Videodisc Systems: Theory and Applications

Jordan Isailović Ph.D.

Prentice-Hall, Inc., Englewood Cliffs, New Jersey 07632

Library of Congress Cataloging-in-Publication Data

Isailović, Jordan (date)
Videodisc systems.

Bibliography: p.
Includes index.
1. Video discs. I. Title.
TK6685.I83 1987 621.388'332 86-3185
ISBN 0-13-941865-2

Editorial/production supervision and interior design: Sophie Papanikolaou
Cover design: 20/20 Services, Inc.
Manufacturing buyer: Gordon Osbourne

Printed in the United States of America

10 9 8 7 6 5 4 3 2 1

ISBN 0-13-941865-2 025

Prentice-Hall International (UK) Limited, *London*
Prentice-Hall of Australia Pty. Limited, *Sydney*
Prentice-Hall Canada Inc., *Toronto*
Prentice-Hall Hispanoamericana, S.A., *Mexico*
Prentice-Hall of India Private Limited, *New Delhi*
Prentice-Hall of Japan, Inc., *Tokyo*
Prentice-Hall of Southeast Asia Pte. Ltd., *Singapore*
Editora Prentice-Hall do Brasil, Ltda., *Rio de Janeiro*

Contents

Preface **ix**

Acknowledgments **xi**

Introduction **1**

0.1 Key component performance 2
0.2 Past, present, and future 4
0.3 Videodiscs and the personal computer 5
References 6

one *Discs and Pickups* **8**

1.1 Introduction 8
1.2 Disc overview 9
1.3 Capacitive pickup 12
1.3.1 Signal on the Disc, 12
1.3.2 Readout Principles, 13
1.3.3 Tracking the Information, 16
1.4 Optical pickup: prerecorded/replicated discs 16
1.4.1 Signal on the Disc, 16
1.4.2 Readout Principles, 18
1.4.3 Tracking the Information: Servo Systems, 22

1.5 Optical pickup: recordable discs 27
1.5.1 Signal on the Disc, 28
1.5.2 Readout Principles, 28
1.6 Features and limitations 28
References 31

two ***Recording and Disc Production*** **33**

2.1 Introduction 33
2.2 Premastering: editing and signal encoding 34
2.3 Optical discs: mass replicas 39
2.3.1 Recording Media, 39
2.3.2 Mastering: Optical Recording, 40
3.3.3 Stamper and Disc Replicas, 44
2.3.4 Metallization, Coating, and Miscellaneous, 46
2.4 Mass replicas: capacitive discs 46
2.4.1 Recording Media, 47
2.4.2 Electromechanical Recording, 47
2.4.3 Stamper and Disc Replication, 48
2.5 Writable (video) disc systems 48
2.6 Servo systems 51
2.7 Disc testing and quality control 53
2.7.1 Premastering, 53
2.7.2 Replicated Discs, 54
2.7.3 Recordable Optical Videodiscs, 56
2.7.4 Evaluation of the Digital Optical Discs, 58
References 62

three ***Signal Processing: Modulation*** **64**

3.1 Justification of signal processing and methods classification 64
3.1.1 Messages and Recorded Signal Characteristics, 66
3.1.2 Analog Modulation Systems, 67
3.1.3 Digital Modulating Systems, 71
3.1.4 Modulation Methods: Overview, 76
3.2 Noise in modulating systems 77
3.3 Signal modification: Shaping the transmitting signal spectrum and waveform 79
3.3.1 Preemphasis, 79
3.3.2 Compressor, 80
3.3.3 Pulse Shaping, 81

References 85
3A: Linear Continuous-Wave Modulation Systems 86
3A.1 Double-Sideband Modulation, 86
3A.2 Double-Sideband Demodulation, 86
3A.3 Conventional Amplitude Modulation, 89
3A.4 Single-Sideband Modulation, 91
3A.5 Vestigial Single-Sideband Modulation, 94
3A.6 Noise in Linear CW Modulation Systems, 96

four *Frequency Modulation* **100**

4.1 Introduction 100
4.2 Spectra of FM signals 102
4.3 Power and bandwidth of FM signals 108
4.4 FM signal in the presence of passband white Gaussian noise 110
4.4.1 Feedback FM Demodulators, 114
4.5 Preemphasis and deemphasis in FM 115
4.6 Pulse frequency modulation 117
4.6.1 Duty Cycle Analysis, 120
4.7 Zero crossings and differential detectors 120
References 122
4A: Phase Modulation 123
4A.1 Spectra of Tone-Modulated PM Signals, 123
4A.2 Output Signal-to-Noise Ratio in PM Systems, 124
4A.3 Pulse PM, 125
4B: Ratio of First and Second Order Sidebands to the Carriers in the FM System 126

five *TV Channels* **127**

5.1 Introduction 127
5.2 Black-and-white TV signal 129
5.2.1 Raster Scan Concept, 129
5.2.2 Spectrum of the TV Signal, 137
5.2.3 Standards for Monochrome TV, 147
5.3 Color TV signal 152
5.3.1 Basic Principles and Requirements of the Composite Signal, 152

5.3.2 NTSC System, 161
5.3.3 PAL System, 180
5.3.4 SECAM System, 197
References 201
5A: Vertical Interval in the NTSC Standard 202
5B: Amplitude Spectra 205

six ***TV Signal Recording*** **210**

6.1 Introduction 210
6.2 Selection of optimal modulation techniques 213
6.3 Signal processing 217
6.3.1 Basic Block Diagram, 217
6.3.2 Encoding, 220
6.3.3 Frequency Spectrum of the Recorded Signal, 234
6.3.4 Decoding (Demodulation), 236
6.4 System parameters 239
6.5 Distortion in FM signals 243
6.5.1 Nonflat Frequency Response, 243
6.5.2 Nonlinear Phase Response, 245
6.5.3 Random Noise, 246
6.5.4 Moiré Patterns, 248
6.5.5 Asymmetry, 252
6.5.6 AM-to-FM Conversion, 256
6.5.7 Crosstalk, 256
6.5.8 Distortion Due to Limited Bandwidth, 257
6.6 Time-base-error correction 260
6.6.1 Time-Base-Error Correction Techniques, 261
6.6.2 Color Correction: Color Burst versus Pilot Signal, 268
6.7 Dropout compensation 270
6.8 Film picture recording on the disc 272
6.9 Frame (picture) number 273
6.10 Special effects and playing modes 274
6.11 Operating controls 278
6.12 Evaluation of picture (color) rendering in the videodisc system 279
References 281
6A: Distortion in FM Signals Due to Non-Flat Frequency Response 283

seven *Extended Play* 287

7.1 Introduction 287
7.2 Increasing disc information capacity 288
7.3 More efficient use of the disc surface 296
7.4 Signal processing 299
7.4.1 Classification of Signal Processing Techniques and Fidelity Criteria, 299
7.4.2 Modulation Bandwidth Reduction Techniques, 303
7.4.3 Spatial Resolution and Frame Rate Reduction Techniques, 306
7.4.4 Picture Interlace Techniques, 310
7.4.5 Variable-Velocity Scanning, 315
7.4.6 TV Signal Bandwidth Reduction, 315
7.4.7 Predictive and Transform Compression, 322
7.4.8 General Remarks, 326
References 330
7A: The SNR Improvement ($1/\rho$) in the FM System with Preemphasis/Deemphasis Circuits 332
7A.1 FM Improvement (dB) from Preemphasis, 333
7B: Color Subcarrier for the NTSC-like Systems with 1/2 Line Offset 334
7C: Audio Carriers with $f_H/4$ Shift 336
7D: Audio Carriers with $3/4f_H$ Shift 338

eight *Audio Signal Recording* 341

8.1 Introduction 341
8.2 Speech perception 342
8.3 Audio plus video 346
8.4 Noise reduction techniques 347
8.4.1 Dynamic Noise Reduction Systems, 347
8.4.2 Noise Reduction Systems with Signal Preprocessing, 351
8.5 Audio signal digitalization 356
8.5.1 Pulse-Code Modulation, 357
8.5.2 Differential Pulse-Code Modulation, 360
8.5.3 Delta Modulation, 362
8.5.4 Comparison of Techniques, 365

8.6 Digital audio in analog video channel 367
8.7 Digital audio—digital channel: Compact disc 370
8.7.1 Digital Audio Modem, 371
8.7.2 Standards, Disc Structure, and Specifications, 373
8.7.3 Signal Processing, 373
8.7.4 Error Correction, 379
8.7.5 Word of Caution in the CD Specs Interpretation, 383
8.8 The CD digital audio modulation in the Laservision videodisc coding formats 383
References 387

nine *Other Applications of Videodisc Systems* **390**

9.1 Introduction 390
9.2 Videodisc programming 390
9.2.1 Program Loading and Operating Modes, 392
9.2.2 Sample of a Specified Program Block Diagram, 393
9.3 Information storage and retrieval 397
9.4 Digital data storage 404
9.4.1 Binary Data Transmission Through TV Channels, 405
9.4.2 Digital Data—Digital Channel, 412
9.5 Codes for optical recording 413
9.5.1 Code Parameters and Evaluation Criteria, 414
9.5.2 The Codes, 422
9.5.3 Discussion, 431
9.6 Compact disc—Read Only Memory 432
9.7 Non-disc formats 434
References 434

Index **438**

Preface

This book is the outcome of the author's experience in research and development in the area of videodisc systems. The idea for the book was born during numerous trips between Belgrade and Los Angeles. At the start of each trip I wondered whether I had with me "everything I needed." Thus the idea developed to write a book that would contain the basis of videodisc systems; that is, the book would serve as an "exterior memory." My initial intention in writing this book was to encompass, in a single text, the broad area joining videodisc systems. I planned to discuss signal processing, modulation theory, TV systems, image processing, and optics, among other things. The book originally contained 16 chapters, but publishing constraints have forced me to organize the material differently. The material has been divided between *Videodisc and Optical Memory Technologies* and this book.

This book considers theoretical as well as practical aspects of videodisc systems. Its purpose is to help engineers, scientists, students, and technicians who need an up-to-date understanding of videodisc systems and their applications.

The Introduction gives the reader a panoramic view of videodisc systems. Principles and descriptions of optical and capacitive playback systems are presented in Chapter 1. The functioning of basic servo systems is also discussed.

In Chapter 2, the videodisc is considered from the recording and production point of view. The basic recording materials and their characteristics are discussed. The basic operations of disc production are described for master replica systems for both optical and capacitive discs. Optical recording and playback in programmable optical videodisc systems are also discussed.

Signal processing and the modulation fundamentals necessary for applications in videodisc systems are presented in Chapter 3. The modulation technique used in all existing videodisc systems—frequency modulation—is presented in Chapter 4. The material in these two chapters, although presented briefly, is necessary for a complete understanding of why and how signal processing is performed in videodisc systems.

For a full understanding of videodisc applications for TV signal recording, it is indispensable to have a basic knowledge of the principles of forming a TV color composite signal and the existing standards (NTSC, PAL, and SECAM); these are the topics of Chapter 5.

Chapter 6 considers the problems of standard TV color signal recording. This is perhaps the most basic chapter in the book. Extended-play techniques are presented in Chapter 7. Although the bandwidth reduction and constant linear velocity (CLV) techniques are used primarily, many other techniques are covered in a systematic manner. This rather extensive coverage is given for reference as well as to serve as background for possible future applications.

The final section of the book covers the principal application areas of videodisc systems (as an analog channel rather than for video signal recording/playback). Chapter 8 contains a thorough and self-contained discussion of audio signal recording. Audio and video techniques used for videodisc recording as well as digital audio recording using videodisc technology are covered. Chapter 9 deals with additional applications of videodisc systems including compact disc Read Only Memory (ROM) systems and non-disc formats.

Jordan Isailović

Acknowledgments

I would like to thank those colleagues to whom I am deeply indebted for encouragement, discussions, inspiration, and support for my book, *Videodisc and Optical Memory Technologies,* published by Prentice-Hall in 1985. They include Paul Obradović, Ray Dakin, Scott Golding-Zlatić, John Raniseski, Ronald Clark, Milan Topalovic, John Winslow, Richard L. Wilkison, Carl Eberly, Michael Michalchik, Richard Allen, all members of Discovision Research and Development Laboratory, Dr. Tim Strand, Dr. Zrilić, Prof. Mysore Lakshminarayana, Paul Thomsen, Dr. William Smith, Nada Orbadović, Lenka Golding-Zlatić, and Shari LaBranche.

I would also like to acknowledge here others who helped me to reorganize and complete *this* book. John Watney and Gene Fifield of Ampex Corporation reviewed part of the manuscript and gave me useful comments and suggestions. In addition, students in my graduate course on this subject taught in the spring of 1985 at California State University, Fullerton, worked on the book's preparation. I thank them for their help—in particular, Hyon Yang.

If anyone has questions on any portion of the book, they may be addressed to the author at the following address. I will try to answer any queries promptly.

P.O. Box 17516
Anaheim Hills, CA 92807

Introduction

A videodisc or videodisk is a thin, flat, circular object of any material, which may contain a record of the video information. The term "thin" is conditional because the thickness varies among different types of videodiscs, so that they can be considered as thick or thin in reference to each other. A blank videodisc is intended for video information recording. It is assumed that the video information on the videodisc can be played back. Typically, the spelling "disk" is associated with computer-generated data.

Three major disc categories are:

1. *Mass replica discs:* videodiscs and digital audio discs
2. *Write-once discs:* multiple readings are assumed
3. *Erasable discs*

Videodisc technology was spawned by the home entertainment industry. The initial aim of videodisc development was to produce a system that records audio/video information on a disc, replicates this disc accurately and inexpensively on plastic, and finally, plays the replicas on home television screens by means of a disc player attachment. Beginning efforts were directed to exploring the development of new types of video equipment which would permit the packaging and marketing of entertainment, educational, and industrial audiovisual programs [1]. The spelling "disc" originates from the consumer video product application.

If the baseband video is encoded only in a 10-MHz bandwidth, the total number of cycles to be recorded for a 30-minute program is approximately $30 \times 60 \times 10^7 = 1.8 \times 10^{10}$ cycles. For the 30-cm disc, the recording area is typically between 5.5 and 14 cm, that is, an area of approximately 521 cm^2. Thus one cycle occupies about 2.9 μm^2, that is, 1.7 $\mu m \times 1.7 \mu m$ if a square aspect ratio is assumed. For the spiral track recording format this would mean about 588 tracks/mm and a spatial frequency of 588 cycles/mm along the tracks. This illustrates the very high recording/reading density required.

Although the spur to the development of videodisc technology has been the entertainment market, it was also recognized that this technology would be exploited to store digital information. Currently, the technology has three main areas of application: prerecorded video playback, prerecorded digital audio playback, and digital data storage—prerecorded or user recorded.

Research and development in high-density optical data storage suggests that a storage density of 160 Mbits/cm^2 is possible. This is 100 times denser than the most sophisticated magnetic storage devices now available.

The need for data base, both analog-imaginary and digital, is growing rapidly. Possibly the most formidable data storage problem has been created by the new generation of electronic imagery satellites [2]. Their high-resolution views of the earth involve as many as 36 million pixels per scene with a 10-bit gray scale and a multispectral capability involving up to eight different spectral bands, which results in a scene size of nearly 3×10^9 bits. The newer satellites can transmit 200 scenes per day, resulting in an annual storage requirement per satellite of up to 4×10^{14} bits. Hospitals need a high-density storage medium for x-rays. Acceptable image quality for the digital storage of medical x-rays can be provided by 4 million pixels with an 8-bit gray scale. A typical hospital with more than 100 beds would require 1.5×10^{13} bits of memory to store the 1/2 million x-rays that it has on file. The average annual x-ray throughput of 30,000 exposures would require an additional 10^{12} bits of storage per year per hospital. There is a wide range of document storage applications involving business records, personnel files, banking records, office files, and library material. With multiple file cabinets at any location, the memory requirement could exceed 10^{14} bits. All over the world, a massive map data base is beginning to be converted into digital form. The resolution required for the highest-detail map is a 1 m^2, and it can be up to 64 bits per resolution cell considering such things as topography, feature classification, place/names, and geopolitical boundaries.

0.1 KEY COMPONENT PERFORMANCE

What makes the videodisc so interesting is the potential combination of three properties: very low cost in high-volume duplication, very high information density, and the possibility of rapid access to any portion of a long recording. In general, videodisc systems offer up to 2 hours of a program on a single disc, color video, stereo sound, digital audio channel (stereo), special features, direct rapid access, and external computer control.

The important considerations for videodisc systems are:

Capacity
Performance
Price
Random access
Interactivity
Capability to combine images, data, and audio playback on one medium

Sometimes, capacity and performance should be considered together because some trade-offs in performance are possible: for example, exchanging playing time for picture quality.

Price analysis can be split into three parts:

Recorder
Discs
Player

Figure 0.1 illustrates a price comparison of the two write-once systems: laservision and pregrooved optical discs. The laservision recorder is, say, five times more expensive than a recorder for pregrooved discs, but pregrooved discs are more expensive than write-once discs, which can be played on any standard laservision player. After some number of recorded discs, the total price is higher for the pregrooved system. The player price is not included in the figure.

As information sciences and communications technologies increasingly touch our daily lives, at home, in education, at work, and in leisure, there is a growing need for creative sensibilities of the human interface and for human usage [3]. Random access and interactivity are very valuable features for the human interface.

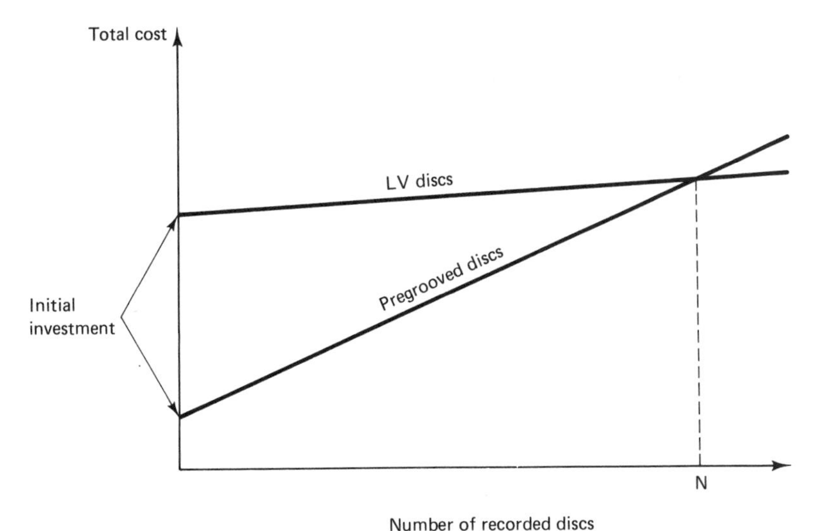

Figure 0.1 Total recorder/discs price for two write-once systems, laservision and pregrooved discs.

The raw material is stored on the disc, but the actual contents are the result of an interaction between the viewer and the system, each of which contributes to the plot, which now becomes a nonlinear event.

The capability of the videodisc to combine images, data, and audio playback has changed the meaning of the term "data base," which now includes whole images.

0.2 PAST, PRESENT, AND FUTURE

The concepts on which videodisc systems are based have found their widest and most dramatic application in recent years, but the ideas are far from new. A brief historical background can be found in the literature (e.g., Refs. 1, 4, 5, and 6). To my knowledge, no one person can be referred to as the inventor of the videodisc in the manner that, for example, Nikola Tesla and Thomas Edison can be named for their inventions of the radio and phonograph, respectively.

Originally, videodisc was conceived of as another method of recording entertainment and educational programs, but with pictures, much as a phonographic recording does for sound. In the 1960s, basic technologies were tested. Optical, capacitive, and pressure pickups were used. Optical, electron beam, and mechanical recorders, respectively, with rates from 200 times slower than real time up to real-time recording are used. The concept of spinning the videodisc at 1800 rpm (NTSC) and 1500 rpm (PAL, SECAM) was developed at that time.

In the 1970s, technical details of various videodisc systems were disclosed through publications, patents, demonstrations, and so on. Playing time up to 1 hour per side was achieved. Optical and pregrooved capacitive discs were the major production types, with flat capacitive discs being tested. Three types of optical videodiscs were used: reflective, thin transparent plastic discs, and film based.

Outside the main stream, three other fields were explored for the optical recording market:

1. Videodisc as a data base was combined with "intelligence" and interactivity was demonstrated.
2. Writable discs were tested.
3. Digital data recording; digital audio discs and discs as a computer peripheral were tested.

These trends continued in the 1980s.

Presently, interactive video [7] and erasable technology are the main areas of development. Also, some approaches considered in the 1970s to be noneconomical are being reconsidered. For example, a two-headed player is very useful for editing, while multichannel systems deliver data at higher rates. The jukebox concept extends data volume. Component recording, including multiplex analog components (MACs) as opposed to composite recording, and high-definition TV are being tested now, but more remains to be done.

Videodisc technology has a long way to go and hopefully, the future will bring improvements in both the technology itself and in systems integration. Improved yield in mass replication and erasable videodisc improvements are the main expectations on the technology side. Interactivity, electronic publishing, and component television are some of the systems integrations expected. A typical component television system consists of a color monitor (in the future, possibly a flat one), at least two speakers, a television tuner packed with a switching mechanism for source selection, an AM/FM receiver, an audio tape deck, an audio turntable, a videocassette recorder, a video camera, a videodisc player (possibly a player/recorder in the future), a video game, or/and a computer [8]. The success of the component TV concept itself has to be proven.

0.3 VIDEODISCS AND THE PERSONAL COMPUTER

In the fight for survival, it looks as if the personal computer is the videodisc's best friend. Personal computers are still an unpersonalized apparatus—with better human interfaces and full sensory richness, recognition abilities, and speech-production improvements yet to come. Even now, the images on the disc can be part of the computer program/system so that the disc is an integral part of the system rather than simply a repository for a limited number of video sequences. [3] That is, the optical storage can be integrated into the more general storage hierarchy of the computer as opposed to being a flexible peripheral. The videodisc is an electronic publication in that the contents are distributed on disc, but read via a conversation with the computer. The computer is a participant in a "semantic" way—not for image processing, although it can control it, but as an active participant in a human–computer conversation. The density and access style of the disc enrich the computer's repertoire and allow it to hold its part of the conversation. Nevertheless, a disc has features not available in print. At least it can be dynamic: an illustration can take on motion and narrate the text. The density of the disc allows it to be encyclopedic, but in addition, the system is a book that "knows" its contents.

Fundamentally, one can equate radio, television, and print as being technically different versions of broadcast communications. They use different channels and have different descriptive modes, but they all follow a one-to-many paradigm. Powerful local computing and storage can change this, transforming a broadcast into a dialogue, a presentation into a conversation. This is a basic change in the notion of publication, with technical and intellectual ramifications [3].

Videotext, teletext, and the possibility of combining different video sources, including that generated by computer, further extend the system's capabilities.

The number of functions that videodisc systems can perform is broad and there is a tendency to classify hardware-players accordingly. Originally, players were classified into two groups: consumer and industrial-type players. Consumer players can perform normal play, slow motion, step forward, step reverse, scan forward and reverse, select audio channels, search for any required picture, and stop at a predefined picture. A remote control unit (RCU) can be provided with the player. Industrial players have built-in microprocessors and as high a level of viewer interaction as is possible. The player operations can also be controlled by digital data stored on the disc, which would be first dumped into local memory.

Lately, players have been roughly classified into levels 1, 2, and 3 in reference to the control capabilities. Level 1 is a player that can play straight linear with use of picture stops and chapters. Remote control can be used to randomly access individual frames and/or chapters; it can be used for catalogs, or used as audio visual teaching materials of foreign languages, calisthenics, video libraries, encyclopedias, etc. Level 2 is a player that uses an internal processor (in conjunction with remote pad) to control play and search functions. Program software (dump) is on audio channel of the disc and is loaded into the microprocessor. This can be used for in-company training, demonstrations, and as information systems for entertainment games. Level 3 is a player with external control of the player through remote port. The external device contains program software for player control (usually through RS 232 interface). The use of light pen, voice recognition, and touch operation of the character display are all possible with this system. Hybrid disc contains video, audio, digital data (used for stop motion audio or data) and is used in systems that may have graphic overlays, etc. Sometimes, essentially anything not in Levels 1, 2, or 3, belongs to level 4.

Two other, relatively simple, level definitions are used. First, Level 0: Linear play; Level 1: Random access; Level 2: Program dump, digital data is in audio; Level 3: Digital data is in video. Second, Level 1: Consumer player; Level 2: Microprocessor built-in (industrial player); Level 3: Level 1 or level 2 combined with outside control. Obviously, classification is arbitrary; up to eight level classifications are used.

REFERENCES

1. J. Isailović, *Videodiscs and Optical Memory Systems,* Prentice-Hall, Englewood Cliffs, N.J., 1985.
2. M. W. Goldberg, "Large Memory Applications for Optical Disk." Topical Meeting on Optical Data Storage, *Tech. Dig.,* Incline Village, Nev., Optical Society of America, January 17–20, 1983, p. MA3-1-3.
3. A. Lippman, "Optical Videodiscs and Electronic Publishing," in J. Isailović, *Videodisc Technology: Recording, Replicating, and Reproduction of Audio, Video and Data Signals,* Lecture Notes, May 1984.
4. A. Korpel, "Laser Applications: Videodisc," in *Laser Applications,* Vol. 4, eds. Joseph W. Goodman and M. Ross, Academic Press, New York, 1980, Chap. 4.

5. P. Rice and R. F. Dubbe, "Development of the First Optical Videodisc," *SMPTE J.*, Vol. 91, No. 3, 1982, pp. 277–284.
6. G. C. Kenney: A Time Line of Videodisc Milestones," The Videodisc Monitor, April 1985, Vol. IV, N.Y., pp. IA-IO.
7. W. M. Bulkeley, "Videodiscs Make a Comeback as Instructors and Sales Tools," *The Wall Street Journal,* Feb. 15, 1985, p. 25.
8. T. S. Perry, "Component Television," *IEEE Spectrum,* Vol. 20, No. 6, 1983, pp. 38–43.

one

Discs and Pickups

1.1 INTRODUCTION

In videodisc systems the recording and disc production can alternate orders. Mass replica discs are made after recording takes place, while the recording on writable discs follows disc production. Except for testing, read during writing, direct read after write (DRAW), and so on, the reading is normally performed after recording and disc production. But for pedagogical reasons, we first discuss the reading process in this chapter, and then, in Chapter 2, recording and disc production will be introduced. Some phenomena that occur before and after reading in the videodisc chain are better understood once reading is understood, examples of this being required pit depth and system frequency response.

For two major pickups, capacitive and optical, three subjects are discussed:

1. The signal on the disc
2. Readout principles
3. Tracking

These three issues are discussed for both read-only discs (RODs) and programmable discs [1]. The second type includes programmable RODs (PRODs) and erasable PRODs (EPRODs).

In order to recover the signal information represented by the disc relief along the spiral track, an accurate means of reading the information is necessary. When a video signal is recorded, the videodisc must be rotated at a speed compatible with the frequency at which the television system is operating (i.e., either 50 or 60 Hz). The 50-Hz PAL system is designed to give 25 pictures per second, or 1500 per minute, on the TV screen. A PAL-system videodisc must therefore be rotated at 1500 rpm. The 60-Hz NTSC system gives 30 TV pictures per second, requiring that an NTSC-system videodisc be rotated at 1800 rpm if one frame per revolution is recorded.

A simplified model of the videodisc reading process is shown in Figure 1.1. An oscillator generates a periodical signal, electrical or optical, with a high carrier frequency (f_0). Through the stylus/sensor, the reading carrier is basically amplitude modulated by the pit pattern on the disc surface. A pit-pattern detector demodulates the carrier with frequency f_0 and thus converts the mechanical pattern relief from the disc surface to the electrical signal, which is then amplified and possibly equalized in the low-noise preamplifier. After signal processing, the original information is recovered.

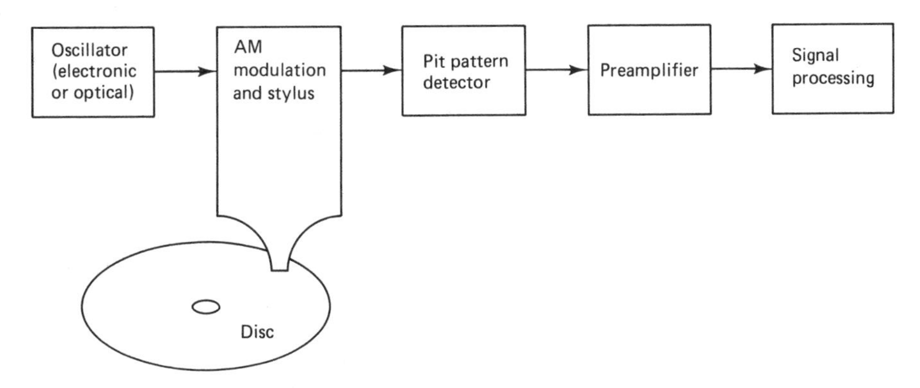

Figure 1.1 Model for the videodisc reading process.

A typical simplified block diagram of the signal processing in the videodisc player is shown in Figure 1.2. The detected signal from the disc goes through the signal conditioner and low-noise preamplifier. A filter separates the audio and video portions, out of which two (stereo) audio channels and baseband video are obtained. A dropout detector controls the output video signal. When the dropout is detected, the video signal from the previous line—delayed by the CCD delay line, for example—is displayed. If a TV set is serving as the display, the modulator places the video and audio signals in TV channel 3 or 4. Otherwise left (L) and right (R) audios must be obtained of two audio channels, for the stereo speakers.

1.2 DISC OVERVIEW

All present videodisc systems use a plastic disc rotating on a turntable. The player picks up information represented by changes in the disc's surface and converts them into signals for a television set [2]. All use pulse frequency modulation for both video and audio signals. Each disc also has a spiral information track of a given pitch, q (i.e., b revolutions per millimeter, b $= 1/q$), which is usually

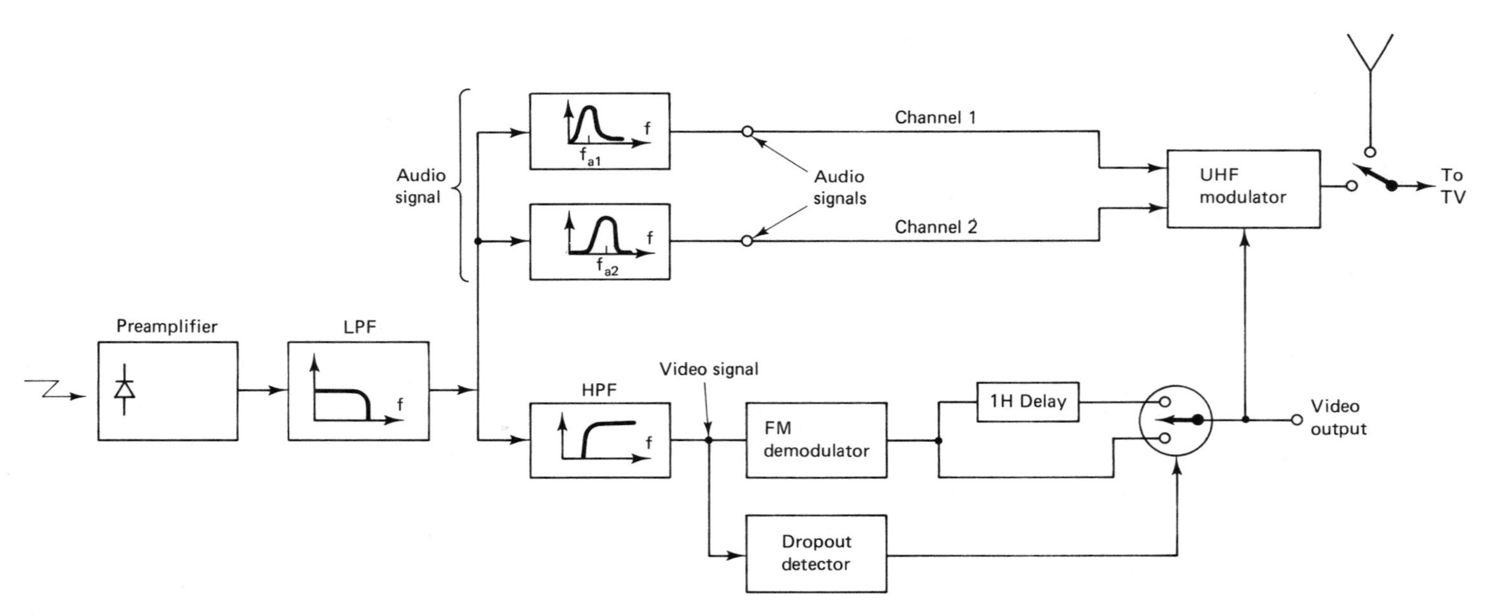

Figure 1.2 Simplified block diagram of signal processing in the videodisc player.

constant over the whole recorded area, and a given tangential density of information elements, a (elements per millimeter), which may or may not be constant. The information on the disc is stored in a spiral track, typically starting at the inside at a fixed diameter and moving to the outside for optical discs, and vice versa for capacitive discs.

There are two types of disc formats, called CAV and CLV. CAV stands for *constant angular velocity,* which means that their speed of rotation is constant, 1500 rpm for PAL/SECAM (n = 25 rotations per second) and 1800 rpm for NTSC (n = 30 rotations per second), if one frame is recorded per revolution. In flat disc systems, this type of disc allows special playing modes, such as still frame, slow motion, and so on. For NTSC, 900 rpm corresponds to two frames, and 450 rpm to four frames recorded per revolution. The tangential information density is not constant for this format, being lower than optimum on the outer radius.

In CLV, *constant linear velocity,* the speed of rotation decreases inversely proportional to the readout diameter. As a result, more information can be stored on the disc. On the other hand, these discs can only be played in a continuous way (i.e., in the normal forward mode). Also, a hybrid mode, the modified CLV (MCLV), is possible. In this method the disc area is divided, in the radial direction, into zones. Every zone may contain equal number of tracks per revolution. The angular velocity varies discretely from zone to zone but is constant in each zone. The mean linear velocity value for each zone is equal.

The two major pickup systems in use are the optical and the capacitive. Although, in principle, reflective, absorbtive, and transmissive optical systems are possible, the reflective one is mainly used today. Prerecorded optical discs are flat*; capacitive discs can be grooved or grooveless (i.e., flat). The capacitive electronic disc (CED) is of the grooved type and the video high-density disc (VHD) is a flat capacitive disc.

In today's systems, the signal on the disc is restricted to two levels. Information can be carried by transition spacing of the signal elements on the disc (for example, pulse frequency modulation (PFM)–analog channel) or on their combination, for example, pulse code modulation (PCM) or delta modulation (DM)–digital channel. When the disc is used as an analog channel, it is usually referred to as a videodisc (Table 1.1). Optical discs are used as an analog and also as a digital channel, whereas capacitive discs are used presently only as analog channels and for playback only (read only).

Consumer application has been the driving force behind research and development of the read-only disc (ROD). The consumer videodisc was the first product. Mastering and mass replication are used for ROD production. Computer application has been the driving force for the archival write-once, or nonalterable, and erasable, or reversible, discs. That is why the latter are usually spelled with a "k": disk.

* Some recordable optical discs are pregrooved.

TABLE 1.1 DISC CLASSIFICATION

	Signal channel				
	Analog: videodisc		*Digital: digital disk/disc*		
Disc type	*Video/audio signal*	*Data (digital)*	*Audio (CD)*	*Data*	*Comment*
Read only	Optical and capacitive	Optical only	Optical	Optical†	Many discs per recorded master through the stamping method
Write once	Flat or pregrooved	Flat or pregrooved	Can be used	Typically pregrooved	Optical: typically, one disc per recording
Erasable	Can be used	Can be used	Can be used	Yes	In general: not compatible with read-only optical discs

† CD ROM is an example of the read-only digital disc.

By analogy with semiconductor memories, a write-once disc can be called a programmable read-only disc (k)—PROD [1]. Similarly, the erasable disc is an erasable PROD (EPROD).

Recording media can be either write-once (used for ROD and PROD) or erasable. It should be noted that the read-only disc does not contain the recording medium. Instead, the recording medium is used during mastering. Some recording media have a very desirable property: the signal can be read during write. This is usually called direct read after write (DRAW). The DRAW property can be used to verify information that has just been recorded. Sometimes, the write-once discs or digital write-once discs are called DRAW discs.

1.3 CAPACITIVE PICKUP

1.3.1 Signal on the Disc

A simplified arrangement of the signal elements on the capacitive discs is shown in Figure 1.3. For the grooved system (CED) the track pitch is typically $q = 2.7\ \mu m$ and is slightly larger than the pit width (γ). The pit length (β) is of the order of $0.25\ \mu m$. Figure 1.3 shows the lower covers of the hypothetical capacitive structure. The upper covers are in the same plane, but the lower covers can be at either of two distances. In practice, the upper covers in the hypothetical structure are replaced by one moving cover.

In Figure 1.4, the perspective sketches of the grooved (a) and flat (b) capacitive discs are shown. A spiral groove is V shaped for two reasons: to guide the diamond pickup systems and to increase capacitance of the signal element, that is, to improve sensitivity and thus signal-to-noise ratio during the

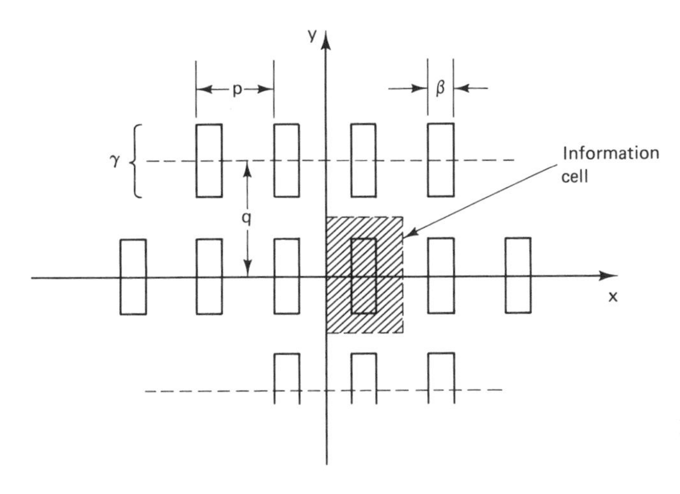

Figure 1.3 Arrangement of the signal elements on a capacitive disc.

playback process. The flat capacitive disc has two types of signal elements (Figure 1.4b): main signal elements and tracking signal elements, recorded in the "guard band" between the main signal tracks.

1.3.2 Readout Principles

The stylus for the capacitive pickup is made of the stylus tip or substrate, diamond or sapphire, and a thin metal electrode on the tip's flat trailing edge. For the grooved disc, the stylus foot is shaped to fit the triangular groove cross section. A typical size is 2 μm across the groove and 4 μm along the groove. The electrode thickness is usually 1000 to 1500 Å. The rest of the stylus shape is determined by mechanical requirements.

The following should be made clear for the capacitive readout [1, Chap. 4].

1. Relation of the recorded signal elements and capacity
2. Detection of capacitive variations

The capacity between the electrode and the conductive surface (Figure 1.5) is a nonlinear function of the distance between them. The capacity decreases as the distance increases. For the peak-to-peak signal elements, typically depth is 850 to 1000 Å, the change in capacitance, ΔC, is on the order of 10^{-16} to 10^{-14} F. The idealized transfer function, T, for the small-signal amplitude is [1, Chap. 4]

$$T(\nu) = \frac{h_c}{h} e^{-k\nu} \tag{1.1}$$

where ν is a spatial frequency, h an average elevation for the electrode edge, and h_c an arbitrary reference height.

The transfer function shows some important properties of the capacitive pickup. For example, shorter wavelengths will be detected with smaller amplitudes than equal-amplitude recorded signals of longer wavelengths. Also, the amplitude

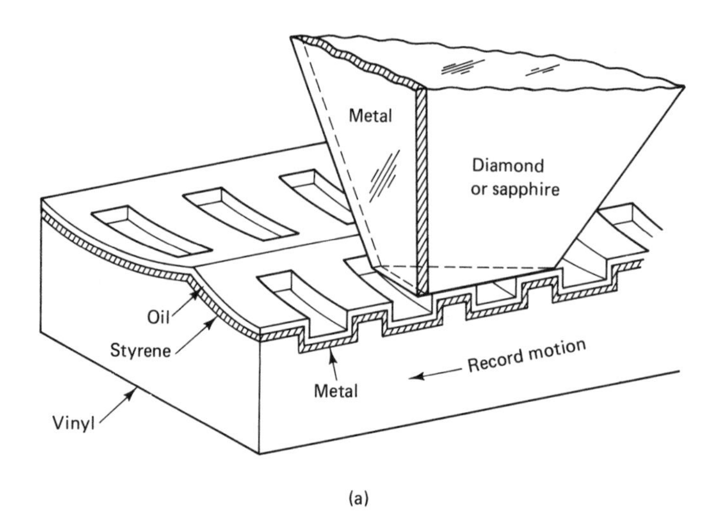

(a)

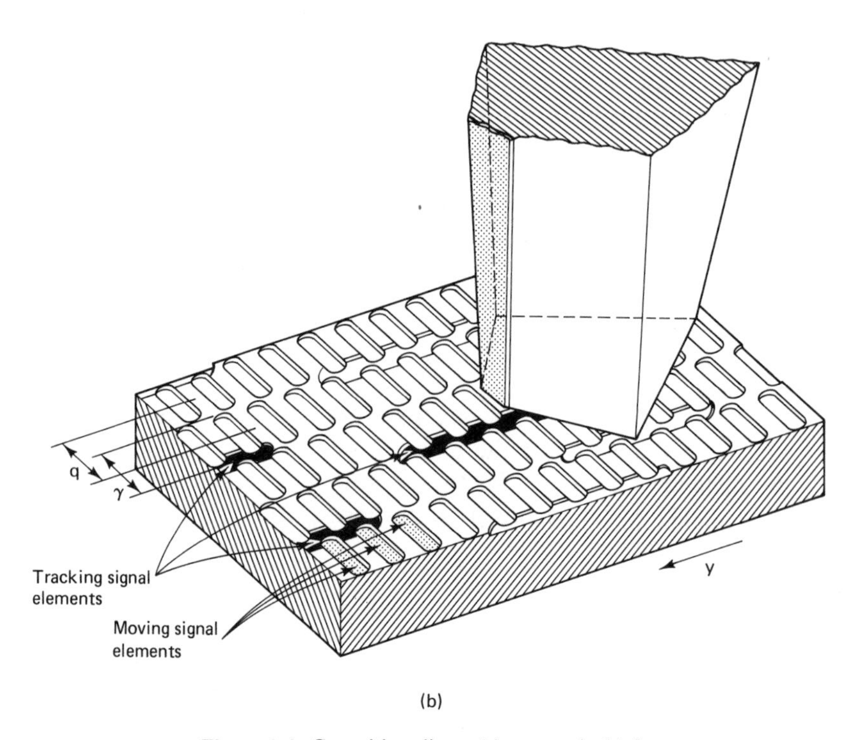

(b)

Figure 1.4 Capacitive discs: (a) grooved; (b) flat.

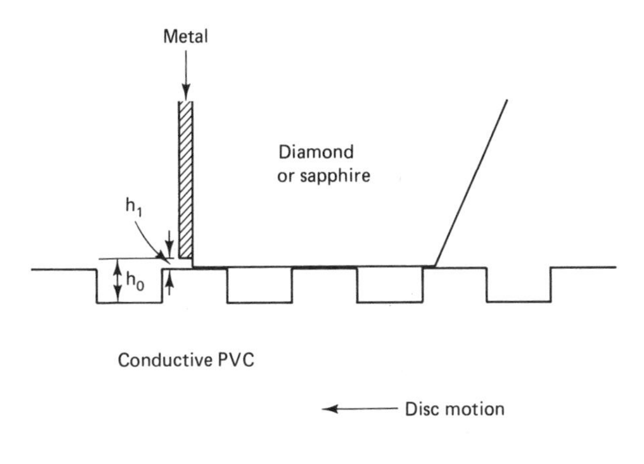

Figure 1.5 Cross section of the capacitive stylus/disc combination (not to scale).

for a given wavelength decreases as h is increased. The shortest resolvable wavelength is a function of the noise in the system and the desired signal-to-noise ratio.

The transfer function of the system does not give a complete picture of its performance. It can even be misleading because it tends to obscure the fact that (capacitive) detection is basically nonlinear. Nonlinearity introduces, for example, harmonics and beat-frequency components in the output signal.

The small change in capacitance experienced by the stylus electrode can be detected only at high frequency [3–6]. Although transmission-line techniques are used, because the circuitry operates at about 1 GHz, its function may be described schematically by the lumped-element circuit in Figure 1.6. The resonant

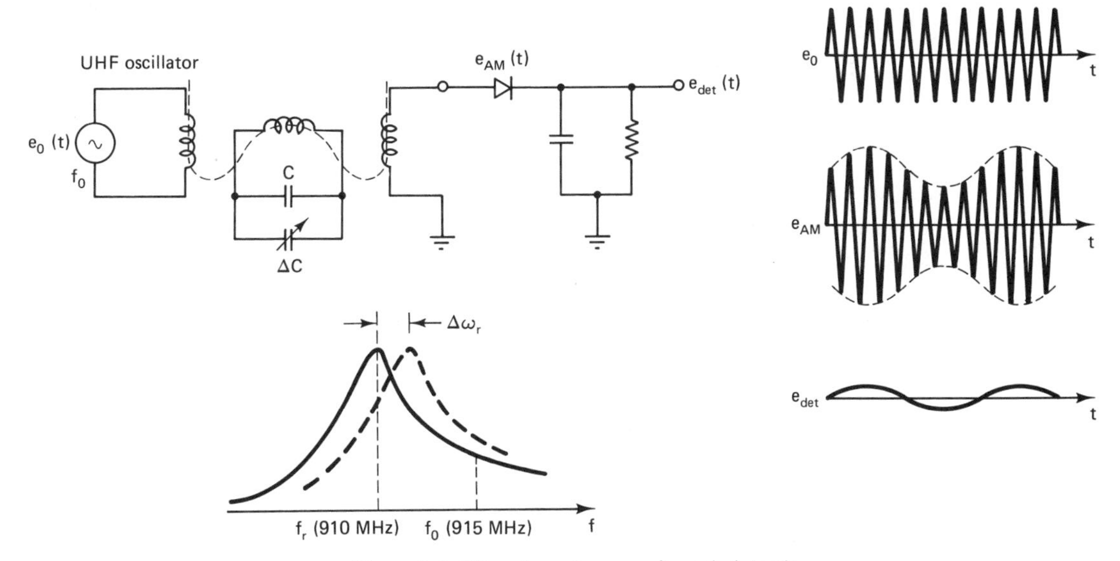

Figure 1.6 Pit pattern (or capacitance) detection.

circuit includes the stylus–conductive disc capacitance. The resonant frequency (f_r) of the circuit changes as the capacitance (ΔC) changes, and the UHF signal (f_0) is amplitude modulated, $e_{AM}(t)$. The original, video FM, is obtained at the output of the AM detector, $e_{det}(t)$.

1.3.3 Tracking the Information

The grooved capacitive system is by far the simplest case of all videodisc systems. A grooved track guides the stylus so that during normal play there is no need for extra radial control. Only during jump (e.g., search) operation is the arm "kicked" radially; but tolerances are now much wider. The stylus weight "focuses" it on the track surface, and because of that, there is stylus/disc wear.

Even if the spindle control is perfect and the rotational speed is constant, the signal recovered from the disc may have significant time-base errors, due, for example, to imperfect centering and warp of the disc [3, p. 212]. A 50-μs peak-to-peak timing error varying at a once-around rate is the result of a centering error of about 0.18 mm. Because a shift of a few nanoseconds with one line of video can cause objectionable variations in hue, chroma phase errors must be reduced at least 70 dB. For horizontal sync instability, 20 to 30 dB correction is sufficient for the typical TV receiver. This can be performed by the "arm stretcher" transducer closed-loop correction system. The error signal basically is derived by comparing reference color subcarrier with that recovered from the disc. The final correction required to assume accurate rendition of color is provided electronically. A phase-locked loop is controlled by the same error signal, which moves the stylus along the groove to maintain constant relative velocity between the stylus and the recorded information on the disc. However, for synchronizing with professional television equipment for keying and mixing purposes, even more stable pictures are needed.

1.3.3.1 Tracking Servo System for Flat Capacitive Disc Readout. In the system with flat capacitive discs, the readout lead can follow the track, move directly over the desired track(s), or be used for any of the special effects during the readout. Electromagnets (with corresponding coils) are used to move the readout lead transversely and longitudinally.

The level of the readout signal varies with the radial positional displacement of the stylus on the disc. On the flat capacitive disc, pilot signals (e.g., f_{p1} = 511 kHz, f_{p2} = 716 kHz) are recorded on the opposite sides of each track carrying the main information signal (Figure 1.4). An output tracking error signal is generated from the pilots.

1.4 OPTICAL PICKUP: PRERECORDED/REPLICATED DISCS

1.4.1 Signal on the Disc

A simplified arrangement of the signal elements on the optical (transmissive or reflective) discs is shown in Figure 1.7. The average track pitch is 1.6 μm (q in

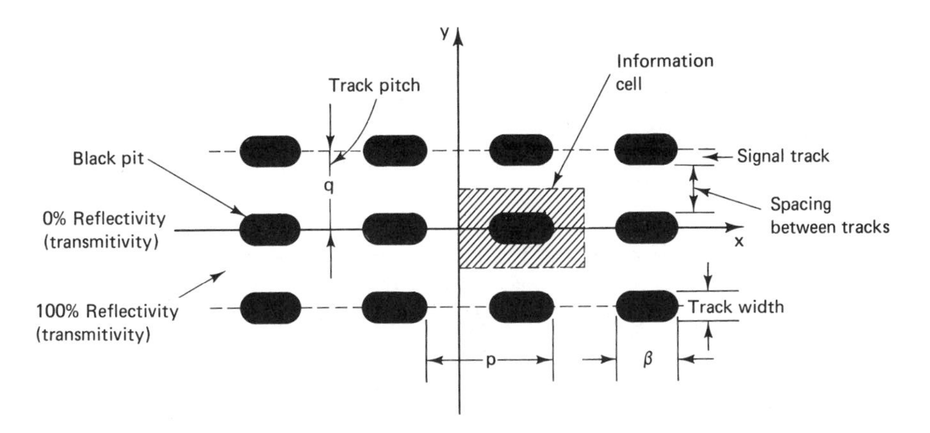

Figure 1.7 Arrangement of the signal element on the optical disc.

Figure 1.7). This means that the period at the end of this sentence would cover over 500 tracks. The effective pit width (γ) is 0.4 μm; the effective pit length (β) varies with radius and signal content and other factors. A single television picture, one frame, requires a surface on the disc of 0.6 mm^2. Ideally, the reflectivity of the disc surface is 100%, for the transmissive disc transmittivity is 100%. Signal pits are black holes; that is, they have 0% reflectivity, 0% transmittivity (i.e., 100% absorptivity). The closest approximations to the ideal optical disc are obtained by scattering effects or by creating phase modulation [1, Chap. 3]. In Figure 1.8, a cross section of the reflective optical disc is shown (not to scale). A reflective (metal) layer is sandwiched between two plastic coatings.

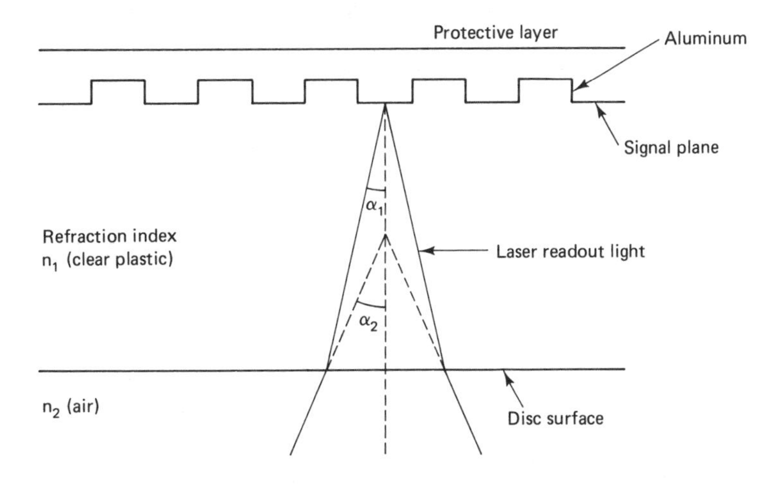

Figure 1.8 Cross section of the optical disc.

1.4.2 Readout Principles

The optical principles used for optical disc pickup are based on light-ray diffraction, a phenomenon that occurs if the object dimensions are the same order of magnitude as the wavelength of light. One well known example is the light diffraction by means of a narrow slit [1].

In order to read the signal information represented by the individual pits along the spiral track, the laser beam must be focused, and reduced, to a spot of light 1 μm in diameter. There is, however, a fundamental diffraction-limited lower limit to the size of the details that can be read (approximately $\lambda/2NA$), depending on the wavelength of the light and the numerical aperture (NA) of the objective lens. The numerical aperture is defined as the product of the refraction index, n, and the sine of the angle between the optical axis and the outermost light ray contributing to the imaging: $NA = n_1 \sin \alpha_1 = n_2 \sin \alpha_2$ (Figure 1.8).

Reflective videodiscs show their greatest utility when they are read through the body of the disc; this mode of use puts the information-bearing surface out of harm's way, a distance many times the depth of focus from the nearest optically effective exposed surface. This has the benefit of holding dirt and scratches out of focus to reduce or remove their effect on signal playback; it has the drawback of requiring compensation elsewhere in the optical system. Focusing a beam of light through a slab of refracting material makes the point of focus move axially and introduces spherical aberration. Above a certain tolerance limit, this aberration degrades the focused spot enough so that the signal is no longer useful.

The basic principles of an optical pickup system are illustrated in Figure 1.9. The ideal optical videodiscs (Figure 1.7) are assumed to be either transmissive or reflective.

Within the approximations of color diffraction theory, transmissive and reflective phase gratings are equivalent when they impart the same spatial phase variations to the readout beam [1]. Discussion can be carried out for either one. Two causes are presented for the reflective optical disc.

1.4.2.1 Pit Depth of $\lambda/4$: Phase Cancellation/Central Detection. For the sake of simplicity, let us assume that the adjacent tracks are separated enough (or equivalent) so that only one track is recorded (Figure 1.10). Also, let's point out that:

1. The pit width (w) is smaller than the reading spot diameter.
2. The output signal is proportional to the total signal from the photodetector; or assume that the photodetector has one photocell (diode).

When the reading spot is in between pits, the output signal is maximal. Consider now the case when the recording beam is focused on the pit. It can be shown that the best resolution is obtained when half of the total area (or, more precisely, energy) of the focused spot is intercepted by the pit. Detail for a cross section

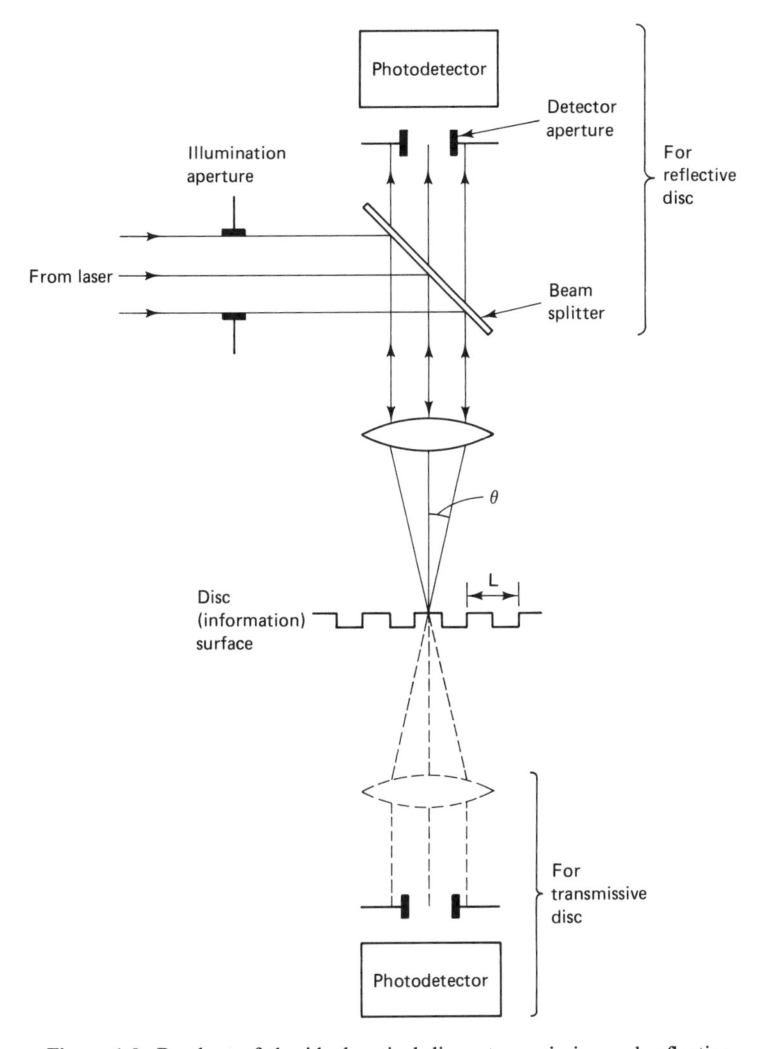

Figure 1.9 Readout of the ideal optical discs: transmissive and reflective.

A-A′ is also shown in Figure 1.10. The difference in optical path length between rays reflected from the surrounding area and the bottom of the pit is $2d$, and the phase difference is

$$\Delta \varphi = \frac{2\pi}{\lambda} \times 2d \tag{1.2}$$

For $d = d_{opt} = \lambda/4$, the phase difference is 180°; that is, the difference in optical path length is $\lambda/2$. To be more precise, it should be

$$d_{opt} = \frac{\lambda}{4n} \tag{1.3}$$

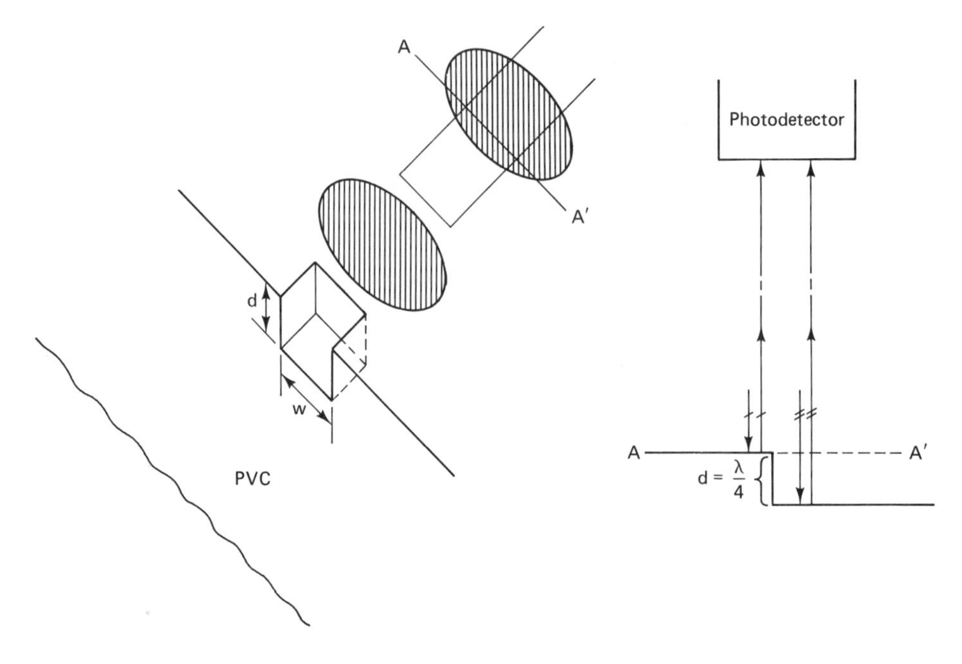

Figure 1.10 One-quarter-wavelength pits.

where λ is the wavelength in air and n is the reflective index of the disc material. For plastics (PVC) $n \simeq 1.5$, hence $d_{opt} = \lambda/6$. With $\lambda = 632.8$ nm, $d_{opt} = 100$ nm $= 1000$ Å.

It makes no difference where in the range the differential phase delay is $3\pi, 5\pi, \ldots, (2n + 1)\pi$: cancellation behavior is periodic with pit depth. Normally, the minimal optical depth is used. Experiments show that a certain minimum depth is required rather than an optimum depth. Practical results do not follow completely previous theoretical conclusions because after some minimal depth the scattering dominates the readout mechanism.

1.4.2.2 Pit Depth of λ/8: Optical Edge Detection. Let's assume that (Figure 1.11):

1. The pit width is larger than the reading spot diameter.
2. A split photodetector (photodiode) is used, and the output signal is proportional to the difference of the diode signals. A split line is normal to the reading direction.

Now, when the reading spot is in between pits or completely in a pit, the output signal is zero because the light intensity is equal on the two halves of the photodiode.

Consider now the case when the reading spot is focused on the edge of a

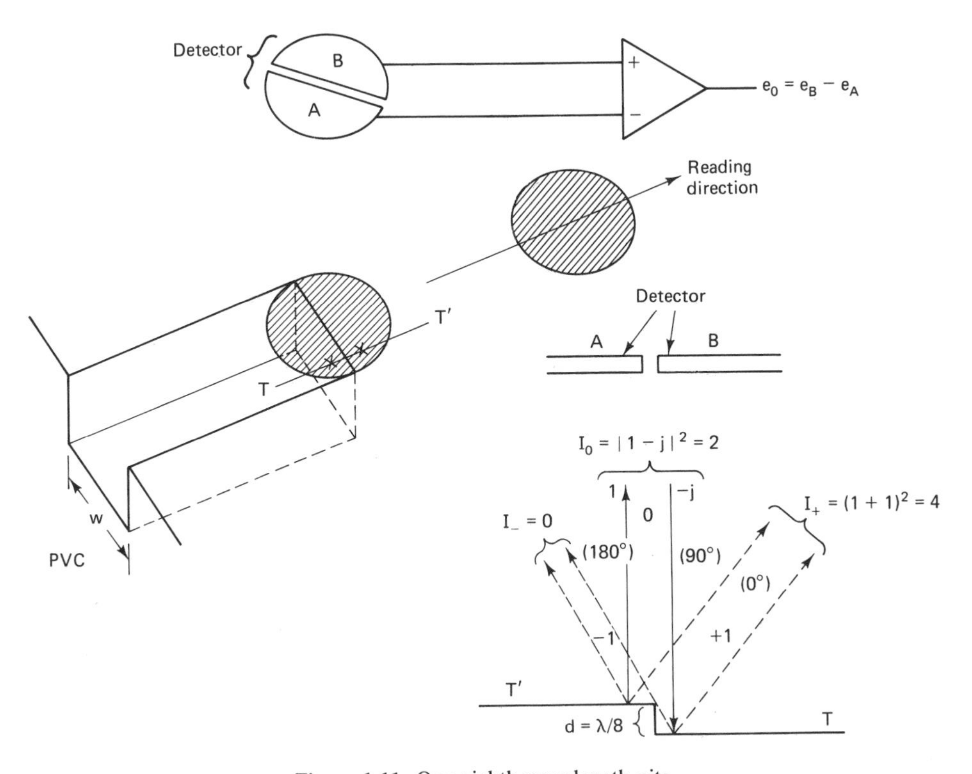

Figure 1.11 One-eighth-wavelength pits.

pit passing under it. Detail for a cross section along the x-axis is also shown in Figure 1.11. Traversing the region at the edge, the wavefronts in one half of the beam are displaced by $\lambda/4$ with respect to those in the other half. In this simplified example, the intensity of the zero-order diffracted light is practically constant (see Figure 1.11). Thus a differential detector should be used. The output signal is proportional to the difference between the -1 and the $+1$ orders of diffracted light.

The transfer function of the optical pickup in the first case, the $\lambda/4$ pits, is practically defined by the optical transfer function of the focusing lens [1, Chap. 5]. Figure 1.12a shows typical curves for the different levels of defocusing. The fictitious maximal resolving spatial frequency (fictitious because no noise consideration is applied), ν_c, is

$$\nu_c = \frac{2\text{NA}}{\lambda} \tag{1.4}$$

Figure 1.12b shows the analog transfer function for the optical edge detector with $\lambda/8$ pits. As with the capacitive system, the modulation transfer function (MTF) gives no information about the inherent nonlinearities in the system or about the accuracy in detecting zero crossings (edges).

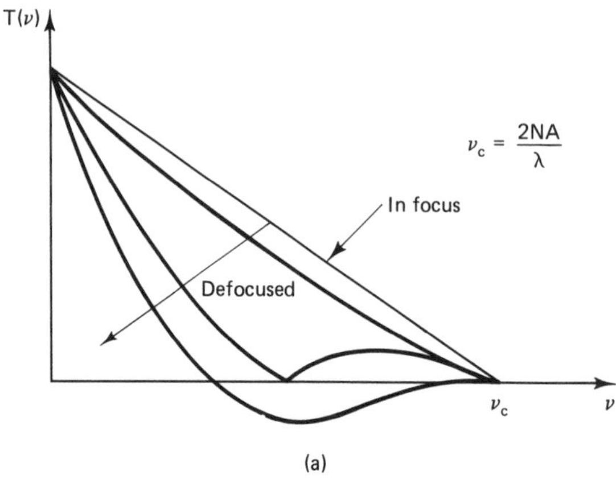

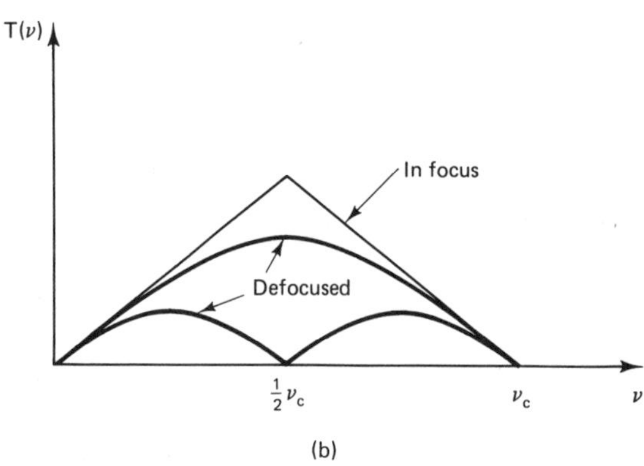

Figure 1.12 Transfer functions of the optical pickup: (a) $\lambda/4$ pits; (b) $\lambda/8$ pits.

1.4.3 Tracking the Information: Servo Systems

The positional accuracy of the pickup stylus with respect to the truly microscopic dimensions of the recorded signal location is a problem that must be resolved by any videodisc technology. A three-dimensional positioning is required: time base—tangential (x), tracking—radial (y), and focus—vertical (z). The spindle control is, at least in principle, similar for all systems.

The most important advantage of the optical system is the contact-free readout of the information. The result is that wear of the disc or readout device is nonexistent. The second advantage is that information on the disc can be protected against the influence of dust, fingerprints, and so on. The price for this is a need for precise servo systems. The existence of practical high-performance optical servos is the most important factor in making possible removable optical storage media. For the consumer market, that is, for the mass production of the players, all designs must be inexpensive to implement and still perform well enough.

The pickup used in optical videodisc player requires servomechanisms for three directions: the vertical servo system, which focuses the light beam onto the information plane of a disc; the radial servo system, which enables the beam to follow information tracks; and the tangential servo system, which adjusts the time-base errors.

Figure 1.13 shows the simplified structure of the optical pickup. The optical system includes a coherent light source, gas laser, or laser diode—usually, a collimator (planoconvex) lens which expands the beam to fill the objective lens; a combination of a polarizing prism (beam splitter) and a quarter-wave plate used to prevent the light reflected by the disc from returning to the light source; two mirrors that can be rotated by piezoelectric or electromagnetic bender motors to move a focused spot in the tangential or radial direction; an objective lens focusing the light beam into a diffraction-limited spot; and finally, a photodetector, usually a quadrant PIN photodiode having four light-sensitive segments. For a uniformly illuminated diffraction-limited lens, the spot width, a, between half-intensity points is given approximately by

$$a = \lambda/2\text{NA} \tag{1.5}$$

where λ is the wavelength of the light and NA is the lens numerical aperture. Practically, this defines the limit in the resolution that can be obtained with the particular system. A special grating before the beam splitter is sometimes used to generate two auxiliary beams of light, needed, for example, for the radial tracking.

The main servo systems in the player will be considered here: spindle, focus, radial, and tangential.

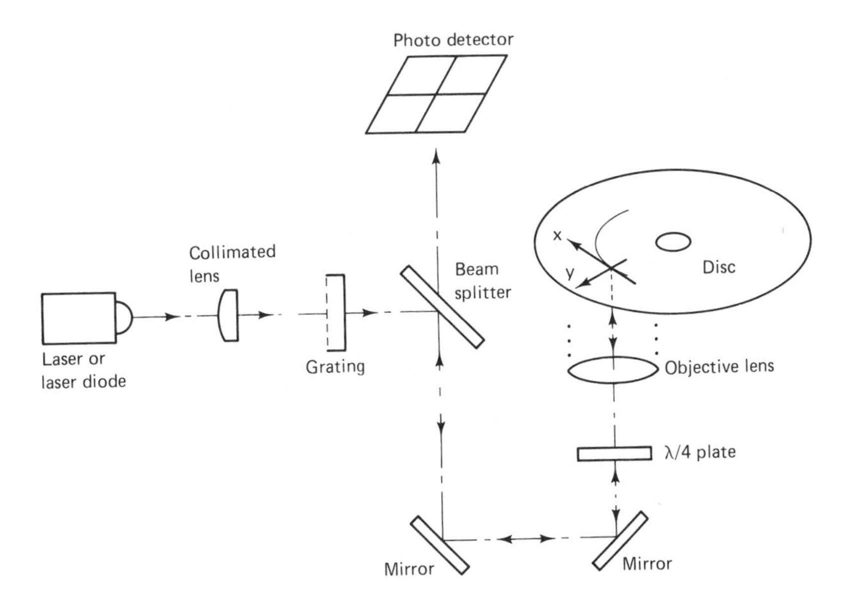

Figure 1.13 Player optics.

1.4.3.1 Spindle Servo. The videodisc is set on an axis driven by an electrical motor (with, e.g., 1798.2 rpm). The variation in the rotating speed is reflected in the phase and frequency change of the signal that is read. This is commonly referred to as *time-base error.* For standard monitors and receivers, the typical requirement is 0.1% stability [1]. For genlocking equipment, greater stability is required.

A system with a tachometer could be used. Such a system would have the tachometer with, say, 525 tick marks on the edge of the disc or on the spindle motor shaft. The tachometer signal is then compared with a reference oscillator, with a quartz crystal, for instance. As an alternative to a tachometric signal, a signal from the disc itself can be used. Such signals can be recorded along with the video during mastering, or the synchronizing video signal can act as the tachometer. These would be the horizontal sync or the color subcarrier [1]. The error signal controls the number of rotations and phase of the electromotor.

1.4.3.2 Focus Servo. Although the action of the rotating disc on its surrounding air creates an air bearing, the disc is not stabilized sufficiently for the required distance between the disc and the read objective. Discs also exhibit a large fundamental error due to warp. The maximal error ranges from ± 250 μm to ± 1 mm. The depth of focus of the focusing lens objective is a function of the reading light wavelength and the lens numerical aperture. The focus accuracy required is typically ± 0.2 to ± 1 μm. This requires the servo correctability to be greater than 1000:1 or 60 dB at the fundamental frequency of rotation. Two problems are obvious:

1. How to move a read objective to follow the undulations of the disc
2. How to generate error signals

The largest amplitudes in vertical movement of the disc occur at the rotational frequency (e.g., 30 Hz) and decrease rapidly for higher frequencies. The objective lens can thus be mounted in a system similar to that of a loudspeaker voice coil and operate according to the electrodynamic principle. Depending on the direction and magnitude of the current through the coil, the lens makes vertical movements.

The amplitude of the output signal (RF) is a (quadratic) function of the defocus Δz [7, p. 1999]. Thus the information recorded on the disc is used for generating an error signal. The acquisition range of this method is only a few times the depth of focus. An acquisition range of up to ± 1 mm is obtainable with methods in which the disc (information) surface is considered to be a plane mirror. The reading or separate beam can be used to measure the distance to the focal plane.

One attractive method is illustrated in Figure 1.14, astigmatic focusing (1,8). Point P has three different images, P', N', and M', for three different positions of the disc: in focus, far, and near. When the beam is focused on the astigmatic lens surface, one of three characteristic spot shapes will be created on the detector surface: circular spot or elliptic spots. The three cases can easily be distinguished

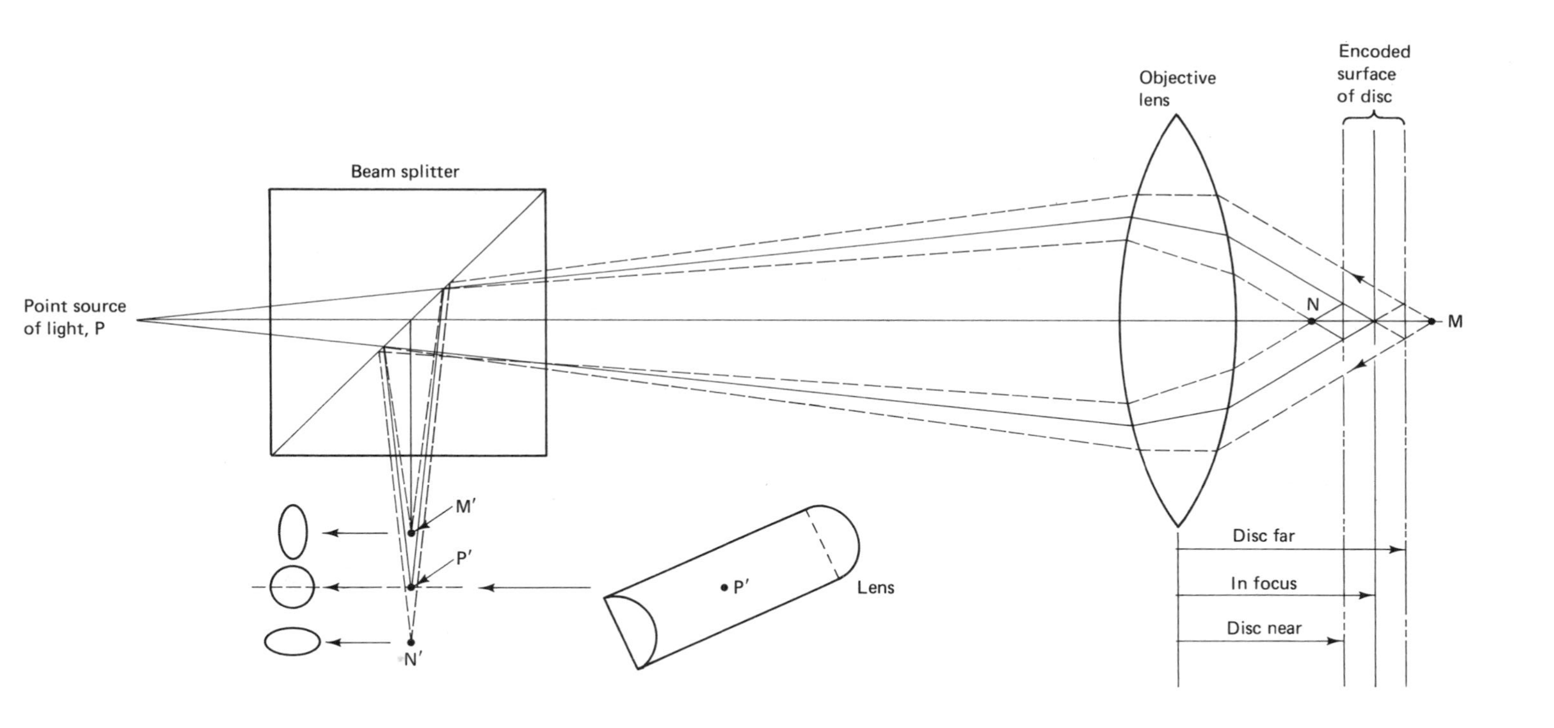

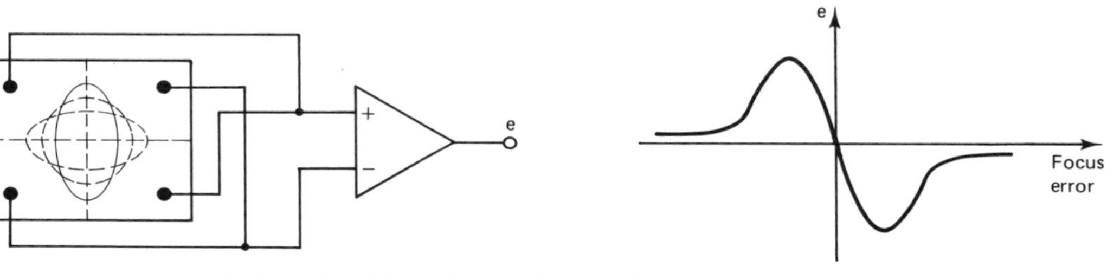

Figure 1.14 Principle of the astigmatic focus system.

by a four-segment photodetector. The error signal is linear for a significant range of focus error and then drops off to zero on each side of focus as the spots get large due to the defocusing effect and drop in intensity.

A focus error signal can be also generated with an auxiliary skew beam generated from the reading beam [7, p. 1998]. A mask technique combined with the asymmetric sensor is also used [8]. The same techniques are used for digital audio and for optical disc mass data storage systems [1]. It should be mentioned that optical disc profiles can be tested by spectral analysis of focusing error signals. Substrates as well as coated discs can be tested this way.

1.4.3.3 Radial Tracking Servo. Components of the optical pickup are usually mounted on a sled driven by a small dc motor and moved radially under the disc. For CAV discs with one frame per revolution (1800/1500 rpm for NTSC/PAL) this means an average linear sled speed of 3 mm/min for NTSC, or 2.5 mm/min for PAL. With a track pitch of 1.6 μm, the scanning light beam has to remain focused on the track with a radial accuracy of 0.1 μm, a requirement that cannot be met by a purely mechanical guidance system. Certain slow corrections are possible, by varying the speed of the drive motor incorporated in a servo control system. However, the eccentricity of the disc can be, for example, on the order of 100 μm, which means that a 60-dB compensation is required (at 30 Hz) for radial tracking of the focused beam. As shown in Figure 1.13, this can be performed by a pivoting mirror.

The tracking error signal can be extracted from the main reading beam, or separate beam(s) can be used. The average radiant energy reflected back from the disc through the objective depends on the position of the focused spot with respect to the tracks. If the focused beam is wobbled across the track on the order of tenths of micrometers, the dc current from the photodiode will contain a component at the wobbling frequency. A phase detector-multiplier, with a tracking position signal and output from photodetector wobbling signal as the multiplier inputs, would generate an error signal. In a $\lambda/8$ pit system, reflective discs ($\lambda/4$ for transmissive disks), the pit edges principle can be used, similar to that for signal readout. To detect edges parallel to the track, the photodiode must also be split parallel to the track. This is usually performed with a four-segment photodetector.

In the $\lambda/4$ system (reflective discs), two auxiliary beams of light are used which are slightly displaced from the centerline of the track, in opposite directions, so that they are partly on and partly alongside the track (Figure 1.15). A spatial grating is used to generate the auxiliary beams (Figure 1.13). Two separate detectors are used, and the error signal is formed as a difference between those two detector outputs. When the radial position of the reading spot is correct, the two auxiliary beam intensity profiles, shown in Figure 1.15, are symmetrically shifted and the error signal is zero. The polarity of the error signal shows the direction of the radial displacement.

When special effects are needed (e.g., a still frame), "a jump back" to the last track may be needed. The jump back of the light beam cannot be accomplished

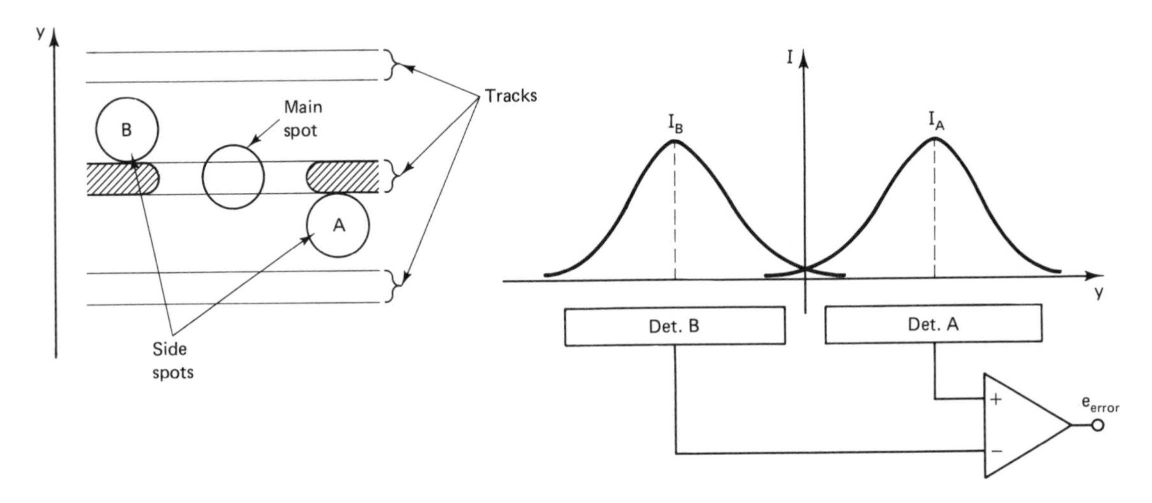

Figure 1.15 Principle of radial tracking.

by a mechanical system such as a radial drive motor and slider. As in the spindle, a pivoting mirror is used to deflect the beam to the next track. This is usually done in the vertical interval so as not to disrupt the picture. Thus the tracking servo incorporates both long-term slow control via the slider and short-term fast control via the mirror. The jump back one track (1.6 μm) is obtained by incorporating an additional pulse into the feedback loop of this system. The same technique can be used to allow for 2× play, 4× play, or 1× reverse simply by changing the amplitude and polarity of the kick-back pulse.

1.4.3.4 Tangential (Time-Base Correction) Servo. In a videodisc player, linear speed of the track, as seen by the objective, is not constant. This is caused by changes in the rotational speed of the motor, by imperfections of the disc, and by unavoidable tolerances in the centering of the disc on the turntable. The eccentricity of the track is the main cause of time-base error (TBE). Time-base errors of over 10 μs are possible. A maximum short-term time error of 5 ns results in satisfactory performance in any TV receiver, so a reduction of over 66 dB is required (at 30/25 Hz).

Time errors can be corrected by a second pivoting mirror. Also, electronic compensation methods (using CCD elements, for example) are used. The time-base error can be detected using a color subcarrier or specially recorded pilot signals, for example.

1.5 OPTICAL PICKUP: RECORDABLE DISCS

This subject is discussed in more detail in Ref. 1, Chap. 7. Both write-once and erasable discs are discussed there. Only brief coverage is included here.

1.5.1 Signal on the Disc

In contrast to mass replica discs, where recording takes place before disc production, here recordable discs are formed first and then the signal is recorded.* The signal on the disc is mainly, as in the cases discussed previously, in the mechanical form. The various forms are exhibited: pits, holes, optical density change, crystalline/amorphous change, and so on. In erasable magneto-optical discs, the signal is placed in the magnetization direction. The erase process is the same as the recording process: data are erased by reversing the direction of the magnetic field applied while the laser heats the surface of the recording film. The recording mechanism of phase-change optical media is based on reversible switching between the amorphous and crystalline status of the materials by heating with a laser. The crystalline/amorphous transition can be reversed simply by applying energy; data recording is then completely erasable.

1.5.2 Readout Principles

In all cases, a signal on the disc modulates one of the light-beam parameters: amplitude, polarization, and so on. If pits are recorded (e.g., dye discs), the pit depth is optimized for the given reflectivity and detection process. The principles are the same as for the mass replica optical discs ($\lambda/4$ or $\lambda/8$ effective depth).

Readout of the ablative thin film (discs) is based on the light amplitude modulation by the recorded pattern. Either reflection or transmission of light can be used. The magneto-optic readout is done by using the principles of either the Kerr or the Faraday effect. In the Kerr effect a linear polarized beam reflected off a vertically magnetized surface will have its polarization partially rotated. An elliptic polarization is created, with the axis of the ellipse depending on the direction of magnetization. In the Faraday effect, the polarized beam passes through the material rather than being reflected from its surface.

Readout of erasable discs with phase-change optical media is provided by the difference in the optical properties between the two states. Because the crystalline form is both more reflective and more opaque than the amorphous material, such a disc could be used in either reflective or transmission playback mode.

1.6 FEATURES AND LIMITATIONS

The several videodisc systems have varying degrees of technological differences, but the following operational features can be found, in general:

1. Freeze frame
2. Frame-by-frame viewing (forward or reverse)
3. Scan (forward or reverse)
4. Slow motion (forward or reverse)

* Some signals, sync for example, can be prerecorded.

5. Frame number display
6. Search (rapid access to a desired frame)
7. Auto stop (stops the normal playing mode at a preselected frame and goes into the freeze-frame mode)
8. Reject (canceling of the operating mode; returning the disc to the load position)
9. Dual audio (selecting either one or both of the audio channels)
10. Programming features:
    a. User programming capability
    b. Self-programming capability (self-programmed discs)
    c. Direct computer interface
    d. External synchronization (to the composite signal and color subcarrier)

For the existing optical and capacitive videodisc system, the parameters are closely related [2].

Basically, there are five types of videodiscs: three optical (reflective, transmissive, absorbtive) and two capacitive (grooved and flat disc). In principle, two modes, CAV and CLV, can be used. Also, signal processing can be performed according to one of the existing standards (NTSC, PAL, SECAM) or in compressed bandwidth form for each of them. Thus total industrial standards are practicably impossible, although detailed standards for the various disc formats do exist. For now, the only total "standards" are: the outside diameter of the disc is approximately 30 cm; the VHD disc is smaller, about 25 cm; the audio bandwidth is 20 kHz; and pulse frequency modulation is used as the video and frequency sharing principle (linear summation) for adding the audio carrier(s). Another example of variation in videodiscs is the difference in spindle speeds. In the CAV mode, the rotational rates used are 1800, 900, or 450 rpm (30, 15, or 7.5 rotations per second). Random access by keyboard control is typically available.

In Table 1.2, some examples are given for optical reflective and capacitive discs.

The recorded wavelength, track width, rate of rotation, and playing time are tied together by a set of equations. None of the systems now available have, simultaneously, a rotational rate of 1800 rpm (corresponding to one TV frame per revolution) and a playing time of 60 minutes per side. A lower rotational rate is mechanically simple to handle. But selecting, for example, the 450-rpm rate dictates a minimum wavelength of about 0.5 μm, a small value. Less rpm allows a relatively large track pitch; this means less crosstalk between adjacent tracks.

Capacitive videodisc systems are limited by mechanical constraints. Pickup sizes, stylus controls, and so on, are dictated by mechanical laws. Optical videodisc systems are limited by optical laws. This is illustrated in Figure 1.16. For the case of round, uniform, equal apertures, defocus of

$$\pm \frac{\lambda}{2(\mathrm{NA})^2}$$

TABLE 1.2 VIDEODISC SYSTEM PARAMETERS

Parameters	*Pickup Methods*			
	Optical: reflective		*Capacitive*	
System designation	*Constant angular velocity (CAV)*	*Constant linear velocity (CLV)*[a]	*CED, grooved discs*	*VHD, flat discs*
Disc rotation mode	Constant rotation rate	Varying rotation rate; constant track velocity	Constant rotation rate	Constant rotation rate
Disc material	Plastic with metal coating	Plastic with metal coating	Conductive plastic with lubricant	Conductive plastic
Disc diameter (cm)	30	30	30	25
Rotation rate (rpm)	1800	Continuously variable 1800–600	450	900
Playing time per side (min)	30	60	60	60
Track pitch (μm, estimate)	1.6	1.5	2.7	1.4
Minimum wavelength (μm)	1.2	1.13	0.55	0.59
Tracking method	Servo to follow signal track	Servo to follow signal track	Groove in disk	Servo to follow between tracks
Vertical tracking	Focus servo	Focus servo	Stylus wear	Stylus wear
TV frames per revolution	1	Continuously changing	4	2
Stop picture(s)	Yes	No	Yes	Yes
Visual picture during scan	Yes	No	Yes	Yes
True slow motion	Yes	No	No	Yes

[a] Digital audio disc, also called compact disc (CD), is of this type; the rotational rate is continuously variable: 500–215 rpm.

takes the system halfway to the first null of response; this range is frequently referred to as the *depth of focus* of the system, d:

$$d = \frac{\lambda}{(\mathrm{NA})^2} \tag{1.6}$$

The cutoff spatial frequency, for a diffraction-limited system and perfect focus, ν_c, is

$$\nu_c = \frac{2\mathrm{NA}}{\lambda} \tag{1.7}$$

It can be seen that the depth of focus drops rapidly with increasing cutoff frequency, and the maximum useful cutoff frequency falls as the required depth of focus increases [1, Chap. 5].

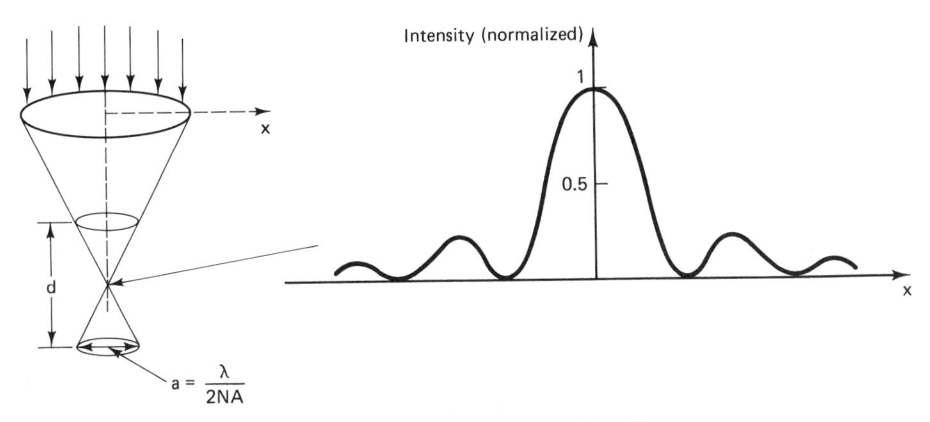

Figure 1.16 The focused spot has a finite diameter.

The smallest focused spot diameter (Figure 1.4)

$$a = \frac{\lambda}{2\text{NA}} = \frac{1}{\nu_c} \tag{1.8}$$

can be made smaller, which means high (pickup) resolution, by increasing NA. That would require more expensive servo controls. For the consumer player this is compromised (price/resolution) and NA is typically 0.4 to 0.6. The wavelength is determined by the laser used: 441.6 nm for a Cd laser or He-Cd, 488 nm for an argon laser, 632.8 nm for He-Ne, 840 nm for laser diodes, and so on. Laser diodes would be preferred over gas lasers, but two problems with them still exist: long wavelength and higher noise than for gas lasers [9, p. 56]. Detailed modeling of the various interactions in the videodisc system would allow one to more readily evaluate the alternatives, while independently varying different parameters [10].

At first glance, the noise-amplitude probability density in the system with the optical readout seems to be quite close to a Gaussian distribution; however, detailed analysis [11] shows that the Gaussian distribution may be taken only as the first crude approximation to the experimentally determined probability density function. The power spectral density is almost uniform (i.e., the noise in the system is white). The slope of this function can be considered to be the result of the MTF of the read lens, and the slope is somewhat lower if the MTF compensation circuit is included in the player. It may also be noted that the experimental work does not show a clear difference in disc noise with or without information recorded on the disc.

REFERENCES

1. J. Isailović, *Videodiscs and Optical Memory Technologies,* Prentice-Hall, Englewood Cliffs, N.J., 1985.

2. J. K. Clemens, "Video Disk: Three Choices," *Spectrum,* Mar. 1982, pp. 38–42.
3. Special Issue on Videodiscs, *RCA Rev.,* Vol. 35, No. 1, 1978.
4. Special Issue on Videodisc Optics, *RCA Rev.,* Vol. 39, No. 3, 1978.
5. Special Issue on Videodisc Systems, *RCA Rev.,* Vol. 43, No. 1, 1982.
6. T. Inoue, T. Hidaka, and V. Roberts, "The VHD Videodisc System," *SMPTE J.,* Vol. 91, No. 11, 1982, pp. 1071–1076.
7. Five papers on video long-play systems, *Appl. Opt.,* Vol. 17, No. 13, 1978.
8. C. Bricot, J. C. Lehureau, and C. Puech, "Optical Readout of Videodisc," *IEEE Trans. Consum. Electron.,* Vol. CE-22, 1976, p. 304.
9. Optical Disc Technology, *Proc. SPIE,* Vol. 329, Jan. 26–28, 1982, Los Angeles.
10. V. B. Jipson and C. C. Williams. "Two-Dimensional Modeling of an Optical Disc Readout." *Appl. Opt.,* Vol. 22, No. 14, 1983, pp. 2202–2209.
11. J. Isailović, "Channel Characterization of the Optical Videodisc," *Int. J. Electron.,* Vol. 54, No. 1, 1983, pp. 1–20.

two

Recording and Disc Production

2.1 INTRODUCTION

In Chapter 1 we discussed playback systems: the performance requirements imposed by low-cost discs and design considerations imposed by an economical player. In this chapter we discuss videodisc recording, mastering, and mass replication.

For read-only videodiscs, the recording is part of the disc production cycle, while the signal recording takes place after writable discs are produced. In both cases, it is desirable to read during recording. The direct-read-after-write (DRAW) function is readily implemented using a secondary beam which forms a read spot that trails the primary writing spot by several micrometers (Figure 2.1). Approximately 1 μs after the data are recorded, the trailing beam reads the data and verifies that they are correct. This represents an important element in the overall strategy to control errors resulting from media microdefects. It should be noted that the DRAW is referred to the recording media, not the disc, property. This makes a great difference for mass replica (prerecorded) discs.

One important feature of the videodisc is the high information density which can be, for read-only discs, replicated in an inexpensive medium such as polymethyl methacrylate (PMMA) or polyvinyl chloride (PVC) plastic, or for writable discs, directly recorded on the disc allowing instantaneous playback. This leads to the following requirements for the recording [1]:

1. Materials and technology for reliable recording with sufficient resolution to make well-defined topographical signal elements approximately 1000 Å deep and a fraction of a micrometer in width and length.

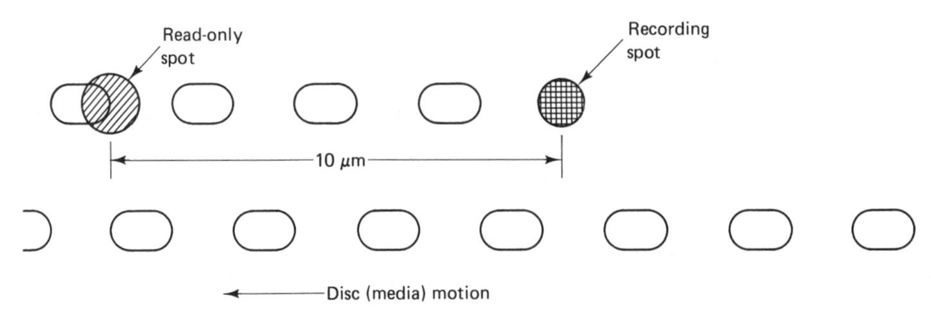

Figure 2.1 Principle of DRAW.

2. Recording in real time, that is, making signal elements at an average rate between approximately 4 and 10 MHz.
3. Recording the signal elements in a uniform spiral (even track spacing).
4. A process for obtaining from the original recording a flaw-free metal master qualified to be the starting point for fan-out to stampers for use in pressing thousands of finished discs.
5. Read after write (if possible).

Three major processes are involved in the disc fabrication. The first is premastering: gathering together all video and audio and proper control information. The second process is mastering, which consists of recording the master and production of the metal mold from the master. The third is the replication process, which produces discs from the mold.

2.2 PREMASTERING: EDITING AND SIGNAL ENCODING

The video and audio software for the recording can originate from different sources. Programs are received for mastering in the form of film (35 mm or 16 mm), slides, or any professional form of videotape. Computer or control software may also be required at this time if level 2 program damps are used. The video signal can be obtained from cameras or film chains and can be electronically generated, for example, as a test signal, character generator, or multilevel coding signal based on a digital data. Signal editing is performed to insert identification number codes in the vertical blanking interval, flags, sync signals—if needed, and so on. Also, special commands and a "program dump" for the player's computer can be added. Materials may be edited together from several sources; color correction, if any, may be accomplished; audio may be added or changed; and frame-by-frame address codes can be inserted in the programming. Each video frame could be uniquely identified for fast electronic recording and synchronization. Thus the duration of a selected scene or program interval can be determined with frame accuracy. If, after recording, videodiscs are used in editing, this frame identification allows interchangeability between editing systems. The IN and OUT points for edits selected on one editing system can be listed by their ID designation and input to another editing system by electronic mail, for example. Also, the frame identification allows precision synchronization of one player to another. This means perfect frame-to-frame matchups at the edit points; hence clean edits that exhibit no video breakup.

Frame number and/or the SMPTE/EBU vertical interval time code (VITC) are used for the frame identification. Both record the indexing information for each field/frame in the video signal during the vertical blanking interval. In laservision (LV) systems, the frame numbers are between 0 and 54,000, and 24 or 40 bits.

The SMPTE/EBU identifies every single video frame by hour, minute, second, and frame: for example, 11 (hours), 26 (minutes), 54 (seconds), 17 (frames). Total number of bits per frame is 90. Most of these 90 bits have a specific value assigned to them. Some indications obtained from VITC are: the end of one frame and the start of another, whether the disc is read forward or reverse, whether the code was recorded in drop-frame of nondrop-frame format (NTSC color signals have an actual frequency of close to 29.97 frames per second, so a generator counting at 30 frames per second would produce an increasing error of 3.6 s—108 frames—every hour), and so on. The VITC is normally recorded on two nonadjacent vertical blanking interval lines in both fields of each frame (lines 10 through 20 in NTSC). Recording the code four times per frame provides a redundancy factor which lowers the possibility of reading errors due to disc dropouts. In conjunction with the included error checking code bites, redundancy thus virtually eliminates reading errors.

The signal encoding is used to produce a properly tailored signal for the videodisc channel. Luminance, chrominance, and audio signals can be encoded in different ways. Composite video can be formed according to one of the existing standards (NTSC, PAL, SECAM) or in some other way, in order, for example, to get an extended play time [1–8].

The signal format of current NTSC and PAL laservision (LV) is a two-level signal (HF) which is frequency modulated, after preemphasis, by the composite (luminance and chrominance) video signal. Addition of the stereo sound signal is achieved by means of pulse-width (duty-cycle) modulation of the HF signal by the two frequency-modulated audio carriers. Figure 2.2a shows a block diagram of the signal path of the encoder. The signal $x_0(t)$ is the frequency-modulated composite video signal. Signals $x_1(t)$ and $x_2(t)$ are the frequency-modulated sound signals. Also, a digital audio channel placed at a normally noisier part of the channel to lower frequencies, can be included. The sum signal is limited, so that a pulse-width and frequency-modulated two-level signal $y(t)$ results. A detailed discussion is given in Chapter 6, and the main parameters of the disc and player for the LV standard accepted in 1978, are given in Tables 6.2 and 6.3, respectively.

The luminance bandwidth can be chosen to be less than specified by TV standards. This will result in (horizontal) resolution reduction. If the line and frame rate are kept the same, the discrete line structure of the spectra will remain the same as for the standard TV signal, so the interlace technique can still be used. The color information can be processed relative to the luminance information, either on a time- or a frequency-sharing basis. But since they must be matrixed together (if placed in separate channels), the amplitudes, frequency responses, and the noise characteristics of the two channels should be matched. An interlace technique allows the color signal to be placed in the luminance band (Figure 2.2b) and recovered in the player. This technique is used in the capacitive systems, because mechanical limitations reduce the channel bandwidth below the bandwidth required for the standard TV signal. In the capacitive electronic

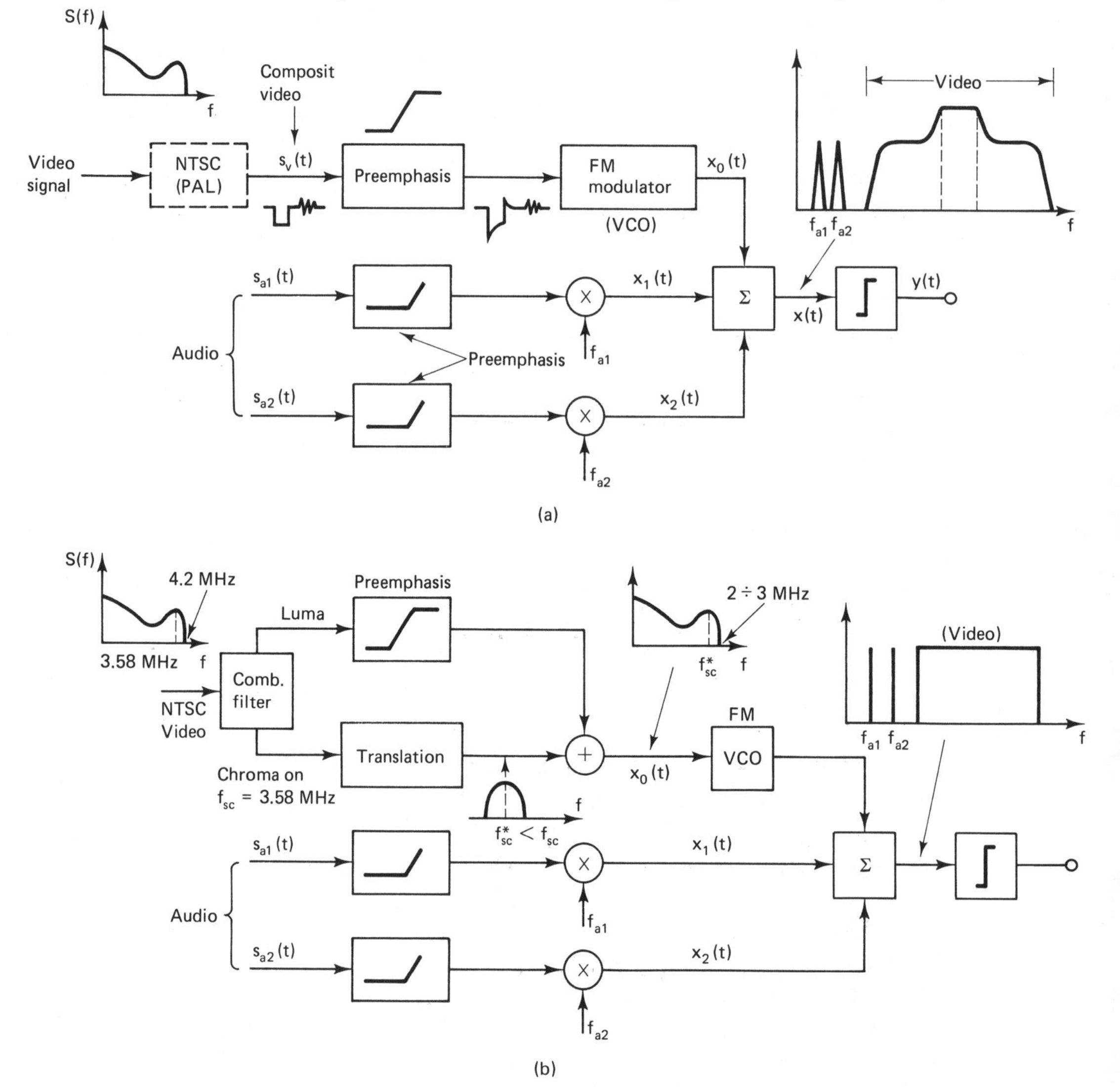

Figure 2.2 TV signal encoding: (a) standard TV signal; (b) NTSC-like compressed system (bandwidth reduction, "buried subcarrier"); (c) color under.

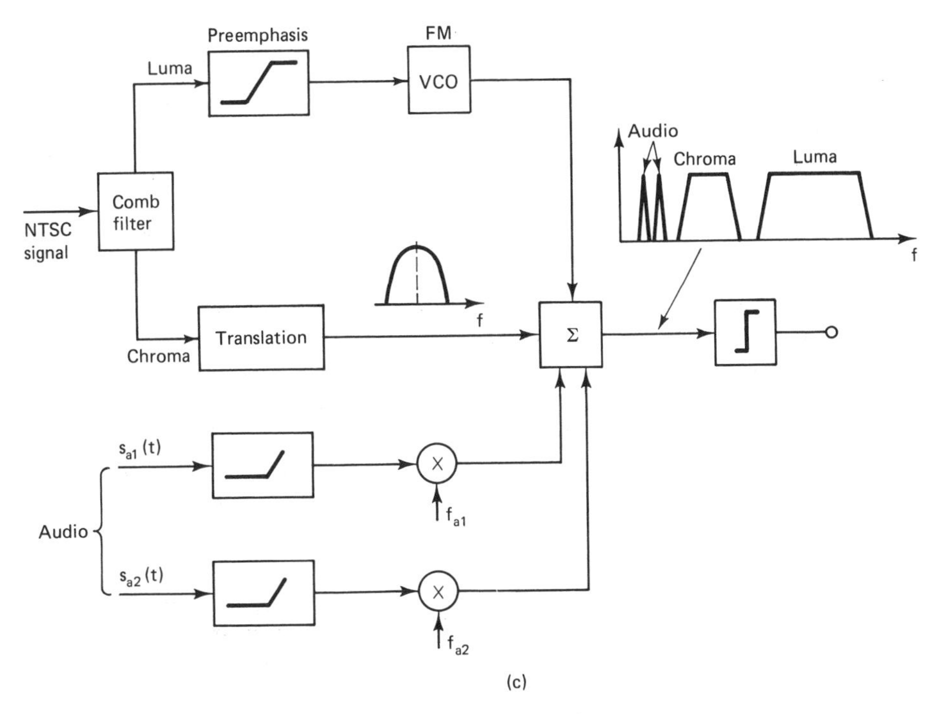

Figure 2.2 Continued

disc (CED) system, the technique is called *buried subcarrier*. The method shown in Figure 2.2b is also used in the optical videodisc systems to get extended play time.

Assuming that it is desirable to maintain the chrominance information in its quadrature-modulated format, there is another approach for encoding a composite video signal. The method consists of separating the chrominance subcarrier from the luminance part and encoding them differently, for example, in a *color under* or *crossband* scheme, where the subcarrier is transposed to a lower frequency [7,8]. The system is used mostly where the recording medium has a rather limited bandwidth.

Figure 2.2c is a block diagram of the record signal processing. A key element is the comb filter, which allows separation of the luminance and color without band limiting. A video preemphasis is employed before FM modulation.

A time-sharing principle can be used to form a signal for the videodisc channel. In a multiplexed analog component (MAC) system bandwidth/time exchange is applied (Figure 2.3). The luminance signal (Y) of the one TV line is compressed in time into one-half of the active line interval, thus occupying a two-times-wider bandwidth. In the same way, the two color-difference signals, for example B-Y and R-Y, are compressed each into approximately one-fourth of the active line interval, thus occupying the same bandwidth as the time-compressed luma component [9] (see also Refs. 10–13). This approach offers at

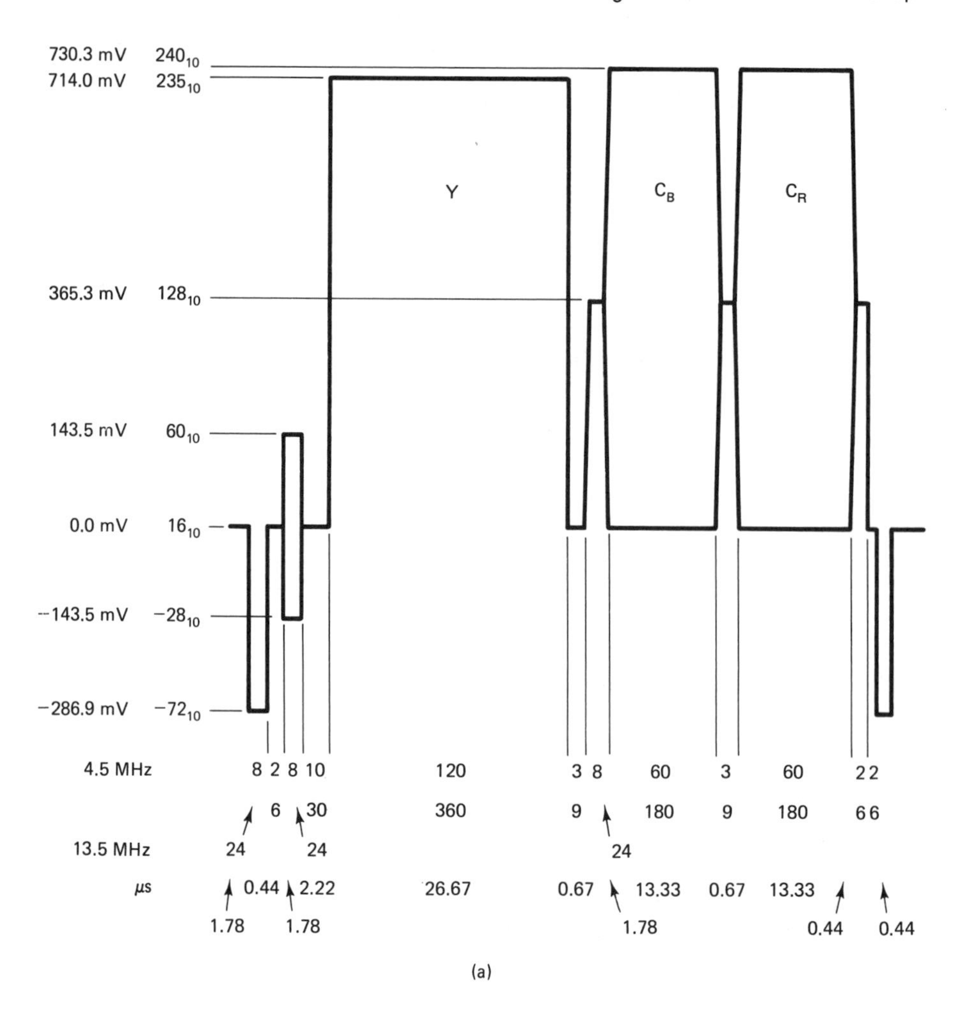

(a)

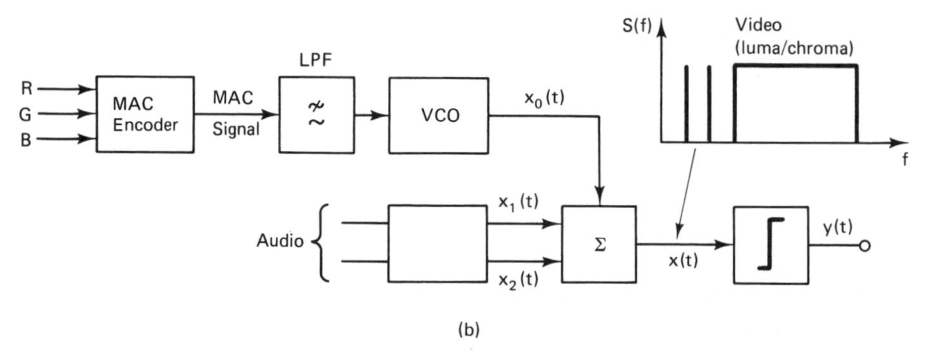

(b)

Figure 2.3 Multiplexed analog components: (a) waveform; (b) block diagram for a videodisc recording.

least two advantages compared with the standard frequency-sharing (interlace) technique. First, there are no luma-chroma beat components introduced by the system nonlinearities. Second, the chroma noise (FM noise) will be less, assuming adequate preemphasis/deemphasis.

Figure 2.3b shows a simplified block diagram of the MAC signal encoding for the videodisc recording. Compared with Figure 2.2a, the difference is that a MAC signal, instead of a composite TV signal, frequency modulates VCO.

2.3 OPTICAL DISCS: MASS REPLICAS

Playback-only videodiscs are pressed from masters in a manner similar to the technique used to make audio records, except for the much higher packing density. Making the master recording requires very sophisticated equipment. The main steps involved in making a master recording and producing discs in quantity form are [1]:

1. Premastering
2. Videodisc mastering
   a. Quality inspection
3. Stamper replication, also called matrix processing
   a. Quality inspection
4. Disc replication
5. Metallization
6. Disc coating
7. Miscellaneous: central hole, etc.
   a. Testing
8. Packaging

Premastering has already been discussed, and we will continue with a discussion of recording media before considering mastering.

2.3.1 Recording Media

Ideally, the recording medium for the optical videodisc master should provide [1, Chap. 2] high resolution (greater than 1000 cycles/mm), high sensitivity, high smoothness, high signal-to-noise ratio (SNR), real-time recording and instant playback, and high freedom from defects. A number of available materials have been screened. These include thin metal films, ablative materials, silver halide materials (both evaporated and in emulsion), and both positive- and negative-working photoresists. There is still room for new materials.

Photoresists and metal films are used mostly as optical recording media. The depth of the photoresist, 1300 Å for example, corresponds to the real depth of the pits on the plastic replica. The depth of the metal film is significantly less, 300 Å, for example. Thin metal films are needed because of the melting process involved in the recording process. The typical minimal hole size is approximately

0.4 μm, obtained after development of photoresist. The corresponding diameters for metal film are 0.6 μm in tellurium (Te) and 0.8 μm in bismuth (Bi). In both cases the edges are distorted approximately 0.1 μm. This can cause both proportional and fixed-length asymmetry. This distortion will cause second harmonics to be generated on playback.

A glass substrate is usually used for the master disc, which is 365 mm in diameter and about 6 mm thick. Substrate discs are cut from twin-ground plate glass, the surface of which contains hundreds of small pits/mm^2. The glass substrate is therefore reground with a fine abrasive to eliminate the deepest pits. Finally, the surface is optically polished until the pit density has been reduced to less than 1 pit/mm^2.

As a substrate material, polymethyl methacrylate (PMMA) can be used. PMMA is a plastic material and is normally produced, at least for this purpose, by casting between two polished plates. Good gross flatness, minimal surface roughness, low surface defects, and long-term stability of the plexiglass material itself are some of the advantageous characteristics of this substrate.

2.3.2 Mastering: Optical Recording

There is more than one way to produce a master disc. Each method has unique advantages. In addition to criteria that are strictly technical, the choice among mastering methods also depends on commercial considerations such as capital and operating costs, yields, process control, and so on.

The limits of an optical recording system are defined by the ability of the recording lens to form the laser beam into a tightly focused spot. For a uniformly illuminated diffraction-limited lens, the focused spot width, w, between half-intensity points can be represented by

$$w = \lambda/2\text{NA} \qquad (2.1)$$

where λ is the wavelength of the light and NA is the lens numerical aperture. In practice, because high numerical aperture lenses are not perfect, the focused spot width is slightly larger than that predicted by equation 2.1. The finite size of the focused spot in an optical recorder for a videodisc affects the response of the system: the response tends to roll off at high spatial frequencies, the spatial frequency being a function of the temporal frequency of the recording signal, the frequency of rotation of the disc, and the radial position of the recording beam on the master disc.

The choice of NA for the focusing lens in the recorder differs from that for the player. Namely, higher NA gives better resolution but smaller depth of focus. Smaller depth of focus requires a higher-quality focus servo and more expensive lenses, which can be justified for mastering recorders. Conversely, consumer products are subjected to strong pressure on product price. For the player, NA is typically 0.4 to 0.6, whereas for master recorders it is 0.65 to 0.75.

Figure 2.4 shows the optomechanical system for mastering. It includes both

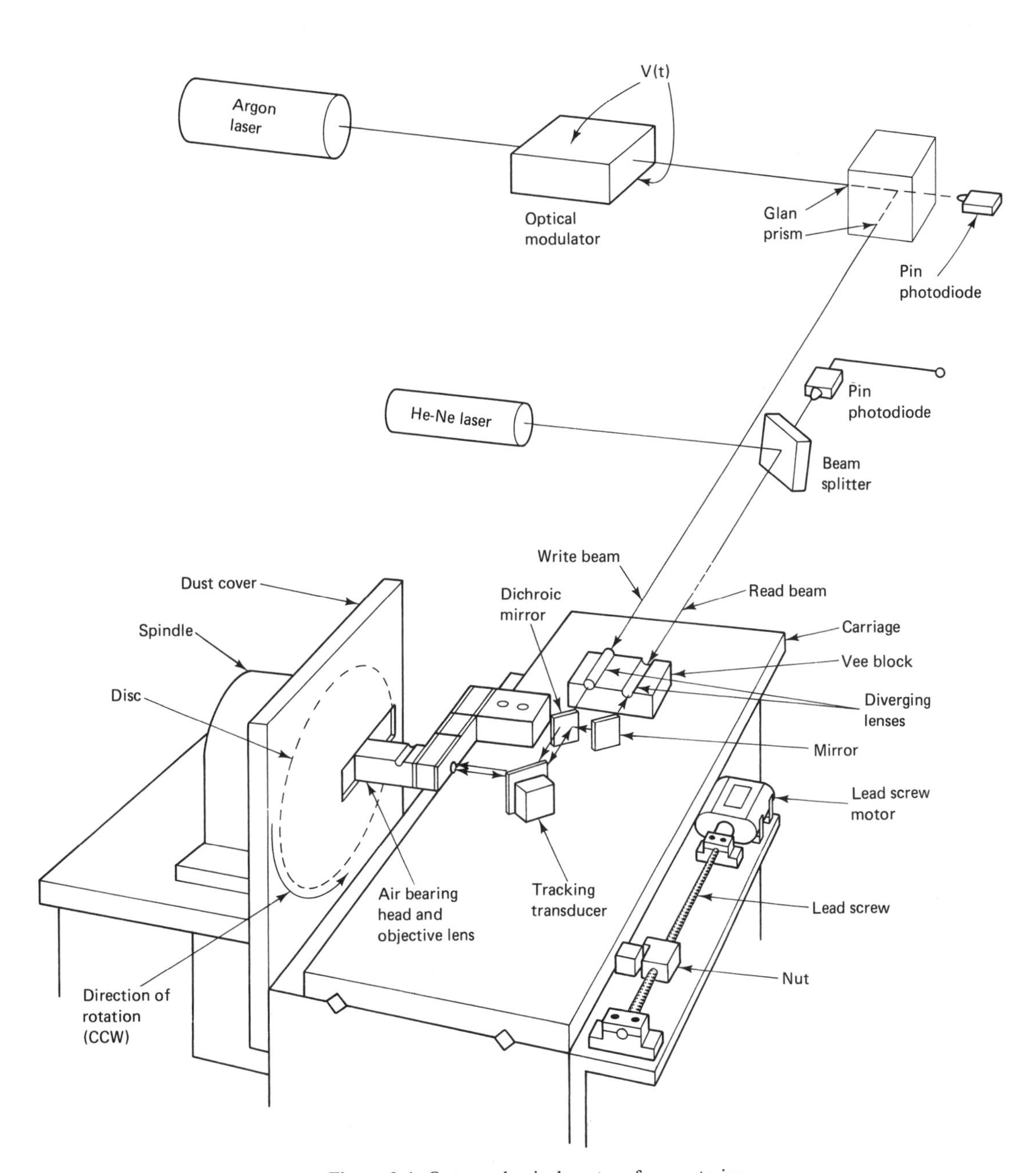

Figure 2.4 Optomechanical system for mastering.

a writing laser and a reading laser, providing a direct-read-after-write (DRAW) capability. An argon laser of several watts generates the light beam for recording the information on the disc. The choice of laser power is, to a great extent, directly affected by the choice of a mastering method. More specifically, it is dependent on the material used to coat the master disc substrate. The light beam is modulated in the electro-optic modulator or acousto-optic modulator. It then passes through the corresponding focusing lenses and mirrors and is led to the disc. The cost and transmission efficiency are advantages of the acousto-optic modulator over an electro-optic modulator. The light of wavelength corresponding to the argon radiation passes through the microscopic lens with a numerical aperture which enables the metal or photoresist surface covering the disc to be illuminated in an area of less than 1 μm in diameter. This ensures the formation of a spiral channel less than 1 μm wide. For the metal film, the laser intensity and the thickness of the disc metal cover–film are so adjusted that at the maximal laser intensity corresponding to the higher level of electrical signal brought to the optomodulator, a hole is made in the metallic film. The light-beam intensity corresponding to a lower level of the electrical signal is not sufficient to make a hole in the metal film.

To keep the microscopic lens at the same, very small distance from the disc, it is supported by an air cushion obtained by an air bearing under a relatively high pressure. The disc itself is in an enclosure protecting it against dust.

The light beam creates a spiral channel less than 1 μm wide. The contents of the channel aperture depend on the content of the signal fed to the optomodulator. The spiral trajectory is obtained by a simultaneous rotation of the disc around a fixed axis and by a horizontal shifting of the microscopic lens by the lead screw.

Exposing a thin photoresist film [14] on a rotating substrate with a focused and intensity-modulated laser beam and subsequent development result in a pit pattern on the disc arranged in a spiral track. After development, the minute areas of exposed photoresist are washed away, leaving a series of holes, or pits if positive resist is used. Many parameters, including those of the modulator signal, the resist film thickness, the applied laser power, the substrate, and so on, determine the final dimensions of the pits. At this point, the developed disc is ready for galvanic processing.

The optimum depth of the pits (i.e., the thickness of the resist film) is determined by the playback system. Reading out a replica in reflection through the bulk material yields, for the central detection, an optimum modulation depth of the signal at a pit depth of

$$d_{opt} \geq \lambda/4n \tag{2.2}$$

For $\lambda = 633$ nm (the assumed wavelength of the laser in the player) and a reflective index for a vinyl disc of $n = 1.54$, the optimum thickness on a master disc is $d_{opt} \geq 103$ nm.

Although the thin-metal-film category sounds broad, functionally it means tellurium or bismuth, or, more precisely, alloys of the group 6A elements in which a focused laser beam can easily produce small holes because of its 450°C melting point and low thermal conductivity.

Sensitivity, a material property, translates into recording speed, and thus is a systems property. Typically, 1-μm spots can be ablated at speeds of 10 million per second with the modulated output of a continuous-wave 10-mW laser. Fast recording requires short laser pulses, hence higher optical power to deposit the same amount of energy. The relationship is not purely reciprocal, however; in general, less energy is needed in a short pulse than in a long pulse because there is less time for the heat to dissipate.

Because the hole-forming process is a thermal—not an optical—effect, the laser wavelength is not critical as long as the light is absorbed. That is significant because it allows writing (recording) with any laser that can deliver the required power at the required speed—a category that includes high-power, single-mode diode lasers which are compact and easy to modulate.

The average power in the spot is of the order of 20 mW. Since the FM carrier frequency is about 8 MHz, 8×10^6 holes are cut per second, and the energy per hole is 2.5×10^{-3} J.

After the master disc has been cut, it must be transformed into a configuration from which replicas can be made. This is done by transforming the essentially "two-dimensional" master record into a "three-dimensional" configuration which can be used to stamp or form inexpensive, plastic replica discs.

The essentially flat hole pattern produced on the thin metal film could not be read if it were replicated by molding. The difference in height between the glass and the uncut metal (200 to 300 Å) is too small to produce much optical interference, or scattering, at visible wavelengths. The height difference can be enhanced by coating with photoresist and exposing through the holes. This exposure will polymerize negative resist at the hole sites and will leave bumps or holes/pits when the unexposed resist is developed away. (Positive resist will yield holes at the hole sites.) This bumpy surface can then be reproduced in metal by the same galvanic technique as that used in audio disc processing.

The master is therefore coated with a layer of negative photoresist material and is exposed through the rear (undersurface) of the disc. An ultraviolet-light source exposes (polymerizes) the photoresist through the information holes. The uncut metal film shields the photoresist where there are no holes. This results in an array of hardened areas which coincide with the initial array of information holes. The unpolymerized photoresist material is then washed away with an appropriate solvent, leaving bumps over the holes. Depending on the photoresist used, the hardening program, and other parameters, the height and profile of these bumps may be tailored to optimize the optical contrast between the bumps and the surrounding flat area when they are illuminated by the optical scanning system of the player. Next, 500 Å of nickel is evaporated over the photoresist, and the discs are then played back on the test player, for example with the 0.45 NA lens.

Photoresist mastering requires two fewer steps than that required for metal film mastering: the disc is coated only once rather than twice; and photoresist exposure takes place during mastering rather than in a separate operation. However, the read-while-write feature can be lost, so that the recorded master disc quality is unknown until the photoresist is developed and a metal film deposited on it.

2.3.3 Stamper and Disc Replicas

After the metal-coated master is tested, it is nickel plated to produce a mother, and then the mother is separated from the master. At this point, the mother can either be used directly as a stamper for injection molding, or it can be plated again, to make an intermediate piece of tooling, which can in turn be plated to make additional submaster stampers.

The production of videodisc stampers from masters is the same as that used in audio record manufacture except that it is highly refined. The same basic processes for metal-to-metal duplication, utilizing nickel as the electroformed metal, and passivation of the nickel surfaces for duplication are used. Thus electroforming of nickel metal molds and molding stampers from the recorded master is called *matrix processing*. Enclosure stamper takes 6 to 8 hours at about 380 μm thickness. Quality inspection is performed after each step for possible defects. Steps for the manufacturing process from the master disc to the metal mold, nickel stamper, are shown in Figure 2.5.

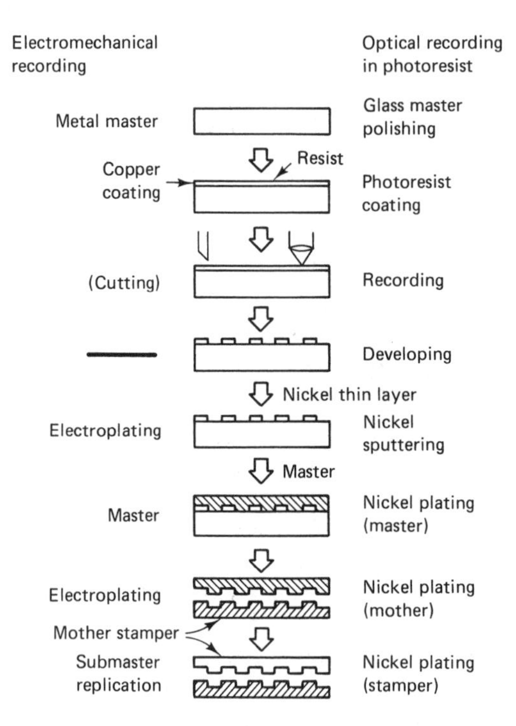

Figure 2.5 Mastering and master replica.

In general, the number of stampers depends on the desired number of replicas. To produce 20,000 to 30,000 replicas or less, the mother can be used as the only stamper. Otherwise, the fan-out process must be used to make additional stampers: multiple metal masters can be made from the original master, so a large number of stampers, and better replicas, can be made from a single recording.

Both sides of the disc can be processed at once, or separately, and then two one-sided discs can be mounted together. Compression molding is the oldest method used for the production of audio records and remains the technique used almost exclusively for 30-cm record production. The stamper is precisely centered and mounted securely into the mold and force is used to form a disc with the recorded safety. Before the halves of the mold cavity are pressed together, forcing the warm PVC to assume the shape of the cavity, the mold is heated. Some time after the mold is closed, it is cooled so as to solidify the PVC and then opened so that the disc can be removed. This technique is used to make thick videodiscs, and has been used, for many years, to make audio LP records. In injection molding, the halves of the mold cavity are first clamped shut, then hot PVC is injected at high pressure to fill the cavity. The mold is cooled to solidify the PVC, opened, and the disc is removed. This technique is used to make 45-rpm audio discs and is being used to make thick videodiscs. One-sided discs may also be pressed by injection molding. Compression and injection molding are thermoforming processes commonly used to replicate thick videodiscs. A third distinct thermoforming process is embossing. A preformed sheet of plastic is placed in a press. A heated stamper that bears the encoding is pressed against the surface of the sheet long enough to transfer the encoding without deforming the sheet. This technique is used to make thin "mailable" audio discs or thin videodiscs, and is a web press technique similar to the method used to print newspapers. The injection compression method combines the merits of the two methods. Recently, a method has been developed in which the resin is cured by ultraviolet light [15].

Besides thermoforming, videodiscs may be replicated by casting and photopolymerization techniques. Thin videodiscs can also be made by a casting process. A thin film of liquid-resin monomer is first applied to a preformed substrate, and then the liquid resin is pressed against an encoded mold surface at relatively low pressure. After contact, the resin polymerizes (either spontaneously or with the help of heat or radiation) and the disc may be separated from the mold. Materials that are pigmented and difficult to cross-link by photographic methods may be polymerized with electron beam radiation [16].

The mold is derived from the photoresist in the same way that the replica is made from the mold. A film of silicone (elastomeric) rubber monomer is applied to a glass disc and this is pressed against the photoresist surface. The elastomer is cured in contact with the photoresist and then the two are separated. The elastic properties of the cured mold make its separation from the photoresist, and the separation of the replica from the mold, relatively easy. A rigid mold would be more difficult to separate, and distortion of the encoding could result.

Because of the low temperatures, pressure, and shrinkage involved, the casting process introduces the least geometrical distortion of any of the replication techniques. The replicas are nearly as round as the masters, and the encoded detail is faithfully reproduced. The flatness, however, depends on the substrate.

In some systems, for example film-based videodiscs, replicas can be made by means of fairly straightforward photographic contact-printing techniques. In this case, the master disc serves the same purpose as the master negative and is used to make copies in the same way that a photographic negative is used to produce contact prints. This technique has the advantage of low cost per copy, plus the ability to make hundreds of inexpensive copies one at a time. This permits replication at the local level for very little cost in equipment.

2.3.4 Metallization, Coating, and Miscellaneous

The reflectivity of the optical (reflective) disc is improved by deposition of a thin metallic film, usually made of aluminum. The metal film follows the shape of the information elements. To protect the reflective surface from dust and damage, a protective overcoating is needed. The protective cover is a thin layer of transparent polymer. After coating, there are a number of miscellaneous operations (items) to be added, depending on the purpose of the videodisc (e.g., for mass or laboratory application).

One important item is certainly the formation of a hole in the middle of the videodisc, of a definite diameter corresponding to the adopted standards. The disc shape is very important (may cause excessive time-lag errors) because, for example, any eccentricity could be reflected in the quality of the reproduced color. The reflective disc does not need special protection and its packaging is simple.

It is necessary to inspect the quality of the videodisc made, and this control is performed for each step in production. Some level of playback testing must be done to verify product quality for consumer acceptance criteria. The control tests are usually performed on a sample basis. Besides visual (subjective) inspection, some objective measurements must be performed, such as: dropouts, carrier level, video and audio signal-to-noise ratio, intermodulation (IM) products, and eccentricity.

2.4 MASS REPLICAS: CAPACITIVE DISCS

The main steps in capacitive videodisc production are:

1. Mastering
2. Matrix/stamper replication
3. Molding
4. Disc coating
5. Packaging

Signal encoding for a buried subcarrier is shown in Figure 2.2b.

2.4.1 Recording Media

In the grooved capacitive videodisc system, signals are recorded in an electroplated copper substrate using a piezoelectrically driven cutterhead [17–19]. Typically, at one-half the real-time rate, a small diamond tool cuts the FM signals. A peak-to-peak signal amplitude is on the order of 875 Å. Nickel replicas made from this substrate serve as stampers to press the vinyl CED (capacitive electronic disc) videodiscs. The purity, uniformity of properties, and extremely low levels of defects are the most critical specifications for the copper substrate. A layer of fine-grained or amorphous copper is plated on approximately 12-mm-thick aluminum discs, with precise control of copper's metallurgical properties.

Optical recording of capacitive videodisc masters is obtained when the photoresist-covered pregrooved substrates are exposed to the focused light beam in the manner similar to the optical disc recording [17, p. 427]. The recording substrates are trapezoidally pregrooved copper-clad discs onto which a dilute solution of photoresist was spun. Masters for the flat capacitive discs are normally optically recorded.

2.4.2 Electromechanical Recording

In pregrooved systems, information elements can be recorded on the pregrooved surface, or both information elements and grooves can be recorded simultaneously. A 0.5-mm-thick bright copper layer electroplated on a smoothly machined flat, thick aluminum disc provides a substrate material into which to machine the groove with a sharp diamond cutting tool. Application of the resist to the surface of the grooved substrate can be done by a spinning process. For the 5555-groove-per-inch (GPI) format, trapezoidal-cross-section substrate grooves can be used to provide the desired slightly cusped groove shape. For the 9541-GPI format, triangular-cross-section substrate grooves are required to provide a desired final groove depth of 0.2 μm [17, p. 64].

In an electromechanical recorder (EMR) a smoothly turning precision turntable is accurately locked to the signal source by a tachometer and speed servo system. A cutterhead is mounted on a sturdy arm with a translation mechanism that moves the cutterhead smoothly a distance of 1 in. every 9524 turns. (Due to shrinkage in master replication and in disc-molding steps, the final disc has a slightly different number, nominally 9541 turns per inch.) On the turntable is a flat metal disc substrate covered by a layer of material that can be smoothly cut by sharp diamond cutting tools or styli. Before recording, the top surface of the material is carefully machined to be flat.

The sharp diamond recording stylus has a tip-face shape that corresponds to the desired cross-sectional shape of the finished groove and cuts the groove at the same time that it is recording the signal. In operation, the depth of the cut is determined by the position of the cutterhead support relative to the machined surface of the recording material. The cut is deeper than the groove, however, so groove depth is controlled by the shape of the recording stylus tip and the amount that it has been translated between turns. The amplitude of the signal recorded is determined by the high-frequency motion imparted to the tip. Signals to drive the cutterhead are provided through an equalizing circuit, which is necessary to compensate for the amplitude and phase characteristics of the cutter.

The cutterhead is an electromechanical transducer [4]. It is characterized by how its mechanical displacements, which cause the signal undulations in the grooves, correspond to electrical signals applied to activate it. Ideally, its motion should be limited to a single direction perpendicular to the recording substrate surface, and its motion should be exactly proportional to the applied signal voltage. The cutter assembly of practical size does not meet these requirements. It is a problem to achieve this in the presence of natural resonances that occur in the signal band, which is approximately 0.5 – 10 MHz, with mechanical accelerations of 1.5×10^7 g, and internal dissipations that cause large temperature rises. Even with the half-rate recording process (mainly used), the standard 875-Å peak-to-peak signal amplitude requires accelerations of several million *g*'s.

2.4.3 Stamper and Disc Replication

Steps for the manufacturing process from master disc to metal mold with the nickel stamper are similar for all types of playback-only videodiscs [1, Chap. 2]. The matrix processing is used. The original construction of capacitive discs consisted of a nonconductive compression-molded disc made from a polyvinyl chloride (PVC)–polyvinyl acetate (PVA) copolymer, the surface of which contained the modulating signal information and grooves. The compounding process can include some additives to improve compound stability and flow characteristics or to form some desired characteristic. The required surface conductivity was obtained by vacuum deposition of 200 Å of a glow-discharge-deposited organic dielectric filter [17, p. 116]. Instead of surface only, the disc itself can be made conductive. For example, carbon of extremely small and uniform particle size is mixed uniformly into the PVC base resin to achieve disc conductivity. In both cases, conductive or nonconductive discs with conductive surface, both sides of the disc are pressed at once.

The capacitive disc employs a stylus in contact with the disc and requires protection. To ensure that the disc is shielded from surface contamination and handling damage, a protective package, or disc "caddy" is used. The disc never leaves the protective package except during playback, when it is automatically extracted by a simple mechanism in the slot-loading player.

For the pregrooved capacitive disc, a lubricant is required to reduce wear of the disc surface, especially the stylus, to a minimum. A thin coat of a lubricant, approximately 200 Å thick, is applied by a relatively simple spray process to both sides of the disc, simultaneously, after which it is respindled.

2.5 WRITABLE (VIDEO) DISC SYSTEMS

The mastering system for the optical read-only videodisc with DRAW properties can be thought of as a writable (programmable) videodisc system, although an expensive one. Desired optical videodisc master recording material characteristics are also desired characteristics for the optical disc media [1, Chap. 7]. Two of those characteristics, at least in general, have more important roles for the programmable optical disc recording than for read-only (prerecorded, playback only) videodiscs: high sensitivity and high immunity to defects.

Until recently, gas lasers He-Cd, Ar^+, and He-Ne, were the only lasers used for optical recording applications. High output power and short wavelengths are their advantages. But programmable (disc) recording systems require an extremely compact design, with significant reductions in the size and cost of record and playback systems. The size, high efficiency, and potential high reliability of diode (semiconductor) lasers make them more favorable for the programmable optical systems. However, the high sensitivity of the recording material is a must, because of lower output power for diode lasers. Optical recording with an approximately 50-ns light pulse (λ = 800 nm) with a total power of 15 mW or less focused into 1-μm^2 write spot is required. This means that media write sensitivity of 75 mJ/cm^2 or better at the diode laser wavelength is required [20]. For the programmable optical disc the recording medium is also a final memory. For the archive storage the recording material should allow permanent recording of the data and should not degrade under ambient conditions or prolonged readout by a lower-power laser.

Optical disc recording materials can be classified into two groups: nonerasable (nonrecyclable) an ... ıerasable optical recording media are: ablative ... resists, and photopolymers. The recording me ... nd there is not an erase-recycling process f ... cation can be accomplished except for photop ... se-change materials are examples of erasable ... le of optical recording and playback in progra ... e same one as in the read-only videodisc sys ... is used for both recording and reading. Thre ... ial, and focus—control the focused beam in al ... . In addition, a motor-servo is required to mai ... stant. The controller design depends on wheth ... address, operational mode, etc., are required) ... numbers, sync pulses, etc., are required).

A recording ... ıser is shown in Figure 2.6. The power of the ... rmine the recording bit rate, disc speed, and sp ... ght output of the laser is first divided, with mo ... channel that records. The recording beam is ... tor, electro-optic or acousto-optic, which modulates the light intensity in response to an input electrical signal. The attenuator following the modulator is used to reduce the light intensity during playback. The modulator could also perform this function, but because of the potential for accidental erasure, a separate attenuator seems the prudent choice [22]. The beam is then expanded to fill a focusing lens (typically a 0.40- to 0.85-

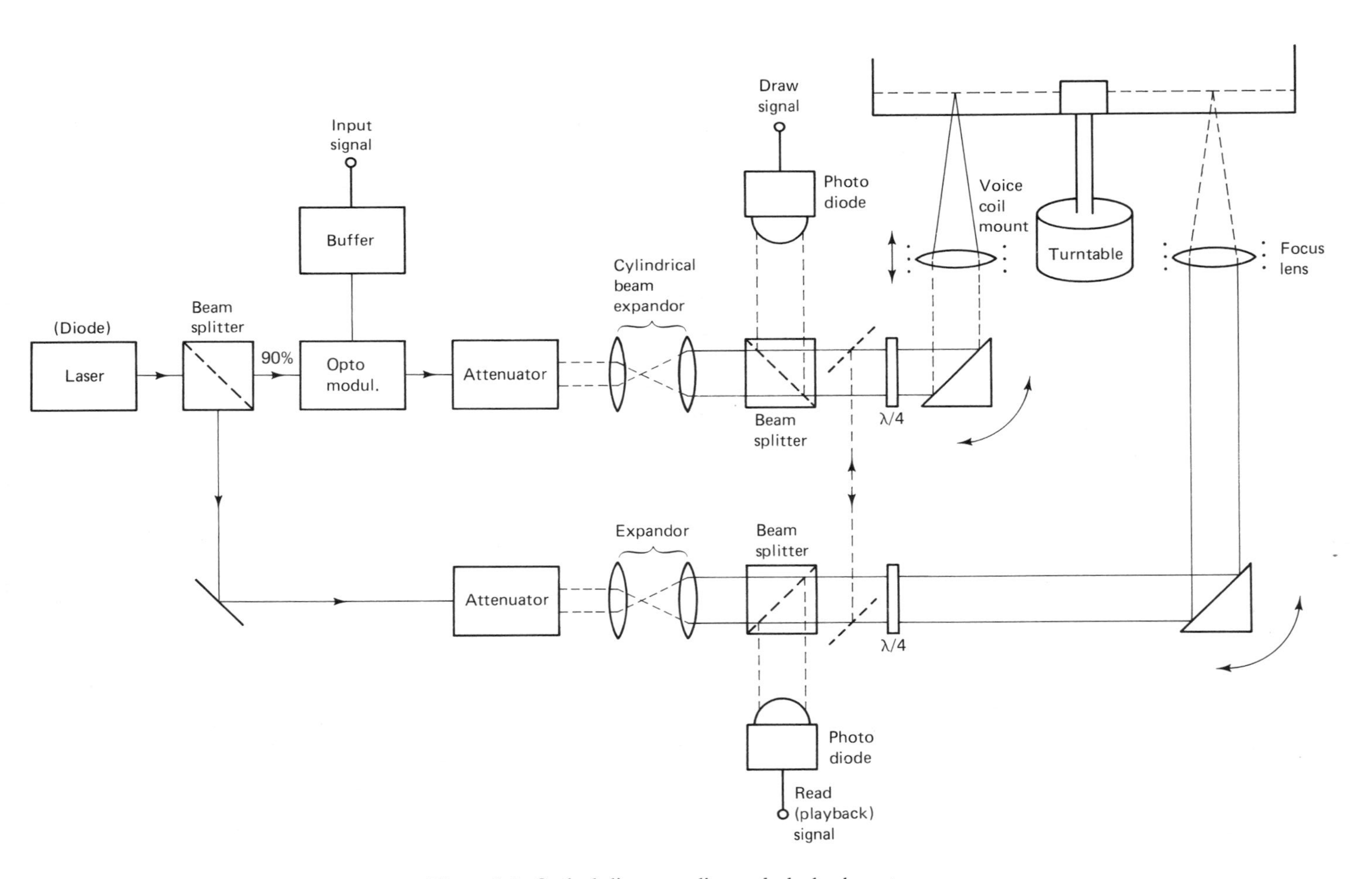

Figure 2.6 Optical disc recording and playback system.

NA microscope objective). A polarizing beam splitter and a quarter-wave plate transmit nearly all the incident light toward the disc, while directing nearly all the reflected light toward the DRAW signal detector. The playback beam passes similar devices, except the optomodulator; a separate focusing lens is shown. But the optics can be rearranged so that record and playback beams are focused with the same lens. Since the read beam is slightly skewed to the optical recording axis, the playback spot trails the recording spot by a few micrometers. In this way the recorded signal is revealed shortly after writing. This is a DRAW signal, typically used to detect any difference between the input and recorded signals, for error-free recording. New real-time information can be recorded immediately, if necessary. Thus the read signal can be used as the DRAW signal during recording, and as a playback signal after recording.

Dust and dirt are unavoidable in any optical disc system. The most widely used approach toward a solution of the problem is to manufacture the active disc surface under the cleanest possible conditions and then immediately cover this sensitive surface with a transparent layer. Recording and playing take place through the protective layer, so dirt on the disc exterior is always out of focus. Two structures providing this protective layer are commonly utilized. In the Philips DRAW system, two recording surfaces are assembled into a sandwich, with their substrates facing out and an air space in between (Figure 2.7). Thus the substrate itself becomes the protective layer. Detail of an ODC's recordable laser videodisc (RLV) is also shown in Figure 2.7a. The RLV is a single-sided disc made by bonding two PMMA substrates together. One of these substrates is coated with a layer of aluminum and a layer of the polymeric DRAW material. The other substrate is clear uncoated plastic. In the case of the metallic or metallic-like thin film recording materials, the oxidation in an air sandwich proceeds from the cavity side of the sandwich structure and the rate of oxidation is determined by the equilibrium vapor pressure within the cavity of the air sandwich [23]. RCA introduced coating of the recording surface with a transparent material (Figure 2.7b). In both cases it is necessary to use optics designed for the appropriate transparent layer. The protection in the air sandwich is afforded without loss of sensitivity. In the overcoated (encapsulated) structures, only a few materials can be utilized without suffering a significant loss in working sensitivity. More complex variants of these structures are considered, such as those employing multiple layers of different polymers and antireflective structures employing a multiplicity of polymer and metallic layers.

The recording layer can be placed on the flat substrate or on the pregrooved layer (e.g., a grooved depth of about 70 nm). The pregrooved layer can be included in the substrate or added on. Pregrooved tracks are used for guidance and pregrooved data for indicating the position of the spot on the optical disc.

2.6 SERVO SYSTEMS

Obviously, the higher the accuracy during recording, the fewer the problems during playback. A ground rule for mass-replicated videodisc systems is: it is

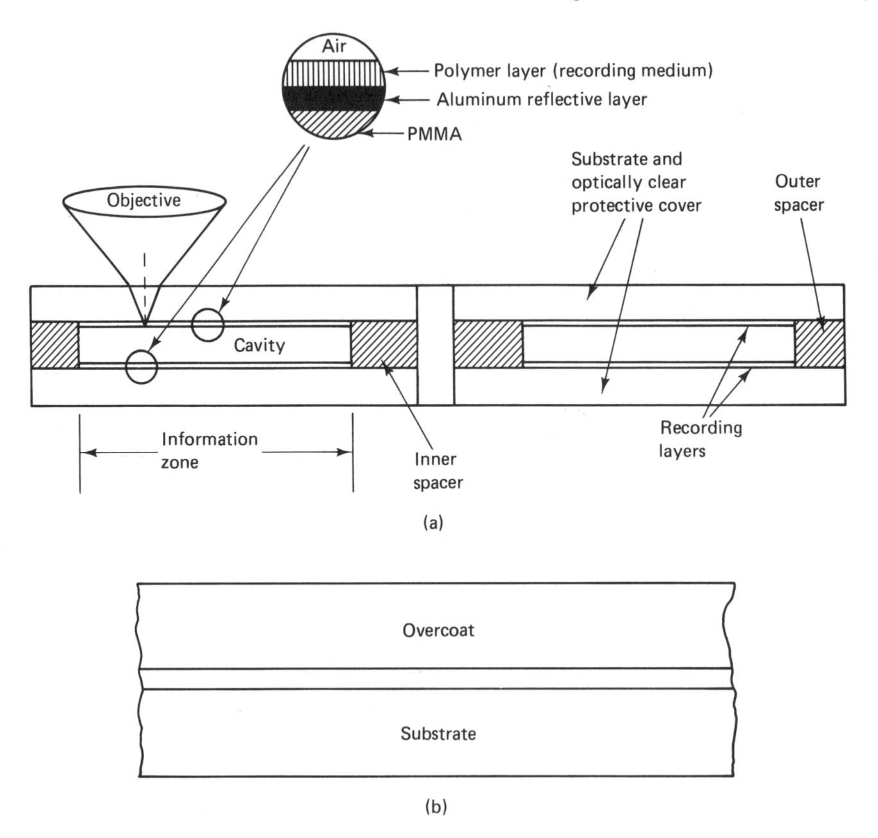

Figure 2.7 Simple optical disc structures: (a) air sandwich; (b) coating with transparent material.

easier to compensate or predistort signals once during recording than in each player. For example, in video signal processing before recording a group delay predistortion is included.

On the other hand, within the constraints of the ultimate system specifications, reduction of player and/or recorder/reader (for writable discs) costs must be found partially in the redesign of optics and servomechanics (e.g., application of solid-state lasers, simplification of the light-path or optical components, different designs for tracking and focusing actuators, etc.).

The carriage drive, which translates the writing head radially across the disc, is similar in principle for all recorders. To produce a constant pitch, the radial translation rate is uniform with time for the CAV recording mode and decreases with radii increase in the CLV mode. The translation is accomplished by a threaded nut on a lead screw driven by a synchronous motor. The reference for the motor drive is derived from the color subcarrier frequency. In the writable optical disc systems, the reference information necessary to control the radial tracking error can be provided by embossing the entire disc substrate with a pattern of circular and concentric grooves. The grooves' location is optically detectable regardless of the presence or absence of recorded data within the grooves. The groove depth is typically about one-fourth of the read/write beam wavelength.

The spindle, also similar in principle for all videodisc systems, must be phase locked to a master timing source for the entire system, typically derived from the color subcarrier frequency. For optical recorders the laser and focus controls are critical. For example, small changes in the profile and intensity of the laser beam can cause harmonic distortion, while large changes can cause even catastrophic distortions.

Write laser beam power is controlled by adjusting the current to the laser tube [1, Chap. 2]. The laser tube current is controlled by an error signal derived from the difference between the output of the photodetector, whose input is a sample of the write beam and a specified power profile. The power increases with radius in the CAV mode and is constant in the CLV mode, since it is proportional to the tangential velocity of the write head.

Each laser has some type of noise "signature." Studying noise in the videodisc channel [24] it is sometimes possible to recognize discs recorded by the same laser. Two main types of laser noise are the relative intensity noise and feedback noise. The first one is an inherent characteristic of the laser, while feedback noise appears when a small portion of reflected light is fed back to the single longitudinal mode laser's facet. Usually, this noise is well suppressed by reducing the feedback light to the laser, using a combination of polarized beam splitter and quarter-wavelength plate. But mainly due to the deviation of the disc's birefringes, it is unavoidable that a small portion of reflected light goes back to the laser's facet [25].

In the mastering process, the write laser beam is focused to a very small spot (0.4 μm) by an objective lens with a high numerical aperture (e.g., NA = 0.75). To maintain this small spot requires that the focus servo keep the lens positioned to within ± 0.1 μm of focus. The principles are the same as for the player's readout focusing servo.

2.7 DISC TESTING AND QUALITY CONTROL

Disc testing and quality control are very important for all types of videodiscs. Procedure differs for the replicated and recordable discs.

2.7.1 Premastering

Premastering consists of four distinct steps [26]: editing, coding, evaluation, and revision. Presently all premastering is done on videotape. In principle, this can be done on the recordable disc.

Video signals originated from different sources, such as videotape, videodiscs, slides, motion-picture film, computer-generated, and so on, and are combined into one program. At some point, certain codes that trigger automatic functions on certain players must be inserted into the program. These codes are:

1. Picture codes that identify the start of each frame and the field dominance of the frame
2. Chapter codes that identify sections of the disc by chapter number for level 1 players
3. Picture stop codes that trigger automatic stops on level 1 players
4. 3/2 pull-down codes that identify the correct field/frame relationship for 24-frame-per-second film segments
5. Programming dumps that load the internal memory of level 2 players
6. Digital audio codes for still-frame audio

For checking how composed material would play on the disc, a disc simulator can be used, although the recordable disc could be even more suitable. After checkings and revisions are completed, the new premaster must be carefully technically evaluated as the original premaster was.

2.7.2 Replicated Discs

Scratches and dust on the substrate for the master disc degrade the quality of the replicated discs. Thus substrate must be checked for the high-quality requirements.

A master videodisc is recorded from the premaster. Direct read during or after writing is a very helpful feature for this stage. Even if the DRAW test is satisfactory, the master disc should be played back. Also, certain signal parameters must be measured, such as carrier-to-noise ratio, signal-to-noise ratio for the video and for the audio, fixed-length distortions, and so on. Special prerecorded test signals can be used for this purpose. Dropout measurements, track "kissing," pit parameters, and so on, must also be tested. A similar procedure is required for every step in the matrix process. A bed master disc or bed stamper cannot result in better or good replicated discs.

In the mass replication, replicated discs are checked statistically [1, Chap. 2]. For example, 1% are inspected visually and short sequences played back, and 0.1% are checked completely similar to the way a master videodisc is checked. In addition to the playback quality on audio and video and the measurements already mentioned, disc flatness and reflectivity for optical systems are tested.

Figure 2.8 shows an output FM signal (RF) when different videodiscs are played on the same player. Also, channel spectra are shown. The amplitude change is caused mainly by a reflectivity index change.

Automatic inspection would be very desirable. At present there is no efficient automatic videodisc test. Because of the optical limitations, or pickup limitations in general, the disc cannot be read, for example, 10 times normal speed or faster. The 1-hour program requires 1 hour of playback time. Tests based on the optical diffraction are used for some testings, as track kissing, for example.

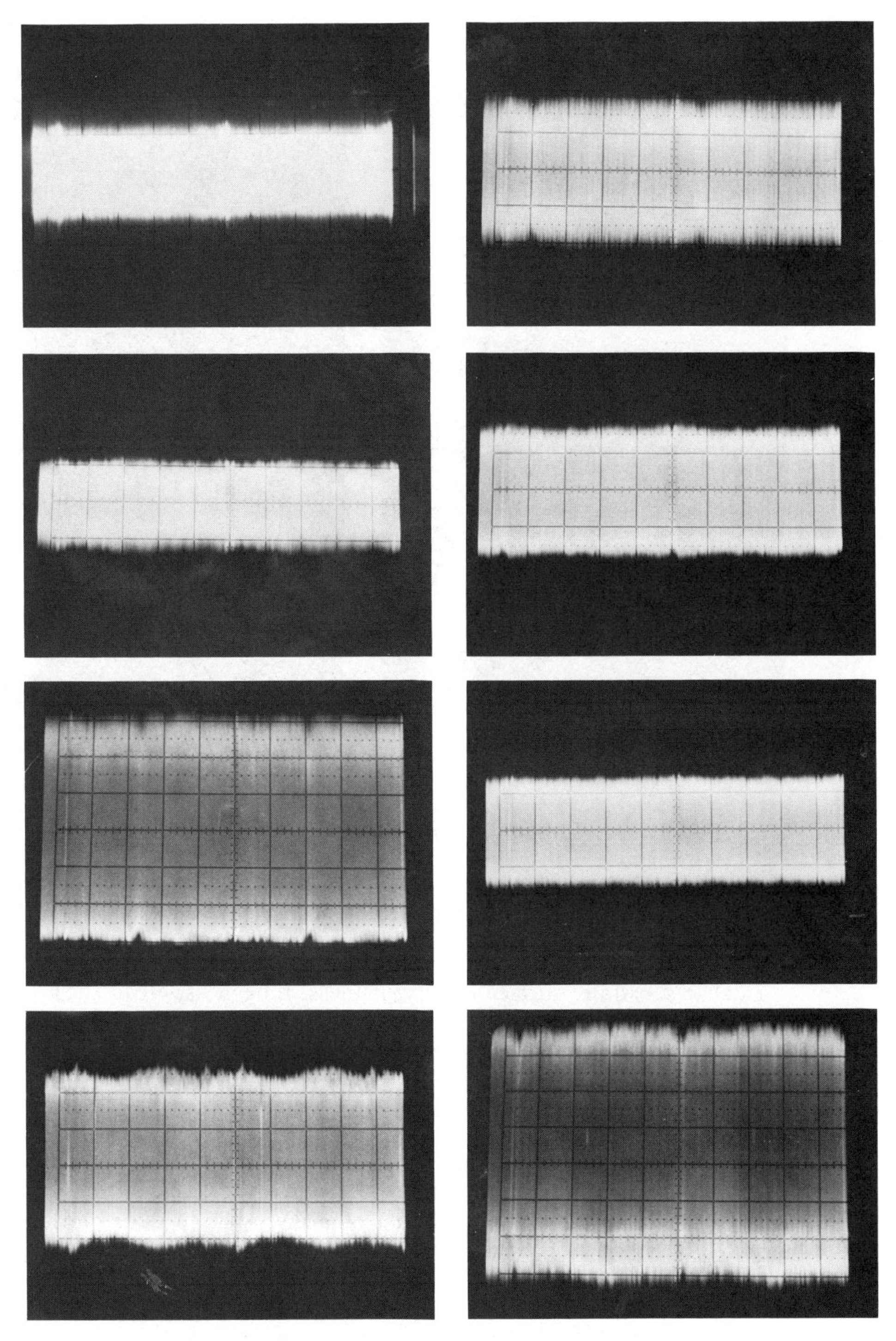

Figure 2.8 Output RF signal and corresponding spectra from the same player and different replicated optical discs, at the same radius.

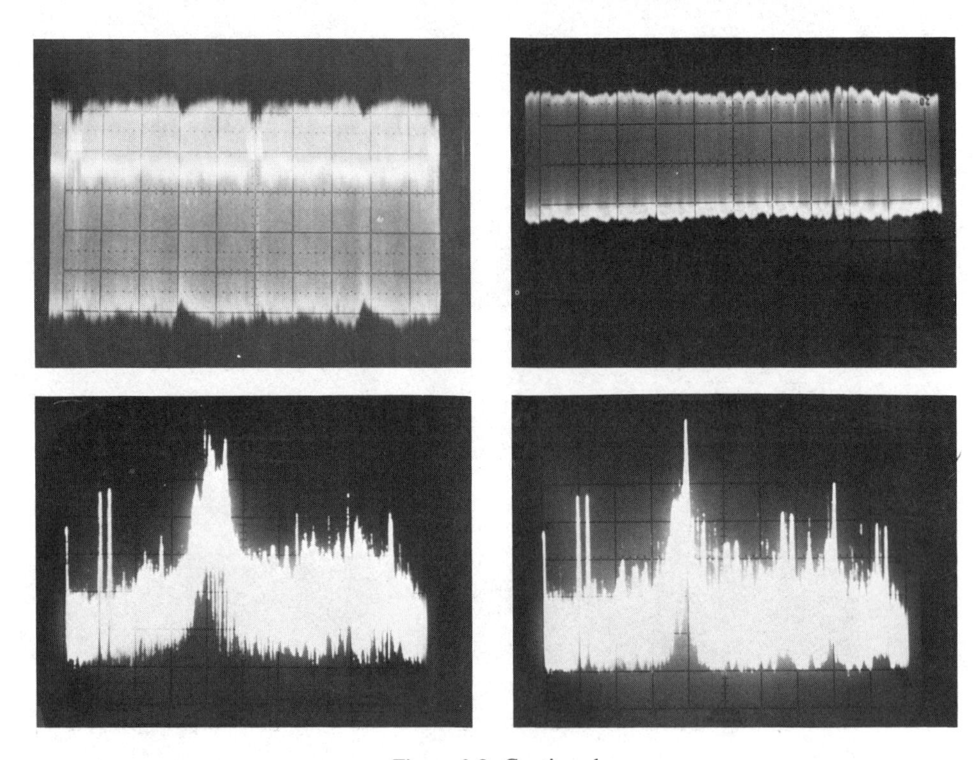

Figure 2.8 Continued

2.7.3 Recordable Optical Videodiscs

Here, the statistical checking is done before recording [27–29]. Special disc testers are available, mainly for the recording media evaluation [30]. In the disc production process, typically the following are tested:

1. Disc substrate: surface quality, flatness, roughness, etc.
2. Media parameters
3. Disc coating: thickness variation, birefringes
4. Reflectivity
5. Grooves parameters (for pregrooved media)

After recording, standard signal parameter measurements are required.

Examples of the disc layouts for the write-once discs are shown in Figure 2.9. Figure 2.10 shows the spectrum of a disc surface during measurement. Approximately 50 Å of gold was sputtered onto the disc surface to allow SEM photos to be taken. Broken portions in Figure 2.9b are caused by overexposure during examination. Lifetime is a parameter that requires long-time testing and examination [31–33]. Under some circumstances, the accelerated degradation of non-hermetically sealed discs can happen [34].

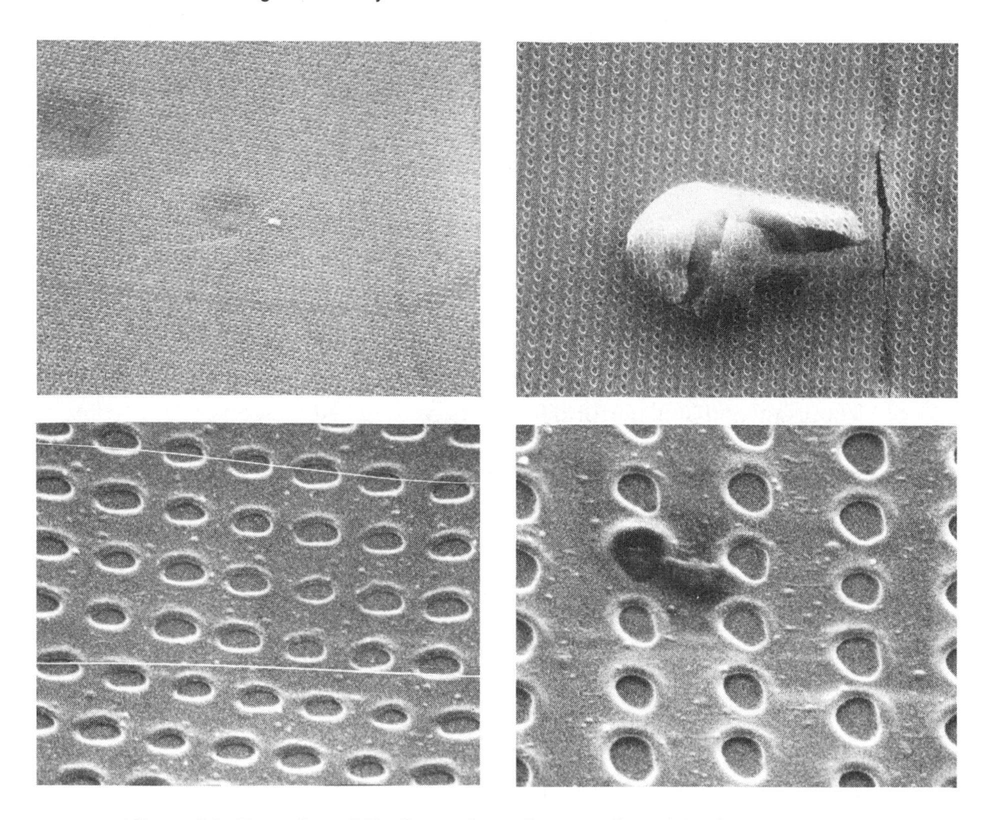

Figure 2.9 Examples of disc layout for write-once discs: (a) 1000×; (b) 2000×; (c) 10,000×; (d) 20,000×.

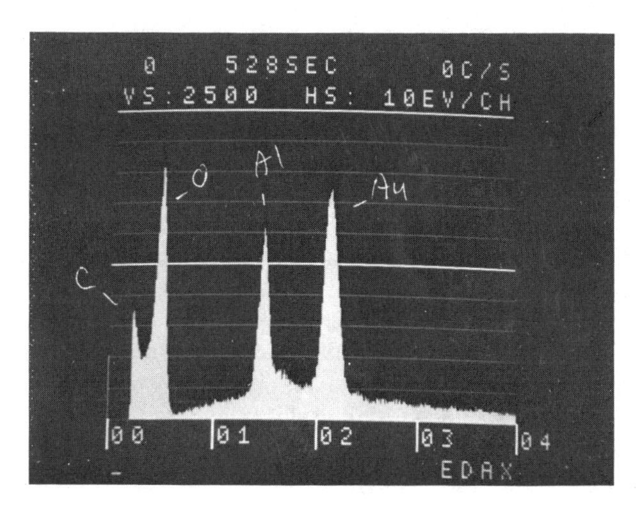

Figure 2.10 Elements present on the disc surface: carbon, oxygen, aluminum, and gold.

For the erasable videodiscs, the rerecorded signal parameters must be measured. For example, the signal-to-noise degradation and, in general, playback quality degradation after erase/record processes are repeated are essential.

2.7.4 Evaluation of the Digital Optical Discs

Using videodisc technology, digital data can be recorded two ways: in analog or digital channels. Digital audio combined with the LaserVision signals and digital data in the video are examples of the first, while CD and CD-ROM are examples of the second. In general, for digital data recording typical specifications for the disc and drive are: capacity, access time, latency, line on recording density, recording method, error rate, etc. Bit error rate is one of the most important and final performance indices used in digital optical recording equipment evaluation. The problem of specifying error rate is familiar to all manufacturers of optical disc systems [35–38]. More difficult is the problem of identifying which part of the system is responsible when the system fails to meet the specified performance levels. The major sources of errors are: linear and nonlinear intersymbol interference (ISI), media defects, noise, adjacent trace pickup, bit shift, external disturbances, and overwrite (for erasable media).

The accurate way to verify a bit error rate would be to read complete disc content during or after recording, and compare with the original data. A less accurate, but useful method would be to check "typical" numbers. Both are time consuming. For example, for the replicated optical discs a "typical" error rate is less than 10^{-12}. To measure, say, at least 10 errors, we should check 10^{13} bits. Assuming an average data transfer of 10^6 bits just to illustrate the problem, it would be necessary for 10^7 s, or almost 4 months to verify specification! Because of this, more practical methods, such as sample testing or margin analysis are used.

Figure 2.11 illustrates the variations of the edge transitions on the Jordan code waveforms caused by noise and other disturbances. Also, the idealized distribution of the edge deviation, the Gaussion one, is shown for linear and logarithmic scale. The distribution is a histogram of the positions of the pulse transitions. In the ideal case of a noiseless channel and no inter-pulse interaction, there is no variation in the signal transitions and the distribution is simply delta function located of the center time window. Namely, to detect digital data, the time window is placed around the expected position transition. In the case of the double density codes, Jordan and Miller codes, the time window is $\pm T_b/4$, where T_b is bit interval. For the measurement purposes the window can be some other time interval.

In real systems, where transitions are packed closely together to maximize the data storage density, the effect of the ISI is to shift the positions of various transitions, from the centers of the data windows, by unique discrete amounts. This is illustrated in Figure 2.12. In the ideal noiseless channel, to periodical pulse trains are shown with one-half and one-third duty cycle. Just by considering first harmonics, it is obvious why in the second case the distribution is bimodal.

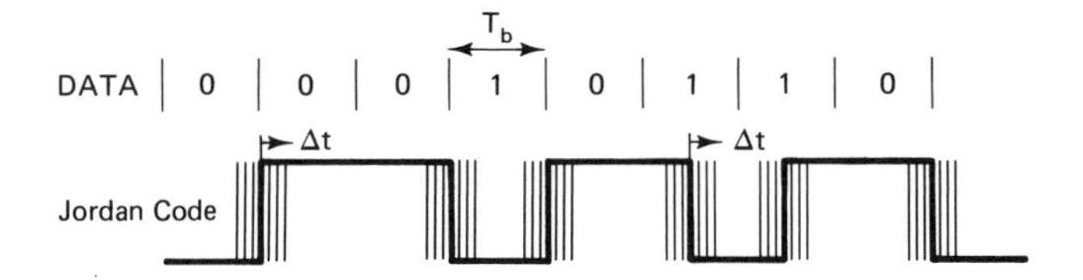

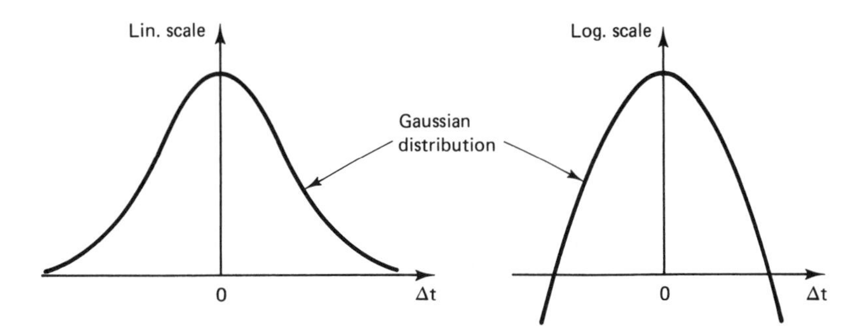

Figure 2.11

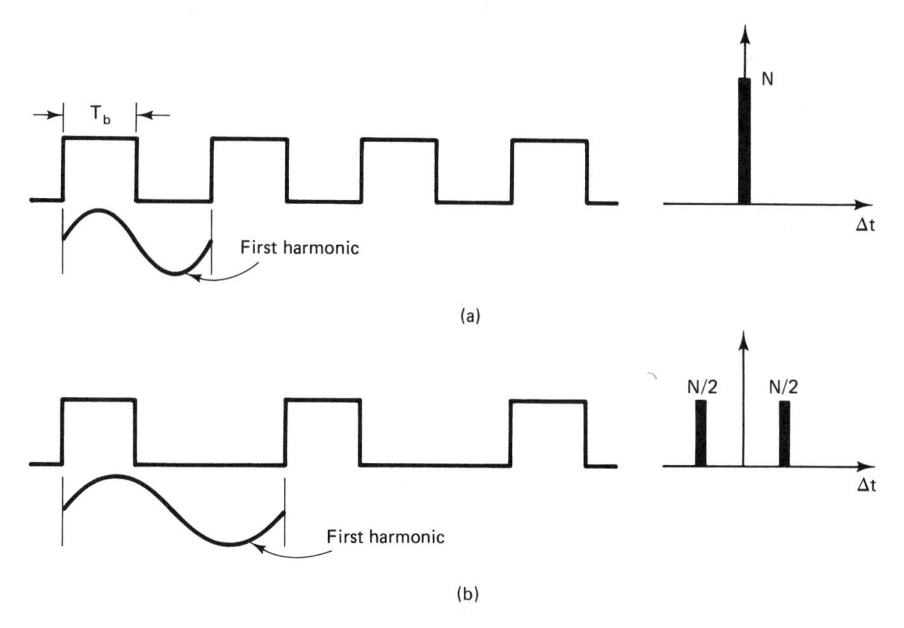

Figure 2.12

Some practical distributions are shown in Figure 2.13. For the simplicity, notation "t" instead "Δ t" is used. If data window is $\pm T_w$, the error rate of the channel is simply the area of the tails of the distribution beyond T_w normalized to the total area of the distribution. For the normalized actual distribution, P(t), the error rate, E, is the value of the integer for times larger than T_w:

$$E = \int_{-\infty}^{-T_w} P(t)dt \pm \int_{T_w}^{\infty} P(t)dt$$

In practice different methods are used to estimate the error rate in a short period of time, rather than waiting a few months [37].

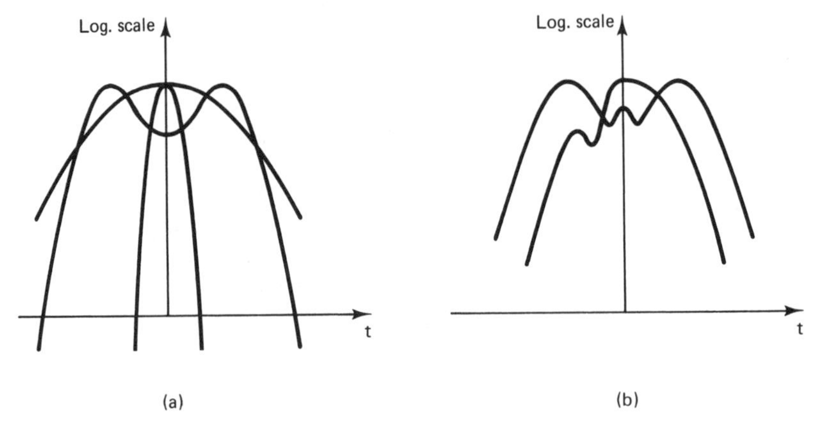

Figure 2.13

Eye Pattern

By these simple methods, the oscilloscope is triggered from the zero crossings, while the read beam signal is displayed (see Figure 2.14). The "eye" opening is correlated to the error rate. This is the simplest, quickest but least accurate technique.

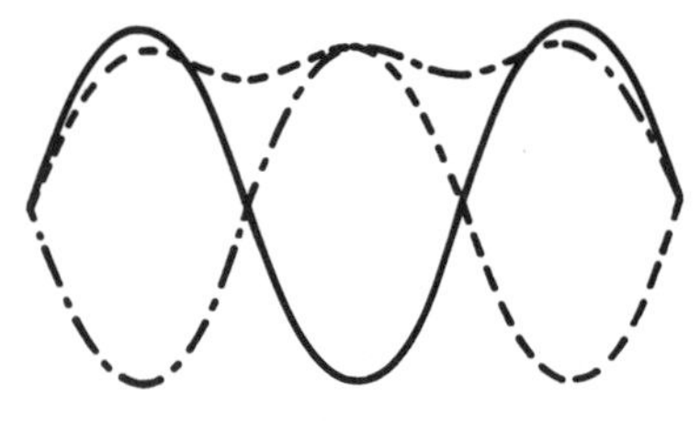

Figure 2.14

Window Sliding

In this method, the error rate is artificially increased by moving the window out of center. By moving the data window in increments and measuring the error rate (number of edges outside the window) each time, it is possible to generate a transition distribution plot similar to the plots in Figure 2.13. However, this method is not capable of producing an exact rate. The data window can be generated using data phase-lock-loop (PLL) or independent-second PLL.

Transition Distribution Histogram

Here, the data window is divided into small time intervals, baskets, or bins. During measurement, the number of transitions is counted for each basket. Typically there are 50–100 baskets. This method produces plots as shown in Figure 2.15. The method is the best of three margin analysis methods listed here. There are many different ways of implementing this technique [38].

REFERENCES

1. J. Isailović, *Videodisc and Optical Memory Technologies,* Prentice-Hall, Englewood Cliffs, N.J., 1985.
2. J. K. Clemens, "Capacitive Pickup and the Buried Subcarrier Encoding System for the RCA Videodisc," *RCA Rev.,* Vol. 39, No. 1, 1978, p. 33.
3. D. H. Pritchard, J. K. Clemens, and M. D. Ross, "The Principles and Quality of the Buried Subcarrier Encoding and Decoding System," *RCA Rev.,* Vol. 42, No. 3, 1981, p. 367.
4. M. D. Ross, J. K. Clemens, and R. C. Palmer, "The Influence of Carrier-to-Noise Ratio and Stylus Life on the RCA Videodisc System Parameters," *RCA Rev.,* Vol. 42, No. 3, 1981, pp. 394–407.
5. R. N. Rhodes, "The Videodisc Player," *RCA Rev.,* Vol. 39, No. 1, 1978, pp. 198–221.
6. T. Inoue, T. Hidame, and V. Roberts, "The VHD Videodisc System," *SMPTE J.,* Vol. 91, No. 11, 1982, pp. 1071–1076.
7. G. C. Kenney and A. H. Hoogendijk, "Signal Processing for a Videodisc System (VLP)," *IEEE Trans. Consum. Electron.,* Vol. CE-24, No. 3, 1978, pp. 453–457.
8. T. Noruse and A. Ohkoshi, "A New Optical Videodisc System with One Hour Playing Time," *IEEE Trans. Consum. Electron.,* Vol. CE-24, No. 3, 1978, pp. 453–457.
9. *SMPTE Doc. T14.22-X2A.0,* "Preliminary Draft of a Standard for a Single-Channel Component Analog Video Waveform," Revision 1, Aug. 30, 1984.
10. IBA, MAC Television System for High-Quality Satellite Broadcasting. Draft.
11. E. F. Morrison, "Videotape Recording: Digital Component versus Digital Composite Recording," *SMPTE J.,* Vol. 91, No. 9, 1982, pp. 789–796.
12. J. D. Lowry, "B-MAC: An Optimum Format for Satellite Television Transmission," *SMPTE J.,* Vol. 93, No. 11, 1984, pp. 1034–1043.
13. G. Clounard and J. N. Barry, "NTSC and MAC Television Signals in Noise and Interference Environments," *SMPTE J.,* Vol. 93, No. 10, 1984, pp. 930–942.
14. W. S. DeForest, *Photoresist: Materials and Processes,* McGraw-Hill, New York, 1975.

15. Y. Okino, K. Sano, and T. Koshihara, "Development in Fabrication of Optical Discs," Optical Disc Technology, *Proc. SPIE,* Vol. 329, 1982, pp. 236–241.
16. M. Michalchik, Private communication.
17. Special Issue on Videodiscs, *RCA Rev.,* Vol. 35, No. 1, 1978.
18. Special Issue on Videodisc Optics, *RCA Rev.,* Vol. 39, No. 3, 1978.
19. Special Issue on Videodisc System, *RCA Rev.,* Vol. 43, No. 1, 1982.
20. A. E. Bell, "Optical Data Storage Technology Status and Prospects," *Comput. Des.,* Jan. 1983, pp. 133–146.
21. G. C. Kenney, et al., "An Optical Disc Replaces 25 Mag Tapes," *IEEE Spectrum,* Feb. 1979, pp. 33–38.
22. R. A. Bartolini, et al., "Optical Disc Systems Emerge," *IEEE Spectrum,* Aug. 1978, pp. 20–28.
23. T. W. Smith, G. E. Johnson, A. T. Ward, and D. J. Luca, "Barrier Coating for Optical Recording Media," Optical Disc Technology, *Proc. SPIE,* Vol. 329, 1982, pp. 228–235.
24. J. Isailović, "Channel Characterization of the Optical Videodisc," *Int. J. Electron.,* Vol. 54, No. 1, 1983, pp. 1–20.
25. T. Gotoh, M. Ojima, A. Arimoto, and N. Chinone, "Characteristics of Laser Diodes and Picture Quality," Optical Disc Technology, *Proc. SPIE,* Vol. 329, 1982, pp. 56–60.
26. R. Daynes and B. Butler (eds.), *The Videodisc Book—A Guide and Directory,* Wiley, New York, 1984.
27. R. Imanaka et al., "Recording and Playing (R/P) System Having a Compatibility with Mass-Produced Replica Disc," *IEEE Trans. Consum. Electron.,* Vol. CE-29, No. 3, 1983, pp. 135–139.
28. K. Sadashige and M. Takenaga, "Optical Disk Technology for Permanent and Erasable Memory Applications," *SMPTE J.,* Feb. 1985, pp. 200–205.
29. M. L. Levene, "Optical Disc Media Parameters and Their Relationship to Equipment Design," Optical Storage Media, *Proc. SPIE,* Vol. 420, 1983, pp. 273–281.
30. R. Hezel, "An Overview of Optical Disk Testers," Optical Storage Media, *Proc. SPIE,* Vol. 420, 1983, pp. 113–120.
31. "Optical Discs System and Applications," *Proc. SPIE,* Vol. 421, 1983.
32. Optical Data Storage, *Proc. SPIE,* Vol. 382, 1983.
33. D. J. Gravesteijn and J. van der Veen, "Organic-dye Films for Optical Recording," *Philips Tech. Rev.* Vol. 41, No. 11/12, 1983/84, pp. 325–333.
34. L. Vriens and B. A. J. Jacobs, "Digital Optical Recording with Tellurium Alloys," *Philips Tech. Rev.* Vol. 41, No. 11/12, 1983/84, pp. 313–324.
35. E. R. Katz and T. C. Campbell, "Effect of Bitshift Distribution on Error Rate in Magnetic Recording," *IEEE Trans. on Magnetics,* Vol. Mag-15, No. 3, May 1979, pp. 1050–1053.
36. G. F. Hughes and R. K. Schmidt, "On Noise in Digital Recording," *IEEE Trans. on Magnetics,* Vol. Mag-12, No. 6, November 1976, pp. 752–754.

37. N. D. Mackintosh, "Evaluate Disk-Drive Performance with Margin Analysis," *Computer Design,* January 1984, pp. 81–88.

38. N. D. Mackintosh, "A Margin Analyser for Disk and Tape Drives," *IEEE Trans. on Magnetics,* Vol. Mag-17, No. 6, Nov. 1981, pp. 3349–3351.

three

Signal Processing: Modulation

3.1 JUSTIFICATION OF SIGNAL PROCESSING AND METHODS CLASSIFICATION

In principle, it is possible through a videodisc system to record/transmit signals in their source shape, similar to their shape at the output of the message-signal transducer. This would be a baseband system. Other possibilities are also available, but these methods require processing of the original signals. In general, some parameter(s) of the auxiliary periodical deterministic signal is so determined that it carries the original signal and in that way is the carrier of the message. This process is called *modulation,* and its purpose is to make the signal suitable for the particular channel [1–7]. Basically, modulation is changing or modifying one wave shape in accordance with another wave shape. Thus, in any modulating scheme we are dealing with two wave shapes: (1) the *carrier,* which is the signal being modulated, and (2) the *information,* or modulating signal. A modified carrier according to the modulating signal is said to be the *modulated signal.*

The inverse process of modulation is *demodulation,* and the original signal obtained is the *demodulated signal.* Thus modulation and demodulation are two inseparable processes in videodisc systems: the first relating to the recorder, the second to the player. The *modulator* is a device that performs modulation and the *demodulator* is a device that performs demodulation; both together are sometimes called a *modem.*

The first reason to introduce modulation in videodisc systems is relatively high noise at low freqencies in recording and playback. The second reason is that in the videodisc channel, frequency sharing can be used to record several independent signals (information).

There are many ways to modulate a carrier. They may be classified according to some common properties. As a result of the modulation, the modulated signal can be continual or in a pulse-wave form.

In continuous modulation systems the carrier is a sinusoidal signal. The characteristic parameters of the carrier are amplitude, frequency, and phase. Depending on which parameter is changing directly proportionally to the modulating signal, we have:

Amplitude modulation (AM). The carrier's amplitude is changing as a function of the modulating signal.
Frequency modulation (FM). The changing parameter is the carrier's frequency.
Phase modulation (PM). The changing parameter is the carrier's phase.

In pulse modulation systems, the modulating signal has a pulse waveform and the carrier is, usually, a periodic pulse train. The characteristic parameters of the pulse carrier are pulse amplitude, pulse duration, and pulse position–pulse location at the particular moment in time. Depending on which pulse parameter is changing according to the modulating signal, we have:

Pulse amplitude modulation (PAM)
Pulse duration (width) modulation (PDM, PWM)
Pulse position modulation (PPM)

In the same group are pulse frequency modulation (PFM) and pulse phase modulation (PPM). Because the modulated signal is discrete, all modulations in this group are based on the principles given by the sampling theorem. To these two groups a third can be added: combined modulation systems. For example, it is possible to form the frequency-modulated signal (FM) and then to use it as a modulating signal to modulate another carrier (FM/FM). In the same way, for example, PWM/FM can be formed; and so on.

From the videodisc system point of view, another approach in the signal processing classification is important. Namely, classification can be done based on whether any parameter of the carrier is changing analog to the modulating signal, or if the carrier is *coded* according to the modulating signal. Thus modulation is either *analog* or *digital.* All methods mentioned hitherto (i.e., AM, FM, PM, PAM, PWM, PPM, PDM) are analog.

Some examples of digital modulations are pulse code modulation (PCM), delta modulation (DM), and differential pulse code modulation (DPCM). Digital-modulated signals can be recorded/transmitted in their physical frequency band, baseband, or in a shifted frequency band: the digital signal modulates some parameters of the carrier. Sometimes, to avoid possible confusion, PCM, DM, and DPCM are called *coded modulations,* and *digital modulations* refer to processes in which digital signals (e.g., a PCM signal) modulate another carrier.

When the digital signal changes the amplitude of the carrier, digital amplitude modulation is obtained. This modulation is usually called *amplitude-shift keying* (ASK), according to the terminology used in the telegraph transmission technology. Digital frequency modulation is obtained when a digital signal modulates the frequency of the carrier; the common name for it is *frequency-shift keying* (FSK). Similarly, digital phase modulation is called *phase-shift keying* (PSK). If the amplitude, frequency, and phase can have two values, it is common to talk about binary modulation: BASK, BFSK, and BPSK. In general, parameters can have M-states ($M \geq 2$ refers to an M-ary modulation: M-ASK, M-FSK, M-PSK). For $M = 4$, the term *quadrature* can be used [e.g., quadrature PSK (QPSK)].

3.1.1 Messages and Recorded Signal Characteristics

Messages to be recorded on the videodisc can be discrete or continuous. Discrete messages appear as strings of separated elements which can have a finite number of different values. Elements called *symbols* belong to a finite set, the *alphabet*. Data signals are one example of discrete messages. *Continuous messages* are a function of time having all possible values in between given limits (L_{max}) and (L_{min}). The word "continuous" is referred here to the fact that the time function can have any value in the given interval; continuity in the mathematical sense is not necessary. Both signals are illustrated in Figure 3.1. It is possible to make discrete signals to be continuous. Amplitude and time can both be made discrete.

Another signal classification is also in two groups: deterministic and stochastic signals. Both can be represented as a time function, although not in the same way. *Deterministic signals* can be given (represented) by a defined time function: for example, a sinusoidal voltage, a stream of square pulses, and so on. Where the mathematical expression for the signal is known, the signal value can be obtained in any moment of the future. For *stochastic signals,* fluctuations in time cannot be known precisely. Stochastic signals and corresponding waveforms are known in the past but are not known in the future. Audio and video signals are examples of stochastic signals.

Although stochastic signals are recorded on the disc, it is sometimes useful

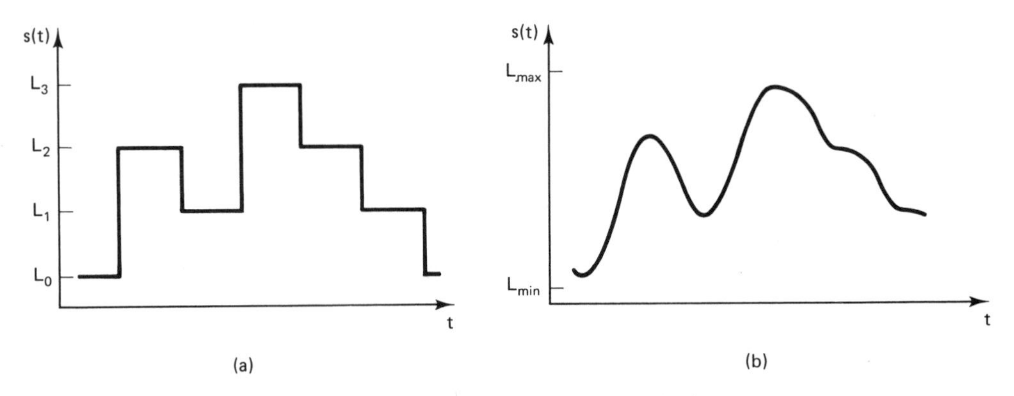

Figure 3.1 (a) Discrete and (b) continuous signals.

to analyze deterministic signals. In the deterministic signal analysis the harmonic analysis of the function representing that signal is used. The harmonic analysis is based on the Fourier series for periodic signals, and Fourier transformation for aperiodic signals.

According to signal classification as continuous and/or discrete, transmission/recording can be classified as analog and/or discrete. In analog systems, there is a correspondence between the value of the message in finite numbers, and electrical signal values, also in finite numbers. In discrete systems the correspondence is between a finite set of symbols and a set of signal values. Coded or digital systems are also discrete.

Band limited continuous signals can be made discrete without loss of information. A continuous message represented by function $f(t)$ can be presented with a discrete set of its instantaneous values—samples. This follows from the sampling theorem. If a continuous function of time $f(t)$ has a spectrum in the interval 0 to f_m, this function is completely defined with its instantaneous values, taken at equidistant points of time that form a set with interval $\Delta t = t_j - t_i = 1/2f_m$. Further, samples can be quantized in the amplitude domain to form the finite set of amplitude values. This quantization is based on the criterion of fidelity, that is, on the usability of the message for the user.

3.1.2 Analog Modulation Systems

In general, linear and exponential modulation methods result in a continuous waveform of the modulated signal, and all pulse methods give discrete waveforms. All continuous and some pulse methods are analog; the remaining pulse methods are digital.

Amplitude modulation (AM) is a linear modulation. The carrier is always sinusoidal and its amplitude is modified, in the modulating process, so that it is time-function proportional to the modulating signal, which carries the message. Depending on which characteristic part of the AM spectra is used, there are several types of AM (Appendix 3A):

1. Double-sideband-suppressed carrier modulation (DSB or DSBC)
2. Conventional amplitude modulation (AM or CAM)
3. Single-sideband modulation (SSB)
4. Vestigial single-sideband modulation (VSB), also called AM with asymmetrical sidebands

Only conventional AM has all components: carrier and both sidebands (higher and lower). Common for all AM is that the modulation process results in simple translation of the fundamental (modulating) signal.

A relative ratio of the spectral densities for the amplitudes of the modulated and modulating signals is conserved during AM. This means that the resulting modulated signal has only spectral components which are obtained by translation of the corresponding spectral component in the modulating signal; this is why

AM is linear modulation. The inverse process, AM signal demodulation, is also linear, which means spectral translation from high frequencies to low frequencies.

In principle, AM can be obtained in many ways: using nonlinear elements (e.g., with the transfer function $a_0 + a_1x + a_2x^2$), switching element parameter variation, a balance modulator, a ring modulator, modulators with phase shift, and so on. AM signal can be demodulated using basically the same process: product demodulation. A detector can also be used. Usually, the terms *demodulation* and *detection* are used with the same meaning. Strictly speaking, this is not correct [8]. *Demodulation* is an inverse operation to modulation in which the modulating signal is reconstructed from the modulation products. *Detection* is a modulating signal reconstruction obtained by an asymmetrically conducting device, without use of a local oscillator (e.g., an envelope detector).

In the process of continuous angular modulation, the carrier is also sinusoidal, but its amplitude is constant and its angle is the parameter carrying the message; the angle of the carrier is modified by the modulating signal. Like AM, angular modulation includes products of two functions, but before multiplication each function goes through an exponential, not a linear transformation. This is the principal difference between AM and angular modulations: AM spectra is obtained by simple translation of the modulating signal spectra and it is limited if the modulating signal spectra are limited; thus there are no other new components besides those obtained by translation, whereas an angularly modulated signal has infinite spectra. This is characteristic of the nonlinear processes, and among them the exponential process has the advantage that demodulation is easily realized. Because the exponential process is essentially angular, modulation is also called *exponential modulation*.

Angular modulations are frequency (FM) and phase (PM) modulations. Using special devices it is possible to convert FM into PM, and vice versa. The same is possible for phase demodulator (PD) and frequency demodulator (FD). The devices are differentiator and integrator because

$$\text{PM} = \text{differentiator} + \text{FM}$$

$$\text{PD} = \text{FD} + \text{integrator}$$

$$\text{FM} = \text{integrator} + \text{PM}, \ \text{FD} = \text{PD} + \text{differentiator}$$

In the process of angular modulation one modulating component generates an infinite number of spectral components with different frequencies. Thus each spectral component of the modulated signal is dependent on all spectral components of the modulating signal. But in practice the bandwidth of the modulated signal is limited because some spectral components make relatively small contributions to the modulated signal.

Angular modulations can be obtained directly (e.g., an FM modulator with a varicap diode) or indirectly using relations between FM and PM. Detection of angularly modulated signals can be done in the same way. A device for FM signal detection is usually called the *discriminator*. The output voltage of the ideal discriminator is linearly proportional to the instant frequency of the input signal, whose amplitude is constant. In pulse modulations, the modulated signal is discrete. Common for all of them is that the pulse-modulated signal has two different states. Conditionally, it can be said that in one state a signal exists, and that in another state there is no signal. Each state has finite duration, and active and passive intervals are interchanging alternatively. Also, the sampling theorem is the basis for pulse modulation.

Pulse modulation can be analog or digital. In analog systems, changing parameter(s) have an infinite number of possible values: this condition is enough and necessary for the system to be analog. For example, the sample can be quantized before modulation. Analog pulse modulations are pulse amplitude, pulse frequency, pulse phase, pulse width, and pulse position modulation.

The main difference in spectra between corresponding pulse and continuous modulations is that each harmonic of the pulse carrier is modulated similarly as the continuous carrier. Also, under the same conditions, a pulse-modulated signal can contain a modulating spectrum in its natural band, the baseband. A baseband signal is present in the modulated signal spectrum whenever in the modulating process a fixed process is included (independent of the modulating signal).

3.1.2.1 Beat Frequencies. If a spectral pair with frequencies f_1 and f_2 is passed through nonlinear system, the output signal contains new spectral components, harmonics mf_1 and nf_2 and combinations $mf_1 \pm nf_2$, for $n, m = 1, 2, 3, \ldots$. For example, if output signal y is related to the input signal x as

$$y = a_0 + a_1x + a_2x^2 \tag{3.1}$$

and if

$$x = \cos \omega_1 t + \cos \omega_2 t \tag{3.2}$$

then

$$y = (a_0 + a_2) + a_1(\cos \omega_1 t + \cos \omega_2 t) + (a_2/2 \cos 2\omega_1 t + \cos 2\omega_2 t) + a_2[\cos (\omega_1 + \omega_2)t + \cos (\omega_1 - \omega_2)t] \tag{3.3}$$

The term *beat frequency* is usually referred to the component $f_1 - f_2$. If one frequency is constant and the other is presenting a modulating signal, the carrier and two sidebands can be filtered out. Also, if one frequency is the carrier, then new carriers, $f_1 + f_2$ or $f_1 - f_2$, can be generated. This is referred to as the *heterodyne technique*.

If a pair of spectral components is linearly superposed, there are no new spectral components, but the resulting waveform can show interesting new properties. Linear superposition of two cosine signals is

$$s(t) = s_1(t) + s_2(t) = S_1 \cos \omega_1 t + S_2 \cos \omega_2 t \tag{3.4}$$

Initial phases are set to zero, for the sake of simplicity. Equation (3.4) can be rewritten so that signal $s(t)$ is given as the sum of odd and even components:

$$s(t) = [S_1 + S_2 \cos (\omega_2 - \omega_1)t] \cos \omega_1 t - [S_2 \sin (\omega_2 - \omega_1)t] \sin \omega_1 t \tag{3.5}$$

Further,

$$s(t) = \sqrt{S_1^2 + S_2^2 + 2S_1S_2 \cos(\omega_2 - \omega_1)t} \cos \frac{\omega_1 t + \arctan[S_2 \sin(\omega_2 - \omega_1)t]}{S_1 + S_2 \cos(\omega_2 - \omega_1)t} \tag{3.6}$$

Thus the sum of two cosine signals is a signal whose amplitude is changing with frequency $(f_2 - f_1)$, and instant phase is a nonlinear function of time (Figure 3.2).

If $S_1 > S_2$ and $f_1 >> f_2$, then $s_1(t)$ can be considered as a carrier and $s_2(t)$ as one AM sideband (SSB). If $S_2 >> S_1$, then

$$s(t) \simeq S_1[1 + (S_2/S_1)\cos(\omega_2 - \omega_1)t] \cos[\omega_1 t + (S_2/S_1)\sin(\omega_2 - \omega_1)t] \tag{3.7}$$

That is, the beat signal presents an amplitude- and phase-modulated signal.

So-called *simple beat* is obtained for

$$S_1 = S_2 = S \tag{3.8}$$

Then $s(t)$ is simply

$$s(t) = 2S\left(\cos\frac{\omega_2 - \omega_1}{2}t\right)\cos\frac{\omega_2 + \omega_1}{2}t \tag{3.9}$$

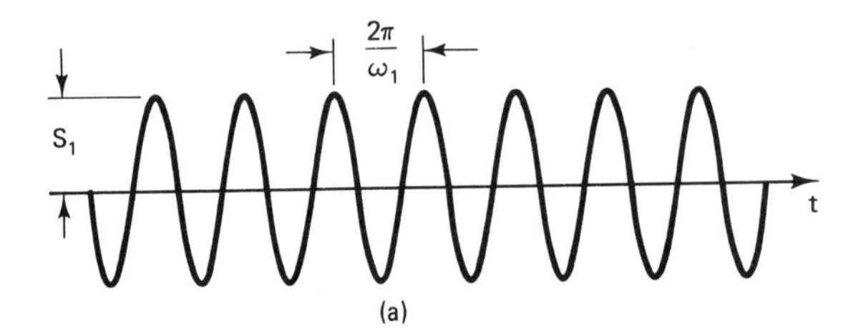

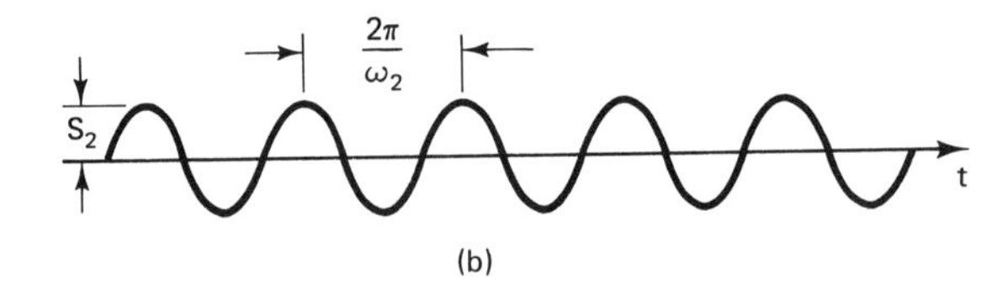

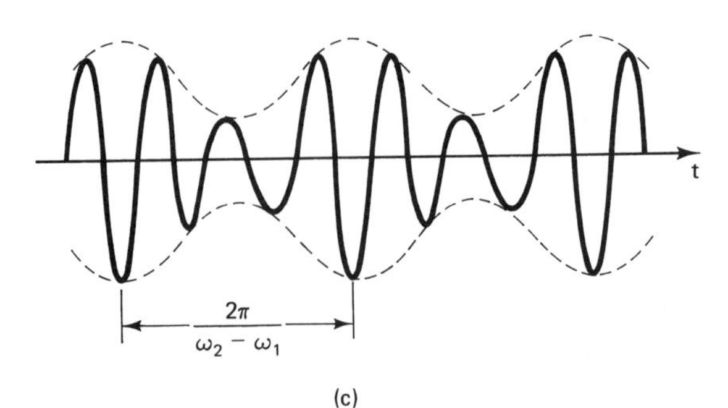

Figure 3.2 Linear beat: (a) and (b) spectral components; (c) resulting waveform.

Figure 3.3 illustrates the simple beat of two deterministic periodical signals when frequencies are close. In the case of simple beat, waveforms oscillate at a relatively low beat frequency $(f_2 - f_1)/2$ between zero and maximal amplitude $2S$. Thus a simple-beat waveform presents an AM signal with two sidebands and a suppressed carrier (DSB).

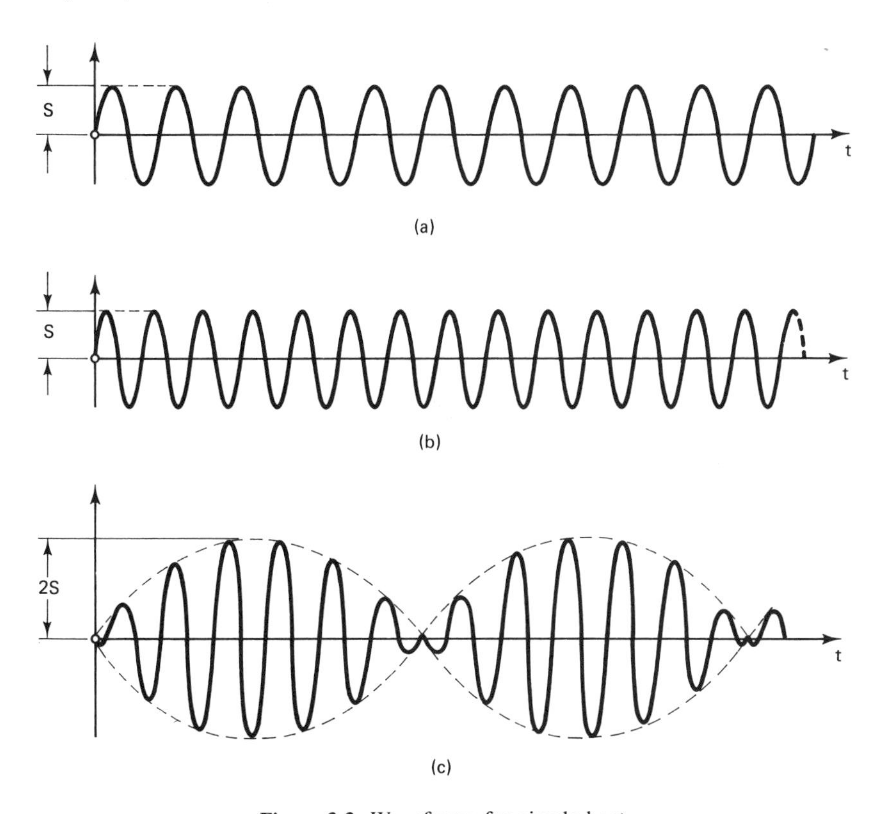

Figure 3.3 Waveforms for simple beat.

3.1.3 Digital Modulating Systems

Digital signals originate in two ways: by analog signal discretization in time and amplitude, or naturally generated as digital signals—computer-generated signals, for example. According to the definition [9], any digital signal is a signal that must have some characteristic discontinuity in time and be described by a finite set of discrete values; each value can be associated with some number. Because of this it is said that in this type of transmission/recording, numbers of digits are transmitted/recorded.

In analog modulating systems, the bandwidth of the modulating signals is a specially important characteristic in the hierarchy system description. In digital systems the corresponding parameter is a digital rate: it describes the quantity of digits in a time unit. For binary systems it is beat rate, and for M-ary systems it is M-ary digit rate. In general, digital rate is used for synchronous systems.

There is more than one shape of binary signal. A common property for all of them is that the important parameter of the signal can have two different values, which are marked 0 and 1, or S and M for *space* and *mark,* respectively. Figure 3.4 shows some binary signals.

The basis for M-ary signal formation is a number system with M different numbers, $M = 2^n$, where n is number of binary digits needed for presentation of each M number (level). M-ary rate R_s is expressed by the number of M-ary digits in seconds. The equivalent binary rate, R_b, is

$$R_b = R_s \cdot n = (1/T_s)\text{ldM} \tag{3.10}$$

where $R_s = 1/T_s$ is the signaling rate, usually expressed in baud.

The simplest method of digital signal transmitting/recording is transmitting/recording in the so-called baseband, that is, in source form as a basic digital signal. Another way is to transmit/record digital signals in the same way as analog signals, using a carrier-deterministic sinusoidal signal.

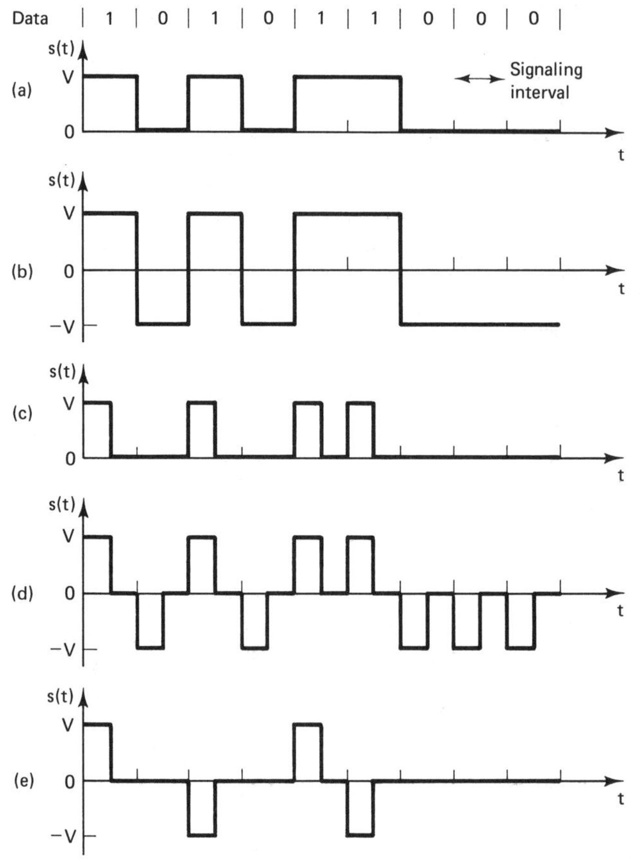

Figure 3.4 Binary signals: (a) unipolar; (b) polar; (c) unipolar with return to zero; (d) polar with return to zero; (e) bipolar with return to zero.

3.1.3.1 Baseband Data Transmission/Recording. Although the digital signal at the output of a discrete source can have almost ideal square shape, at the receiving end digital signals are distorted. Because of the limited channel bandwidth, the edges are not sharp, the pulses are wider, and their edges overshoot and ring. The pulse overlapping interferes in the receiver decision making process and is a source of errors. The residual effect of all other transmitted bits (symbols) on the mth bit being decoded is called *intersymbol interference* (ISI).

It is obvious that the problem could be overcome by system bandwidth extension at the penalty of admitting more noise to the system. But for the videodisc system this is not desired even if extra noise is disregarded, because a valuable bandwidth should be saved as much as possible. An overall pulse shaping that would yield zero ISI is a better solution. In practical systems some amount of residual ISI will inevitably occur due to imperfect filter realization, incomplete knowledge of channel characteristics, and changes in channel characteristics. Hence an equalizing filter is often inserted between the receiving filter and the A/D converter to compensate for changes in the parameters of the channel.

The design of a baseband system consists of specifying the generated and received pulse shape $p_g(t)$ and $p_r(t)$ and the receiving end transmitting filter transfer functions $H_F(f)$ and $H_T(f)$ to minimize the combined effects of intersymbol interference and noise in order to achieve a minimum probability of error for a given data rate and signal levels in the system.

For zero ISI, $p_r(t)$ should satisfy

$$p_r(nT_s) = \begin{cases} 1 & \text{for } n = 0 \\ 0 & \text{for } n \neq 0 \end{cases} \tag{3.11}$$

where T_s is the signaling interval (period). To meet this constraint, the Fourier transform $P_r(f)$ of $p_r(t)$ needs to satisfy a condition:

$$\sum_{k=-\infty}^{\infty} P_r(f + k/T_s) = T_s \qquad \text{for } |f| < 1/2T_s \tag{3.12}$$

This condition is called the *Nyquist (pulse shaping) criterion.*

A raised cosine frequency characteristic is most commonly used when the bandwidth available for transmitting data at a rate of R_b bits/s is between $R_b/2$ and R_b hertz. A raised cosine frequency spectrum consists of a flat amplitude portion and a roll-off portion that has a sinusoidal form. The pulse spectrum $P_r(f)$ is specified in terms of a parameter as

$$P_r(f) = \begin{cases} T_s \cos\dfrac{2\pi}{4\beta}\left(|f| - \dfrac{R_b}{2} + \beta\right), & R_b - \beta < f \leq R_b + \beta \\ 0, & f > R_b/2 + \beta \end{cases} \tag{3.13}$$

where $0 < \beta < R_b/2$. The corresponding pulse shape $p_r(t)$ is

$$p_r(t) = \frac{\cos 2\pi\beta t}{1 - (4\beta t)^2} \frac{\sin \pi R_b t}{\pi R_b t} \tag{3.14}$$

Plots of $P_r(f)$ and $p_r(t)$ for three values of the parameter β are shown in Figure 3.5. It can be seen that the bandwidth occupied by the pulse spectrum is $B = R_s/2 + \beta$; the minimum value of B is $R_s/2$ and the maximum value is R_s. Larger values of B imply that more bandwidth is required for a given symbol (bit) rate R_s. However, a larger value of B leads to faster decaying pulses, which means that synchronisation will be less critical and modest time errors will not cause large amounts of ISI. The phase characteristic is zero; $P_r(f)$ is real nonnegative.

For $\beta = R_b/2$ (really "raised cosine") oscillations in the response are not only small, but additional zero crossings are obtained at $t = \pm\frac{3}{2}T_s, \pm\frac{5}{2}T_s, \pm\frac{7}{2}T_s, \ldots$, in addition to the original zero crossings at $\pm T_s, \pm 2T_s, \ldots$. Also, the relative amplitude of response at points $t = \pm\frac{1}{2}T_s$ is 0.5.

According to the second Nyquist criterion for binary signals, the time period between instances when signal (current, voltage) has (passes through) an average value—or some other specified value—should be the same as the corresponding interval at the transmitting side. If this criterion is satisfied, the state duration will not be distorted.

When signal identification is obtained by simple integration, the surface form during one signaling interval is an important parameter. The third Nyquist criterion states when ISI influence to this parameter is eliminated; if $A(mT_s)$ is the area in the mth synchronizing interval, the analytic formulation for this criterion is

$$A(mT_s) = \int_{(2m-1)T_s/2}^{(2m+1)T_s/2} p_r(t)\, dt = A_0\delta_{m0}, \qquad m = 0, \pm 1, \pm 2, \ldots \tag{3.15}$$

where δ_{m0} is the Kronecker symbol,

$$\delta_{ij} = \begin{cases} 1, & i = d \\ 0, & i \neq d \end{cases} \tag{3.16}$$

and $A_0 = A(0)$.

Digital signaling schemes require accurately shaped waveforms. The shape of a pulse can be specified either by its amplitude as a function of time or by its Fourier transform. If a pulse shape is specified by its transform in the frequency domain, then, in principle at least, we can design filters to shape the pulse. This design becomes difficult in a practical sense because of the need for linear phase characteristics of filters. An alternative method is to generate the pulse waveform directly. Such a method should be capable of generating a signal composed of many overlapping pulses that are superimposed. One of the most commonly used methods of direct waveform generation makes use of a binary transversal filter [4].

3.1.3.2 Digital Carrier Modulation Schemes. Similar to analog modulation systems, the spectra of the digital signal can be transposed. The deterministic signal, the sinusoidal carrier, is defined by its three parameters: amplitude, frequency, and phase. Thus ASK, FSK, and PSK can be obtained.

In the case of AM (ASK), the modulating signal is obtained from the carrier

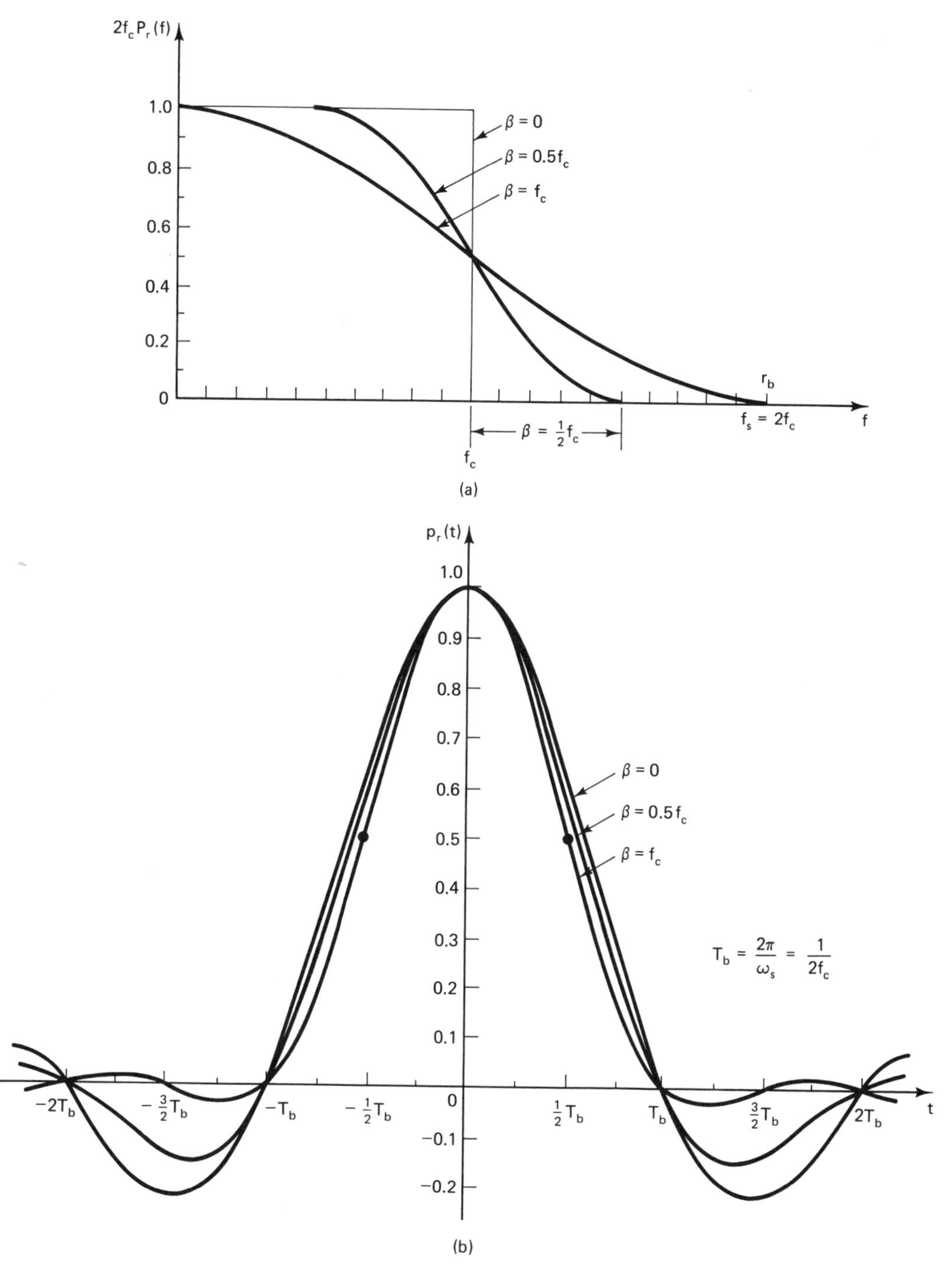

Figure 3.5 Raised cosine frequency characteristics: (a) $P_r(f)$ for three values of β; (b) pulse response $p_r(t)$ for three values of β.

by coherent demodulation—for any kind of AM: SSB, DSB, and so on. In the case of conventional AM, an envelope detector can be used.

FSK systems are simple in realization. System performance depends greatly on the demodulation process. Frequently used methods for FM demodulation are:

Limiter-discriminator
Zero-crossing detection
Differential detector
Coherent demodulation
Demodulation with two filters and two envelope detectors

The main drawback of FM systems is the bandwidth required. *M*-ary FSK is another choice, but the system complexity is higher than for binary FSK.

In its nature, phase modulation is, like FM, a nonlinear process. But when digital signals are transmitted by phase modulation, it is possible (under some conditions) to show that phase-modulated carrier presents two quadrature AM signals with two sidebands. Sometimes the term "PM-AM" is used for this case. The linear nature of the modulating process follows from the properties of the modulated signal spectra. Coherent demodulation is usually used. For differential PM demodulation a local carrier is not needed.

M-ary PSK systems are also used. Quadrature PSK (QPSK) is the simplest *M*-ary PSK.

3.1.4 Modulation Methods: Overview

According to the previous discussion, modulation methods can be classified as follows:

1. Continuous wave: analog
   a. Amplitude modulation (AM, DSB, SSB, VSB)
   b. Frequency modulation (FM)
   c. Phase modulation (PM)
2. Pulse modulation: analog
   a. Pulse amplitude modulation (PAM)
   b. Pulse frequency modulation (PFM)
   c. Pulse phase modulation (PPM)
   d. Pulse width modulation (PWM)
3. Pulse modulation: digital (baseband)
   a. Pulse code modulation (PCM)
   b. Delta modulation (DM)
   c. Differential pulse code modulation (DPCM)
4. Digital carrier systems: binary data
   a. Amplitude modulation [amplitude shift keying (ASK)]
   b. Frequency modulation (FSK)
   c. Phase modulation (PSK)

5. Digital carrier systems: multilevel coding
   a. MASK
   b. MFSK
   c. MPSK
   d. Combined amplitude and PSK
   e. Quadrature PSK (QPSK)
   f. Offset QASK (OQASK)
   g. OQPSK
   h. Timed frequency modulation (TFM)

Other classifications are also possible.

3.2 NOISE IN MODULATING SYSTEMS

Signals transmitted by any communication system will always be received with a certain amount of noise present, as the latter cannot be eliminated entirely from the system. Its presence tends to impede the reception of the wanted signal and is usually the limiting factor in its detection. Hence, in evaluating system performance or when comparing different communication systems, the most generally used criterion is the output signal-to-noise ratio (S/N) of the system.

In many cases the most used definition of signal-to-noise ratio is that of *average* signal power to *average* noise power present in the system, although sometimes other definitions may be used. Although the signal is random in nature, for different transmitted messages, the parameters that best describe the signals are different. Difficulties appear especially when signal has to be defined in order that some system can be verified by measurement in reference to S/N ratio. Because of all that, conventionally, it is taken that the signal S in the expression for the S/N ratio at the receiver output is always a test signal. For every kind of transmitted message the test signal is strictly defined. For instance, in sound and music transmission the test signal is sinusoidal, and the signal S in the S/N ratio is a valuable average power P_s of these test tones at the receiver output, while noise N is presented by the average noise power P_n. For example, in television, the signal S is defined as "peak-to-peak" of the picture signal, and noise N is characterized by its effective (rms) voltage across the useful band.

Attention should also be paid to S/N ratio measurements. Average noise power can easily be measured. However, when the average signal power is measured, the sum of the average signal plus noise power will be measured, because noise cannot be separated or eliminated. In practice, the noise is much smaller than the signal, and measured power can be considered as signal power.

Modulation systems are not equally immune to noise. In AM and pulse AM the ratio cannot be improved by increasing system bandwidth. But in FM and pulse position modulating systems the transmission quality can be improved, to some limit, by bandwidth increasing. This improvement is even more efficient in digital modulation methods.

The S/N ratio is useful for comparison with its value at the input and output of the receiving system. Typical graphs of input and output S/N ratios are shown in Figure 3.6, assuming the same input noise bandwidth for the various systems [10]. At low values of $(S/N)_i$ the AM systems appear to be the best, but for $(S/N)_i$ above the 10-dB threshold, FM is superior to AM systems. To achieve larger values of $(S/N)_o$, however, PCM appears to offer the best advantage. As a comparison, the ideal system shown is still about 8 to 10 dB better than the AM or PCM system. Obviously, when S/N ratios are determined in some systems, noise from other than random noise sources should be taken into account. One example is noise caused by nonlinear distortion. Total output noise should be included in the system's measure of quality.

System bandwidth and S/N ratio are the two most important parameters of system performance. Corresponding parameters in the digital systems are *digit rate* and *error probability*. In analog systems, quantitative noise influence is judged by the S/N ratio. The noise influence in the digital system is as follows. The digital signal receiver operates on a decision basis. It has to make decisions, based on the input signal in the given time interval, as to which symbol from the limited set was sent. Symbol discrimination is based on some characteristic signal parameter. If the volume of that parameter is changed because of the noise superposed to the signal, the decision will be wrong. This effect is quantitatively judged by the degree of error, which represents the probability that the decision is wrong. As in different analog modulator systems, different S/N ratios are required; in different digital systems, a different probability of error is required. The usual value for the probability of error is 10^{-3} to 10^{-9}.

The term *optimization* is usually used when both intersymbol interference (ISI) and noise influence are considered. The transmitting and receiving filters

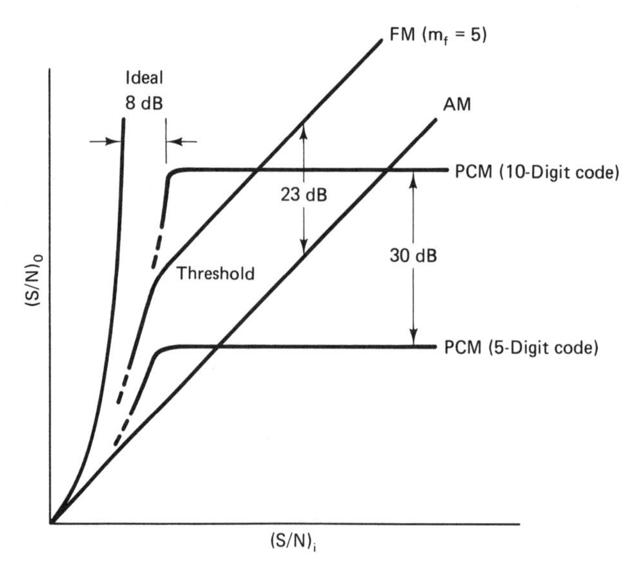

Figure 3.6 Typical graphs of input and ouput (S/N) ratios.

are chosen to produce zero ISI and minimize the probability of error for a given transmitted power. The zero-ISI condition is met if $P_r(f)$ has the form given in equation (3.13), and

$$P_g(f)H_T(f)H_C(f)H_R(f) = K_c e^{-j2\pi f t_d} P_r(f) \tag{3.17}$$

where t_d is the total time delay, $P_g(f)$ the transform of the basic pulse output of the pulse generator, K_c a normalizing constant, and $H_c(f)$ the channel transfer function. The optimal transmitting and receiving filters are [4, Chap. 5]

$$H_R^2(f) = \frac{K[P_r(t)]^2}{H_c(f)G_n^{1/2}(f)} \tag{3.18}$$

$$H_T^2(f) = \frac{K_c^2 P_r(f) G_n^{1/2}(f)}{K P_g(f)^2 H_c(f)} \tag{3.19}$$

where K is a positive constant. The filter phase responses are arbitrary as long as equation (3.17) is satisfied. For the same error probability, the M-ary signaling schemes requires less bandwidth than do binary systems.

3.3 SIGNAL MODIFICATION: SHAPING THE TRANSMITTING SIGNAL SPECTRUM AND WAVEFORM

In many applications, a spectrum and waveform of the signal should be carefully shaped to match the channel characteristics in order to arrive at physically realizable transmitting and receiving filters. Besides spectrum, waveform, or combined shaping, adaptive shaping methods are also attractive.

3.3.1 Preemphasis

The preemphasis/deemphasis method is widely used in FM systems. The spectral density of the random noise power at the FM receiver output is proportional to f^2: noise is higher for the higher spectral components of the modulating signal. On the other hand, many real signals, such as audio, have considerably more energy at low frequencies than at high frequencies. This means that when real messages are transmitted through an FM system, a valuable bandwidth $B = 2(\Delta f + f_{\max})$ is not used efficiently enough: On the average, only part of the available bandwidth is used, while noise exists in the whole bandwidth. This results in a poor S/N ratio due to noise at the higher part of the transmitted signal. This can be avoided using the following procedure: namely, the modulating signal can be processed before FM modulation, so that high-frequency components are boosted compared with low-frequency components. In this manner, high-frequency components will, on the average, cause higher-frequency deviation; that is, the S/N ratio will be improved. At the receiver, after the discriminator, an inverse process—deemphasis—is performed and the original spectral density is obtained; otherwise, the signal would be distorted. The process of deemphasis reduces the signal back to a normal unity-gain level at higher frequencies and correspondingly reduces the noise level to below a unity-gain system. Therefore, the signal is effectively amplified, transported through a noisy channel, and then attenuated. The process attenuates the signal, but more important, it also attenuates the channel noise.

As an example, the preemphasis curve for the 625 standard TV signal is shown in Figure 3.7. The relative deviation Δf_{0s}, in reference to the deviation of 8 MHz caused by the 1-V "peak-to-peak" TV signal, is shown as a function of the frequency f of the TV signal spectra. This curve is obtained from the expression

$$H(j\omega) = (0.282)\frac{1 + jf/0.313}{1 + jf/1.565}; \qquad f \text{ in MHz} \tag{3.20}$$

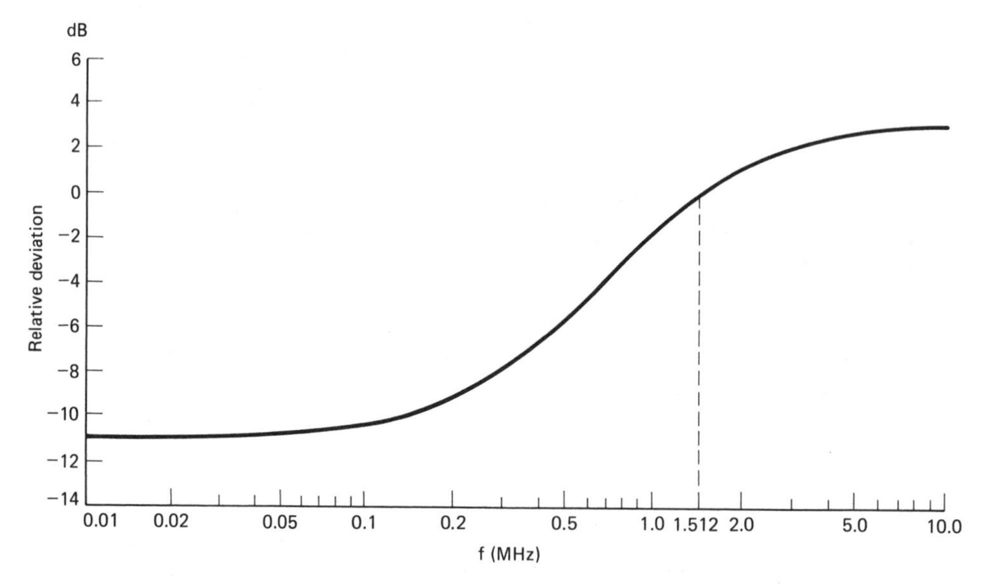

Figure 3.7 Preemphasis curve for the TV signal, 625-line standard.

The deemphasis function is inverse to this. This preemphasis/deemphasis transfer functions can be obtained by the circuits shown in Figure 3.8.

3.3.2 Compressor

The compressor/expander technique, used, for example, in PCM systems, is an example of signal amplitude modification (preprocessing). Similar to the frequency domain for real signals, on the average, lower amplitudes are more occupied than are higher levels. Obviously, if uniform quantization is performed for those signals, the signal-quantizing noise ratio will be larger for large signals and smaller for small signals. If, statistically, small signals are dominant, uniform quantization is not the optimal solution. For the same number of the quantizing levels (steps) it is better to have small steps for small signals and higher steps for larger signals;

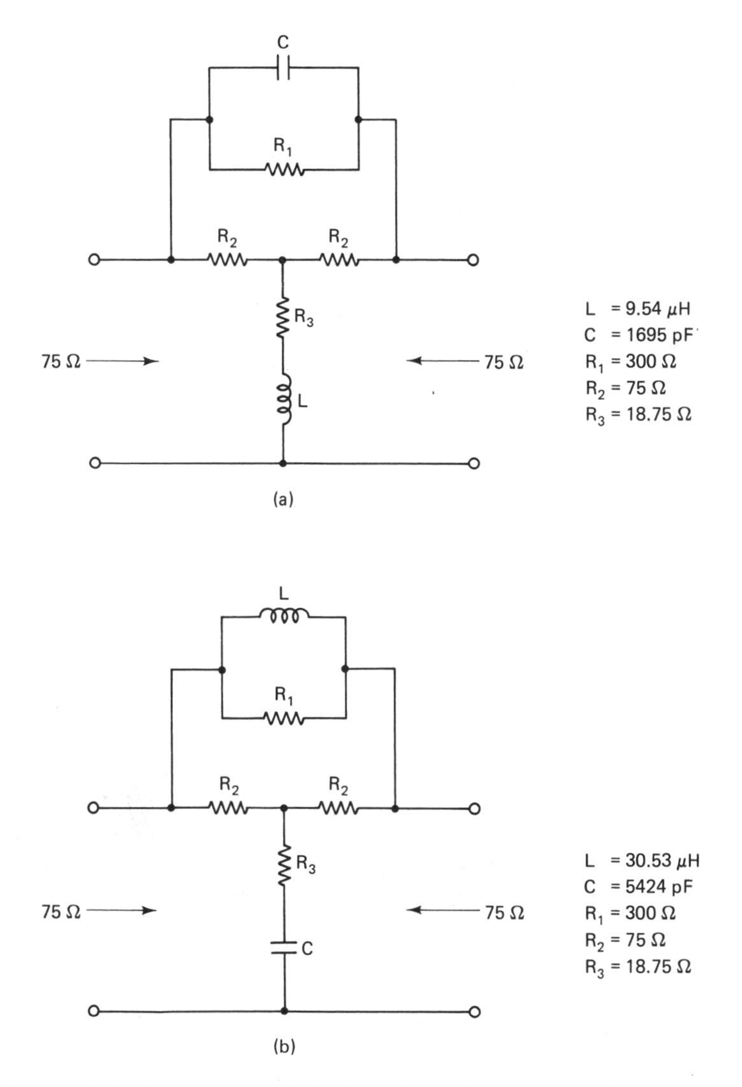

Figure 3.8 (a) Preemphasis and (b) deemphasis circuits for the TV signals from the 625-line system.

the signal/quantizing noise will be greatly improved for small signals and insignificantly reduced for large signals. Input–output characteristics for the compressor and the expander are shown in Figure 3.9. The first characteristic is performed in the decoder. Obviously, these two characteristics should be well matched.

3.3.3 Pulse Shaping

The spectrum of the transmitted signal in a PAM system will depend on the signaling pulse waveform and on the statistical properties of sequences of transmitted

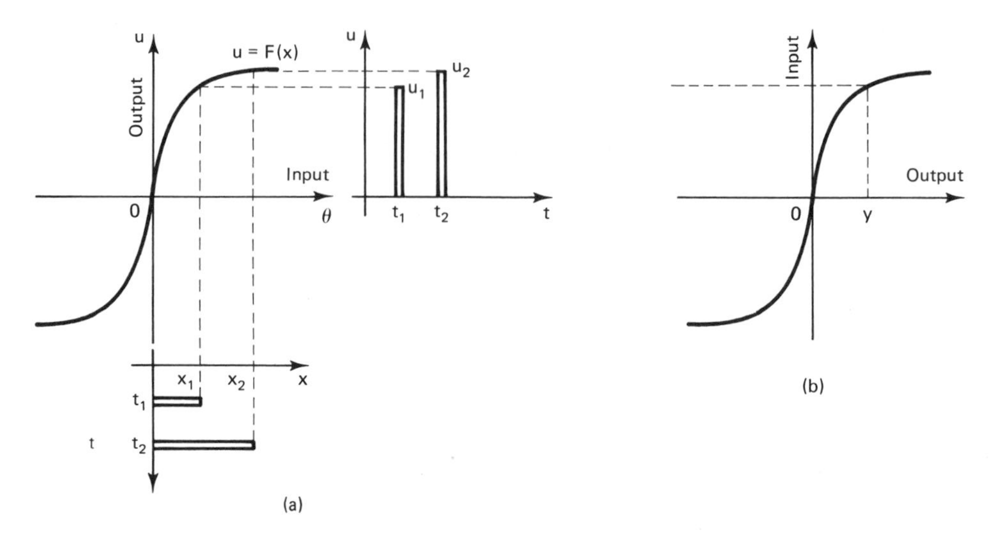

Figure 3.9 (a) Compressor [$u = f(\theta)$] and (b) expander characteristics.

digits. The pulse waveform in a PAM system is specified by the ISI requirements. Although it might be possible to shape the transmitted signal spectrum by changing the transmitted pulse shape, such changes might lead to increased ISI. An easier way to shape the transmitted signal spectrum is to alter the statistical properties of the transmitted bit or symbol sequence [4, Chap. 5].

The transfer function in pulse modulation systems should satisfy Nyquist criteria, in order that ISI is eliminated. Unknown exact channel characteristics and their variation in time, among other reasons, indicate that almost permanent correction of the transfer function is needed. This can be done through special circuits—correctors—for which a *transversal filter* is frequently used. The basic idea for its construction was given as early as 1940 [11]. A binary transversal filter is shown in Figure 3.10.

A shift register is used as the delay line [4]. The pulse waveform $p_t(t)$ which is to be amplitude modulated is sampled at intervals of Δt. If the sample values outside the time interval $-8\Delta t$ to $8\Delta t$ are supposed to be small enough, and for purposes of simplicity are ignored, the transversal filter for this waveform will consist of a 17-bit shift register with 17 outputs, each of which can have two possible levels, 0 or 1. These outputs are alternated b_{-8} to b_8 and summed by an analog summing network. Successive bits d_k are inserted into the shift register once every T_b seconds. The output due to successive input bit overlap and the summing device performs the superposition of individual pulses to form the composite waveform. The digitally generated waveform has zero errors at sampling times 0, $\pm T_b$, $\pm 2T_b$,... . The spectrum of the staircase waveform is centered around the sampling frequency, $1/\Delta t$, and its harmonics. By choosing a suitably large sampling frequency, the "noise" due to sampling can be separated from

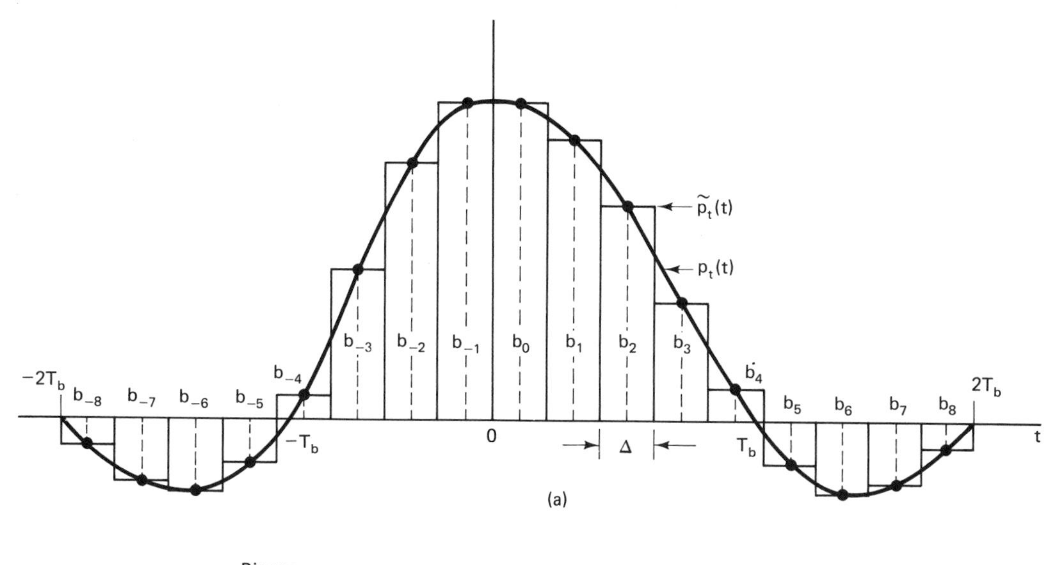

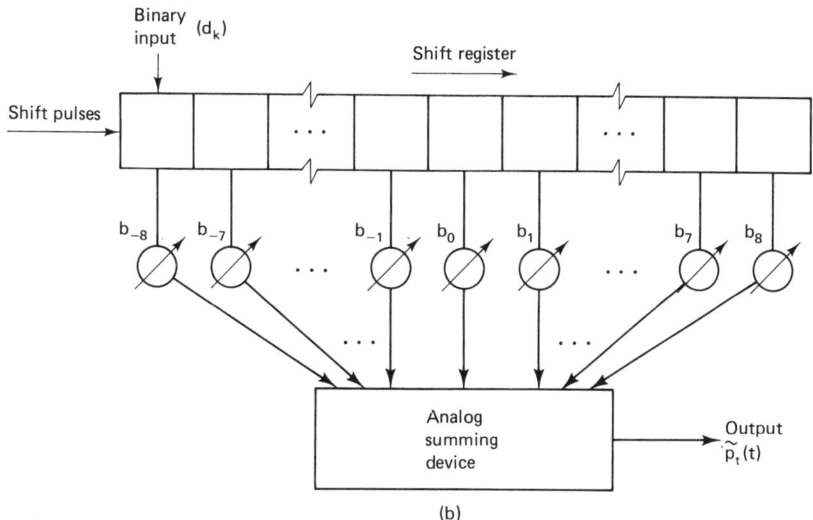

Figure 3.10 Binary transversal filter.

the spectrum of the desired pulses. Then a low-pass filter can be used to smooth out the waveform.

In general, transversal filters are used as correctors in two ways. A filter can be placed at the transmitter and used to predistort the signal so that conditions for zero ISI are satisfied when the signal reaches the receiver, or the filter can be placed at the receiver and used to compensate for the distortion. The process of correcting channel-induced distortion is called *equalization*.

An adjustable equalizer in the form of a transversal filter is shown in Figure

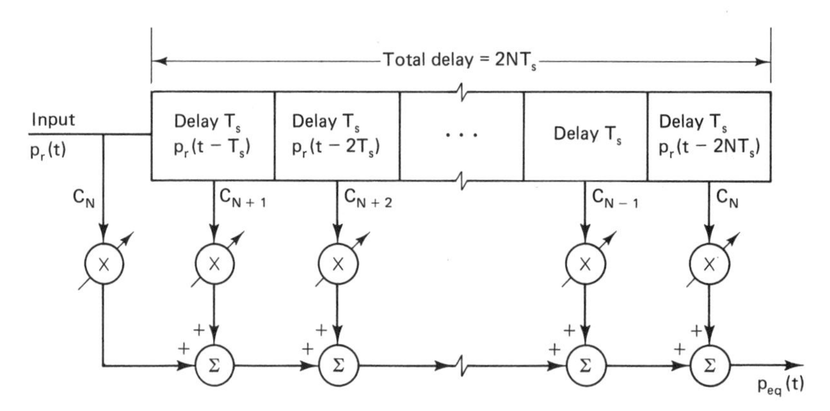

Figure 3.11 Transversal equalizer.

3.11. A delay line is tapped at T_S-second intervals. Each tap is connected through a variable-gain device (C_i) to a summing amplifier. The input to the equalizer is $p_r(t)$, which is known, and the output is $p_{eq}(t)$. One example, for received and equalized pulses, is shown in Figure 3.12.

If we specify the value of $p_{eq}(t)$ for $2N + 1$ points as

$$p_{eq}(k) = \begin{cases} 1 & \text{for } k = 0 \\ 0 & \text{for } k = \pm 1, \pm 2, \ldots, \pm N \end{cases} \tag{3.21}$$

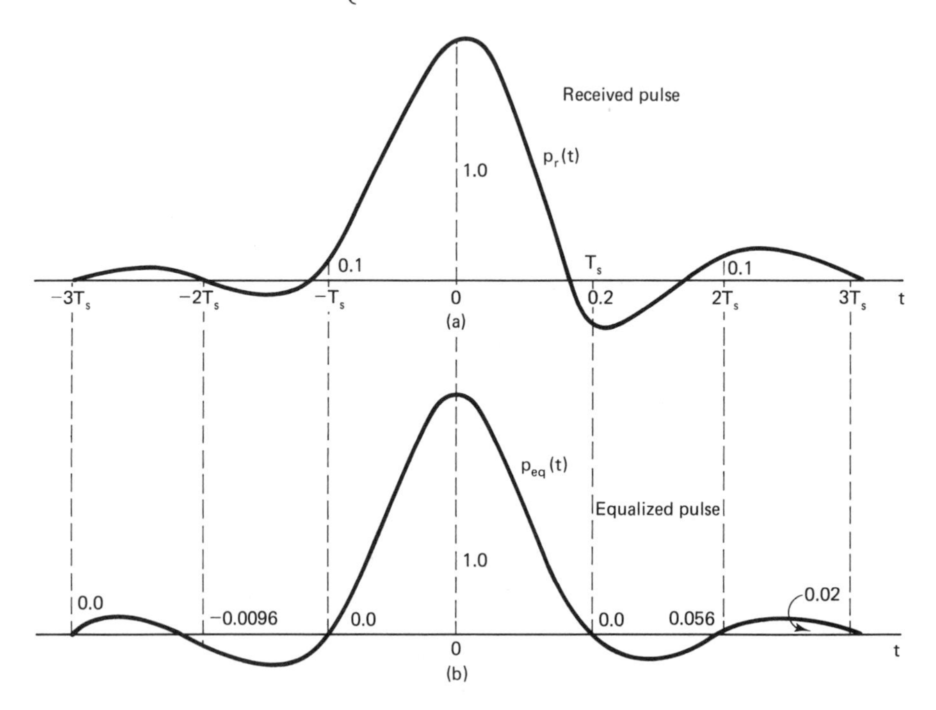

Figure 3.12 Input and output waveforms of the equalizer.

the equalizer is called a *zero forcing equalizer* since $p_{eq}(k)$ has N zero values on either side. Thus equalizer is optimum in that it minimizes the peak intersymbol interference. The main disadvantage of a zero forcing equalizer is that it increases the noise power of the input to the A/D converter [4]. But this effect is normally more than compensated for by the reduction in the ISI.

The essential function of the receiver in the digital carrier modulated systems is to determine which of two (for binary systems) known waveforms $s_1(t)$ or $s_2(t)$ was present at its input during each signaling interval. The optimum receiver distinguishes between $s_1(t)$ and $s_2(t)$ from the noise version of $s_1(t)$ and $s_2(t)$ with minimum probability of error. A matched filter receiver [where the impulse response $R(t)$ is matched to the signal $s_1(t)$ and $s_2(t)$] or correlated receiver can be used.

Appropriate shaping of the input data symbols allows one to generate an entire class of shift-keying-type signals, whose spectral properties are more desirable in some applications. For example, minimum shift keying (MSK) has constant envelope and signal RF bandwidth requirements.

Lately, a new approach has been taken in noise reduction, especially for audio noise reduction. Basically, adaptive, level-dependent signal processing is performed. This is discussed in greater detail in Chapter 8.

REFERENCES

1. H. S. Black, *Modulation Theory,* Bell Laboratories Series, D. Van Nostrand, Princeton, N.J., 1953.
2. M. Schwartz, *Information Transmission Modulation and Noise,* 2nd ed., McGraw-Hill, New York, 1970.
3. P. F. Panter, *Modulation, Noise and Spectral Analysis,* McGraw-Hill, New York, 1965, 1959.
4. K. S. Shanmugam, *Digital and Analog Communication Systems,* Wiley, New York, 1979.
5. W. D. Gregg, *Analog and Digital Communications: Concepts, Systems, Applications, and Services in Electrical Dissemination of Aural, Visual, and Data Information,* Wiley, New York, 1977.
6. I. S. Stojanović, *Osnovi Telekomunikacija* [Telecommunication Fundamentals], Gradjevinska Knjiga, Belgrade, 1977.
7. G. Lukatela, *Statistička Teorija Telekomunikacija* [Statistical Telecommunication Theory], Gradjevinska Knjiga, Belgrade, 1978.
8. *List of Definitions of Essential Telecommunication Terms,* ITV, Geneva, June 1957, Definitions 02.38 and 02.39.
9. CCITT, Fifth Planary Assembly, *Green Book,* Vol. IV-2: *Line Transmission,* Geneva, 1973, p. 353, Definition 2002.
10. F. R. Connor, *Noise,* Edward Arnold, London, 1973.
11. H. E. Kallmann, "Transversal Filters," *Proc. IRE,* July 28, 1940, pp. 302–310.

APPENDIX 3A: LINEAR CONTINUOUS-WAVE MODULATION SYSTEMS

Essentially, the linear continuous-wave system is a multiplication process; the modulating signal, $g(t)$, and the carrier are multiplied together. This operation is equivalent to a symmetrical translation of the baseband spectrum through a distance ω_c, the carrier angular frequency. The modulated carrier is represented by

$$s(t) = A(t) \cos \omega_c t \tag{3A.1}$$

The carrier amplitude $A(t)$ is linearly related to the message signal $g(t)$. Depending on the relationship between $g(t)$ and $A(t)$ in the frequency domain, there are different types of linear modulation systems.

3A.1 Double-Sideband Modulation

The modulated waveform is

$$s(t) = A_c g(t) \cos \omega_c t, \qquad \omega_c = 2\pi f_c \tag{3A.2}$$

The spectrum of DSB signal is

$$S_c(f) = F\{s(t)\} = \tfrac{1}{2}A_c[S(f - f_c) + S(f + f_c)] \tag{3A.3}$$

where F{ } denotes the Fourier transform.

The required bandwidth is

$$B = 2f_g \tag{3A.4}$$

where f_g is the message signal bandwidth (Figure 3A.1). For tone modulation with

$$g(t) = a \cos \omega_m t \tag{3A.5}$$

the DSB signal is (Figure 3A.2)

$$s(t) = \tfrac{1}{2}Aa \cos [(\omega_c + \omega_m)t] + \tfrac{1}{2}Aa \cos [(\omega_c - \omega_m)t] \tag{3A.6}$$

The average DSB signal power is

$$P = \lim_{T\to\infty} 1/T \int_{-T/2}^{T/2} A_c^2\, g(t) \cos^2\omega_c t \, dt = P_c P_g \tag{3A.7}$$

where $P_c = \tfrac{1}{2}A_c^2$ is the average carrier power, and

$$P_g = \lim_{T\to\infty} 1/T \int_{-T/2}^{T/2} g^2(t)\, dt \tag{3A.8}$$

is the normalized average signal power.

3A.2 Double-Sideband Demodulation

For an ideal channel, the received signal $s_r(t)$ is

$$s_r(t) = a_c g(t) \cos \omega_c t \tag{3A.9}$$

where a_c/A_c is the channel attenuation. The output of the multiplier (with a local

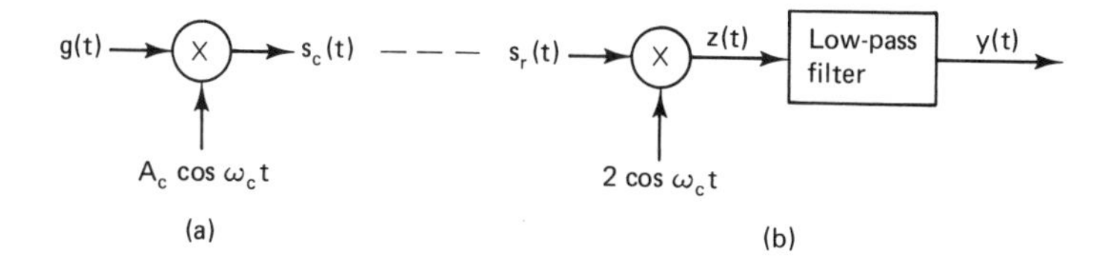

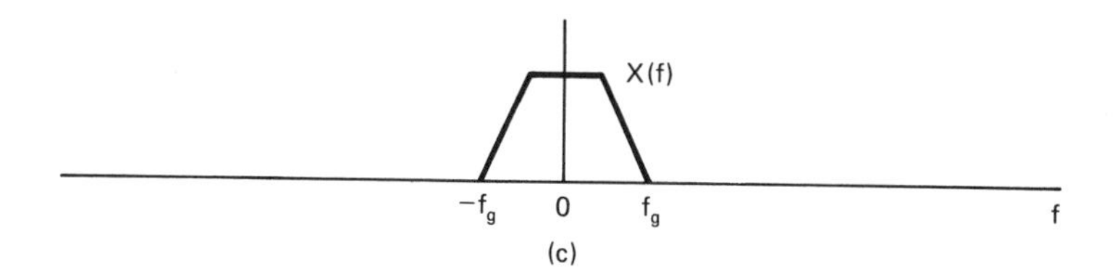

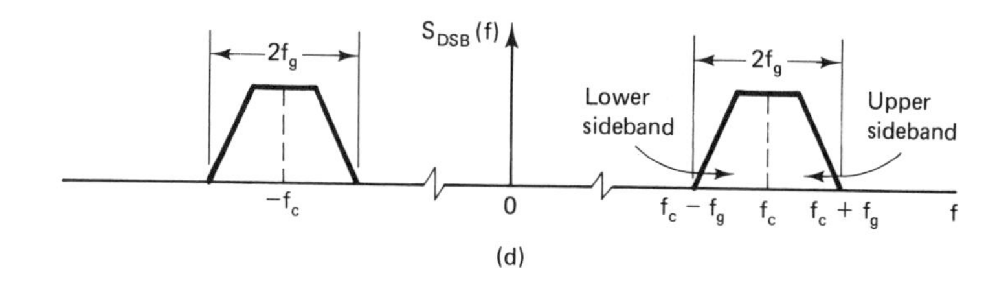

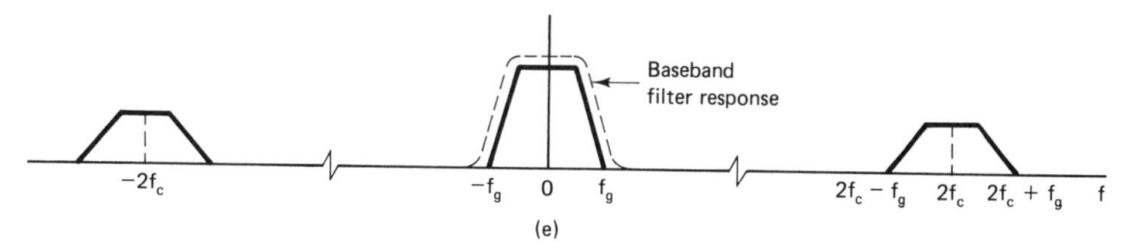

Figure 3A.1 Double-sideband modulation: (a) modulator; (b) synchronous (or coherent) demodulator; (c) message spectrum for an arbitrary $g(t)$; (d) $S_{DSB}(f)$; (e) $Z(f)$.

carrier) in the case of synchronous (coherent) demodulation is

$$z(t) = s_r(t)2 \cos \omega_c t = a_c g(t) + a_c g(t) \cos 2\omega_c t \tag{3A.10}$$

The spectrum of $z(t)$ is (Figure 3A.1)

$$S_z(f) = F\{z(t)\} = a_c S(f) + \tfrac{1}{2}a_c[S(f - 2f_c) + S(f + 2f_c)] \tag{3A.11}$$

For $f_g < 2f_c - f_g$, there is no overlapping of spectral components and proper filtering can be done, so the output signal is

$$y(t) = a_c g(t) \tag{3A.12}$$

When the local oscillator signal has a frequency offset $\Delta\omega$ and phase offset θ, the output will be

$$y(t) = a_c g(t) \cos (\Delta\omega t + \theta) \tag{3A.13}$$

For $\Delta\omega = 0$ and $\theta = \pi/2$, the signal is lost entirely.

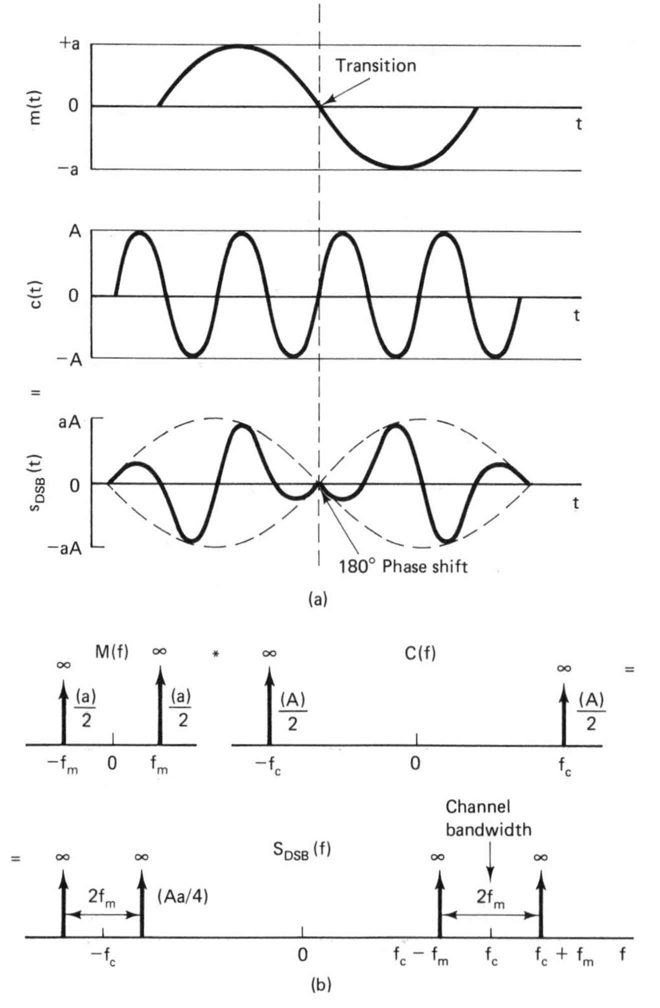

Figure 3A.2 Tone DSB.

For $\theta = 0$, $y(t) = a_c g(t) \cos \Delta\omega t$, which results in serious signal distortion. For instance, in the case of tone modulation and $\Delta\omega = 0.1\omega_c$ (10% offset):

$$y(t) = \tfrac{1}{2}a_c[\cos(\omega_m + 0.1\omega_c)t + \cos(\omega_m - 0.1\omega_c)t] \tag{3A.14}$$

Because usually $f_c >> f_g$ or $f_g << f_c$, even a small percentage error in f_c will cause a deviation Δf that may be comparable to or larger than f_g.

In Figure 3A.3, a method for adding a small carrier signal to the DSB system of the modulator and its extracting methods are shown. Another method to generate a coherent carrier for demodulation is to use a squaring circuit and a bandpass filter centered to $2f_c$. The frequency divider should be used, too, because $g^2(t)$ has a nonzero dc component and hence a discrete frequency component of $2f_c$ can be extracted from $s_r(t)$, although $g(t)$ may have zero dc value.

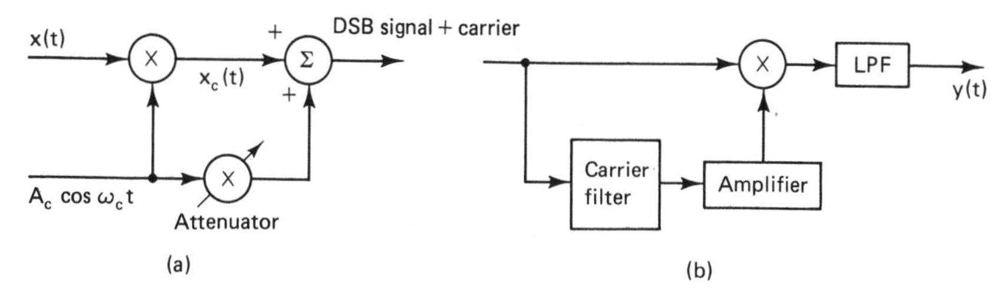

Figure 3A.3 Pilot carrier DSB system: (a) transmitter; (b) receiver.

3A.3 Conventional Amplitude Modulation

AM signal = DSB signal + large carrier. The AM signal has the form (Figure 3A.4)

$$s(t) = A_c[1 + g(t)] \cos \omega_c t = A(t) \cos \omega_c t \tag{3A.15}$$

where $A(t)$ is the envelope of the modulated carrier, and the modulation index

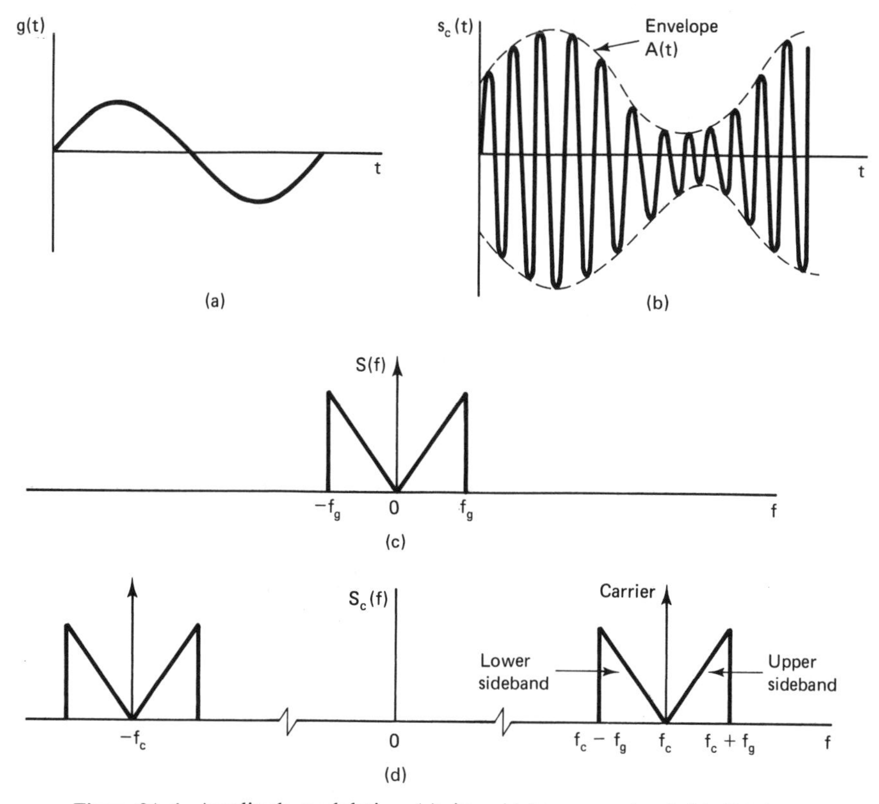

Figure 3A.4 Amplitude modulation: (a) sinusoidal message signal; (b) AM signal; (c) message spectrum for an arbitrary $x(t)$; (d) modulated signal spectrum.

m of the AM signal is

$$m = \frac{[A(t)]_{\max} - [A(t)]_{\min}}{[A(t)]_{\max} + [A(t)]_{\min}} \tag{3A.16}$$

When m exceeds 1, the carrier is overmodulated and the envelope is distorted (Figure 3A.5). For easy signal recovery (using simple demodulation schemes), the modulated signal amplitude has to be small:

$$|g(t)|_{\max} << A_c \tag{3A.17}$$

and the dc component has to be zero:

$$\lim_{T\to\infty} 1/T \int_{-T/2}^{T/2} g(t)\, dt = 0 \tag{3A.18}$$

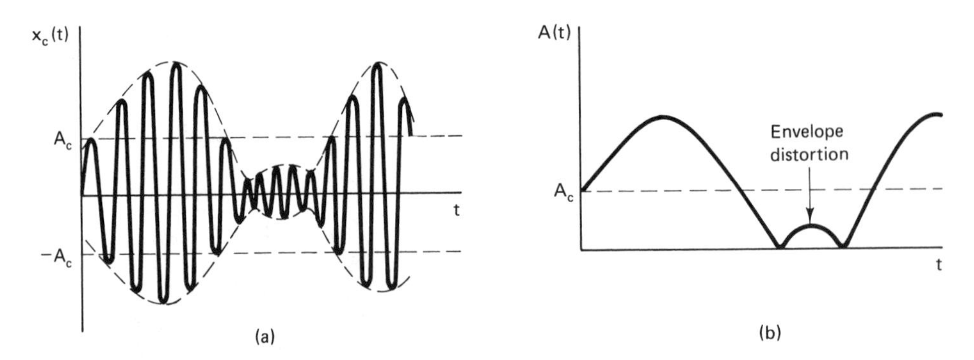

Figure 3A.5 Envelope distortion of an AM signal: (a) modulated signal; (b) envelope $A(t)$.

The spectrum of the AM signal (Figure 3A.4) is

$$S_c(f) = F\{s(t)\}$$
$$= \tfrac{1}{2}A_c[S(f - f_c) + S(f + f_c)] + \tfrac{1}{2}A_c[\delta(f - f_c) + \delta(f + f_c)] \tag{3A.19}$$

The bandwidth of the AM signal is

$$B = 2f_g$$

The baseband message signal $g(t)$ can be recovered from the AM signal $s_r(t)$ using a simple envelope detector (Figure 3A.6). Under ideal operating conditions, $f_c >> f_g$ or $f_g << f_c$, the discharge time constant RC is adjusted so that the maximum negative rate of the envelope will never exceed the exponential discharge rate; the output of the demodulator is

$$z(t) = K_1 + K_1 g(t) \tag{3A.20}$$

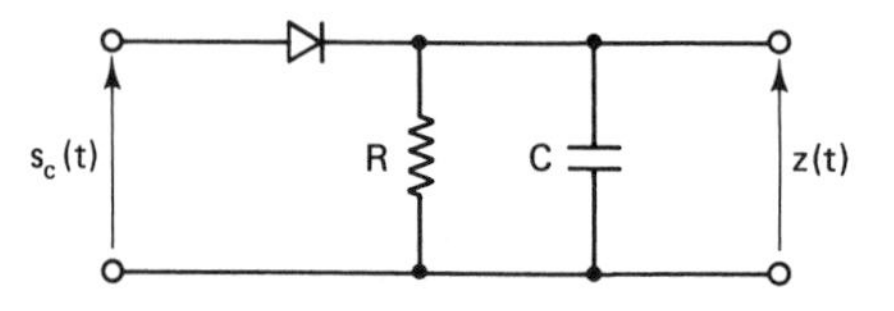

Figure 3A.6 Envelope detector of AM signal.

where K_1 is a dc offset due to the carrier and K_2 is the gain of the demodulator circuit.

AM is not suitable for transmitting message signals having significant low-frequency components, because a coupling capacitor or a transformer should be used to remove the dc offset. This was one of the reasons for the assumption that the dc value of $g(t)$ is zero.

3A.4 Single-Sideband Modulation

The primary motivation for considering SSB modulation is to get a 50% bandwidth savings comparing to DSB or AM: only half the spectrum of $s_{DSB}(t)$ is necessary to convey all the information in its entire spectrum, because the other half reflected about f_c can be constructed by symmetry (Figure 3A.7). The bandwidth of the SSB signal is

$$B = f_g \tag{3A.21}$$

Difficulties in practical implementation:

1. The modulator calls for an ideal bandpass sideband filter.
2. The demodulator requires a synchronous carrier.

SSB can be used for message signals with little or no low-frequency content (e.g., voice and music).

The SSB signal can be generated by the phase-shift (or quadrature) method, which does not require a sideband filter (Figure 3A.8). An example for tone modulation:

$$g(t) = G \cos(\omega_m t + \theta_0), \qquad \omega \leq \omega_g \tag{3A.22}$$

The SSB upper sideband signal:

$$\begin{aligned} s(t) &= (A_c/2)G \cos[(\omega_c + \omega_m)t + \theta_0] \\ &= (A_c/2)[g(t) \cos \omega_c t - \hat{g}(t) \sin \omega_c t] \end{aligned} \tag{3A.23}$$

where

$$\hat{g}(t) = G \sin(\omega_m t + \theta_0) \tag{3A.24}$$

is the quadrature message signal component and is referred to as the *Hilbert transform* of $g(t)$.

An arbitrary information signal $g(t)$ can be expressed as a function of its spectrum by the inverse Fourier transform (Figure 3A.9a):

$$\begin{aligned} g(t) &= \int_{-\infty}^{\infty} S(f)e^{j\omega t}\, df = \int_{-\infty}^{0} S(f)e^{j\omega t}\, df + \int_{0}^{\infty} S(f)e^{j\omega t}\, df \\ &= g_-(t) + g_+(f) \end{aligned} \tag{3A.25}$$

Similarly,

$$S(f) = S_+(f) + S_-(f) \tag{3A.26}$$

where, for example,

$$S_+(f) = \tfrac{1}{2}[1 + \text{sgn}(f)]S(f) \tag{3A.27}$$

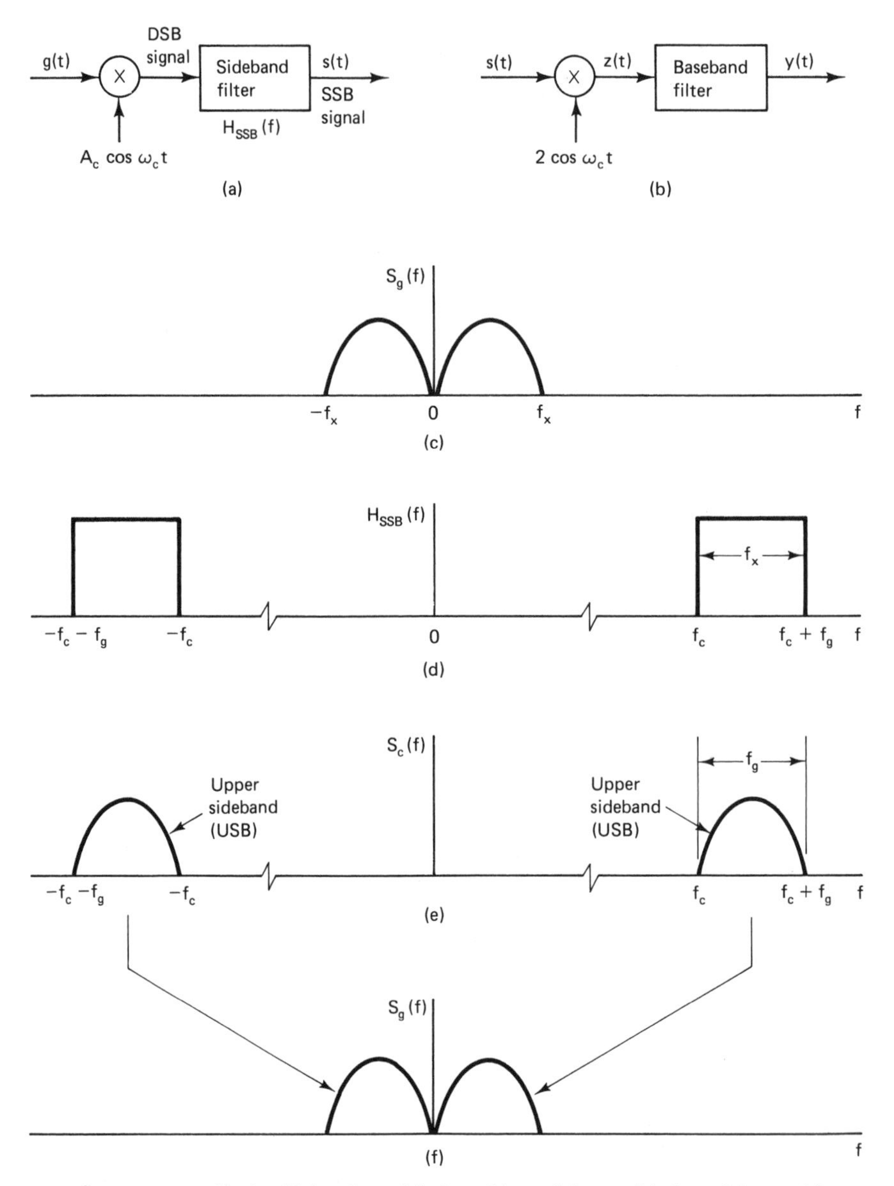

Figure 3A.7 Single-sideband modulation: (a) modulator; (b) demodulator; (c) signal spectrum; (d) ideal sideband filter; (e) transmitted signal spectrum; (f) reconstructed signal.

Using this type of decomposition, it follows that (Figure 3A.9)

$$s_{\text{USB}}(t) = (A_c/2)[g(t) \cos \omega_c t - h(t) * g(t) \sin \omega_c t] \qquad (3A.28)$$

and

$$s_{\text{LSB}}(t) = (A_c/2)[g(t) \cos \omega_c t + h(t) * g(t) \sin \omega_c t] \qquad (3A.29)$$

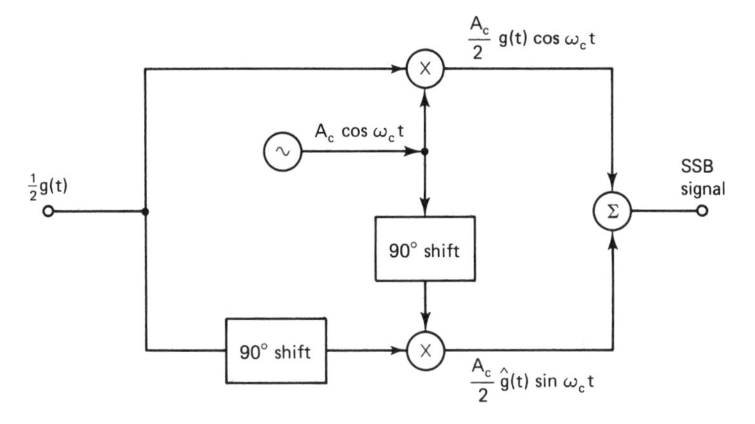

Figure 3A.8 Phase-shift SSB modulator.

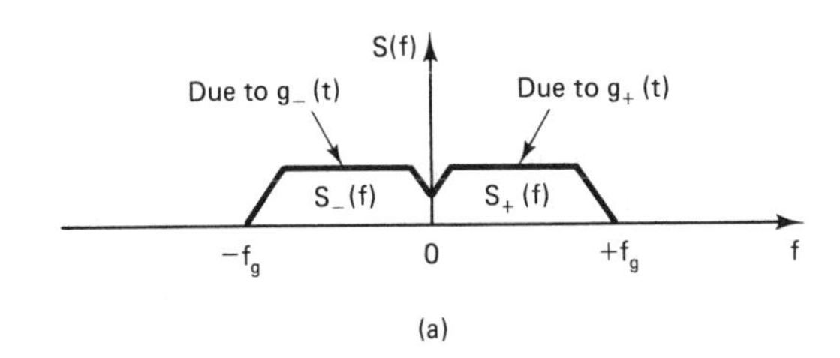

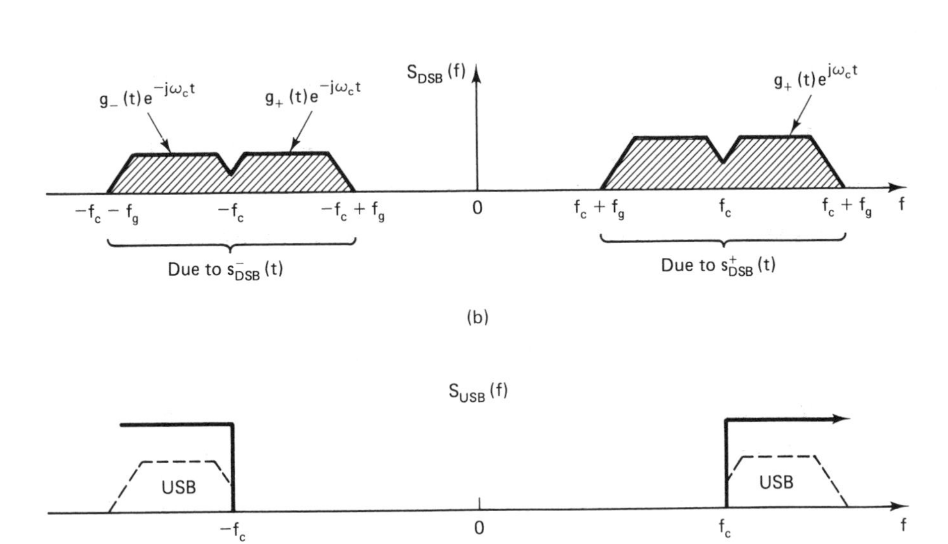

Figure 3A.9 (a) Decomposition of $S(f)$; (b) spectral decomposition of $S_{DSB}(f)$; (c) VSB spectrum obtained by an ideal zonal sideband filter.

where $*$ is the convolution symbol and

$$h(t) = \int_{-\infty}^{\infty} - j \operatorname{sgn}(f) e^{j\omega t}\, df = 1/u(t)$$

The function $H(f) = -j \operatorname{sgn} u(f)$ is referred to as a *quadrature* or *Hilbert transform filter,* and it delays harmonics by 90° with no attenuation.

3A.5 Vestigial Single-Sideband Modulation

This modulation method comprises the bandwidth conservation of SSB, and the good performance for messages with significant low-frequency content of DSB (or AM). VSB modulation is obtained by filtering a DSB or an AM signal in such a fashion that one sideband is passed almost completely while only a trace of the other sideband is passed (Figure 3A.10). The essential requirement of the VSB filter $H_{VSB}(f)$ is that it must have odd symmetry about f_c and a relative response of $\frac{1}{2}$ at f_c. The transition interval of the VSB filter is of width 2α hertz, and the transmission bandwidth of the VSB signal is

$$B = f_g + \alpha \tag{3A.30}$$

The bandwidth is about 125% of SSB or 62% of DSB rather than 50%, due to the imperfect cutoff of one sideband. This yields DSB transmission of low-

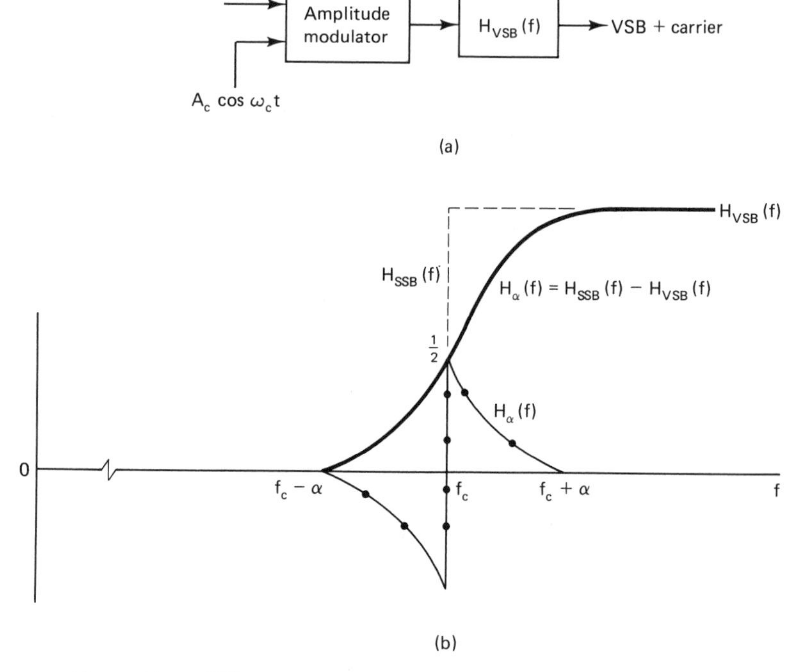

Figure 3A.10 VSB modulation: (a) modulator; (b) filter characteristics.

frequency components of $g(t)$ (those near f_c) and SSB transmission of higher-frequency components of $g(t)$. Reception of such a composite signal would require amplitude equalization in the receiver. The vestigial sideband characteristic shown in Figure 3A.10 does not require equalization. The output of the VSB filter is given by the convolution:

$$s_{\text{VSB}}(t) = h_v(t) * s_{\text{DSB}}(t) \tag{3A.31}$$

where the Fourier transform of $h_v(t)$ is (Figure 3A.10)

$$F[h_v(t)] = H_{\text{VSB}}(f) = H_{\text{SSB}}(f) - [H_\alpha(f)] \tag{3A.32}$$

$H_\alpha\ (f)$ represents the difference between the SSB and VSB filters, and it is required to have odd symmetry about f_c. If the input to VSB filter is an AM signal, $A_c[1 + g(t)] \cos \omega_c t$, the output signal, can be expressed as

$$\underset{\substack{\uparrow\\ \text{VSB}\\ +\\ \text{carrier}}}{s(t)} = \underbrace{\tfrac{1}{2}A_c \cos \omega_c t}_{\textit{carrier}} + \underbrace{\underbrace{\tfrac{1}{2}A_c[g(t) \cos \omega_c t - \hat{g}(t) \sin \omega_c t]}_{\textit{SSB signal}} - \tfrac{1}{2}A_c g_\alpha(f) \sin \omega_c t}_{\textit{VSB signal}} \tag{3A.33}$$

where $\frac{1}{2}A_c g_\alpha(t) \sin \omega_c t$ is the response of $H_\alpha(f)$ to $A_c g(t) \cos \omega_c t$.

The VSB signal can be demodulated using a synchronous demodulator. If a sufficiently large carrier component had been added to the VSB signal at the modulator, the VSB signal can be demodulated with a small amount of signal distortion using envelope demodulation. The VSB signal can be expressed as

$$s(t) = \tfrac{1}{2}A_c[1 + g(t)] \cos \omega_c t - \tfrac{1}{2}A_c\gamma(t) \sin \omega_c t \tag{3A.34}$$

where

$$\gamma(t) = \hat{g}(t) + g_\alpha(t)$$

If $\gamma(t) = 0$, an AM signal is obtained, and with $\gamma(t) = \hat{g}(t)$, an SSB + carrier signal is obtained. Adding a carrier to the VSB signal, equation (3A.34) becomes

$$s(t) = A_c\{[1 + g(t)] \cos \omega_c t - \gamma(t) \sin \omega_c t\} \tag{3A.35}$$

or

$$s(t) = R(t) \cos [\omega_c t + \phi(t)] \tag{3A.36}$$

where $R(t)$ is the envelope given by

$$R(t) = A_c[1 + g(t)] \left\{1 + \left[\frac{\gamma(t)}{1 + g(t)}\right]^2\right\}^{1/2} \tag{3A.37}$$

The envelope is distorted, but for $|\gamma(t)| << 1$, the distortion is negligible: $R(t) \simeq A_c[1 + g(t)]$. For $\gamma(t) = \hat{g}(t)$, the SSB + carrier signal, the quadrature component cannot be ignored.

It can be easily shown that the VSB signal is of SSB structure; that is, it has the quadrature carrier form. The output of the VSB filter,

$$s_{\text{VSB}}(t) = h_v(t) * s_{\text{DSB}}(t) \tag{3A.38}$$

can be expressed as

$$\begin{aligned} s_{\mathrm{VSB}}(t) &= A_c \int_{-\infty}^{\infty} h_v(x)[g(t-x)\cos\omega_c(t-x)]\,dx \\ &= A_c\left[\int_{-\infty}^{\infty} h_v(x)\cos\omega_c x g(t-x)\,dx\right]\cos\omega_c t \\ &\quad + A_c\left[\int_{-\infty}^{\infty} h_v(x)\sin\omega_c x g(t-x)\,dx\right]\sin\omega_c t \\ &= g_i(t)c(t) + g_q(t)g(t) \end{aligned} \tag{3A.39}$$

where $h_i(t) = \mathrm{h}_v(t)\cos\omega_c t$ and $h_q(t) = h_v(t)\sin\omega_c t$ for simplicity. $H_i(j\omega)$ and $H_q(j\omega)$ are the Fourier tranform of $h_i(t)$ and $h_q(t)$, respectively.

$$H_i(j\omega) = \tfrac{1}{2}\{H_v[j(\omega-\omega_c)] + H_v[j(\omega+\omega_c)]\} \tag{3A.40}$$

$$H_q(j\omega) = \tfrac{1}{2}j\{H_v[j(\omega-\omega_c)] - H_v[j(\omega+\omega_c)]\} \tag{3A.41}$$

For the sake of comparison, the SSB signal is

$$s_{\mathrm{LSB}}(t) = (A_c/2)[g(t)c(t) \mp (h(t) * g(t))g(t)] \tag{3A.42}$$

In Figure 3A.11, phasor diagrams for ideal continuous wave (CW) modulation by a single tone are illustrated (carrier is present in SSB).

3A.6 Noise in Linear CW Modulation Systems

The detector input signals are:

AM: $A_r[1 + mg(t)]\cos(\omega_c t + \phi_c)$
DSB: $A_r g(t)\cos(\omega_c t + \phi_c)$
SSB: $(A_r/2)[g(t)\cos(\omega_c t + \phi_c) + [h(t) * g(t)]\sin(\omega_c t + \phi_c)]$

The detector noise input, $n(t)$, is a filtered version of channel output noise, and it can be expressed as

$$\begin{aligned} n(t) &= V(t)\cos[\omega_c t + \phi_n(t)] \\ &= n_c(t)\cos\omega_c t - n_s(t)\sin\omega_s t \end{aligned} \tag{3A.43}$$

The input noise is assumed to be white with a power spectral density (PSD) of $n_0/2$ watts/Hz, and also

$$\overline{n_c^2}(t) = \overline{n_s^2}(t) = \overline{n^2(t)} \tag{3A.44}$$

The noise power in bandwidth B is (Figure 3A.12)

$$N_0 = \int_0^B n_0\,df = n_0 B \tag{3A.45}$$

DSB: The output signal at the post-detection filter is

$$A_r g(t) + n_c(t) \tag{3A.46}$$

The output signal component is

$$E[A_r g(t)] = A_r^2 S_g \tag{3A.47}$$

where E[ ] denotes mathematical expectation.

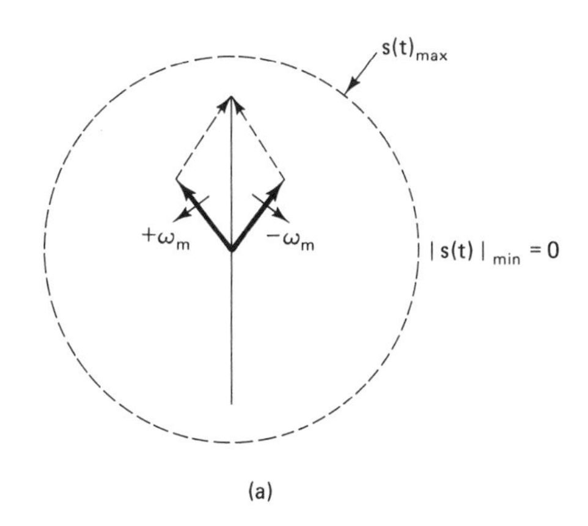

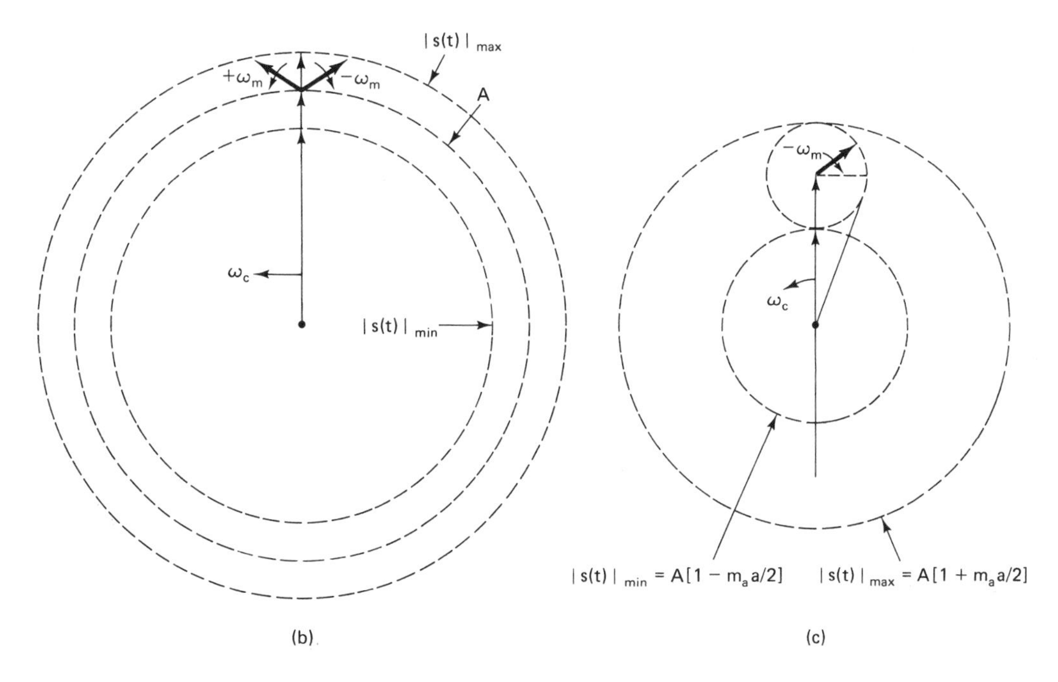

Figure 3A.11 Phasor diagrams for ideal CW modulation by a single tone: (a) DSB; (b) AM; (c) SSB(LSB).

The output noise component is

$$E[n_c^2(t)] = 2n_0 f_g \tag{3A.48}$$

where f_g is the signal base bandwidth. The signal-to-noise ratio at the output is

$$(S/N)_o = \frac{A_r^2 S_g}{2n_0 f_g} = \frac{S_r}{n_0 f_g} \tag{3A.49}$$

where $S_r = \frac{1}{2}A_r S_g$ is the average power at the receiver input. The signal-to-noise

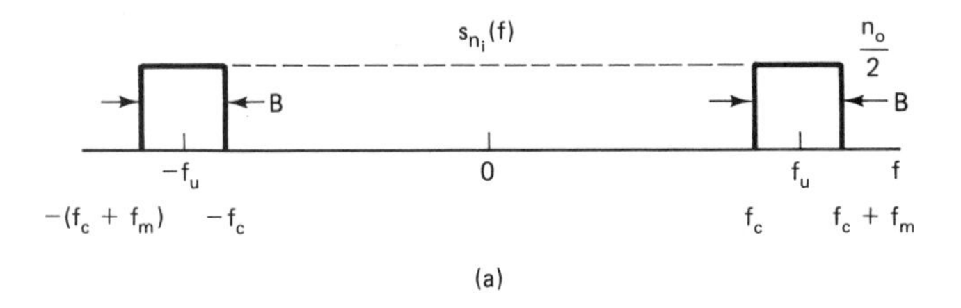

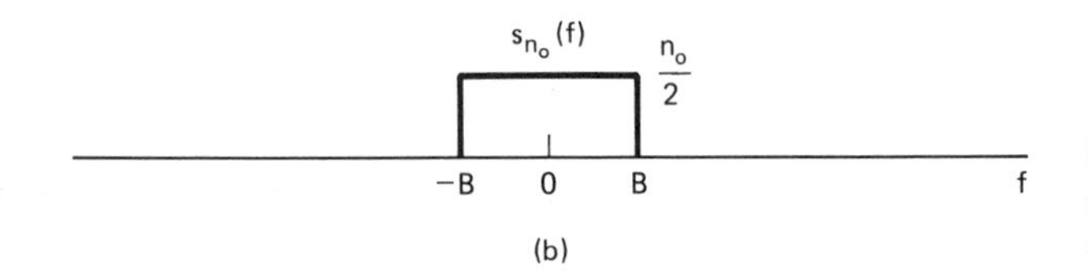

Figure 3A.12 Noise power spectral densities for the SSB system: (a) predetection band; (b) postdetection band.

ratio at the input of the detector is

$$(S/N)_i = \frac{S_r}{2n_0 f_g} \tag{3A.50}$$

and

$$\frac{(S/N)_o}{(S/N)_i} = 2 \tag{3A.51}$$

which is the detection gain.

SSB and VSB:

$$(S/N)_o = \frac{S_r}{n_0 f_g} \tag{3A.52}$$

which is the same as DSB; however, the transmission bandwidth for the SSB system is $B = f_g$, which is half the transmission bandwidth of DSB modulation.

AM (Figure 3A.13): The input to the detector is

$$\begin{aligned} y(t) &= \{A_r[1 + mg(t)] + n_c(t)\} \cos \omega_c t - n_s(t) \sin \omega_c t \\ &= R(t) \cos [\omega_c t + \theta(t)] \end{aligned} \tag{3A.53}$$

where

$$R(t) = \sqrt{\{A_r[1 + mg(t)] + n_c(t)\}^2 + [n_s(t)]^2}$$

and

$$\tan \theta(t) = \frac{n_s(t)}{A_r[1 + mg(t)] + n_c(t)}$$

Case 1: The signal power at the receiver input is considerably higher than the noise power.

The demodulator output, after the post-detection filter, is

$$z_0(t) \approx A_r mg(t) + n_c(t) \tag{3A.54}$$

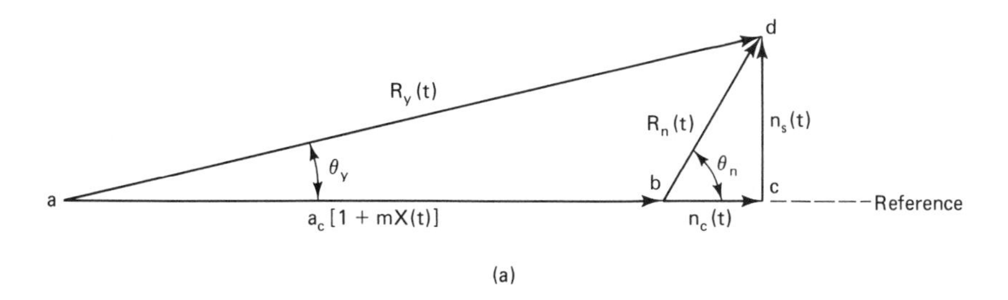

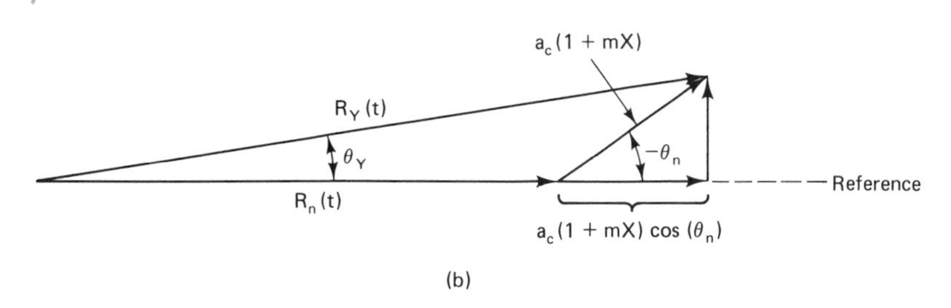

Figure 3A.13 Phasor diagram for an AM signal plus noise: (a) $a_c^2 >> E[n^2(t)]$; (b) $a_c^2 << E[n^2(t)]$.

The output signal-to-noise ratio is

$$(S/N)_o = \frac{A_r^2 m^2 E[g^2(t)]}{E[n_c^2(t)]} \tag{3A.55}$$

where the receiver power is

$$S_r = (A_r^2/2)(1 + m^2 S_g) \tag{3A.56}$$

The power efficiency of the AM signal is

$$\frac{m^2 S_g}{1 + m^2 S_g} \leq \frac{1}{2} \tag{3A.57}$$

For a given value of average transmitted power, $(S/N)_o$ for AM is at least 3 dB poorer than $(S/N)_o$ for SSB and DSB using coherent demodulation.

For a given peak power the difference is 6 dB compared to DSB, since the peak powers are proportional to A_c^2 and $4A_c^2$. In terms of equal peak powers, SSB is 2 to 3 dB better than DSB and 8 to 10 dB better than AM.

Case 2: The signal power is much lower than the noise power (Figure 3A.13b).

The message term is captured by the noise and it is meaningless to talk about output signal-to-noise ratio. The loss of message signal at low predetection signal-to-noise ratios is called the *threshold effect*. The threshold effect does not occur when coherent demodulation is used, and if AM is used for transmitting digital information, it is necessary to use synchronous demodulation when $(S/N)_i$ is low in order to avoid threshold effects.

four

Frequency Modulation

4.1 INTRODUCTION

Presently, this modulation is used in all videodisc systems. *Frequency modulation* (FM) is a form of angle modulation, also referred to as an *exponential modulation*. Its principal point of superiority over amplitude modulation lies in its ability to exchange bandwidth occupancy in the transmission medium for improved noise performance. Under certain conditions, the signal-to-noise ratio is improved by 6 dB for each 2:1 increase in bandwidth occupancy.

As might be expected, against these advantages there are certain disadvantages. A wide frequency band is required in order to obtain the advantage of reduced noise and decreased power. An FM system is more complicated than DSB. Although FM systems are tolerant to certain types of amplitude nonlinearity, it is necessary to pay special attention to the phase nonlinearity, such as AM-to-PM conversion which is commonly encountered in up and down conversion.

In the expression $s(t) = A_c \cos \theta(t)$, FM is by definition that type of modulation in which the instantaneous frequency is equal to the constant frequency of the carrier plus a time-varying component that is proportional to the magnitude of the modulating signal. A frequency-modulated signal has the form [1–4]

$$s(t) = A_c \cos [\omega_c t + \phi(t)] = \text{Re}\,\{A_c e^{j[\omega_c t + \phi(t)]}\} \tag{4.1}$$

Where Re $\{\cdot\}$ denotes the "real part of" a complex number. The instantaneous phase of $s(t)$ is defined as

$$\theta_i(t) = \theta(t) = \omega_c t + \phi(t) \tag{4.2}$$

and the instantaneous angular frequency is

$$\omega_i(t) = d\theta_i/dt = \omega_c + d/dt(\phi(t)) \tag{4.3}$$

It would be erroneous to write instantaneous angular frequency in FM signal analogous to modulated amplitude ("instantaneous amplitude") or modulated phase ("instantaneous phase"):

$$\omega = \omega_c[1 + m_f g(t)] \tag{4.4}$$

where $g(t)$ is the message signal. If $s(t)$ is the frequency-modulated current

$$i(t) = I_c \cos(\omega_c[1 + m_f g(t)]t + \phi_0) \tag{4.5}$$

and it is supplied through an inductance L, the voltage across the inductance is

$$\begin{aligned} v(t) &= L(d/dt)i(t) \\ &= -\mathrm{L}\mathrm{I}_c\omega_c[1 + m_f g(t) + m_f t g(t)] \sin\{\omega_c[1 + m_f g(t)]t + \phi_0\} \end{aligned} \tag{4.6}$$

and we face a physical absurdity since $v \to \infty$ as $t \to \infty$.

An alternative definition of instantaneous frequency of $s(t)$ would be that it is the ratio of the number of zeros of $s(t)$ in interval of time τ; or as $\frac{1}{2}$ of the mean density of zero crossing. If $s(t)$ has consecutive zeros at $t = t_1$ and $t = t_2$, then

$$\theta(t_2) - \theta(t_1) = \pi \tag{4.7}$$

If $\phi(t)$ changes slowly compared to $s(t)$, then

$$\theta(t_2) - \theta(t_1) = \theta'(t)(t_2 - t_1) \tag{4.8}$$

and

$$t_2 - t_1 = \pi/(\omega_c + \phi'(t)) \tag{4.9}$$

If $\phi(t)$ is nearly constant during the time interval τ, then $\frac{1}{2}$ of the mean density of zero crossings is

$$f_i = \frac{1}{2\pi}[\omega_c + \phi'(t)] \tag{4.10}$$

which is essentially the same result as in equation (4.3). It should be noted that the term *instantaneous frequency* is used in the literature for both f_i and ω_i. But there should be no confusion because $\omega_i = 2\pi f_i$.

The instantaneous frequency deviation, $d\phi/dt$, is the difference between the instantaneous frequency of the modulated wave and the frequency of the unmodulated carrier, and is proportional to the message signal:

$$d\phi/dt = m_\omega g(t) \qquad 0 < m_\omega < \infty \tag{4.11}$$

where m_ω is the frequency deviation index (constant) expressed in (rad/s)/V. In terms of $g(t)$, the angle function for FM is

$$\theta(t) = \int_{t_0}^{t} \omega(t)\, dt = \omega_c t + \int_{t_0}^{t} m_\omega g(t)\, dt - \omega_c t_0 \tag{4.12}$$

and the frequency-modulated signal is

$$s(t) = A_c \cos\left[\omega_c t + m_\omega \int^{t} g(t)\, dt + \phi_0\right] \tag{4.13}$$

The constant $\phi_0 = \omega_c t_0$ represents the initial phase angle of the unmodulated carrier and usually will be considered to be zero. The lower limit of the definite interval in equation (4.13) is omitted because it contributes nothing to the argument.

Figure 4.1 shows typical modulated waveforms for two different message waveforms for $m_\omega = 0.1\omega_c/V$. For the sake of comparison, AM and PM waveforms are included. It can be seen that for tone modulation it is impossible to distinguish visually between FM and PM waveforms.

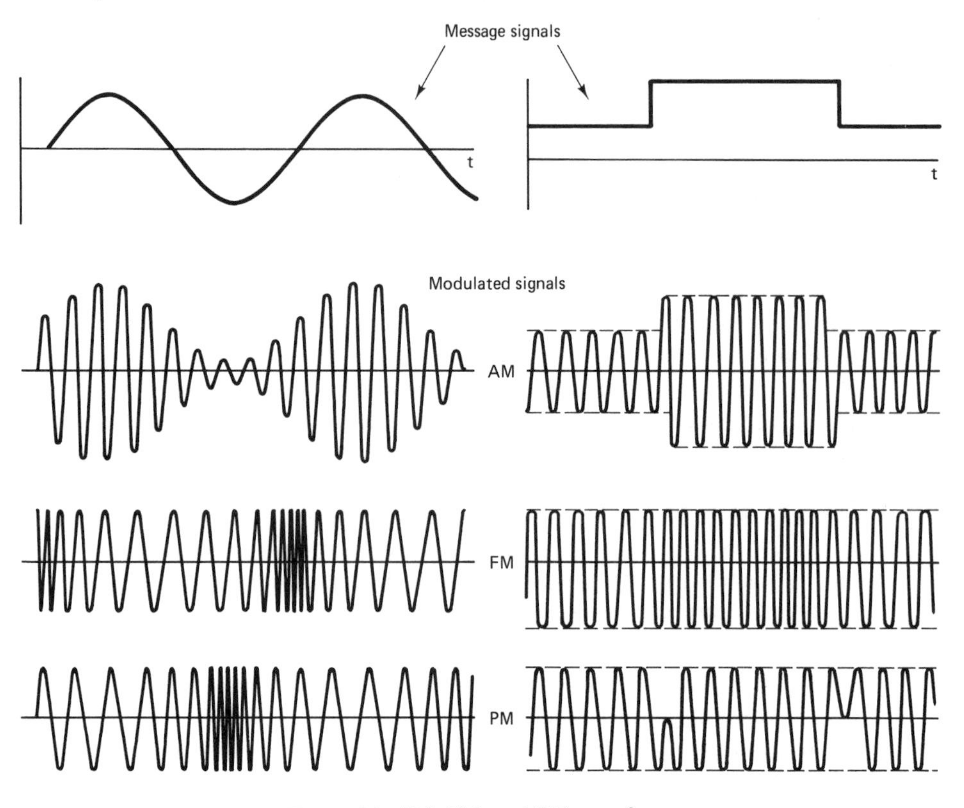

Figure 4.1 AM, FM, and PM waveforms.

The instantaneous frequency and phase deviation are illustrated in Figure 4.2 for $m_\omega = 0.1\omega_c/V$, for a pulse type of modulating (information) signal.

4.2 SPECTRA OF FM SIGNALS

In terms of the spectrum of $g(t)$, the spectrum $S(f)$ of the FM signal is not readily obtained, due to the fact that $g(t)$ is contained in a trigonometric function, linearity and superposition are not valid, and it is difficult to apply the Fourier transform. Simple expressions for $S(f)$ exist only for a few modulating signals: periodic function and certain pulse forms. In order to gain insight into the frequency behavior of FM signals, the spectra for sinusoidal message signals will be examined.

For a *sinusoidal,* sometimes called a *tone, signal,*

$$g(t) = A_m \cos \omega_m t \tag{4.14}$$

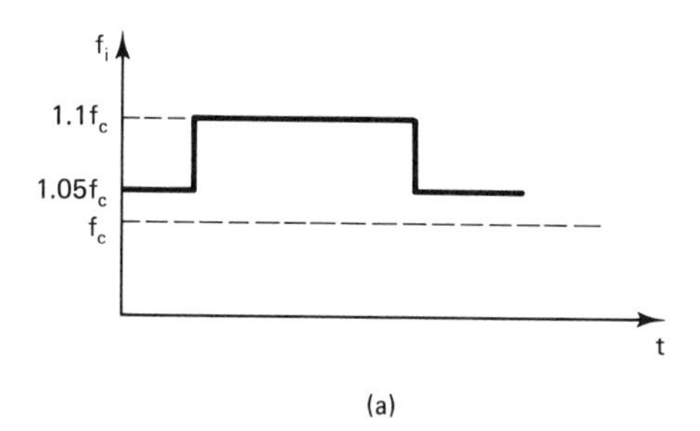

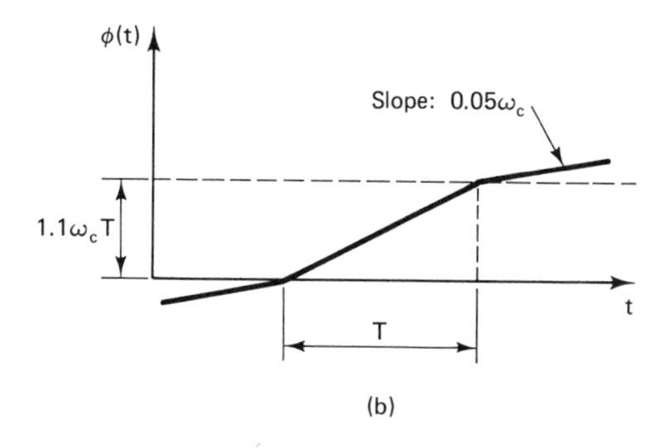

Figure 4.2 FM: (a) instantaneous frequency; (b) instantaneous phase deviation.

the instantaneous angular frequency of the modulated signal is

$$\omega_i(t) = \omega_c + m_\omega A_\omega \cos \omega_m t \tag{4.15}$$

or

$$\omega_i(t) = \omega_c + \Delta\Omega \cos \omega_m t \tag{4.16}$$

where $\Delta\Omega$ is a maximum angular frequency deviation. The maximum frequency deviation is constant, independent of ω_m.

The instantaneous phase of the modulated signal is

$$\theta(t) = \omega_c t + (\Delta\Omega/\omega_m) \sin \omega_m t \tag{4.17}$$

where $\phi_0 = 0$ is assumed.

The frequency-modulated carrier is then given by

$$s(t) = A_c \cos (\omega_c t + \beta \sin \omega_m t) \tag{4.18}$$

where the modulation index β is

$$\beta = \frac{m_\omega A_m}{\omega_m} = \frac{\Delta\Omega}{\omega_m} \tag{4.19}$$

The parameter β represents the maximum phase deviation produced by the modulating tone and is defined only for tone modulation.

The frequency-modulated carrier can be expressed as

$$s(t) = A_c \operatorname{Re} \{\exp [j(\omega_c t + \beta \sin \omega_n t)]\} \tag{4.20}$$

Using a Fourier series, the next identity can be obtained:

$$e^{j\beta \sin \omega_m t} = \sum_{n=-\infty}^{\infty} J_n(\beta) e^{jn\omega_m t} \tag{4.21}$$

where $J_n(\beta)$ are Bessel functions of the first kind:

$$J_n(\beta) = \left(\frac{1}{2\pi}\right) = \int_{-\pi}^{\pi} e^{j(\beta \sin x - nx)}\, dx \tag{4.22}$$

By substituting equation (4.21) in (4.20), the following expression for the FM signal with tone modulation can be obtained:

$$s(t) = A_c \sum_{n=-\infty}^{\infty} J_n(\beta) \cos\,(\omega_c + n\omega_m)t \tag{4.23}$$

The spectrum of $s(t)$ can easily be obtained from this expression. The values of $J_n(\beta)$ are well tabulated in the literature (see, e.g., Ref. 5, Chap. 9), and a short listing is given in Table 4.1. Also, a plot of Bessel functions of the first kind as a function of argument is shown in Figure 4.3. In Appendix 4B very

TABLE 4.1 BESSEL FUNCTIONS $J_n(\beta)$

β	$J_0(\beta)$	$J_1(\beta)$	$J_2(\beta)$	$J_3(\beta)$	$J_4(\beta)$	$J_5(\beta)$	$J_6(\beta)$
0.0	1.0000000	0.0000000	0.0000000	0.0000000	0.0000000	0.0000000	0.0000000
.1	0.9975016	.0499375	.0012490	.0000208	.0000003	.0000000	.0000000
.2	.9900250	.0995008	.0049834	.0001663	.0000042	.0000001	.0000000
.3	.9776262	.1483188	.0111659	.0005593	.0000210	.0000006	.0000000
.4	.9603982	.1960266	.0197347	.0013201	.0000661	.0000026	.0000001
0.5	0.9384698	0.2422685	0.0306040	0.0025637	0.0001607	0.0000081	0.0000003
.6	.9120049	.2867010	.0436651	.0043997	.0003315	.0000199	.0000010
.7	.8812009	.3289957	.0587869	.0069297	.0006101	.0000429	.0000025
.8	.8462874	.3688420	.0758178	.0102468	.0010330	.0000831	.0000056
.9	.8075238	.4059495	.0945863	.0144340	.0016406	.0001487	.0000112
1.0	0.7651977	0.4400506	0.1149035	0.0195634	0.0024766	0.0002498	0.0000209
1.1	.7196220	.4709024	.1365642	.0256945	.0035878	.0003987	.0000368
1.2	.6711327	.4982891	.1593490	.0328743	.0050227	.0006101	.0000615
1.3	.6200860	.5220232	.1830267	.0411358	.0068310	.0009008	.0000986
1.4	.5668551	.5419477	.2073559	.0504977	.0090629	.0012901	.0001523
1.5	0.5118277	0.5579365	0.2320877	0.0609640	0.0117681	0.0017994	0.0002280
1.6	.4554022	.5698959	.2569678	.0725234	.0149952	.0024524	.0003321
1.7	.3979849	.5777652	.2817389	.0851499	.0187902	.0032746	.0004721
1.8	.3399864	.5815170	.3061435	.0988020	.0231965	.0042936	.0006569
1.9	.2818186	.5811571	.3299257	.1134234	.0282535	.0055385	.0008965
2.0	0.2238908	0.5767248	0.3528340	0.1289432	0.0339957	0.0070396	0.0012024
2.1	.1666070	.5682921	.3746236	.1452767	.0404526	.0088284	.0015875
2.2	.1103623	.5559630	.3950587	.1623255	.0476471	.0109369	.0020660
2.3	.0555398	.5398725	.4139146	.1799789	.0555957	.0133973	.0026534
2.4	.0025077	.5201853	.4309800	.1981148	.0643070	.0162417	.0033669
2.5	−0.0483838	0.4970941	0.4460591	0.2166004	0.0737819	0.0195016	0.0042246
2.6	− .0968050	.4708183	.4589729	.2352938	.0840129	.0232073	.0052461
2.7	− .1424494	.4416014	.4695615	.2540453	.0949836	.0273876	.0064518
2.8	− .1850360	.4097092	.4776855	.2726986	.1066687	.0320690	.0078634
2.9	− .2243115	.3754275	.4832271	.2910926	.1190335	.0372756	.0095032

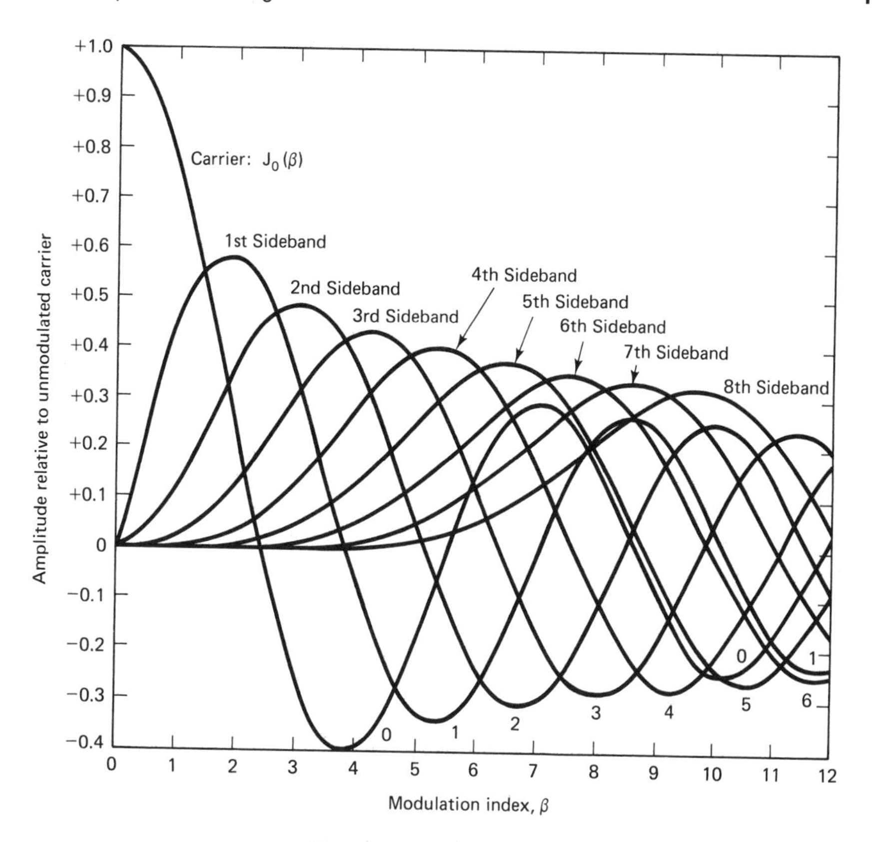

Figure 4.3 Plot of $J_n(\beta)$.

useful curves for the videodisc formats design are given: $J_1(\beta)/J_0(\beta)$ and $J_2(\beta)/J_0(\beta)$.

The next two properties of Bessel functions can also be useful:

$$\sum_{n=-\infty}^{\infty} J_n^2(\beta) = 1 \tag{4.24}$$

and

$$J_{-n}(\beta) = (-1)^n J_n(\beta) \tag{4.25}$$

Two examples of modulated signal spectrum are shown in Figure 4.4.

Several important properties of the FM signal spectrum can be pointed out using equation (4.23). When the carrier is modulated by one tone signal, the FM spectrum consists of a carrier component plus an infinite number of sideband components of frequencies $f_c \pm nf_m$, $n = 1, 2, 3, \ldots$. The amplitudes of these spectral lines are listed below:

Carrier: $J_0(\beta)A_c e^{j\omega_c t}$
Sidebands: $J_n(\beta)A_c e^{j(\omega_c \pm n\omega_m)t}$

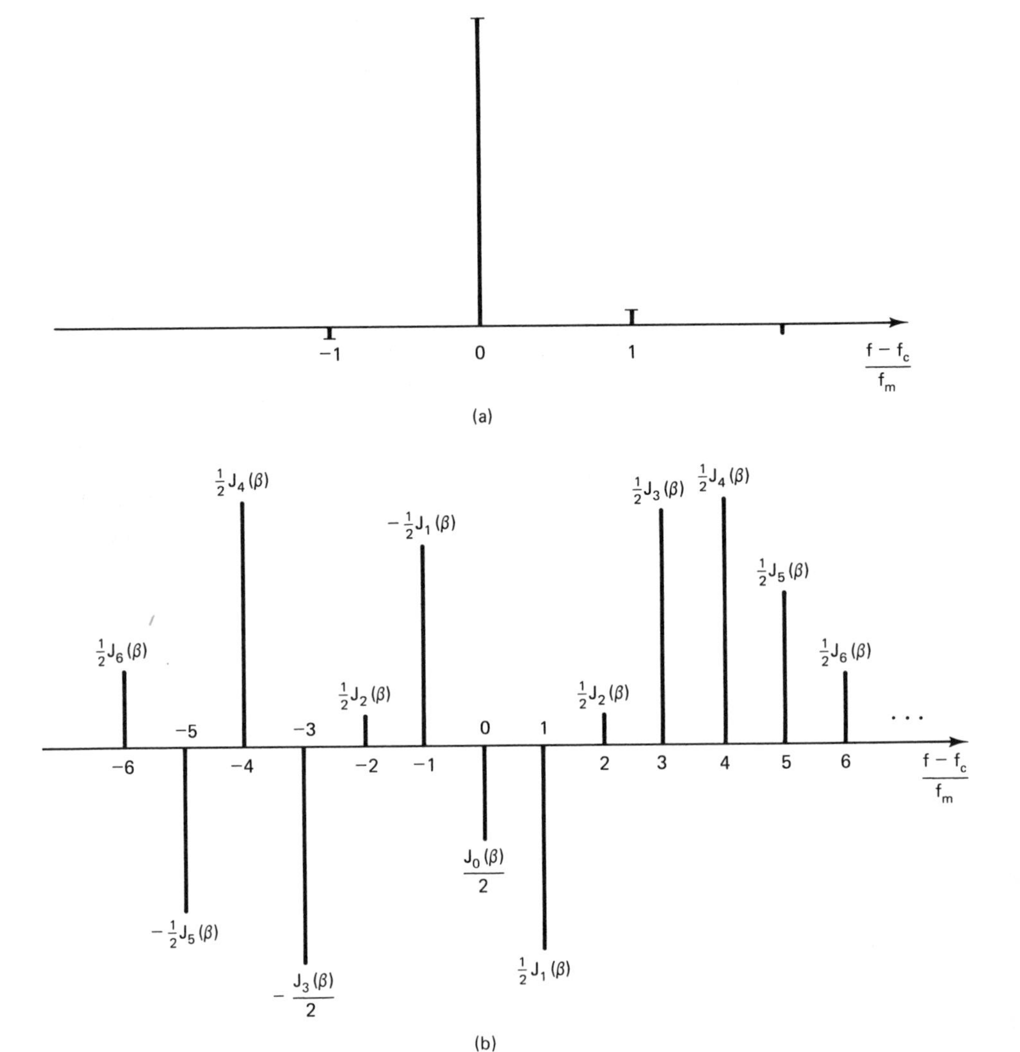

Figure 4.4 Spectrum of FM signal: $A_c = 1, f_c >> f_m$: (a) $\beta = 0.1$; (b) $\beta = 5$. Negative values are shown plotted downward.

In comparison, the spectrum of an AM signal with tone modulation has only three spectral components, of frequencies f_c and $f_c \pm f_m$. The relative amplitude of the FM carrier depends on $J_0(\beta)$ and its value depends on the modulating signal; in AM the residual carrier amplitude does not depend on the value of the modulating signal. With an FM tone-modulated signal, the odd-order lower sidebands are reversed in phase.

For $\beta << 1$ only J_0 and J_1 are significant (Table 4.1) and the FM spectrum approximate with a carrier and two sideband components. This is similar with an AM spectrum with the exception of the phase reversal of the lower sideband component. The number of significant spectral components is a function of β, and as β increases the number of significant sidebands increases while the total average power of the FM signal (the sidebands plus carrier) remains constant, because the signal amplitude is constant, equal to A_c. This means that a large value of β implies a large bandwidth since there will be many significant sideband components.

Note that only the amplitude of the carrier is determined by $J_0(\beta)$; its behavior is determined from the zero-order Bessel function $J_0(\beta)$. From Figure 4.3 it can be seen that when β is increased the value of $J_0(\beta)$ drops off rapidly; for $\beta = 2.404$ the amplitude is zero. The zero-order Bessel function is oscillatory with decreasing peak amplitude, and the spacing between zeros approaching asymptotically the constant value π. This is obvious from the approximation:

$$J_0(\beta) \simeq \cos(\beta - \pi/4)/\sqrt{\pi\beta/2} \tag{4.26}$$

When the carrier is modulated by two tones, the instantaneous angular frequency is

$$\omega_i(t) = \omega_c + \Delta\Omega \cos\omega_1 t + \Delta\Omega_2 \cos\omega_2 t \tag{4.27}$$

The FM signal is

$$\begin{aligned} s(t) &= A_c \operatorname{Re}\{e^{j[\omega_c t + \beta_1 \sin\omega_1 t + \beta_2 \sin\omega_2 t]} \\ &= A_c \operatorname{Re}\left\{\left[\sum_{n=-\infty}^{\infty} J_n(\beta_1)e^{jn\omega_1 t}\right]\left[\sum_{m=-\infty}^{\infty} J_m(\beta_2)e^{jm\omega_2 t}\right]e^{j\omega_c t}\right\} \end{aligned} \tag{4.28}$$

where the modulation index of the first and the second modulating signals is

$$\beta_i = \Delta\Omega_i/\omega_i, \qquad i = 1, 2 \tag{4.29}$$

From equation (4.28) the sideband amplitudes can be expressed

$$s(t) = A_c \sum_{n,m=-\infty}^{\infty}\sum C_{nm} \cos(\omega_c + n\omega_1 + m\omega_2)t \tag{4.30}$$

The carrier and three types of sideband terms follow from equation (4.30):

Carrier: $J_0(\beta_1)J_0(\beta_2)A_c \cos\omega_c t$
Sidebands due to ω_1: $J_n(\beta_1)J_0(\beta_2)A_c \cos(\omega_c \pm n\omega_1)t$.
Sidebands due to ω_2: $J_m(\beta_2)J_0(\beta_1)A_c \cos(\omega_c \pm m\omega_2)t$
Beat frequencies at $\omega_c \pm n\omega_1 \pm m\omega_2$: $J_n(\beta_1)J_m(\beta_2)A_c \cos(\omega_c \pm n\omega_1 \pm m\omega_2)t$

It can be seen that the amplitude of each spectral component is given by the product of two Bessel functions, which means that in two-tone modulation the amplitudes of the spectral components are generally reduced comparing with single-tone modulation at the same modulation index. It should also be pointed out that in the case of two-tone FM, the beat frequencies are obtained. This is caused by the nonlinear nature of the FM process. The conclusion is that, in general, when the carrier is frequency modulated by multitone signals not harmonically related, the spectral components

$$\omega_c \pm n\omega_1 \pm m\omega_2 \pm p\omega_3 \pm \cdots \pm q\omega_q \pm \cdots$$

exist for all combinations of n, m, p, ..., q ... tacking values {0, 1, 2, ...}. Signal energy is now distributed among the carrier and a large number of sidebands.

4.3 POWER AND BANDWIDTH OF FM SIGNALS

The previous analysis explains the wideband nature of FM (and PM), and why it is often called a wideband modulation. It has been seen that for the simplest case, modulation by a single tone of frequency f_m, there are an infinite number of harmonics in an FM signal, spaced f_m apart. Using Parseval's theorem (the average power is proportional to the sum of the individual Fourier components of the modulated signal); from equation (4.23) the average power is

$$P = \frac{A_c^2}{2} \sum_{n=-\infty}^{\infty} J_n^2(\beta) = \frac{A_c^2}{2} \tag{4.31}$$

where equation (4.24) is the substitute. The same result could be derived from equation (4.1) because FM is a constant-amplitude (envelope) modulating system and the power of an FM signal is identical to the average power of an unmodulated carrier of the same amplitude.

The average power in the bandwidth B containing carrier and k pairs of the sidebands is

$$P_k = \frac{A_c^2}{2} \sum_{n=-k}^{k} J_n^2(\beta) \tag{4.32}$$

Now the problem is to find the transmission bandwidth of FM signal, B, for a given modulation index β. The transmission bandwidth means the bandwidth containing spectral components with the main part of energy (e.g., 99%). This means that we have to find the smallest value of k for which $P_k \geq 0.99P$. The value of k can be found by evaluation of the $J_k(\beta)$ (from tables, graphs, or by computation) relative to some criterion, say $J_k(\beta) \geq J_0(\beta)/100$.

After checking for different values of β, it can be shown that a practical guide for general bandwidth requirements is Carson's rule:

$$B = 2f_m + 2\,\Delta F = 2f_m(1 + \beta) = 2\,\Delta F(1 + 1/\beta) \tag{4.33}$$

where ΔF is the maximum frequency deviation and f_m is the highest modulating frequency. For β small ($f_m > \Delta F$), the required bandwidth for an FM signal is $B \approx 2f_m$, as with an AM signal. For β large the required bandwidth for an FM signal is $B \simeq 2\Delta F$.

If the intention is to include more energy or to include smaller sidebands such as 10 dB below $J_0(0)$ [equation (4.33) corresponds to approximately −20-dB sidebands], the bandwidth is given by another approximation:

$$B = 2f_m(\beta + 2) = 2\Delta F(1 + 2/\beta) \tag{4.34}$$

Based on the value of β, FM signals are classified into two categories: narrowband FM (NBFM) signal, for $\beta << 1$, and wideband FM (WBFM) signal, for $\beta >> 1$. The corresponding bandwidths are

$$B = \begin{cases} 2f_m, & \beta << 1 \\ 2\Delta F, & \beta >> 1 \end{cases} \tag{4.35}$$

For narrowband FM frequency and phase deviations are small and the following approximations can be used: $\cos \phi(t) \cong 1$ and $\sin \phi(t) \simeq \phi(t)$.

This yields

$$\begin{aligned} s(t) &= A_c \cos [\omega_c t + \phi(t)] \\ &= A_c [\cos \omega_c t \cos \phi(t) - \sin \phi(t) \sin \omega_c t] \\ &= A_c \cos \omega_c t - A_c \phi(t) \sin \omega_c t \end{aligned} \tag{4.36}$$

For one-tone modulation $\phi(t) = \beta \sin \omega_m t$, and

$$s(t) = A_c \cos \omega_c t + \tfrac{1}{2}\beta A_c \cos (\omega_c + \omega_m)t - \tfrac{1}{2}\beta A_c \cos (\omega_c - \omega_m)t \tag{4.37}$$

This means that the NBFM signal consists of only two sidebands and thus has the same bandwidth as AM or DSB. But there is a distinct difference between NBFM and AM: the lower sideband is 180° out of phase in the NBFM case. From equation (4.37) it can be seen that sidebands are in phase quadrature with the carrier.

It is of interest to examine the behavior of instantaneous frequency and the spectra as f_m is varied. For single-tone modulation with $g(t) = \text{AM} \cos \omega_m t$ and for

$$m_\omega A_\omega = \text{const.} \tag{4.38}$$

the instantaneous frequency is

$$f_i = f_c + \Delta F \cos \omega_m t \tag{4.39}$$

where $\Delta F = m_\omega A_\omega / 2\pi$.

The FM signal is then a constant-amplitude continuous-phase signal. The instantaneous frequency is changing from $f_c - \Delta F$ to $f_c + \Delta F$. The maximal frequency deviation is proportional to the modulation signal amplitude.

On the other hand, the FM signal can be expressed as the sum of signals with constant frequencies ($f_c \pm nf_m$) and constant but dependent on β amplitudes, $J_n(\beta)$, where $\beta = \Delta F/f_m$. But while the maximal frequency deviation on ΔF (sidebands) is independent of f_m, the number of significant components (sidebands) is dependent on f_m through β. As β is large, the energy is spread on a large number of sidebands, but spacing between spectral components is small. For a very large modulation index the bandwidth is equal to $2\Delta F$, centered around f_c. Figure 4.5 shows the FM spectrum of a single-tone modulation.

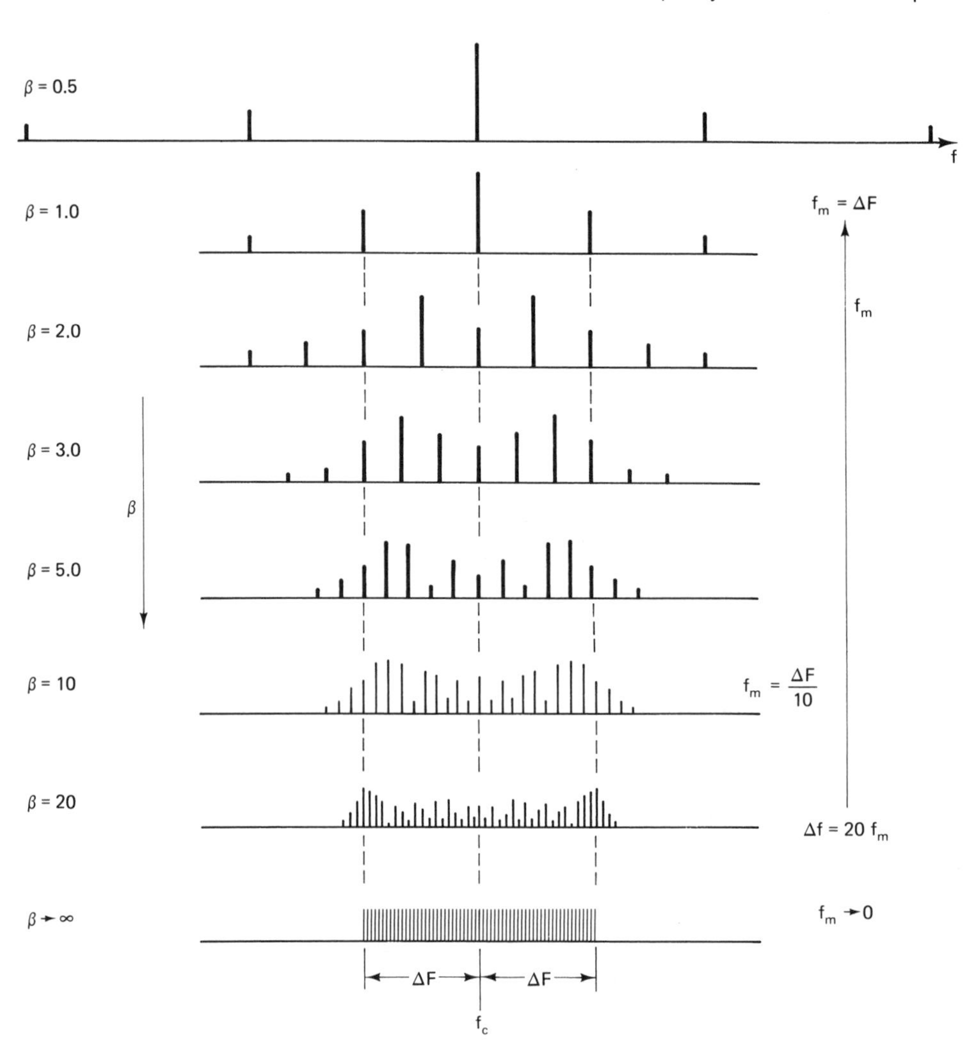

Figure 4.5 FM spectrum of single-tone modulation.

4.4 FM SIGNAL IN THE PRESENCE OF PASSBAND WHITE GAUSSIAN NOISE

The simplified block diagram of the demodulation side of an FM system is shown in Figure 4.6. The noise plus signal supplied to the input of the FM demodulator is given as

$$\begin{aligned} v(t) &= A_c \cos[\omega_c t + \phi(t)] + n(t) \\ &= A_c \cos[\omega_c t + \phi(t)] + n_c(t) \cos \omega_c t - n_s(t) \sin \omega_c t \qquad (4.40) \\ &= A_c \cos[\omega_c t + \phi(t)] + V(t) \cos[\omega_c t + \theta_n(t)] \end{aligned}$$

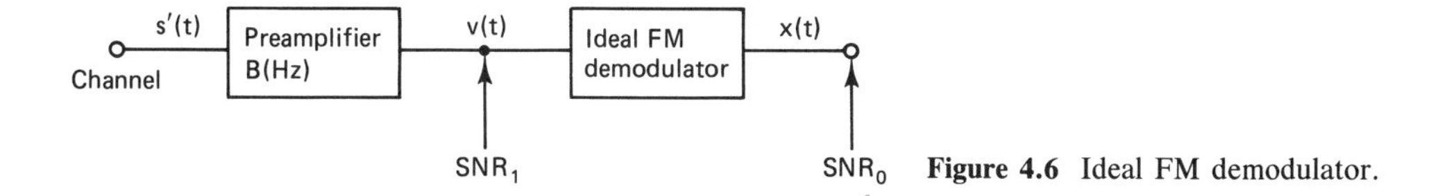

Figure 4.6 Ideal FM demodulator.

where $n(t)$ is bandlimited, zero-mean Gaussian noise with power spectral density $N_0/2$ in watts per cycle, and n is a constant; $V(t)$ and $\theta(t)$ are the envelope and phase of $n(t)$. This expression can be rewritten as (Figure 4.7)

$$v(t) = R(t) \cdot \cos [\omega_c t + \theta(t)] \tag{4.41}$$

where

$$\theta(t) = \phi(t) + \theta_v(t)$$

and

$$\tan\theta_v(t) = \frac{V(t) \sin (\theta_n - \phi)}{A_c + V(t) \cos (\theta_n - \phi)}$$

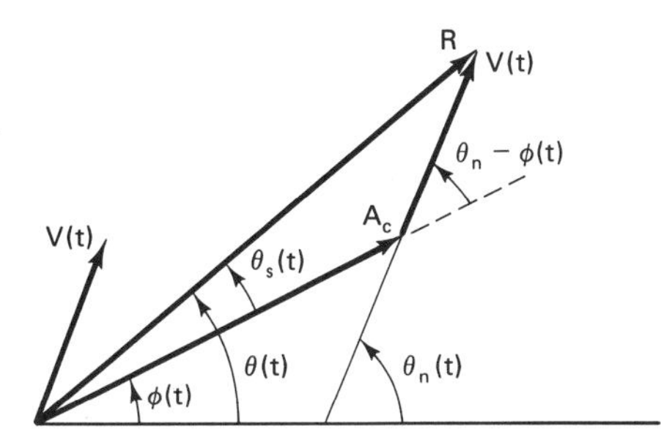

Figure 4.7 Phase diagrams of an FM signal corrupted by additive noise.

For a large signal-to-noise ratio, $A_c^2 >> E\{n(t)\}$, $A_c >> V(t)$ most of the time and the next approximation can be obtained:

$$\theta_v(t) \simeq [V(t)/A_c] \sin (\theta_n - \phi), \tag{4.42}$$

or

$$\theta(t) = \underbrace{\phi(t)}_{\text{signal term}} + \underbrace{[V(t)/A_c] \sin (\theta_n - \phi)}_{\text{noise term}} \tag{4.43}$$

The FM detector output is

$$\begin{aligned} z(t) &= k_d(d/dt)[\theta(t)] \\ &= k_d k_f g(t) + k_d(d/dt)\{V(t)/A_c \sin[\theta_n(t) - \phi(t)]\} \end{aligned} \tag{4.44}$$

For the sake of simplicity we will take $\phi(t) = 0$, because this assumption does not cause any error in the calculation of in-band noise power at the output. Then

$$z(t) = g(t) + (k_d/A_c)(d/dt)[n_s(t)] = g(t) + n_1(t) \tag{4.45}$$

for $k_p k_f = 1$.

If we assume that

$$n_s(t) = \sum_{n=1}^{\infty} A_n(f_n) \sin [(\omega_n - \omega_c)t + \theta_s] \tag{4.46}$$

or

$$n_s(t) = \sum_{m} A_n(f_m + f_c) \sin [\omega_m t + \theta_s] \tag{4.47}$$

where $f_m = f_n - f_c$, then

$$(d/dt)n_s(t) = \sum_{n} \omega_m A_n(f_m + f_c) \cos (\omega_m t + \theta_m) \tag{4.48}$$

The mean-square value is

$$N_s = \omega^2 A_n^2(f + f_c) \tag{4.49}$$

and the two-sided spectral density of $n_1(t)$ is

$$N_1 = \frac{k_d}{A_c} \omega^2 N_0, \qquad \text{for } |f| < B/2$$

where $A_n(f) = N_0$ is the one-sided noise spectral density of the intermediate frequency (IF) noise $n(t)$. In Figure 4.8 the input and output noise PSD in an FM system are shown.

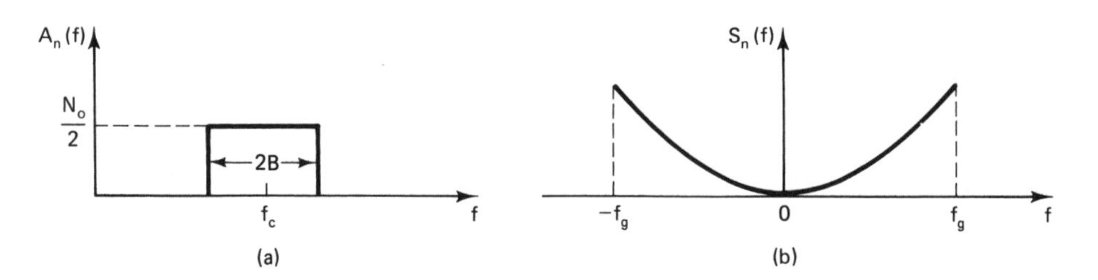

Figure 4.8 FM noise spectrum: (a) input noise; (b) detected output noise.

The signal power is

$$E[x_0^2(t)] = E[g(t)] = P_g \tag{4.50}$$

and the noise power is

$$E[n_0(t)] = \int_{-fg}^{fg} (k_d^2/A_c^2)\omega^2 n_0 df = \tfrac{2}{3}(\mathrm{k}_d^2/\mathrm{A}_c^2)\pi^2 n_0 f_g^3 \tag{4.51}$$

The signal-to-noise ratio at the output of an FM system is given by

$$\frac{S}{N} = \frac{3A_c^2 P_g}{2(2\pi k_d)^2 n_0 f_g^3} \tag{4.52}$$

or

$$S/N = 3(f_\Delta/f_g)P_g(P_r/n_0 f_g) \tag{4.53}$$

because

$$f_\Delta = k_f/2\pi = 1/(2\pi k_d);\ k_f k_d = 1 \tag{4.54}$$

In an FM system S/N can be increased by increasing the modulator sensitivity k_f without increasing the total power. But by increasing k_f, the transmission bandwidth is increased. Thus in an FM system, it is possible to trade-off bandwidth for signal-to-noise ratio.

When $V(t) >> A_c$, the instantaneous phase of FM signal corrupted by noise can be approximated by

$$\theta(t) = \omega_c t + (A_c/V(t)) \sin [\phi(t) - \theta_n(t)] \tag{4.55}$$

In this case the message has been "captured" by the noise. Because of this effect one cannot make an unlimited exchange of bandwidth for signal-to-noise ratio, so the FM system should be made to operate above this threshold, which is always the case in videodisc systems.

When the noise begins to take over the signal, at the value (threshold) dependent on β, the output SNR decreases 30 dB per decade of SNR (Figure 4.9) below the threshold of full improvement. In the region where $(S/N)_i < 0$ dB, the output SNR is proportional to $(S/N)_i^2$.

The threshold is defined as the minimum carrier-to-noise ratio yielding an FM improvement, which is not significantly deteriorated from the value predicted by the usual signal-to-noise formula, assuming small noise. To illustrate the threshold property, consider the operation on the curve $\beta = \beta_1$ (Figure 4.9), for low input carrier-to-noise ratio X_1(dB), and low output signal-to-noise ratio, Y_1(dB). The bandwidth expansion ratio, eq. 4.34 is:

$$B_{FM}/f_{max} = 2(\beta_1 + 2)$$

Let us try to increase the SNR_0 by the usual method of increasing the transmission bandwidth. Suppose we go to $\beta = \beta_2$: transmission bandwidth is $B_{FM} = 2 f_{max} (\beta_2 + 2)$. From Figure 4.9 we see that at $(S/N)_i = X_1$(dB), the $(S/N)_0$ is Y_2(dB), where $Y_2 < Y_1$. That is, increasing the transmission bandwidth has actually

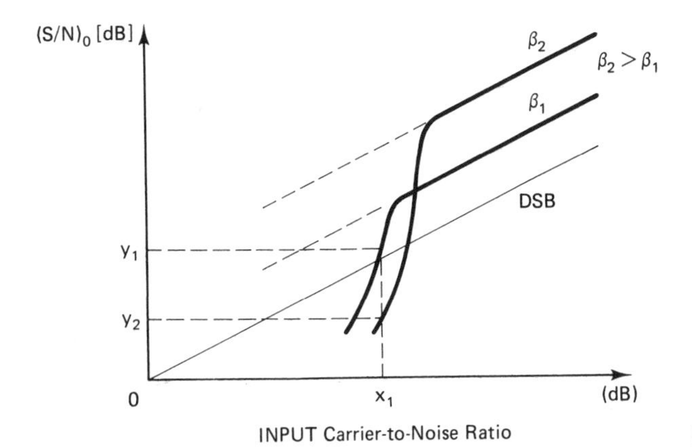

Figure 4.9 $(S/N)_o$ for FM systems.

reduced the SNR and puts us deeper into the threshold. This is because for $\beta = \beta_1$ and $(S/N)_i = X_1$(dB), we are already in the threshold. By increasing the FM bandwidth further, we are increasing the channel noise proportionately. It can be seen that the SNR_0, for $\beta = \beta_2$ and $(S/N)_i = X_1$(dB), is inferior even to the DSB or SSB.

Some reduction of the FM threshold can be achieved by feedback loop implementation with optimized filter characteristics. For example, the filter characteristics can be optimized relative to the spectral density and/or the autocorrelation of the information signal $g(t)$ and the noise. The threshold can be lowered several decibels.

4.4.1 Feedback FM Demodulators

The FM signal can be demodulated using a feedback system. This type of demodulator, referred to as a *feedback demodulator,* (FMFB demodulator), performs better than a discriminator in the presence of noise. Among demodulators in this group is the phase-locked-loop (PLL) demodulator.

The PLL basic functioning can be seen from Figure 4.10. The system consists of a phase comparator made of a multiplier and a low-pass filter and voltage-controlled oscillator (VCO).

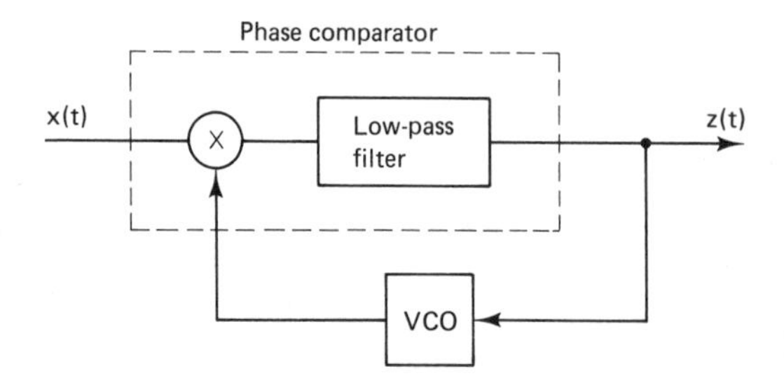

Figure 4.10 Block diagram of a PLL.

The input signal is assumed to be

$$x(t) = X \sin [\omega_c t + \theta_x(t)] \tag{4.56}$$

In the absence of control voltage, the VCO is tuned to operate at an angular frequency of ω_c, but in general the VCO output is

$$y(t) = Y \cos [\omega_c t + \theta_y(t)] \tag{4.57}$$

The multiplier output is

$$x(t)y(t) = \tfrac{1}{2}XY [\sin (\theta_x - \theta_0) + \sin (2\omega_c t + \theta_x + \theta_y)] \tag{4.58}$$

The low-pass filter output is given by

$$z(t) = \tfrac{1}{2}XY \sin [\theta_x(t) - \theta_y(t)] \tag{4.59}$$

Now it can be seen that sin and cos functions for $x(t)$ and $y(t)$ are required so that the comparator output is zero when $\theta_x = \theta_y$.

When the input signal is locked, the phase error $\theta_x - \theta_y$ will be small and

$$z(t) = \tfrac{1}{2}XY[\theta_x(t) - \theta_y(t)] \tag{4.60}$$

The instantaneous frequency deviation of the VCO is proportional to the control voltage:

$$\omega_y(t) = \omega_c + k_v z(t) \tag{4.61}$$

where k_v is the sensitivity of the VCO in (rad/s)/v.

If now $x(t)$ is an FM signal and if, say, θ_x begins to increase, the phase comparator output will be positive, which will force the VCO to increase the frequency. Since the VCO frequency deviation is proportional to $z(t)$, $z(t)$ then represents the frequency deviation of the input FM signal, which means that $z(t)$ is the demodulated output.

4.5 PREEMPHASIS AND DEEMPHASIS IN FM

The noise power spectrum of the output of the FM discriminator is emphasized at the higher frequencies: for a large carrier-to-noise ratio it is proportional to f^2. If most of the energy of the modulating signal is concentrated in the lower-frequency range, then a poor $(S/N)_o$ ratio at the higher frequency of modulating signal will be obtained. This different statistical property of signal and noise can be utilized in order to improve the $(S/N)_o$ ratio of the high-frequency end of the baseband. For this purpose specially designed filters at the transmitting and receiving ends of the channel should be used (Figure 4.11). The transfer function for these filters, known as preemphasis/deemphasis, are chosen such that the output signal-to-noise ratio $(S/N)_o$ is maximized.

For the sake of elimination of linear distortion, $H_p(f)$ and $H_c(f)$ should be chosen to satisfy

$$H_p(f)H_c(f)H_d(f) = ke^{-j\omega f_g} \qquad \text{for } |f| < f_g \tag{4.62}$$

where $H_c(f)$ is the channel transfer function.

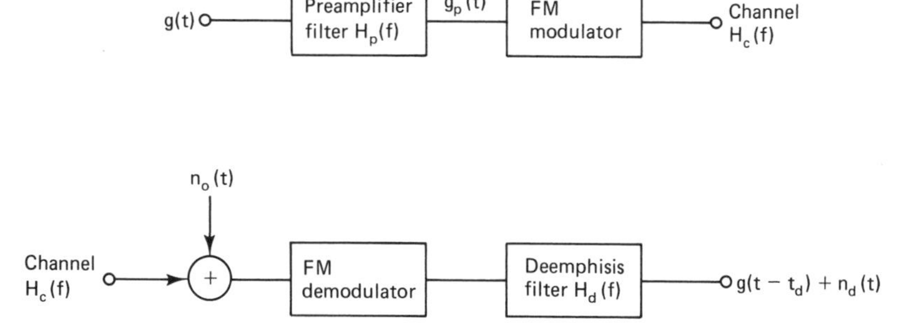

Figure 4.11 Preemphasis/deemphasis filtering.

Obviously, H_p should be proportional with the noise spectral density $S_{n0}(f)$ and inversely proportional to the signal spectral density, $S(f)$:

$$|H_p(f)|^2 = k_1 S_{n0}(f)/S(f)^{1/2} \tag{4.63}$$

while

$$|H_d(f)|^2 = (1/k_1)\,[S(f)/S_{n0}(f)]^{1/2} \tag{4.64}$$

The constant k_1 is defined by

$$\int_{-fg}^{fg} S(f)\,df = \int_{-fg}^{fg} S(f)|H_p(f)|^2\,df \tag{4.65}$$

This constraint, that is, equal power requirement, ensures that the bandwidth of the FM signal remains the same.

A simple RC circuit as shown in Figure 4.12 can be used to emphasize the high frequencies in some FM systems. The corresponding transfer function is

$$H_p(j\omega) \simeq \frac{R_1}{R}\frac{1 + j\omega\,\tau}{1 + j\omega\,\tau_1} \tag{4.66}$$

where the time constant $\tau = RC$, $R >> R_1$, $\tau_1 \cong R_1C$.

The break frequencies are $\omega_1 = 1/RC$ and $\omega_2 = 1/R_1C$ and signals in the range between f_1 and f_2 are emphasized.

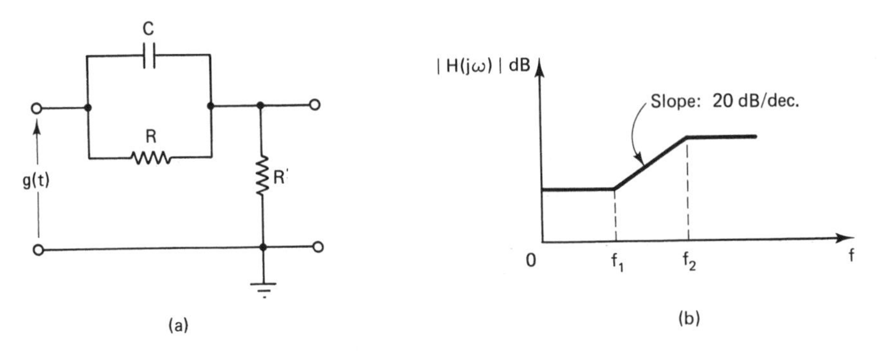

Figure 4.12 Preemphasis: (a) network; (b) asymptotic response.

The corresponding deemphasis network is shown in Figure 4.13, and its transfer function is the inverse of those of the preemphasis circuit:

$$H_d(j\omega) = 1/(1 + j\omega\tau) = 1/(1 + j\omega/\omega_1) \tag{4.67}$$

where $\omega_1 = 1/\tau$ and $\tau = RC$.

For white noise the spectral distribution of the noise power (one-sided)

$$S_{n0}(f) = N_o \tag{4.68}$$

and the output power without preemphasis is

$$(N)_o = (k_d/A_c)^2 N_0 \int_{-fg}^{fg} f^2\,df = (k_d/A_c)(\tfrac{2}{3})N_0 f_g^3 \tag{4.69}$$

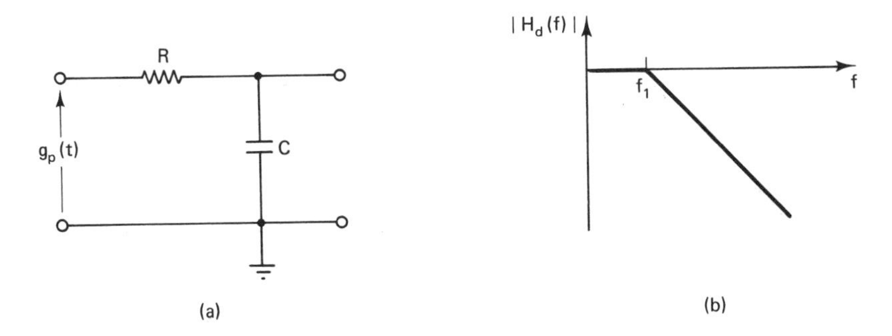

Figure 4.13 Deemphasis: (a) network; (b) asymptotic response.

With preemphasis, the total noise power at the output of an ideal low-pass filter of bandwidth f_g is

$$\begin{aligned}(N)_{o,d} &= 2(k_d/A_c)^2 N_0 \int_{fg}^{fg} f^2/|H_p|^2\, df \\ &= 2(k_d/A_c)^2 N_0 (f_g - \frac{1}{2\pi\tau}\arctan \omega_g\tau)\end{aligned} \tag{4.70}$$

The improvement factor ρ is given by

$$\rho = \frac{(N)_o}{(N)_{o,d}} = \frac{x^3}{3(2\pi\tau)^2\,(x - \arctan x)}; \qquad x = \omega_g\tau \tag{4.71}$$

If the message signal has the power spectral density flat or quasi-flat (as in the case of a large number of multiplexed signals), then, to maintain the same $(S/N)_d$ for the entire bandwidth;

$$|H_d|^2 = 1/K_1\omega^2 \tag{4.72}$$

and to avoid signal distortion,

$$|H_p(f)|^2 = K_1\omega^2 \tag{4.73}$$

where k_1 is chosen to satisfy the equal power constant. For an arbitrary phase response, we can take

$$H_p(j\omega) = jK_1\omega \qquad \text{(differentiator)} \tag{4.74}$$

and

$$H_d(j\omega) = 1/jK_1\omega \qquad \text{(integrator)} \tag{4.75}$$

This makes the entire system become a PM system.

4.6 PULSE FREQUENCY MODULATION

The FM can be utilized not only by the continuous-wave (CW) technique but also by the pulse-modulation technique. The exponential form of the Fourier series for the periodic pulse train (Figure 4.14) is given by

$$f(t) = A_c(\tau/T_c) \sum_{n=-\infty}^{\infty} (\sin x/x)e^{jn\omega_c t} \tag{4.76}$$

where $x = n\omega_c\tau/2$, $\omega_c = 2\pi/T_s$ is the fundamental angular frequency or pulse repetition frequency, T_c the pulse period, T the pulse width (duration), and A_c the constant pulse amplitude.

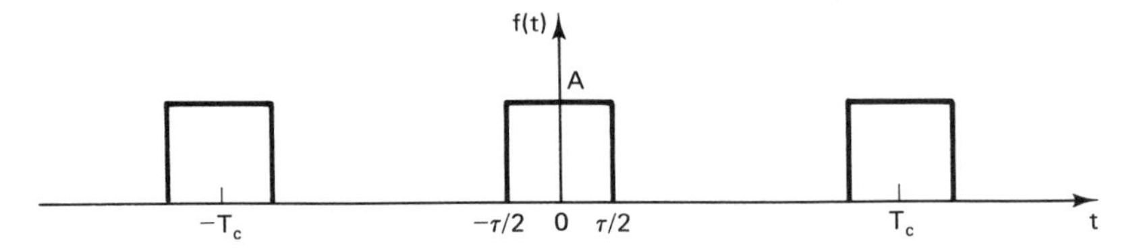

Figure 4.14 Infinite pulse train.

Equation (4.76) can be rewritten as [2]

$$f(t) = \frac{A_c}{2\pi j} \sum_{n=-\infty}^{\infty} \frac{1}{n} (e^{jn\omega_c\tau/2} - e^{-jn\omega_c\tau/2})e^{jn\omega_c t} \tag{4.77}$$

The individual effects of the leading and the trailing edges can now be noticed. The pulse FM can be carried out in two basically different ways: pulse duration can be dependent of the modulating signal or it can be kept constant and equal τ. Consider a single-tone case only; that is, consider a case when the normalized modulation signal is $\sin(\omega_m t + \phi)$. The first type of modulation can be taken into account by substituting $\omega_c\tau/2$ in equation (4.77) by:

$$\omega_c\tau/2 + \beta \sin(\omega_m t + \phi) \quad \text{and} \quad \omega_c\tau/2 - \beta \sin(\omega_m t + \phi) \tag{4.78}$$

for the leading and trailing edges respectively. The modulation index is $\beta = \Delta\omega/\omega_m$. The expression for a modulated pulse train in the first type of modulation is then

$$f(t) = \frac{A_c}{2\pi j} \sum_{n=-\infty}^{\infty} \frac{1}{n} \{e^{jn[\omega_c\tau/2 + \beta\sin(\omega_m t + \phi)]} - e^{-jn[\omega_c\tau/2 - \beta\sin(\omega_m t + \theta)]}\} e^{jn\omega_c t} \tag{4.79}$$

Using the same identity as for CW FM,

$$e^{ja\sin x} = \sum_{k=-\infty}^{\infty} J_k(a)e^{-jkx} \tag{4.80}$$

where $J_k(a)$ is the Bessel function of the first kind and of order k, the next expression can be obtained:

$$f(t) = \frac{A_c\omega_c\tau}{2\pi} + \frac{A_c\omega_c\tau}{\pi} \sum_{n=1}^{\infty} \text{sinc}(n\omega_c\tau/2) \Bigg(J_0(n\beta) \cos n\omega_c t$$
$$+ \sum_{k=1}^{\infty} J_k(n\beta) \{\cos[(n\omega_c + k\omega_m)t + k\phi] + (-1)^k \cos[(n\omega_c - k\omega_m)t - n\phi]\}\Bigg) \tag{4.81}$$

In the second type of pulse FM the displacement of both edges from their unmodulated position at any instant of time t depends on the value of modulated voltage at time t. Now the modulation can be taken into account by the substitution of term $\omega_c\tau/2$, in the expression for unmodulated pulse train, by the expressions

$$\omega_c\tau/2 + \beta \sin(\omega_m t + \phi) \quad \text{and} \quad \omega_c\tau/2 - \beta \sin[\omega_m(t - \tau) + \phi] \tag{4.82}$$

for the leading and trailing edges, respectively.

The expression for the modulated pulse train for the second type of modulation is then

$$f(t) = \frac{A_c}{2\pi j}\sum_{n=1}^{\infty}\frac{1}{n}\{e^{jn[\omega_c\tau/2+\beta\sin(\omega_m t+\phi)} - e^{-jn[\omega_c\tau/2-\beta\sin(\omega_m t-\tau)+\phi]}\}\, e^{jn\omega_c t} \tag{4.83}$$

After simplification similar to that in the first case, this becomes

$$\begin{aligned} f(t) &= \frac{A\omega_c t}{2a} + A_c(\Delta\omega/2\pi)\,\tau\,\frac{\sin c\,(\omega_m\,\tau/2)}{\tau/2}\cos\left(\omega_m t + \phi - \frac{\omega_m\tau}{2}\right) \\ &+ \frac{A_c\omega_c\tau}{\pi}\sum_{n=1}^{\infty}(J_o(n\beta)\sin c\,(n\omega_c\tau/2)\cos n\omega_c t \\ &+ \sum_{k=1}^{\infty} J_k(n\beta)\left\{\frac{\sin(n\omega_c + k\omega_m)\,\tau/2}{n\omega_c\,\tau/2} + \cos\left[(n\omega_c + k\omega_m)t + n\phi - n\omega_m\frac{\tau}{2}\right]\right. \\ &+ \left.(-1)^K\frac{\sin(n\omega_c - n\omega_m)\,\tau/2}{n\omega_c\,\tau/2}\cos[(n\omega_c - k\omega_m)t - n\phi + k\omega_m\frac{\tau}{2}]\right\} \end{aligned} \tag{4.84}$$

These two final expressions can be compared with the expression for the spectrum of a CW FM:

$$\begin{aligned} f(t) = A_cJ_0(\beta)\cos\omega_c t + A_c\sum_{n=1}^{\infty}J_n(\beta)\{&\cos[(\omega_c + n\omega_m)t + n\phi] \\ &+ (-1)^n\cos[(\omega_c - n\omega_m)t - n\phi]\} \end{aligned} \tag{4.85}$$

The following conclusions can be carried out.

In the first type of pulse FM, the low-pass filter cannot be used for the demodulation because the dc component of the pulse spectrum does not have a sideband of the modulation frequency. But in the second type of pulse FM, the modulating signal can be recovered by a low-pass filter, without harmonic distortion. The distortion can come from the lower sidebands of harmonics of the pulse repetition frequency, which fall into the demodulating filter passband.

The nth harmonic at the pulse repetition frequency is frequency modulated and the modulation index is $n\beta$. Based on this fact, the pulse FM signal of the first type can be demodulated using a bandpass filter. The filter bandwidth is centered around one of the harmonics ($n\omega_c$) and it should be wide enough to extract sidebands. In the second type of pulse FM, this method of demodulation will yield to some distortion, because the upper and lower sidebands of the nth pulse-repetition-frequency harmonic are not equal in amplitude.

The amplitude of the carrier and sidebands decreases by the factor Sinc x = $\sin x/x$, $x = n\omega_c\tau/2$, as the order of the pulse repetition frequency increases.

4.6.1 Duty Cycle Analysis

The previous conclusions can be made clearer when the duty cycles for both types of pulse FM are compared. For the first type of pulse modulation, the time of occurrence of the leading and the trailing edges is given by

$$\omega_c(t + \tau/2) + \beta \sin(\omega_m t + \phi) = 2n\pi \quad (4.86)$$

$$\omega_c(t - \tau/2) + \beta \sin(\omega_m t + \phi) = 2n\pi \quad (4.87)$$

For simplicity, it can be used: $\phi = 0$.

Substituting $t = t_1$ in equation (4.86) and $t = t_2$ in equation (4.87), the duration of nth pulse is

$$t_2 - t_1 = \tau - (\beta/\omega_c)[\sin \omega_m t_2 - \sin \omega_m t_1] \quad (4.88)$$

After simplification this becomes

$$t_2 - t_1 \approx \tau/(1 + \Delta\omega/\omega_c \cos \omega_m \bar{t}) \quad (4.89)$$

where $\bar{t} = \frac{1}{2}(t_1 + t_2)$. Differentiating for equations (4.86) and (4.87) and averaging, the instantaneous angular frequency ω_i is

$$\omega_i = \omega_c(1 + (\Delta\omega/\omega_c) \cos \omega_m \bar{t}) = 2\pi/T_i \quad (4.90)$$

The instantaneous duty cycle is

$$\alpha_i = \frac{t_2 - t_1}{T_i} = \frac{\tau}{T_c} = \alpha_i = \text{const.} \quad (4.91)$$

For the second type of pulse FM, the time of occurrence of the leading and the trailing edges is given by

$$\omega_c(t + \tau/2) + \beta \sin(\omega_m t + \phi) = 2n\pi \quad (4.92)$$

$$\omega_c(t - \tau/2) + \beta \sin[\omega_m(t - \tau) + \phi] = 2n\pi \quad (4.93)$$

Following the same procedure as for the first type, the second type of pulse FM is

$$t_2 - t_1 = \tau \quad (4.94)$$

and

$$\omega_i = \frac{2\pi}{\tau_i} = \omega_c[1 + (\Delta\omega/\omega_c)\cos \omega_m(\bar{t} - \tau/2)] \quad (4.95)$$

The duty cycle is

$$\alpha_i = \alpha_c[1 + (\Delta\omega/\omega_c)\cos \omega_m(\bar{t} - \tau)] \quad (4.96)$$

That is, in the second type of pulse FM, the duty cycle is modulated.

4.7 ZERO CROSSINGS AND DIFFERENTIAL DETECTORS

Another two practical and quite often used methods of FM detection are the zero-crossing detector and the differential detector. The idea of zero-crossing detection is that instantaneous frequency is the mean density of zero crossings.

Figure 4.15 shows a block diagram of a zero-crossing detector. The input signal is hard-limited and a rectangular train is obtained. After the signal is differentiated and the alternate train of positive and negative pulses is generated, the rectifier is used to generate a pulse train of the same polarity. If the pulse width is too short, a one-shot is used to generate constant-width pulses. Whenever the input frequency is constant, the average value at the pulse train is constant, but when the frequency is changing, the mean value of the pulse train changes. This is true only because the duty cycle is not constant. The pulse width is constant, and therefore the pulse train average value is proportional to the frequency changes of the input signal. The modulated signal can be extracted by passing the pulse train through a low-pass filter.

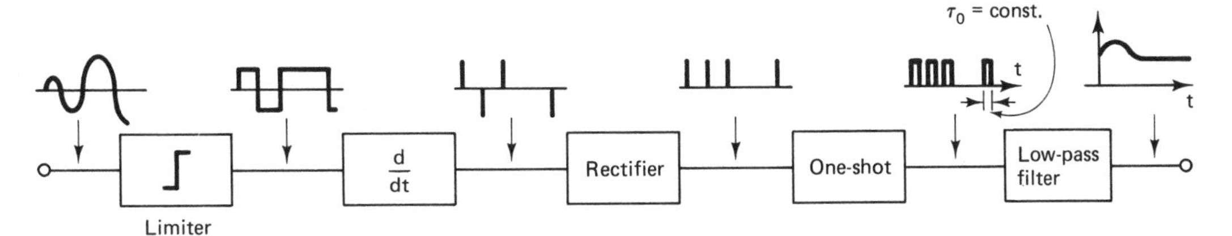

Figure 4.15 Zero-crossing detection.

In Figure 4.16 a binary example is shown: a binary "1" ("mark") is transmitted as a half period of frequency f_0 and a binary "0" ("space") as a 1-period of frequency f_0. The presence of a mark or a space is detected by counting the number of zero crossings in the modulated signal during $T_0 = 1/f_0$-second intervals. In the absence of channel impairments, there are only two possible measurements of zero crossings per T_0 seconds: one zero crossing for a space and two for a mark. This binary FM signal (code) is sometimes called a *Manchester or FM code* or *double-frequency code*.

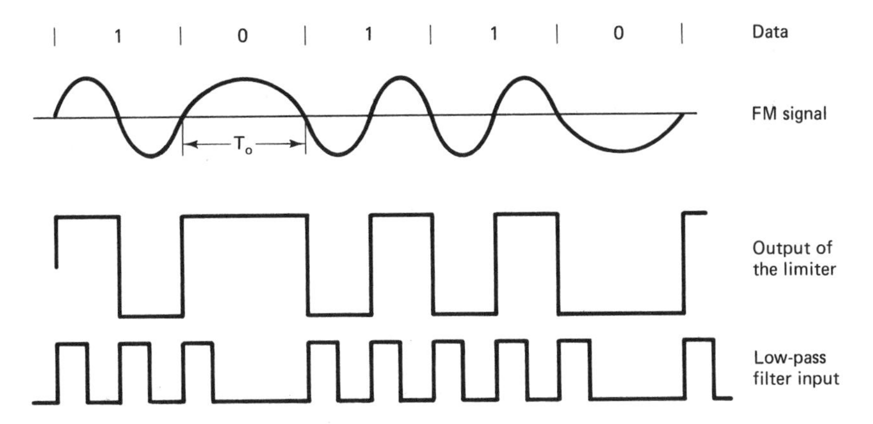

Figure 4.16 Zero-crossing detection in the Manchester code.

The block diagram of differential or product detectors is shown in Figure 4.17. This detection can be regarded as a particular kind of a discriminator or as a phase comparator detecting the phase change in an FM signal.

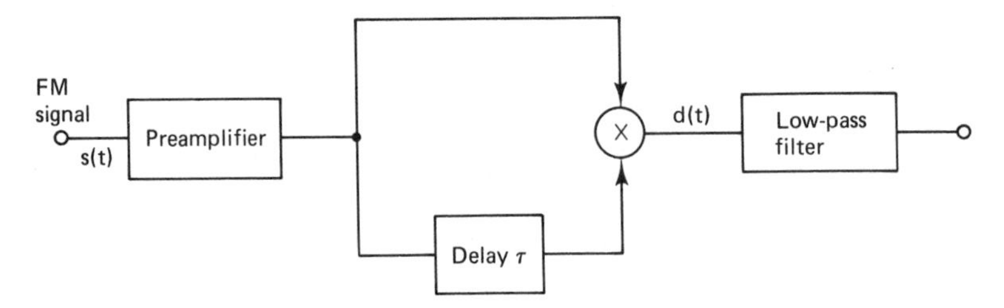

Figure 4.17 Block diagram of differential or product detectors.

If the input signal after the preamplifier is an FM signal,

$$s(t) = A_c \cos\left(\omega_c t + m_f \int^t g(t)\,dt + \phi_0\right) \tag{4.97}$$

the output of the mixer may be calculated as the product of $s(t)$ and its delayed replicas $s(t - \tau)$:

$$d(t) = (A_c^2/2)\cos\left[\omega_c\tau + m_f \int_t^{t+\tau} g(t)\,dt\right]$$

$$+ (A_c^2/2)\cos\left[2\omega_c t - \omega_c\tau + \int^t + \int^{t+\tau} + \theta_0\right] \tag{4.98}$$

For small τ and $\omega_c\tau = \pi/2$, by passing this signal through a low-pass ideal filter, the double-frequency component will be eliminated, and the output signal is

$$d_0(t) = (A_c^2/2)\sin\,[m_f g(t)\tau] \approx (A_c^2/2) m_f \tau_g(t) \tag{4.99}$$

This means that the output signal is proportional to frequency deviation over the range in which the approximation $\sin m_f\tau g(t) \approx m_f\tau g(t)$ is valid.

REFERENCES

1. M. Schwartz, *Information Transmission Modulation and Noise,* 2nd ed., McGraw-Hill, New York, 1970.
2. P. F. Panter, *Modulation, Noise and Spectral Analysis,* McGraw-Hill, New York, 1965, 1959.
3. K. S. Shanmugam, *Digital and Analog Communication Systems,* Wiley, New York, 1979.
4. W. D. Gregg, *Analog and Digital Communications: Concepts, Systems, Applications, and Services in Electrical Dissemination of Aural, Visual, and Data Information,* Wiley, New York, 1977.

5. M. Abramowitz and I. Stegun (eds.), *Handbook of Mathematical Functions with Formulas, Graphs and Mathematical Tables,* Dover, New York, 1972, Chap. 9.

APPENDIX 4A: PHASE MODULATION

Most of the principles of FM also apply to PM. A difference is that the peak phase deviation is controlled by the modulating signal in PM, and therefore the peak frequency deviation varies with the frequency of the modulating signal as well as its amplitude.

In phase modulation the instantaneous phase deviation of the carrier is proportional to the message signal:

$$\phi(t) = k_p g(t) \tag{4A.1}$$

where k_p is the phase deviation constant (rad/V). For unique demodulation of the PM signal, $\phi(t)$ must not exceed $\pm 180°$ since there is no physical distinction between the angles 225° and $-45°$, for instance. The PM signal is

$$s(t) = A_c \cos [\omega_c t + k_p g(t)] \tag{4A.2}$$

where A_c is a constant amplitude.

4A.1 Spectra of Tone-Modulated PM Signals

For sinusoidal (or a tone) modulating signal,

$$g(t) = A_m \sin \omega_m t \tag{4A.3}$$

The PM signal is

$$s(t) = A_c \cos (\omega_c t + \beta \sin \omega_m t) \tag{4A.4}$$

where a modulation index $\beta = k_p A_m$ is defined only for tone modulation, represents the maximum phase deviation produced by the modulating tone, and is independent of modulation frequency.

The instantaneous frequency $\omega_i(t)$ is

$$\omega_i(t) = \omega_c + \omega_m \beta \cos \omega_m t = \omega_c + \Delta\Omega \cos \omega_m t \tag{4A.5}$$

where $\Delta\Omega$ is the maximum angular frequency deviation of the modulated carrier frequency.

For a small phase deviation (narrowband PM) the PM signal can be expressed as

$$s(t) \simeq A_c \cos \omega_c t - \frac{A_c\beta}{2} \cos (\omega_c - \omega_m)t + \frac{A_c\beta}{2} \cos (\omega_c + \omega_m)t \tag{4A.6}$$

which differs from an AM signal in the sine of the lower sideband.

For large phase deviation (wideband PM), β is not small and the PM signal can be expressed as

$$s(t) = A_c \sum_{n=-\infty}^{\infty} J_n(\beta) \cos (\omega_c + n\omega_m t) \tag{4A.7}$$

This expression is the same as the corresponding expression for the FM signal, so that the power and bandwidth consideration are the same as for the FM signal.

4A.2 Output Signal-to-Noise Ratio in PM Systems

The input signal to the detector in the presence of noise is

$$\begin{aligned} Y(t) &= A_c \cos[\omega_c t + \phi(t)] + V_n(t) \cos[\omega_c t + \theta_n(t)] \\ &= R_n(t) \cos[\omega_c t + \psi(t)] \end{aligned} \tag{4A.8}$$

where, for a high input carrier-to-noise ratio,

$$\psi(t) \approx \phi(t) + \frac{V_n(t)}{A_c} \sin[\theta_n(t) - \phi(t)] \tag{4A.9}$$

where $V_n = \sqrt{2 \cdot S'(f_n)\Delta f} << A_c$; $S'(f)$ is the IF power spectral density in watts per cycle, and Δf is a narrow bandwidth.

The output of a phase discriminator is

$$z(t) = K_d \psi(t) = K_d K_p g(t) + K_d \frac{V_n(t)}{A_c} \sin[\theta_n(t) - \phi(t)] \tag{4A.10}$$

The output mean-square value due to the discrete noise component is

$$\overline{\theta^2(t)} = K_d^2 \frac{V_n^2}{2A_c^2} \tag{4A.11}$$

The mean-square value at the total output noise power $(N)_o$ is

$$(N)_o = K_d^2 \frac{1}{A_c^2} \int_{B_{PM}} S'(f_c + f)\, df = \left(\frac{K_d}{A_c}\right)^2 \cdot 2N_0 f_m \tag{4A.12}$$

for white noise with spectral density N_0, and $B_{FM} = 2f_m$.

The two-sided density of the output noise power $S_{N_o}(f)$ is

$$S_{N_o}(f) = \left(\frac{K_d}{A_c}\right)^2 S'(f_c + f) \tag{4A.13}$$

The output signal power is

$$S_o = K_d^2 (\Delta\Omega)^2 \overline{g^2(t)} \tag{4A.14}$$

and the $(S/N)_o$ of a phase discriminator is

$$\left(\frac{S}{N}\right)_o = \frac{(\Delta\Omega)^2 A_c^2\, \overline{g^2(t)}}{2N_0 f_m} \quad \text{for white noise} \tag{4A.15}$$

The signal-to-noise improvement ratio is

$$\frac{(S/N)_o}{(C/N)_i} = (\Delta\Omega)^2 \tag{4A.16}$$

where $(C/N)_i$ is the input carrier-to-noise ratio.

4A.3 Pulse PM

There are two cases: constant and modulated duty cycle. In both cases the expressions for the PM pulse train are similar to the corresponding expression for the FM pulse train.

For a constant duty cycle, the pulse PM spectrum is

$$F(t) = \frac{A_c\omega_c t}{2\pi} + \frac{A_c\omega_c\tau}{\pi}\sum_{k=1}^{\infty}\frac{\sin k\omega_c\tau/2}{k\omega_c\tau/2}\Bigg(J_0(k\omega_c\tau_d)\cos k\omega_c t$$
$$+ \sum_{n=1}^{\infty} J_n(k\omega_c\tau_d)\,\{\cos[(k\omega_c + n\omega_m)t + n\phi] \tag{4A.17}$$
$$+ (-1)^n \cos[(k\omega_c - n\omega_m)t - n\phi]\}\Bigg)$$

In the pulse FM, $k\omega_c\tau_d$ is replaced by $k\beta$. Each of the pulse-repetition-frequency harmonics is phase modulated; the maximum deviation is $k\omega_c\tau_d$. The amplitudes of the pulse-repetition-frequency harmonics and their sidebands decrease with k according to the term $\sin(k\omega_c\tau/2)/(k\omega_c\tau/2)$. There is no sideband accompanying the zero, and hence modulation cannot be recovered by means of a low-pass filter.

The spectrum of the pulse PM when the duty cycle is modulated is similar to the corresponding spectrum of the pulse FM, except that $\omega_r\tau_d$ is substituted for β. The conclusions are similar. Modulation can be recovered by means of a low-pass filter. The upper and lower sidebands of the same order are not equal in amplitude.

APPENDIX 4B: RATIO OF FIRST AND SECOND ORDER SIDEBANDS TO THE CARRIERS IN THE FM SYSTEM

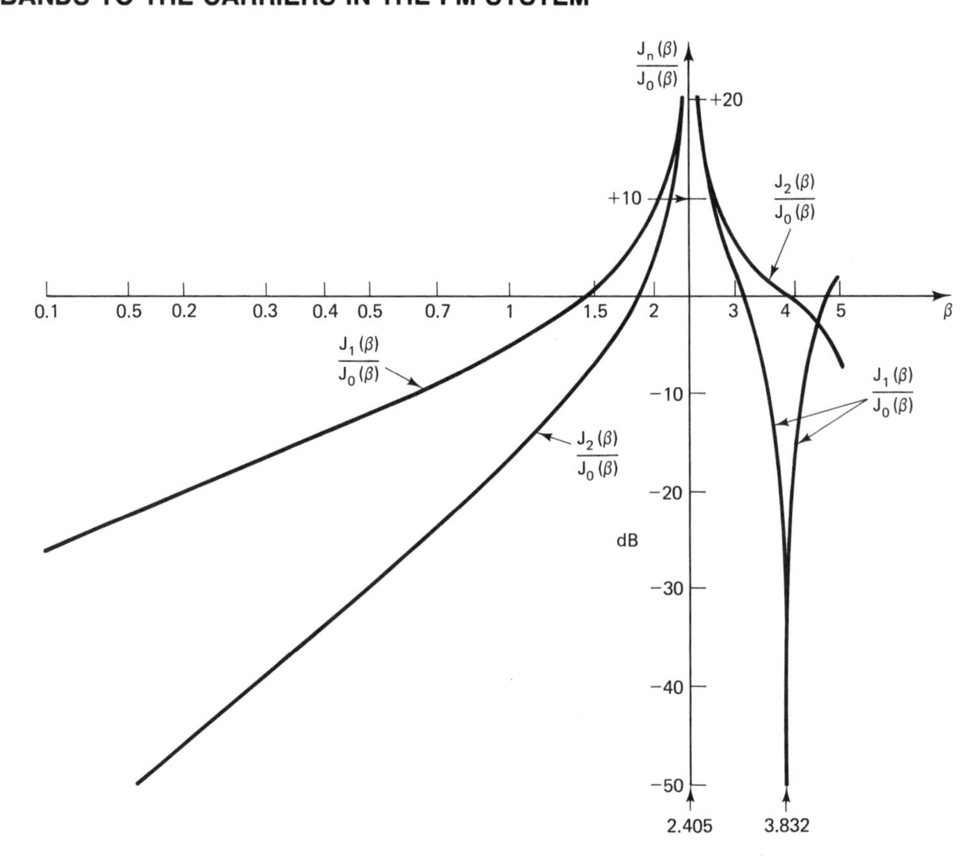

five

TV Channels

5.1 INTRODUCTION

Television can be considered as an extension of two fields: the science of radio communication and the science of film (movie) projection. The fundamental principles of radio are kept, but in the case of television, there is a parallel problem of converting space and time variations of luminosity and color into electrical information and, after processing (e.g., transmitting or recording), reconverting the electrical information into an optical image. The information to be reproduced is optical in character, and at any instant there are an infinite number of pieces of information existing simultaneously, the brightness and color existing in each point of the scene to be reproduced. In other words, the information is a function of more variables, time, and space. The process known as *scanning* is introduced to express this information in the form of a single-valued function of time.

From the point of view of the image formation in the human eye, the fundamental phenomena are similar to the case of film projection. The phenomenon known as *persistence* is a basic phenomenon. A film that causes the same impression as watching the real scene is made from separate still pictures. They are made in successive, but discrete instances of real happenings. To get the impression of a real scene, at least 24 still pictures should be shown to the eye. Basically, if an optical pulse hits the eye, there is the impression of its existence about $\frac{1}{24}$ s after the pulse. Typical movements are satisfied by $N = 24$, but some are not: for example, the wheels of fast-moving wagons.

The object of an ideal television system is to create on a distant light-emitting screen a replica of what the eye would see if it were viewing the original scene directly. The system must produce a picture of realistic color, adequate brightness, and good definition. In its design, account must be taken of the properties and limitations of the human eye. The basic criteria to be met are that the reproduced images be acceptable to the human eye and that the technical details of the system not be obstructively evident to the viewer [1, Sec. 20]. The importance of picture qualities judged by the eye are as follows:

1. *Sharpness, or pictorial clarity*. The eye is very sensitive to edges in the picture and it will be disturbing if the image is out of focus or the details or edges of the objects are not clear and sharp.
2. *Contrast*. The contrast between light areas, dark areas, and the related background illumination: background lighting can introduce contrast changes not present in the original scene.
3. *Continuity of motion*. Motion is created by a succession of still-picture frames. The illusion of motion is created, in part, by the fact that the human eye briefly (for about $\frac{1}{24}$ s) retains any illumination image impeding on it.
4. *Flicker*. Flicker can cause eyestrain, and it makes the reproduced picture unpleasant to view. Flicker is a function of picture brightness, and even when continuity of motion is preserved, the picture may flicker. To eliminate flicker, the pictures must be presented at a rate considerably greater than that required for continuity of motion.
5. *Color value*. Color values, if present, must be accepted as realistic. Accuracy is not necessary in reproduction as compared with the original scene, because colors observed are greatly influenced by the surroundings.

The television picture is reconstructed in time sequence by line scanning, and the components contained in the TV waveforms are:

Horizontal line synchronizing pulses
Field synchronizing pulses
Equalizing pulses
Setup (black) level (generally not used currently)
Picture elements
Color sync (burst)
Bruh blanking
Color hue (tint)
Color saturation (vividness)

The last four components are included in the color composite video signal, and the first five components are included in any TV signal: color or black and white. Synchronizing pulses are introduced by a specific scanning process in which two-dimensional pictures are analyzed and composed (by the one-dimensional signal). These components are discussed below.

5.2 BLACK-AND-WHITE TV SIGNAL

The first four components in the composite TV signal will now be considered. A television system contains an electro-optical conversion device, a transmission channel, and an image-reproducing device. Ideally, this system would be capable of preserving all the details contained in a scene that the eye is capable of utilizing. The characteristics of the system are determined by all components: optical lens system, photographic film (if included), pickup devices, amplifiers, and picture-reproduction devices, usually cathode-ray tubes. The detail that can be reproduced in a system is limited primarily by the pickup and reproducing devices. In the optical system, the resolving power also depends on the size of the aperture through which the scene is viewed. The scanning beam in the pickup device is a limitation with an effect similar to the aperture area. This is a limitation on both pickup and reproducing devices and depends on the cross-sectional area of the scanning beam, the nature of the electron distribution over it, and the separation of individual scanning lines.

For the first approximation in the determination of system requirements and properties, it will be assumed that aperture distortion and its equivalent in scanning-beam area can be neglected and that the picture elements which can be resolved have the same angular spacing as that the eye would see when viewing the scene directly.

5.2.1 Raster Scan Concept

In present TV systems, two basic factors define TV signal waveforms; that is, these two factors define signals in TV systems in time as well as in spectral domain. The original scene content is the first factor. In black-and-white TV systems, the luminance distribution is defined by scene content, and it can have any level between black and white. A scanning process is the second factor. The conversion of a multidimensional process, the scene as viewed in this case, to a one-dimensional process, the TV signal, is managed by a scanning mechanism. This mechanism is the portion of the entire television system that is especially important from the standpoint of the necessity of formulation of standards of methods and performance. Once the scanning system is standardized, the performance of the system is limited.

Scanning is that process which makes possible the conversion of optical information as functions of space and time into electrical information expressed as a function of time only. It is this conversion process that makes possible the use of a signal transmission channel. The scanning process is such that the picture area is traversed at repeating intervals, giving in effect a series of single pictures, in much the same manner as motion pictures are presented. This rate must be chosen sufficiently high so that neither discontinuity of motion nor flicker is apparent to the eye.

Figure 5.1 shows four examples for the continuous scanning process, for

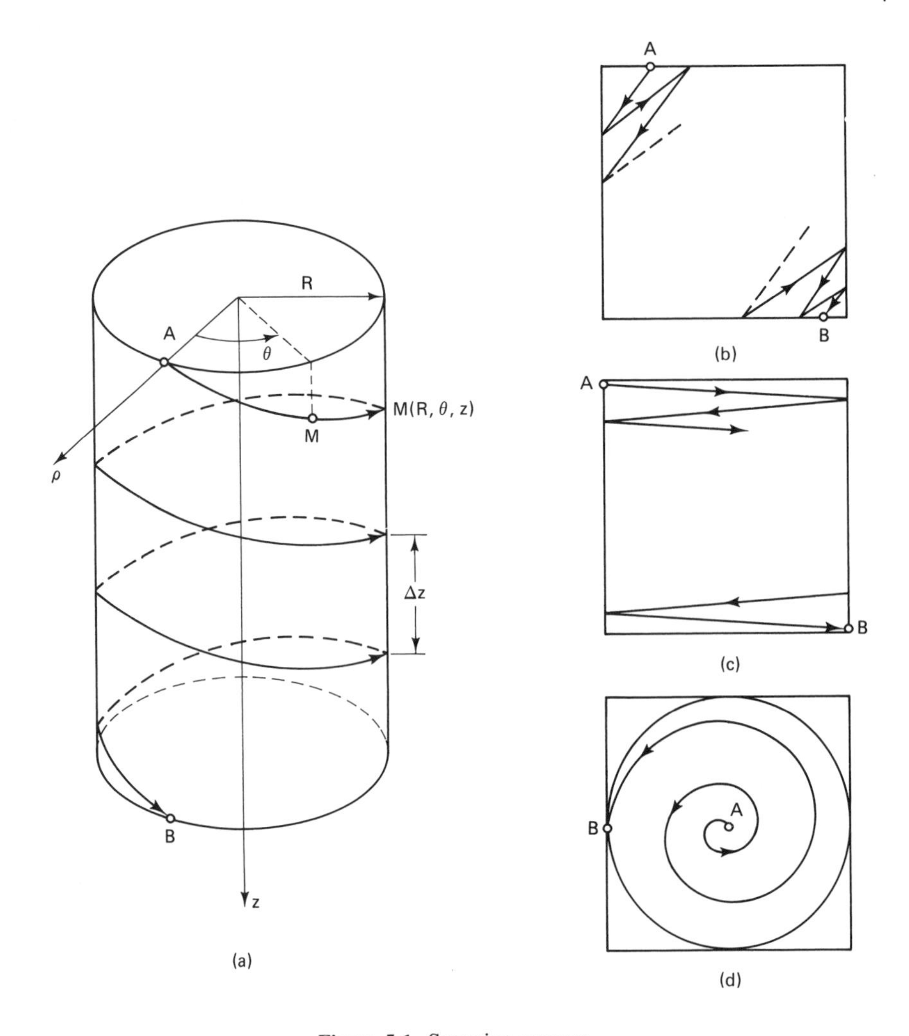

Figure 5.1 Scanning process.

illustration only. In all cases scanning starts at point A and finishes at point B; then the beam returns to point A. Figure 5.1a shows a hypothetical case where the optical system has a 360° view. In the next two examples, rectangular pictures are scanned in which the beam always goes normal: on the one diagonal (Figure 5.1b) and on the vertical line (Figure 5.1c). A spiral scanning of a round picture is shown in Figure 5.1d. Figure 5.2 shows the synchronizing pulses, $s_s(t)$, for the scanning processes in Figure 5.1. These are both synchronizing and blanking pulses. The waveform $s_s(t)$ is periodical with the frame period, T_F. In each period a time interval, τ, is reserved for a beam retrace to the starting point, A. The level of $s_s(t)$ during this period is lower than the black level that will ensure nonvisibility of the beam during the retrace period even in the presence of noise.

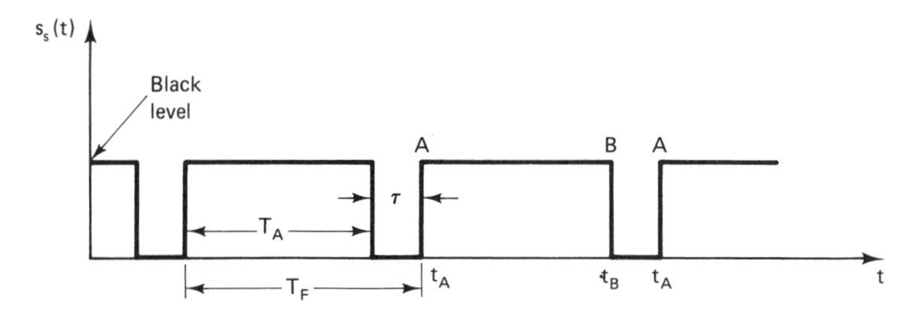

Figure 5.2 Blanking (synchronizing in the same time) pulses for the scanning process in Figure 5.1.

The video signal is generated during active ($T_A = T_F - \tau$) periods and is not shown in Figure 5.2.

The duration of the repetition period is influenced by the flicker. This effect is independent of motion in the picture. For a given level of brightness, flicker becomes less pronounced as T_F is decreased. Flicker is a function of brightness, and T_F should be lower for lower brightness. This is less critical for smaller areas. In any case, between 50 and 60 changes a second is enough for most applications.

The scanning method chosen might depend on the geometry of the image to be reproduced. In motion-picture standards, the picture aspect ratio (width to height) was adopted as 4:3 because it has been found that observers prefer a rectangular picture with slightly greater width than height. It seems that this phenomenon is based on the fact that the fovea of the eye, the area of greatest resolution, is somewhat wider than it is high. Compatibility with the motion picture is important and the 4:3 aspect ratio has also been adopted for TV. Once the geometry of the image has been specified, the track pattern for the scanning beam is determined to cover the entire area. The most practical TV system was found to be one with linear scanning of the image. It is common practice to place a starting point in the top left-hand corner, and scan from left to right and from top to bottom.

Figure 5.3 shows the paths of scanning beams in the active and retrace modes and the corresponding waveforms for a hypothetical system with nine active lines. The retrace times involved, both horizontal and vertical, are due to physical limitations on practical scanning systems and are not utilized for transmitting a video signal but may be employed for the transmission of auxiliary information. The synchronizing signal (Figure 5.3e) is simplified and is called *composite sync* because it contains both line and frame synchronization pulses. The number of active or useful lines, N_{LA}, depends on the number of vertical elements that can be resolved by the eye, N_v:

$$N_v = k_1 N_{LA} \tag{5.1}$$

Here [2, p. 26]

$$N_v = \frac{1}{\alpha \rho} \tag{5.2}$$

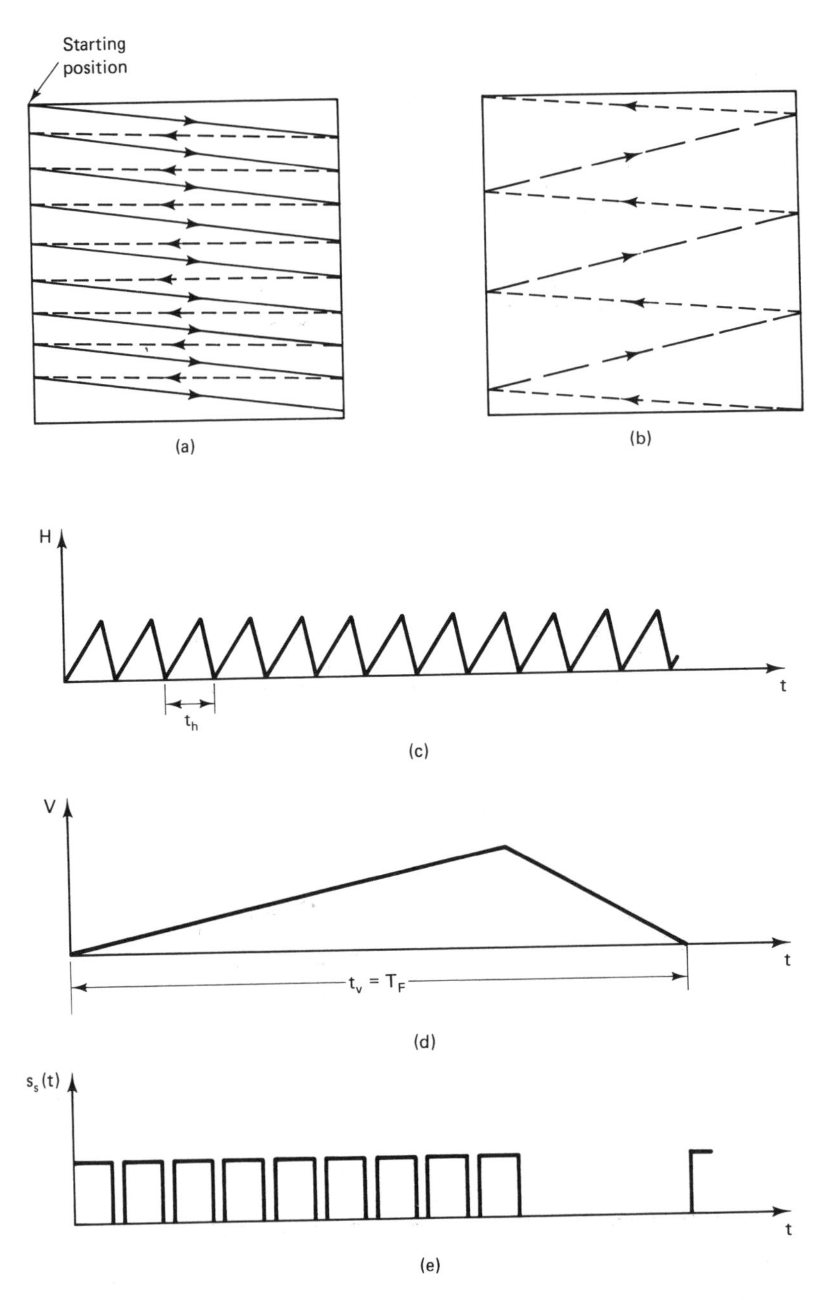

Figure 5.3 Linear scanning: (a) path of the scanning beam covering the picture area; (b) path of the scanning beam during the vertical retrace; (c) waveforms of electric or magnetic fields producing linear (constant velocity) scanning; (d) composite sync (blanking).

where α = minimum resolvable angle of the eye, rad
$\rho = D/H$ = viewing distance/picture height

and k_1 is a constant obtained either from subjective measurements or from a more complex theoretical analysis (e.g., $k_1 = 0.7$).

If the total number of lines, N_L, is included, then

$$N_v = k_1 \times r_1 \times N_L \tag{5.3}$$

where

$$r_1 = N_{LA}/N_L = \frac{\text{number of active lines}}{\text{number of lines required, including retrace}}$$

TV systems should provide sufficient picture detail in the reconstructed picture: for a reasonable amount of detail, the picture should have between 100,000 and 400,000 elements. This means that for 25 pictures a second, there are 10^7 different illuminations each second. The image detail perceived by the eye is determined by the "resolution" capability of the image-reproducing system—the number of basic picture elements that can be reproduced and discerned.

An electron beam, scanning across a photosensitive surface of various pickup devices, can be made to produce in an external circuit a time-varying electrical signal whose amplitude is at all times proportional to the illumination on the surface directly under the scanning beam: the picture signal. The upper-frequency limit of the amplifier for such signals depends on the velocity of the scanning beam.

In Figure 5.4, the picture signal is shown between two successive line, or horizontal, sync pulses. The picture signal changes between black and white levels. The blanking level is lower than the black level, also referred to as the

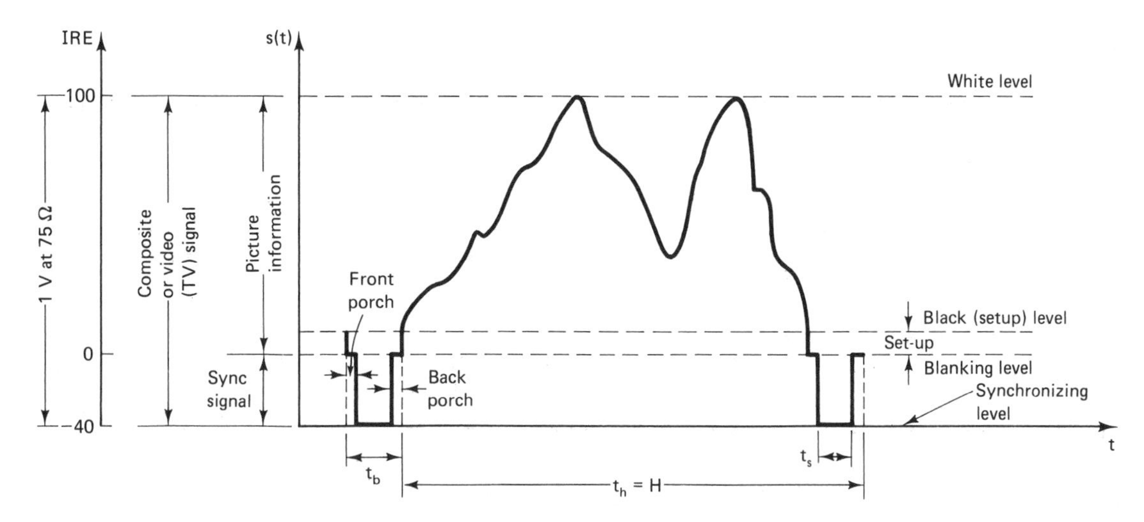

Figure 5.4 TV (composite) signal.

setup level, to ensure sync separation. If the blanking level is marked as 0 and the reference white level as 100 units, the synchronizing level is approximately 40% of that span and is lower than the blanking level. The reference black level used to be about 7.5% higher than the blanking level, but this difference now seems to be disappearing. Absolute values are specified by the CCIR (Comité Consultant International des Radiocommunications). The video signal is 1 V into a 75-Ω termination, that is, 1 V between the synchronizing and reference white levels, 0.7 V between the blanking and reference white levels in the same condition, and 0.3 V between the synchronizing and blanking levels; between the blanking and reference black levels the signal used to be 0.05 V.

The time relationship of the blanking and synchronizing pulses is also depicted in Figure 5.4 ($t_s < t_b$). The leading edge of a blanking pulse precedes the leading edge of the horizontal synchronizing pulse. The region between the leading edges is termed the *front porch,* and the region between the trailing edges is termed the *back porch*. The average level of the TV signal varies as the picture content varies. These dc variations are lost when the signal is passed through any *RC*-coupled amplifier, which in practice means that information about the black level is lost. Special clamping circuits are used to provide a reference for reinserting the black level (the dc level). This circuit sets the synchronizing pulses at a constant dc level, independent of picture-brightness variations.

5.2.1.1 Interlacing of Scanning Lines: Field and Frame. As we can see, the picture repetition rate, $N_F = 1/T_F$, is defined by flicker, while the line rate, or number of lines per frame (picture), N_L, is defined by the desired vertical resolution. The principle of interlacing the scanning lines allows a system to maintain the same resolution and flicker characteristics as those of a noninterlaced system, with a bandwidth requirement only half that of the noninterlaced system. The process gives a vertical scanning frequency, now a field frequency of twice the true picture-repetition, or frame, frequency, and is illustrated in Figure 5.5 for 11 horizontal lines. One frame thus has two fields.

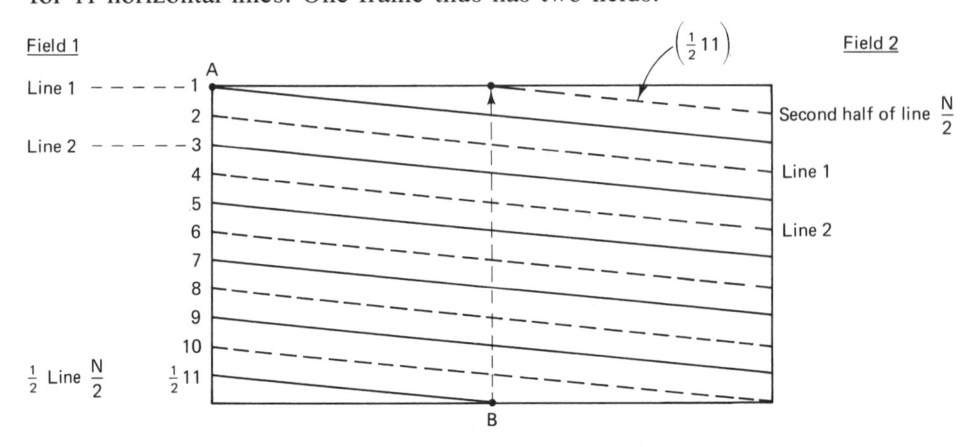

Figure 5.5 Interlaced scanning pattern (raster).

Interlaced scanning is achieved by making the line-scanning (horizontal) rate an odd multiple of one-half the vertical (field-scanning) rate. That is, the interlacing is achieved when the number of lines per frame is an odd number, thus requiring each field to have an even number of lines plus one-half line. The half line left over at the end of a field scan displaces the next field downward by a full line, and interlacing is achieved. The actual picture does not cover the entire raster area. Figure 5.6 shows the actual size of the picture and blanked parts of the raster.

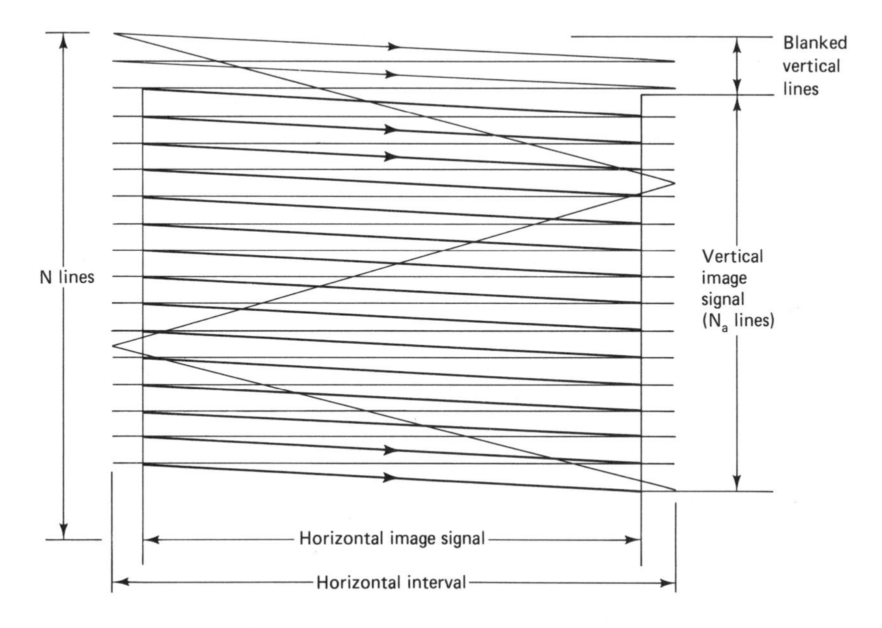

Figure 5.6 Simplified raster scan with marked image size.

The vertical sync pulse is the most complex of the pulses, as the position of the pulse controls interlacing or noninterlacing. The pulse itself can either be a solid negative pulse or a series of shorter negative pulses. In general, it can be as follows:

1. Solid vertical pulse, no interlacing
2. Solid vertical pulse, 2:1 interlacing
3. Series of vertical pulses, 2:1 interlacing

Broadcasting TV and most closed-circuit TV use a vertical pulse system in group 3. The vertical pulses and the surrounding equalizing pulses are characterized by being fed to a clock with twice the frequency of the line rate, and the pulse itself is broken down into, for example, six pulses (can also be five or seven), each of one-half line time. These are often referred to as *serrated field pulses*.

Figure 5.7 shows more detail. First, the time markers with twice the line frequency are given for reference. Then vertical pulses for odd and even fields are shown. Four characteristic parts can be distinguished. Equalizing pulses (six in this example) with duration marked as *l* precede the serrated field-synchronizing pulses for the field duration marked as *m*. Equalizing pulses follow, with duration marked as *n*. The purpose of the initial pulses (*l*) is to assure proper field interlacing and to maintain horizontal line sync. The usefulness of the last group of equalizing pulses (*n*) is not clear in light of present-day circuitry. The fourth part is a series of horizontal pulses without picture information, following the equalizing pulses (*n*). This allows proper reference and settling before a new field is scanned. The entire vertical blanking period has duration *j*.

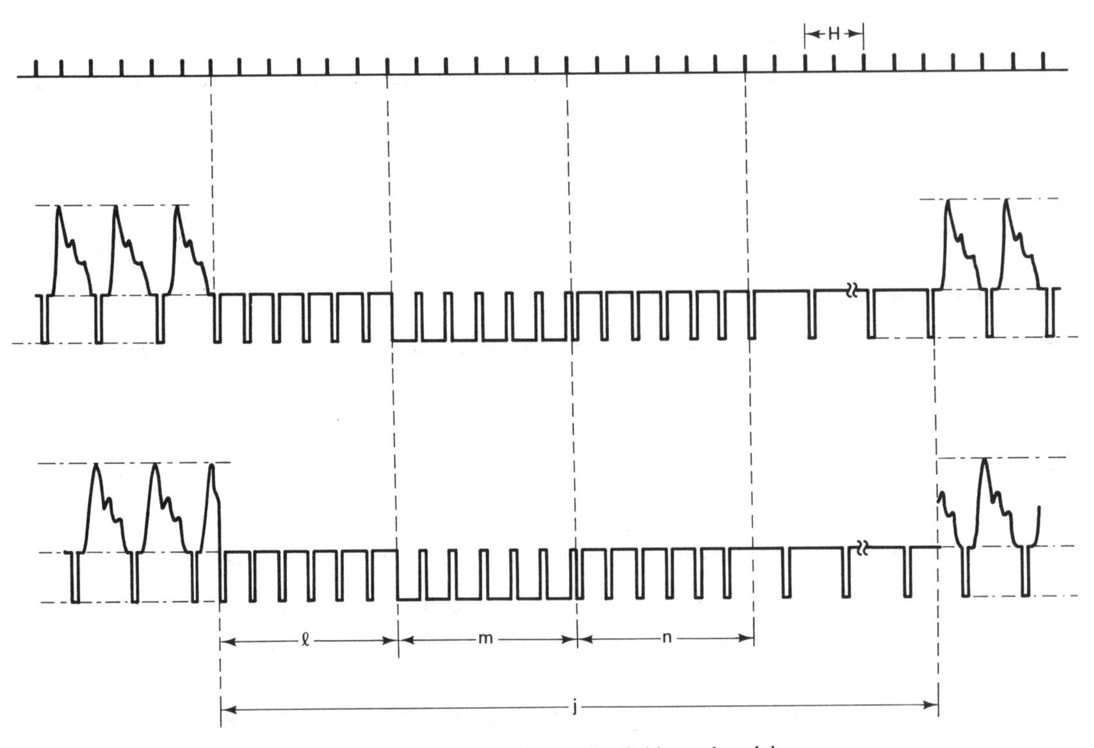

Figure 5.7 Pulses for field synchronizing.

The characteristic time intervals used in Figures 5.7 and 5.4 differ from system to system in the picture analysis. For one frame (two fields) there are systems with: 405 lines, Great Britain; 525 lines, the United States, Canada, and Japan; 625 lines, Europe; and 819 lines, France. In all systems 50 fields per second are used except that in the system with 525 lines, 60 fields are used. Table 5.1 contains characteristic values specified by the CCIR. More details for the vertical interval, NTSC TV standard, are given in Appendix 5A.

TABLE 5.1 CHARACTERISTIC VALUES SPECIFIED BY CCIR

		PAL	*NTSC*
N	Number of lines per frame	625	525
f_F	Frame frequency (pictures/second)	25	29.97
f_f	Field frequency (number of fields per second)	50	59.94
$\frac{1}{2}N$	Number of lines per field	$312\frac{1}{2}$	$262\frac{1}{2}$
h/b	Aspect ratio	$\frac{3}{4}$	$\frac{3}{4}$
B	Video bandwidth (MHz)	5	4.2
H	Line period (μs)	64	63.5566
f_L	Line frequency (Hz)	15,625	15,734
a	Line-blanking interval (μs)	11.8–12.3	10.5–11.4
d	Line synchronizing pulse (μs)	4.5–4.9	4.2–5.1
T_f	Field period V (ms)	20	16.683
j	Field-blanking period	18–22H + 12	25H
l	Duration of first equalizing pulse sequence	2.5H	3H
m	Duration of field synchronizing pulse sequence	2.5H	3H
n	Duration of second sequence of equalizing pulses	2.5H	3H
T_{Fh}	Field synchronizing pulse period	0.5H	0.5H

5.2.2 Spectrum of the TV Signal

The Fourier transform F of the function $f(t)$ is its complex spectrum:

$$F(j\omega) = \int_{\infty}^{\infty} f(t)e^{-i\omega t}\,dt \tag{5.4}$$

The absolute value $|F|$ is an amplitude spectrum, and the arg $F = \theta$ is a phase spectrum of the function $f(t)$. If $f(t)$ is not periodic, we are talking about a continuous complex spectrum F, spectral amplitude density $|F|$, and spectral phase density θ; $|F|^2$ is the power (energy) spectrum of $f(x)$. Obviously, there is not a one-to-one relationship between $f(x)$ and its energy spectrum, although $f(x)$ determines $F(j\omega)$. The information lost when only the energy spectrum can be given is of precisely the same character as that which is lost when the autocorrelation function is given instead of the original function. When $f(x)$ is real, as it would be if it represents a physical waveform, the energy spectrum is an even function and is therefore fully determined by its values of positive frequencies; to stress this fact, the term *positive-frequency energy spectrum* may be used.

The inverse Fourier transform of the spectrum is a function $f(t)$ itself:

$$f(t) = \frac{1}{2\pi}\int_{-\infty}^{\infty} F(j\omega)e^{j\omega t}\,d\omega \tag{5.5}$$

Functions $f(t)$ and $F(j\omega)$ are Fourier transform pairs. More detail is provided in Appendix 5B.

Mathematical analysis of the video signals is very valuable. It is needed for a deeper knowledge of the TV process as well as for the TV system design.

For the case of simplicity, we will assume that the analyzed picture is static. The corresponding video signal is then periodic.

Suppose that the scanning process in Figure 5.1a is performed by a system without inertia (i.e., $\tau = 0$, Figure 5.2). If the scanning velocity is constant (vector) and if the starting point, A, is fixed, then distance l from point A, can be used to define a position of the scanning spot (beam) at the picture surface instead of cylindrical coordinates (R,θ,z). The cylindrical picture at Figure 5.1a can be repeated periodically parallel with the axis z.

The brightness, B, at point $M(R, \theta, z) = M(l)$ can be expressed as the Fourier series:

$$B(l) = \sum_{-\infty}^{\infty} A_n e^{j(2n\pi/L)l} \tag{5.6}$$

where L is the distance between points A and B. Fourier complex coefficients A_n are

$$A_n = \frac{1}{L}\int_0^L B(l)e^{-j(2n\pi/L)}\, dl \tag{5.7}$$

If A_n is expressed in terms of magnitude A_n and phase, ϕ_n,

$$A_n = |A_n|e^{j\phi_n} \tag{5.8}$$

$$B(l) = \sum_{n=0}^{\infty} A_n \cos \frac{2n\pi}{L}(1 + \phi_n) \tag{5.9}$$

If the conversion light-electrical signal is ideal, the output electrical signal, s, is proportional to $B(x)$:

$$s = kB(l), \qquad k = \text{const.} \tag{5.10}$$

If the velocity v along the scanning line is constant, then

$$d = vt \tag{5.11}$$

A one-dimensional picture signal in the time domain is then

$$s(t) = kB(t) = k\sum_{n=0}^{\infty} |A_n| \cos\left(\frac{2n\pi}{L/v}t + \phi_n\right) \tag{5.12}$$

or

$$s(t) = k\sum_{n=0}^{\infty} A_n \cos\left(\frac{2n\pi}{T}t + \phi_n\right) \tag{5.13}$$

where $T = L/v$ and for this case, $T = T_F = T_A$. This means that for assumed conditions (still picture, linear scanning, zero width of the scanning beam, etc.) the spectrum is discrete and the frequency of the spectral components is

$$f_n = nf_F = n/T_F \tag{5.14}$$

That is, the scanning process defines occupied frequencies. The energy distribution or the amplitude of the spectral components is a function of the picture content. These two conclusions are valuable, although they are obtained based on the idealized model. In a more realistic case the influence of the interlace, inertia or finite time for the retrace of the scanning beam, and motion in the picture should be considered.

If the flicker effect is noticeable, the interlace technique can be helpful. The low-frequency content varies little for alternate scanning fields and the lowest-frequency component of any appreciable magnitude is twice the frame frequency:

$$f_n = n \frac{1}{T_F/2} = n(2f_F) \tag{5.15}$$

Thus only the lowest-frequency component that the system must be capable of preserving is represented by scanning frequency, $2f_F$, and not by frame frequency f_F. Details in the picture with vertical dimension larger than $2\Delta z$ (Figure 5.1) will produce individual spectral lines with distance $2f_F$ between the two adjacent lines.

For real, inertial systems, retracing time is not zero. In Figure 5.2 this duration is marked as τ. The repetition period is $T_F = T_A + \tau$, and the distance between two neighboring spectral lines corresponds to the repetition frequency: $f_F = 1/T_F$. The energy content will be slightly redistributed, and spectral components will be slightly changed compared to the ideal case without inertia.

The positions of the spectral components of the picture signal obtained by the scanning process are defined by the picture shape and the scanning procedure, but the energy distribution through the spectral components is a function of the picture content, that is, of the details in the picture. This means that in the analysis of a scene which contains motion, dynamic picture spectral components can change amplitudes only. Effectively, this means that this amplitude modulation will cause side components around each nf_F (or $2nf_F$) component.

Finally, it should be pointed out that the previous analysis was based on the Fourier analysis, more precisely, on the Fourier series. As a consequence of Fourier series convergence, it follows that the amplitude of the spectral components at frequency $f_n = nf_F$ decreases with an increase in n, that is, with an increase in frequency. Generally, this is true only outside some bandwidth when the frequency goes to infinity. As an example, consider a picture with periodical details (100% contrast) normal to the scanning direction (in Figure 5.1), with period T_0 about 100 times smaller than T_F; no interlace is assumed, but inertia is, say, $\tau \simeq T_F/5$. Figure 5.8a shows picture signals $s_p(t)$ for one scanning period. Corresponding spectrum is shown in Figure 5.8b. It can be seen that the amplitude of the spectral components decreases after $f = f_0$.

The spectrum shown in Figure 5.8b can be obtained, for example, using the convolution theorem. That is, the picture signal, $s_p(t)$, can be expressed as

$$s_p(t) = s_s(t)s_0(t) \tag{5.16}$$

where $s_s(t)$ is the signal shown in Figure 5.2, with unit amplitude, and

$$s_0(t) = a + b \cos \omega_0 t \tag{5.17}$$

The spectrum in Figure 5.8b is obtained by convolving these two spectra. Notice

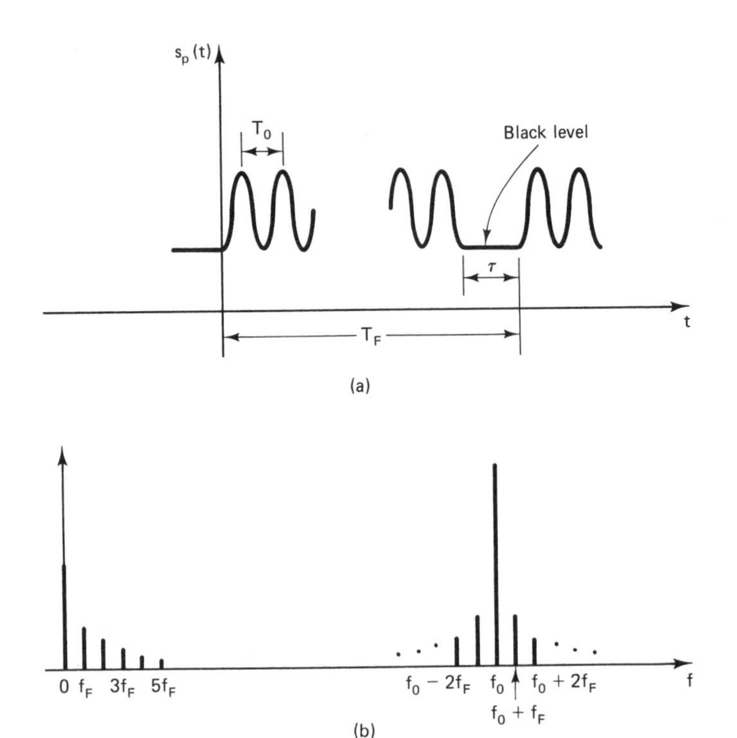

Figure 5.8 High spectral components in the picture: (a) picture signal; (b) corresponding spectra.

that components around zero (positive spectrum is given) are caused by a dc component in $s_0(t)$.

If the cylindrical picture is obtained from a rectangular one, as in the case of phototelegraphy or facsimile, corresponding spectrum would have discrete components:

$$f_{nm} = nf_L + mf_F \tag{5.18}$$

where $f_L = 1/T_L$ is the time period in which the beam passes through 360°. The peak would be at nf_L. If this picture is scanned once, the spectrum would, of course, be continuous. Figure 5.9 shows the envelope of the corresponding spectra (both positive and negative spectra are supposed); folded-over areas are marked. Discrete components (dashed lines) are referred to the periodical scanning.

5.2.2.1 Rectangular Raster. Using the same procedure for a rectangular raster (Figure 5.3) as the one shown for a circular picture, a similar analysis of the standard TV signal for a monochrome still picture can be done [3]. Suppose that the picture width is $2b$ and the height $2h$, and that coordinates x, y have a 0 (0, 0) point at the center of the picture (Figure 5.10). The brightness is a function of x and y:

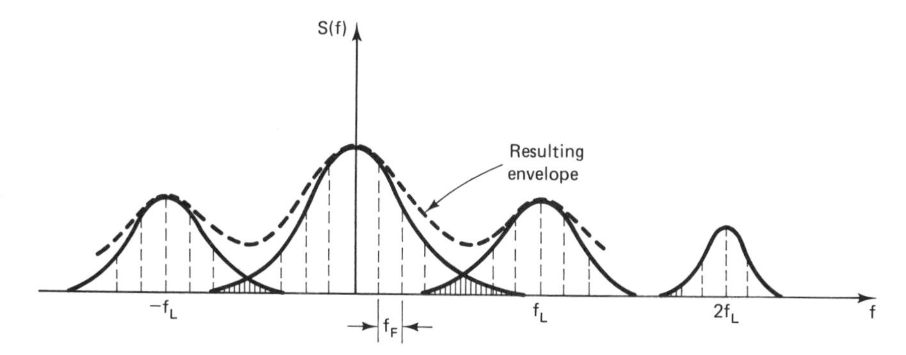

Figure 5.9 Spectra of the phototelegraphic signal.

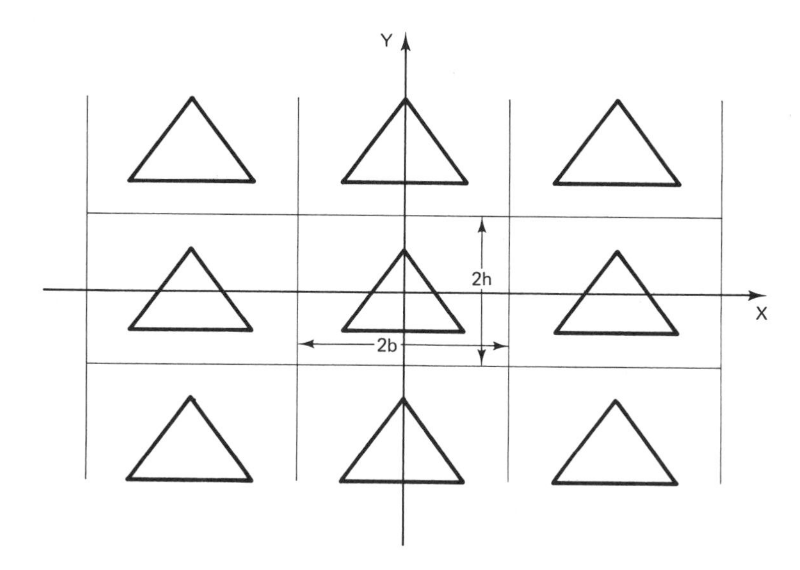

Figure 5.10 Two-dimensional extension of the still picture.

$$B(x, y) = \sum_{n=0}^{\infty} \sum_{m=-\infty}^{\infty} C_{nm} \cos \left[\pi \left(\frac{nv_x}{b} + \frac{mv_y}{h} \right) + \phi_{nm} \right] \tag{5.19}$$

If the analyzing beam has a constant speed in the x, y directions, $x = v_x t$ and $y = v_y t$, and if the electrical signal is proportional to $B(x, y)$,

$$i(t) = K_1 \sum_{n=\infty}^{\infty} \sum_{m=\infty}^{\infty} C_{nm} \cos \left[\pi \left(\frac{nv_x}{b} + \frac{mv_y}{h} \right) + \phi_{nm} \right] \tag{5.20}$$

If line frequency, f_h, and frame frequency, f_F, are

$$f_h = v_x/2b \tag{5.21}$$

and

$$f_F = v_y/2h \tag{5.22}$$

the occupied frequencies are

$$f_{nm} = nf_h + mf_F \tag{5.23}$$

The general shape of the spectra for the coordinates specified is shown in Figure 5.11.

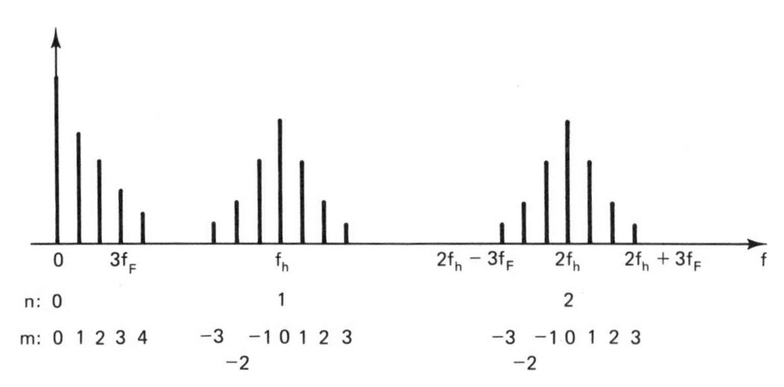

Figure 5.11 General shape of the video signal obtained by analyzing a still picture.

From equation (5.23) it can be seen that spectrum of the scanned picture is composed of the harmonics of the line frequency (nf_h). Around each of them are satellite components (sidebands) disposed symmetrically. The distance between two neighboring sideband lines corresponds to the frame repetition frequency, f_F.

Discussion of the hypothetical case, cylindric scanning, is valid for this case, too. The spectrum shown in Figure 5.11 can be obtained using the convolution theorem: convolving video signal, horizontal blanking, and vertical blanking pulses. It should be ensured that the convolution theorem is properly used.

When movement exists in a scene, each spectral component, f_{nm}, is amplitude dependent (Figure 5.12). This part of the spectrum is not discrete (around each component). If, by electronic methods, movement (AM part of the spectra) is eliminated, a continuous trace will remain on the screen. To reduce AM spectrum, a comb filter can be used (Figure 5.13a) with tooth distance equal to the field frequency (or the frame frequency if there is no interlace). This filter can be realized with a delay (line) equal to the field (frame) period (Figure 5.13b). The same effect can be obtained with a three-dimensional low-pass filter. For example, a screen with long persistence will give a three-dimensional low-pass filter as output.

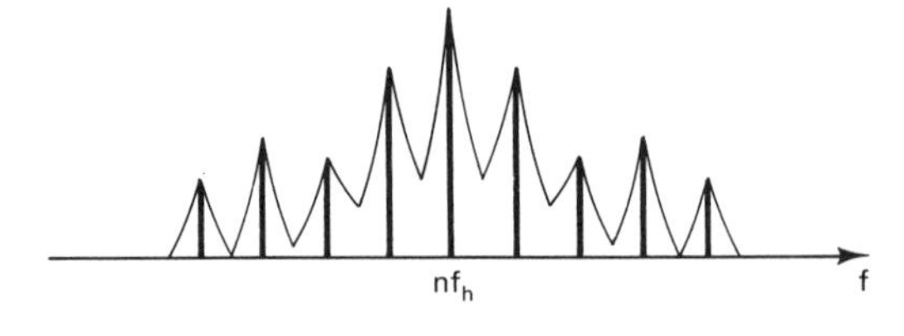

Figure 5.12 Blown-up detail of a luminance signal spectrum for a typical scene with movement (an envelope in the spectra of the video signal).

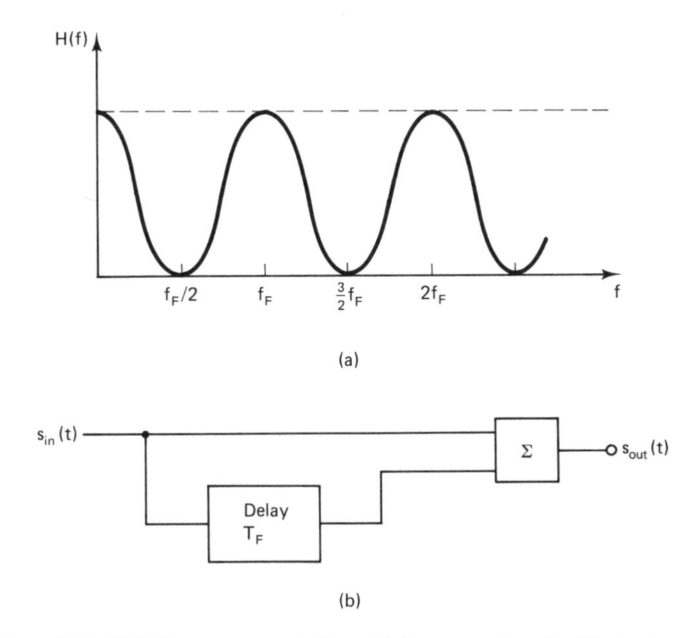

Figure 5.13 Field frequency comb filter: (a) frequency domain; (b) circuit model.

5.2.2.2 System Limitations and Drawbacks. Some of the system limitations and drawbacks can be presented with ideal assumptions, and some of them are introduced by nonideal characteristics of the components in the system. Confusion in signal components can exist under some conditions, although ideal characteristics are assumed. Namely, if the picture content gives slowly decreasing sideband components, overlap will occur (Figure 5.14) which is similar to the aliasing effect.

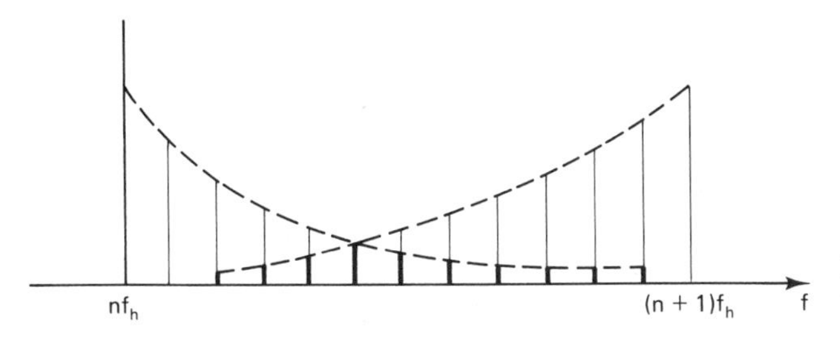

Figure 5.14 Overlap in the spectra of the video signal.

In Figure 5.15 it is shown that reproduction of detail comparable with that of raster scanning lines depends on accurate registration of the scanning lines with such detail.

In the spectral analysis of the video signal it was assumed that the analyzing pulse is a Dirac function but that the beam cross section has finite dimensions which are equivalent to those of a Dirac delta function passed through a linear

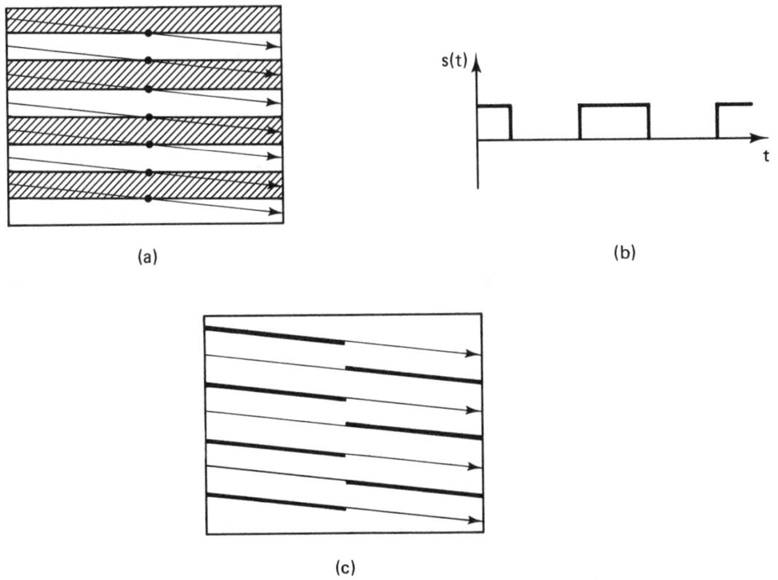

Figure 5.15 Misrepresentation of vertical detail: (a) picture analyzed; (b) waveforms of the signal obtained by a scanning process; (c) picture reproduced.

filter. Also, transmission of the electron beam $T(\xi, \eta)$ is not necessarily constant over the surface of the beam. The output brightness, B_o, is

$$B_o(x, y) = \iint T(\xi, \eta) B(x - \xi, y - \eta)\, d\xi\, d\eta \tag{5.24}$$

or

$$B_o(x, y) = \sum_{-\infty}^{\infty} \sum_{-\infty}^{\infty} Y_{nm} A_{nm} e^{j\pi(n\xi/b + m\eta/h)}\, d\xi\, d\eta \tag{5.25}$$

where

$$Y_{nm} = \iint T(\xi, \eta) e^{j\pi(n\xi/b + m\eta/h)}\, d\xi\, d\eta$$

and $B(x, y)$ is expressed by the Fourier presentation already obtained.

For a symmetrical beam,

$$Y_{nm} = \iint T(\xi, \eta) \cos \pi \left(\frac{n\xi}{b} + \frac{m\eta}{h} \right) d\xi\, d\eta \tag{5.26}$$

This means that Y_{nm} is real and that a symmetrical beam does not introduce phase distortion, only amplitude distortion.

For a square beam with dimensions $(\Delta x, \Delta y)$,

$$Y_{nm} = 4\, \Delta x\, \Delta y\, \text{sinc} \left(n \frac{\pi\, \Delta x}{b} \right) \text{sinc} \left(m \frac{\pi\, \Delta y}{h} \right) \tag{5.27}$$

For a circular beam,

$$Y_{nm} = \left[\frac{J_n[\pi(n\,\Delta r/b)]}{\pi(n\,\Delta r/b)} \right]^2 \tag{5.28}$$

where J_n is a first-order Bessel function. Thus the electron beam acts as a low-pass filter in the three-dimensional domain, which is equivalent to the comb filter in the time-frequency domain.

Figure 5.16 shows the plots of $Y_{n,o}$ for different shapes of electron beams. In practice, the first zero of the admittance corresponds to much higher frequencies (e.g., three times) that really transmitted. The fact that the effective frequency response of the system is a function of the size and shape of the scanning spot, of both the pickup and the reproducing device, is sometimes referred to as an *aperture effect*.

An important parameter of the TV system is the maximal frequency needed in the system, f_m. This is related to the number of vertically, r_v, and horizontally, r_h, resolving elements.

It would be reasonable that r_v is equal to the number of active lines:

$$N_{LA} = k_v N_L \tag{5.29}$$

where N_L is the total number of lines in the frame, and k_v is a factor of line occupancy. Usually, 18 to 22 lines in each field are lost, that is, not displayed on the screen.

But this can be approximately true for a test picture made of horizontal black and white lines with 100% contrast. Because of the vision property and the finite scanning beam (among other reasons), the number of vertically resolvable black and white elements is reduced by the factor K (Kell's factor, $K \approx 70\%$).

$$r_v = K k_v N_L \tag{5.30}$$

In practice, the number of used lines is about 30% higher than the theoretical

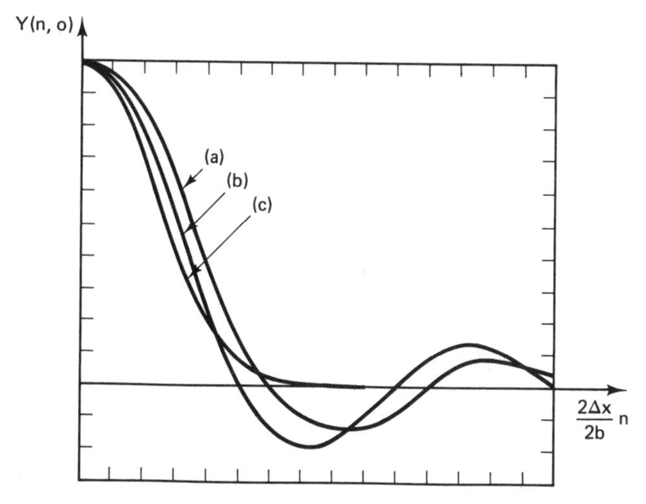

Figure 5.16 Admittance of the electron beam: (a) rectangular beam (uniform density distribution); (b) circular beam (uniform density distribution); (c) circular beam (Gaussian density distribution).

number. Kell's factor is the ratio of the ideal number of lines per vertical cycle and the experimentally obtained number of lines per vertical cycle.

If the screen height is h, the number of vertically resolving elements per unit is

$$r_{v0} = r_a/h = Kk_v(N_L/h) \tag{5.31}$$

The total number of lines in 1 s is $N_L N_F$, where N_F is the number of frames, and the time for one line is

$$H = T_L = 1/N_L N_F \tag{5.32}$$

The active duration is

$$T_{LA} = k_h T_L \tag{5.33}$$

because one (horizontally) resolvable element here means that either a black or a white element, in 1 s would be $2f_m$ just-resolvable elements. The number of the resolvable elements along the active line is

$$r_h = 2f_m T_{LA} \tag{5.34}$$

or, per unit of length,

$$r_{h0} = r_h/b = 2f_m T_{LA}/b \tag{5.35}$$

If the ratio of horizontal and vertical resolution is

$$\rho = r_{h0}/r_{v0} \tag{5.36}$$

then

$$f_m = \tfrac{1}{2}\rho K N_F N_L \frac{b}{h}\frac{k_v}{k_h} \tag{5.37}$$

As an example, consider the European standards of $N_L = 625$, $N_F = f_F = 25$ Hz, aspect ratio $b/h = \frac{4}{3}$, $k_v = 0.935$, $k_h = 0.815$, and $K = 0.7$; then

$$f_m \approx 5.2\rho \quad \text{MHz} \tag{5.38}$$

For $\rho = 1$, $f_m \simeq 5.2$ MHz, and $\rho = 0.96$ (high vertical resolution), $f_m = 5$ MHz. This frequency, 5 MHz, is a CCIR standard. For specified $f_m = 5$ MHz and $\rho = 1$, it would be $N_L = 618$. This number is not suitable as a standard because for a system with interlace, the number of lines should be odd; and second, the stability of the dividers in the synchronizing block, $N_L = 625$ ($625 = 5 \times 5 \times 5 \times 5$) is specified as a standard.

For videodisc systems, from the capacity point of view, it is desirable that the upper-frequency requirement be kept as low as possible. For a given requirement in resolution this is possible only by lowering the scanning speed. If a required vertical resolution defines a particular number of scanning lines, the result is a lowering of picture repetition rate, which may be beyond the limits set by flicker. A particular system design should make a logical compromise among these conflicting requirements.

The analysis of signals produced by the bar patterns gives no information regarding the low-frequency requirement of a video amplifier used to amplify TV signals. From Figure 5.17 it can be seen why an amplifier must have good

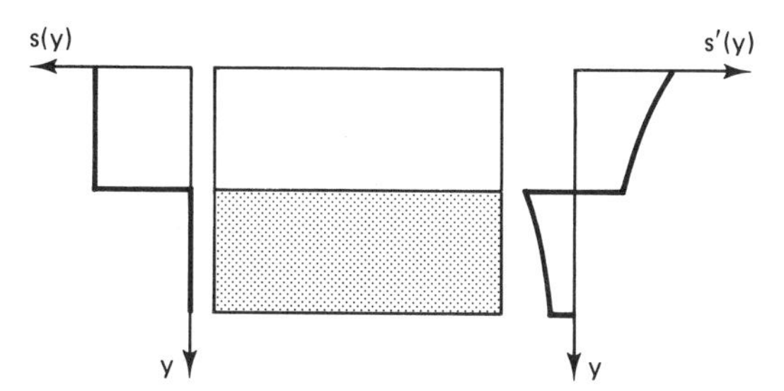

Figure 5.17 Distorted low-frequency response.

time-wave response with negligible phase distortion of square-wave response at field frequency. This, in general, requires good single-frequency response at much lower frequencies than the field frequency for most practical amplifiers. The left-hand side of the picture contains an abrupt black-and-white edge, an ideal signal for a vertical line; on the right-hand side a distorted signal is shown which will reproduce an image of distorted brightness. The top and bottom would look more white than they should and the middle more black. A typical bandwidth for a TV signal is 10 Hz to 5 MHz (for a 625-line system).

5.2.3 Standards for Monochrome TV

Only general aspects will be discussed here. Standards refer to three domains: time (signal shape), voltage, and spectral.

5.2.3.1 Video Signal. Figure 5.4 shows a video signal composed of the picture signal and horizontal sync pulses; Figure 5.7 contains vertical pulses. The sync pulses are rectangular to avoid the influence of possible nonlinearities in the transmitting system. Care is always taken to be sure that the picture signal is in a linear region of the transfer characteristics, but it can happen that sync pulses will be in the nonlinear region. If pulses are rectangular, only the amplitude can be distorted (changed). This can easily be corrected.

Standards for horizontal blanking intervals t_b (Figure 5.4) compromise two opposite demands. First, as much time as possible should be used for the picture signal (i.e., t_b should be as short as possible). Second, it is useful to have a broad blanking interval because it is easier to realize a deflection system in the receiver if it is possible for the beam retrace to be longer.

Similarly, the width of the horizontal sync pulses, t_s, in Figure 5.4, should be less than t_b because part of the blanking interval is used for other information (as in color TV) and for dc component restoration. On the other hand, the sync pulses should not be too narrow. The main part of the energy should be contained in the harmonics under 1 MHz, to be sure that any receiver can reproduce them. Also, in the presence of noise, it is easier to separate wider sync pulses from the picture signal than to separate narrow pulses.

Maximal tolerance for the line period $T_L = H$ depends on the maximal frequency in the system, f_m. One picture element (just resolvable) corresponds to one-half of the period $1/f_m$, so that

$$\Delta T_L < \frac{1}{2f_m} \tag{5.39}$$

where ΔT_L is the change in the period of the horizontal sync. For example, for European systems with $f_m = 5$ MHz (and $T_L = 64$ μs), ΔT_L should be less than 0.1 μs, or the frequency of the horizontal sync should have a stability accuracy greater than 0.16×10^{-2}.

Equation (5.39) expresses a fundamental law of communications, in general: the interchangeability of time and bandwidth. If a total number of units of information is to be transmitted over a channel in a given interval by a specified method, a specific minimum bandwidth is required. If the time available for transmission is reduced by a factor, or if the total information to be transmitted is increased by the same factor, the required bandwidth is increased by this factor. If the picture detail in both the horizontal and vertical directions is to be doubled without changing the frame rate, this represents an increase in total picture elements by a factor of 4 [2, Chap. 2].

The accuracy needed for the vertical interval is even higher. Error in interlacing in the receiver should not exceed 10%. For the European standards (25 frames, 50 fields, 20-ms period of the fields) this means that the beginning of adjacent fields can differ by approximately 3 μs, or the corresponding accuracy for the vertical interval should be greater than 0.15×10^{-3}.

Duration of the vertical blanking interval is between 18 and 20 lines (Figure 5.7) and is related to the conditions for conversion of the picture from a movie tape to a TV picture. The beginning of the vertical sync interval is delayed about 1 to 2 μs from the beginning of the vertical blanking interval. Equalizing pulses have the same duty cycle as horizontal pulses, and half of the duration of horizontal sync pulses.

Voltage levels of the video signal are shown in Figure 5.4. In the spectral domain the maximal frequency, f_m, of the video (TV) signal is specified for each TV standard.

5.2.3.2 Broadcast Video Channel. The broadcast video channel contains both video and audio signals, translated in the frequency domain. Separate carriers are used, f_0 for the video and f_{0T} for the audio signals. The difference between these two carriers is specified for each TV standard. The video signal modulates the amplitude of the carrier, f_0, while the audio signal controls the frequency of the carrier, f_{0T}.

Characteristics of the video signal transmission are deeply influenced by the modulation polarity. The modulation is said to be positive if the amplitude of the modulated signal (f_0) is greater when the brightness in the picture element is greater. Otherwise, it is negative (Figure 5.18). The impulse error visibility, synchronizing an automatic gain control, depends on the type of modulation.

Pulse errors would be reproduced in the system with negative modulation as black spots, and as white spots in the opposite system. Negative modulation

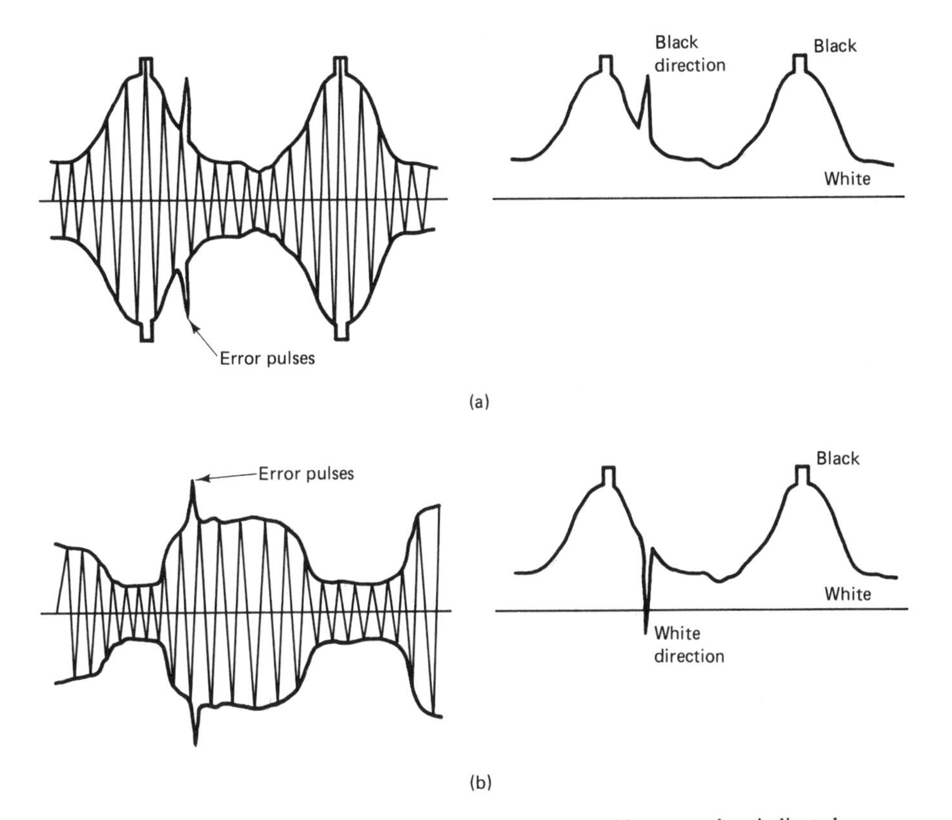

Figure 5.18 (a) Positive and (b) negative modulation with error pulses indicated.

has an advantage because the subjective visibility is higher for white than for black spots.

The influence of pulse errors is greater in systems with negative modulation, which is an advantage of other systems. This is because in negative modulation error pulses go in the same direction as sync pulses and the system can mistake them for sync pulses. Because of this, in systems with negative modulation, the more complicated synchronizing methods should be used.

Automatic gain control (AGC) is simple in a system with negative modulation. The reference signal for the AGC is obtained from the reference levels in the incoming signal. In the video signal and then in the modulated signal, reference levels are sync and black (blanking) levels. It is easier to measure higher levels, which makes negative modulation favorable. Simple peak detectors can register levels corresponding to sync pulses in this system, so it is enabled only during sync pulses. In positive modulation, both reference levels correspond to low amplitudes.

Broadcast systems for the transmission of video signals with corresponding audio signals are divided into four bands, usually referred as bands I, II, IV, and V. Windows in the frequencies are for other purposes. Table 5.2 combines

TABLE 5.2 FREQUENCY AND WAVELENGTH OF EUROPEAN BANDWIDTHS

Name	*Frequency (MHz)*	*Wavelength (m)*
Band I (VHF)	41–68	7.32–4.41
Band II (VHF)	174–223	1.72–1.35
Bands IV and V (UHF)	470–960	0.638–0.3125

the names and borders of the bandwidths in Europe. Although two carriers for the video signal, f_0, and for the audio signal, f_{0T}, are used, the channel (broadcast) is specified as a unit. First, total channel bandwidth, B_o, and distance, $f_{0T} - f_0$, are specified. For example, for the European standards (625 lines, 25 frames) $B_o = 7$ or 8 MHz and $f_{0T} - f_0 = 5.5$ MHz. For the audio channel, FM is used and frequency deviation Δf is specified. For example, for the 625-line system $\Delta f = \pm 250$ kHz, which means that the sound is of high quality.

For the video signal, VAM is used. The reason for this is obvious: to save bandwidth. Corresponding spectrum can be formed (shaped) either in the transmitter or in the receiver. Idealized filter characteristics for the first case are shown in Figure 5.19 and for the second case in Figure 5.20. Bandwidth for the audio signal is the same in all cases.

In Table 5.3 TV channels are listed for the European system with 625 lines; here $B_0 = 7$ MHz from $f_0 - 1.25$ to $f_0 + 5.75$ MHz, and $f_{0T} - f_0 = 5.5$ MHz. In the IV and V bands the channel (high-frequency) width is 8 MHz between $f_0 = -1.25$ MHz and $f_{0T} = +1.25$ MHz, where $f_{0T} - f_0 = 5.5$ MHz. The first

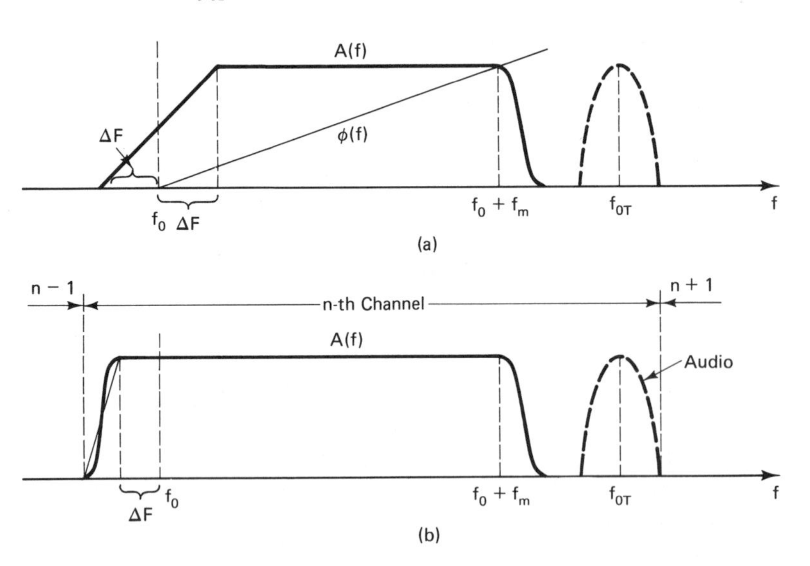

Figure 5.19 Filter characacteristics when the spectrum is formed in the transmitter: (a) filter in the transmitter; (b) filter in the receiver.

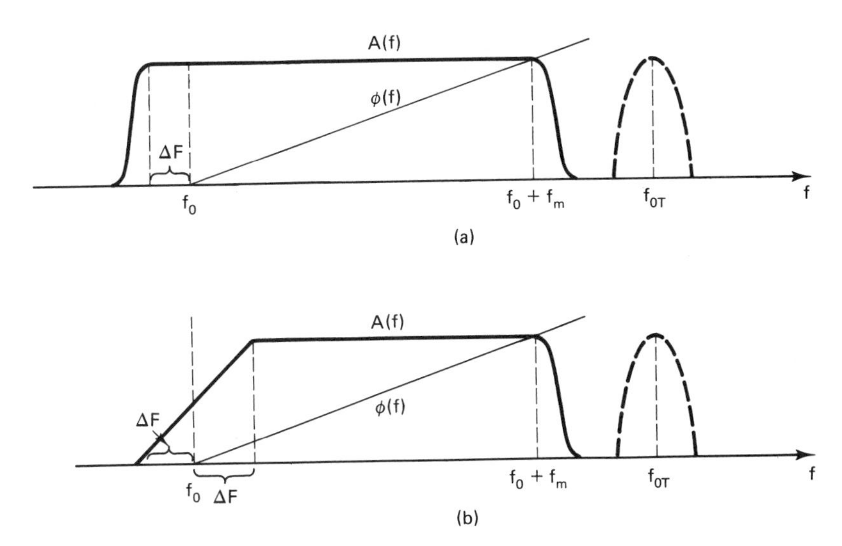

Figure 5.20 Filter characteristics when the spectrum is formed in the receiver: (a) filter in the transmitter; (b) filter in the receiver.

channel in this band, 21, is from 470 to 478 MHz, and further channels are regular; for example, channel 25 is from 504 to 512 MHz. Table 5.4 lists all of the U.S. TV VHF and UHF channels and their frequency allocations for the picture carrier, and the sound carrier. Channel 14, 470–476 MHz, represents the beginning of the UHF allocations of US TV channels. There are a total of 70 UHF channels, 14-to-83, each having a 6-MHz bandwidth between $f_0 - 1.25$ MHz and $f_{0T} + 0.25$, where $f_{0T} - f_0 = 4.5$ MHz. Channel 83 is the highest-

TABLE 5.3 EUROPEAN-SYSTEM TV CHANNELS

Band	*Channel number*	*Frequency band (MHz)*	f_0 *(MHz)*	f_{0T} *(MHz)*
I	1	41–47	Not used	—
	2	47–54	48.25	53.75
	3	54–61	55.25	60.75
	4	61–68	62.25	67.75
III	5	174–181	175.25	180.75
	6	181–188	182.25	187.75
	7	188–195	189.25	194.75
	8	195–202	196.25	201.75
	9	202–209	203.25	208.75
	10	209–216	210.25	215.75
	11	216–223	217.25	222.75
	(12)	(223–230)	(224.25)	(229.75)

TABLE 5.4 U.S. TELEVISION CHANNEL ALLOCATIONS

Band	*Channel number*	*Frequency band, MHz*	f_{oT} *(MHz)*	f_{oT} *(MHz)*
VHF	1*	—		
Low	2	54–60	55.25	59.75
Band	3	60–66	61.25	65.75
	4	66–72	67.25	71.75
	5	76–82	77.25	81.75
	6	82–88	83.25	87.75
VHF	7	174–180	175.25	179.75
High	8	180–186	181.25	185.75
Band	9	186–192	187.75	191.75
	10	192–198	193.25	197.75
	11	198–204	199.25	203.75
	12	204–210	205.25	209.75
	13	210–216	211.25	215.75
UHF	n	$(386 + 6n)–(392 + 6n)$	$387.25 + 6n$	$391.75 + 6n$
	[14–83]	(470–890)		

* The 44- to 50-MHz band was television channel 1 but is now assigned to other services.

frequency channel, 884–890 MHz. There are no gaps in the UHF band of the TV allocations. However, Channels 70 to 83 are also allocated for land mobile radio. For television, these UHF channels are used for special services. Channel 37 is not available for TV assignment.

5.3 COLOR TV SIGNAL

Color TV compromises two fields, colorimetry and engineering. The science of colorimetry is the measurement and specification of color in numerical terms, and its practical application to televised images is included here. Engineering is involved through the methods by which the color characteristics of the image are converted and packaged into the standard composite video waveform. The signal that contains information about luminance, hue, and saturation is called the *composite video signal*.

5.3.1 Basic Principles and Requirements of the Composite Signal

In black-and-white TV, the luminance signal, Y, should be transmitted (Figure 5.21a). The idealized spectral characteristic is also indicated in Figure 5.21a, and it should match the corresponding characteristic of the eye. Based on the three color properties of the eye, that is, on the theory that vision is a three-color process, to produce a picture of realistic color, adequate brightness, and good definition, a three-color system is needed, as shown in Figure 5.21b. The relative sensitivity of the three colors should be fixed so that their output voltages are

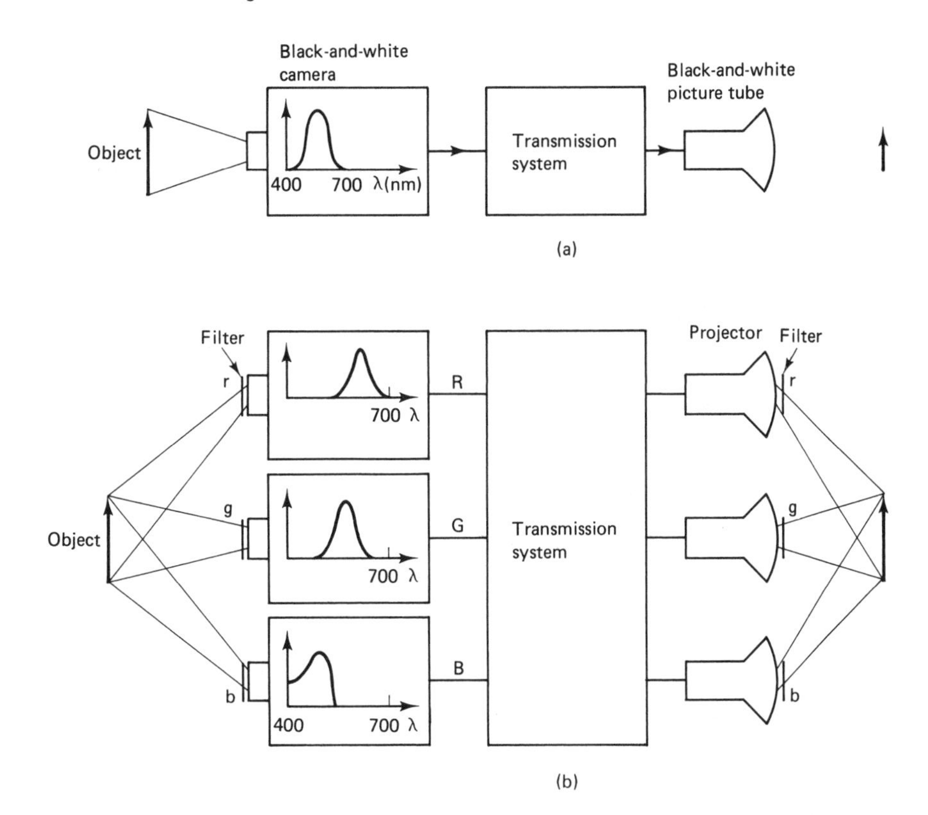

Figure 5.21 Block diagrams: (a) black-and-white system; (b) three-color system.

equal when a white surface is analyzed. To reproduce a particular color contained in the object, corresponding light of the primaries is needed:

$$A = \int \overline{a}(\lambda)I(\lambda)\rho(\lambda)\, d(\lambda); \qquad A = \text{red } (R), \text{ green } (G), \text{ or blue } (B) \qquad (5.40)$$

where $I(\lambda)$ is a continuous-spectrum function of the normalized light with which the object is lighted, $\overline{a}(\lambda)$ is a weighted coefficient (function) of the primary sources and is characterized by object reflection, and $\rho(\lambda)$ is a weighted function characterized by the eye.

To make the color system compatible, the specifications of the standard monochrome signal had to be retained. This means that such things as the channel width, the 4:3 aspect ratio, the number of scanning lines per frame, the horizontal and vertical scanning rates, and the video bandwidth had to remain the same within narrow tolerances. That is, the compatibility condition is that the color television signal must produce a normal black-and-white picture on a monochrome receiver and do so without modification of the receiver circuitry. A reverse compatibility is that a color receiver must be able to produce a black-and-white picture from a normal monochrome signal.

Transmission systems for three-color information fall under two general classifications: simultaneous and sequential systems. In *simultaneous systems,* information involving all three colors is being transmitted at all times, whereas in sequential systems the separate units of color information are transmitted on a time-share basis with alternate switching of the components of information.

Such switching may be made to occur at arbitrary assigned rates, but field-rate color switching, line-rate color switching, and picture-element-rate switching are the only ones used in practice [2, Chap. 2]. These are referred to, respectively, as *field-sequential, line-sequential,* and *dot-sequential systems.*

A preferred signal arrangement had to be developed that resulted in compatible operation within the existing monochrome framework and that made efficient use of the existing channel capacity. To achieve that, certain important features of human color vision were exploited to pack the combination of brightness information and color information within a 6- or 7-MHz channel allocation. First, the color acuity of the eye decreases as the size of the viewed object is decreased and thereby occupies a small part of the field of vision. That is, the theory that vision is a three-color process is true only when the object viewed is relatively large. On a television screen this condition would refer to those objects that are produced by video frequencies of zero to 0.5 MHz. Second, objects that occupy a very small part of the field of vision produce a noncolor sensation, only one of brightness. Therefore, no color information need be transmitted in this spatial frequency range (corresponding to video frequencies about 1.5 MHz). Objects of a size to occupy an intermediate part of the field of view (video frequencies from about 0.5 to 1.5 MHz) are perceived by a two-color vision process consisting of those colors producible by mixing only two primaries of orange and cyan. Blues and yellows are among the first colors to lose their color and become indistinguishable from gray within this range.

5.3.1.1 Three Primaries and Their Characteristics. In colorimetry, three monochrome primaries—red, green, and blue—have wavelengths of 700 nm, 546.1 nm, and 435.8 nm, respectively. If these sources are approximate in the cathode tube, red and blue lights would have very low brightness. To get one lumen of these, red or blue, many watts of electromagnetic radiation are needed. In practice, the span of the colors realized with three primaries and the luminance of the color reproduced are compromised.

According to FCC (Federal Communications Commission, USA) standards, instead of extreme red, orange-red ($x = 0.670$, $y = 0.330$) is used; instead of monochrome blue, a less saturated blue ($x = 0.140$, $y = 0.080$) is used; and instead of monochrome green, a less saturated yellow-green ($x = 0.210$, $y = 0.710$) is specified.

According to FCC standards, reference white is standard C: $x = 0.31$, $y = 0.316$, or $u = 0.201$, $v = 0.307$. Then, the unit luminances of the TV primaries (FCC) are $l_R = 0.2999 \simeq 0.30$, $l_Y = 0.587 \simeq 0.59$, and $l_B = 0.114 \simeq 0.11$. When the system reproduces reference white, $R = G = B$. Standard C is reference white for NTSC; for PAL, reference white is D_{6500}.

Reference white, that is, normalizing light in colorimetry, defines the proportion of the primaries and their equations of the luminance. Coordinates of the light emitted by phosphor in CRTs differs significantly from FCC primary conditions. The reason for this is the intention of CRT manufacturers to increase luminance in the picture reproduced. That difference introduces corresponding colorimetry distortion in the picture. Some of the new reference luminance sources are D_{5500}, cloudy day; D_{6500}, average day; and D_{7500}, sunny day.

The influence of the reference source can be seen in the next example. Consider two simplified systems such as the one shown in Figure 5.21b, the only difference being different reference light sources. If the systems are designed properly, the colors reproduced will be correct in both systems. But if the same color is transmitted through both systems, signals in the input of one group's channels are not equal to signals in the second group. Suppose now that signals in one group are changed to equalize with the second group of signals. Obviously, the colors reproduced will be different. However, if the amplification in the corresponding transmitting channels are properly changed, equal reproduced colors can be obtained. This change basically means reference color equalization, and it means that reference light can be considered as a relative specification in the corresponding channel of the system. Normalized luminance for the same color is

$$L = 0.3R + 0.59G + 0.11B \tag{5.41}$$

where R, G, and B are TV primaries.

A color camera converts colorimetric variables to electrical variables with linear correspondence if the spectral characteristics of the camera and the receiver phosphor are matched; otherwise, nonlinear crosstalk will exist.

If the output of the three color cameras is

$$E_R = kR \qquad E_G = kG \qquad E_B = kB \tag{5.42}$$

these voltages have the same properties as the quantities R, G, and B. This means that they should be equal when normalized light C is transmitted. So the information about luminance is converted in the signal E_Y given by

$$E_Y = 0.3E_R + 0.59E_G + 0.11E_B \tag{5.43}$$

This signal, E_Y, is called the *luminance signal*.

5.3.1.2 Constant Luminance Principle. All standard color television systems in the world are now based on the requirement that the luminance of a given picture area be essentially unaffected by the presence of the signal carrying the color information for that area. In compatible systems, correct representation of the gray levels in the luminance signal is enough. If any other data about luminance are transmitted through other signals, distortion will be introduced in the picture reproduced. This means that in an ideal compatible system, when white color is transmitted, the only existing signal should be E_Y. Obviously, this condition would not be satisfied if any of the E_R, E_G, or E_B signals is transmitted, because they are not equal to zero when white color is transmitted.

As a solution to this problem, a set of signals termed *color-difference signals,* designated as $E_R - E_Y$, $E_B - E_Y$, and $E_G - E_Y$, are introduced. If

$$E_Y = 0.3E_R + 0.59E_G + 0.11E_B \tag{5.44}$$

then

$$E_{R-Y} = E_R - E_Y = 0.7E_R - 0.59E_G - 0.11E_B \tag{5.45}$$

$$E_{B-Y} = E_B - E_Y = -0.3E_R - 0.59E_G + 0.89E_B \tag{5.46}$$

$$E_{G-Y} = E_G - E_Y = -0.3E_R + 0.41E_G - 0.11E_B \tag{5.47}$$

All these are valid for the NTSC system, where C is reference white.

The human eye is less sensitive to the hue and saturation changes than to the luminance changes, which means that the bandwidth for the color difference signals can be drastically reduced compared with the bandwidth of the luminance channel. Two of three color difference signals are needed because the third can be obtained from the E_Y and these two signals. In TV systems signals, E_{R-Y} and E_{B-Y} are used. Taking this combination of the signals instead of E_{R-Y} and E_{G-Y} or E_{G-Y} and E_{B-Y}, simplification in the receiver is obtained. Namely, from equations (5.45) to (5.47) it follows that

$$E_{G-Y} = -0.51E_{R-Y} - 0.19E_{B-Y} \tag{5.48}$$

$$E_{R-Y} = -1.96E_{G-Y} - 0.373E_{B-Y} \tag{5.49}$$

$$E_{B-Y} = -2.68E_{R-Y} - 5.26E_{G-Y} \tag{5.50}$$

and the second alternative would need amplification instead of attenuation. This is not surprising if we remember that characteristic of the luminance (Y) channel is much more matched with the green channel (the middle of the visible spectra) than with the characteristics of the red and blue channels. As can be seen from Table 5.4, maximal and average arithmetic values of the E_{G-Y} signal are smaller than that of the E_{R-Y} and E_{B-Y} signals. Whereas the E_R, E_G, and E_Y signals are always positive, the color difference signals can be both positive and negative and should have different zero levels. These data must be identified in the decoder.

Because, theoretically, the color difference signal contains information about dominant wavelength and saturation, that is, about chrominance characteristics of color, they are also referred to *chrominance signals*.

TABLE 5.4 COLOR SIGNALS FOR SATURATED COLORS WITH MAXIMAL AMPLITUDES

Color	E_R	E_G	E_B	E_Y	E_{R-Y}	E_{G-Y}	E_{B-Y}
Yellow	1	1	0	0.89	0.11	0.11	−0.89
Cyan	0	1	1	0.7	−0.7	0.3	0.3
Green	0	1	0	0.59	−0.59	0.41	−0.59
Magenta	1	0	1	0.41	0.59	−0.41	0.59
Red	1	0	0	0.3	0.7	−0.3	−0.3
Blue	0	0	1	0.11	−0.11	−0.11	0.89
Average arithmetic value					0.47	0.27	0.59

Basically, the reason for the constant luminance principle is the vision property that luminance fluctuations are more noticeable than color fluctuations. The subjective signal-to-noise ratio is greater if the noise in the chrominance channels is manifested as a color fluctuation only. The influence of the waveform of the luminance and chrominance components is also reduced.

5.3.1.3 Gamma Correction. First consider one example. Suppose that the exponent of the cathode-ray tube is $\gamma = 2.2$ and that the system transmits (yellow-orange) color C_1, defined as

$$C_1 = 0.9R + 0.4G + 0B \tag{5.51}$$

If the whole system except the CRT is linear, the color reproduced would be

$$C_r = (0.9)^{2.2}R + (0.4)^{2.2}G \tag{5.52}$$

The coordinates of the original color are

$$r = 0.69 \quad \text{and} \quad g = 0.31 \tag{5.53}$$

but the color reproduced has coordinates

$$r_r = 0.86 \quad \text{and} \quad g_r = 0.14 \tag{5.54}$$

Obviously, the color reproduced (red-orange) differs from the original color.

As a first approximation, a TV system regarded overall ought to be linear; the light emitted from the receiver should be directly proportional to the light falling upon the camera target plate. However, the receiver CRT does not emit light in direct proportion to the voltage applied between the grid and the cathode. This is chiefly due to the nonlinearity of the beam current versus grid voltage characteristic, rather than the light output versus beam current characteristic, which is largely linear. The luminance, L, of the screen is roughly a parabolic function of the grid-to-cathode input voltage, V_g: $L = kV_g^{2.2}$.

In general terms the nonlinear relationship between light output and voltage input may be expressed by writing $L = k(V_g)^\gamma$. In a more general sense, the words *system gamma* are used to express the relationship between the light output at the receiver and the light input at the camera (i.e., to describe the overall "light characteristic" of the entire system) [4, Chap. 2].

The gamma value of the color cinescope is standardized: $\gamma = 2.2$. This requires the reciprocal value, 0.4545, in all three channels in a TV system in order to correct for the influence of gamma. Figure 5.22 shows a plot of both the transmitter and receiver characteristics. The places in the TV chain where gamma should be corrected depend on the properties of the TV system and the eye in terms of contrast [5, Chap. 5]. The system properties depend on many factors, including the contrast properties of the analyzing (color) tube and ambient illumination.

The human eye is able to distinguish very small changes in luminance level. Figure 5.23 shows staircase changes that give a subjective impression of the approximately equal contrast (*a*) and corresponding normalized changes of the cathode-ray tube with $\gamma = 2.2$. The small changes of luminance in the darker

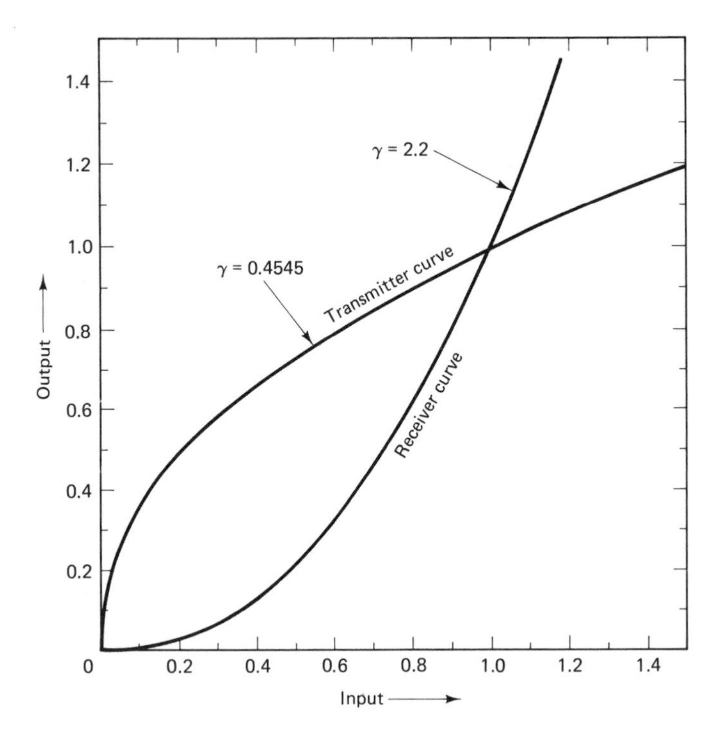

Figure 5.22 Gamma curves used in color TV.

area are as important as large changes in the lighter area. This is because the eye has a variable γ factor. For the low average light level (e.g., in movies), $\gamma = \frac{2}{3}$, and $\gamma = 1$ for high levels.

Gamma correction should be done at the beginning of the chain to correct for the characteristics of the cinescope on the end of the chain, in order to compress contrast; this is an advantage considering the noise introduced later in the chain. The noise is not changed considerably this way, but suitable redistribution is obtained to get less "visibility" of the noise. Namely, when gamma correction is performed at the beginning of the TV chain (i.e., immediately after the color camera) the noise introduced after correction (in practice, the main noise) will be uniformly distributed in the signal. In the cathode-ray tube the noise will be compressed in the lower (black) part and expanded in the higher (white) part of the transfer characteristic.

From an economic point of view, the correction at the beginning of the TV chain is also suitable. It is better to have one gamma amplifier in the transmitter than to have complicated nonlinear amplifiers in each receiver. Control of the error in the reproduced picture is simple, too.

The main drawback of this method of gamma correction after the camera is that the constant luminosity principle is not quite satisfied. The gamma correction

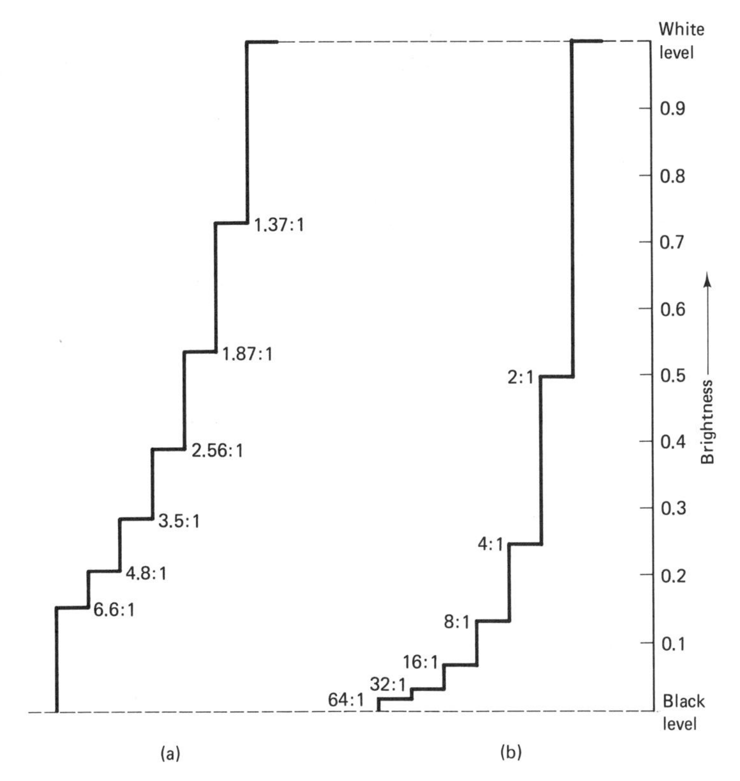

Figure 5.23 Gamma correction: (a) subjective impression of the equal contrast; (b) after taking $\gamma = 2.2$.

in present systems is performed separately for all three primaries and the luminance signal is

$$E'_Y = 0.3E'_R + 0.59E'_G + 0.11E'_B \tag{5.55}$$

where $E'_P = E_P^{1/\gamma}$, E_P is primary (P is R, G, or B), and E'_Y is a linear combination of the corrected primaries. Obviously, $E'_Y = E_Y^{1/\gamma}$ only when white color is transmitted; otherwise, $E'_Y \neq E_Y^{1/\gamma}$.

It has been found that in many modern receivers, CRTs have a gamma value on the order of 2.8 rather than 2.2. Since the normal camera gamma correction factor employed is 1/2.2, the overall system clearly has a gamma of approximately $2.8 \times 0.4545 = 1.27$; that is, the system has rising characteristics such that

$$\text{light output} = k(\text{light input})^{1.27} \tag{5.56}$$

It has been suggested that a rising gamma such as this is desirable when the ambient lighting level in which the receiver is being viewed is low, but that an overall gamma of unity is to be preferred when the ambient lighting is bright. This is in agreement with eye-γ changes mentioned previously.

5.3.1.4 Interleaving Process. One of the requirements of the composite color signal is to occupy the same bandwidth $B = f_m$ as that of monochrome (black-and-white) television. Putting a chrominance signal in the allotted channel of 6 or 7 MHz created a difficult problem, since this chrominance signal had to be transmitted along a luminance signal and had to be included without objectionable interference to the luminance signal. This was accomplished by proper placement of the chrominance signal within the band of the video frequencies and by limitation of the bandwidth of this signal.

The chrominance and luminance signals are included within the f_m (MHz) video band by an interleaving process (Figure 5.24). This process is possible because the energy of the luminance signal concentrates at specific intervals in the frequency spectrum. The spaces between these intervals are relatively devoid of energy, and the energy of the chrominance signal can be concentrated in these spaces.

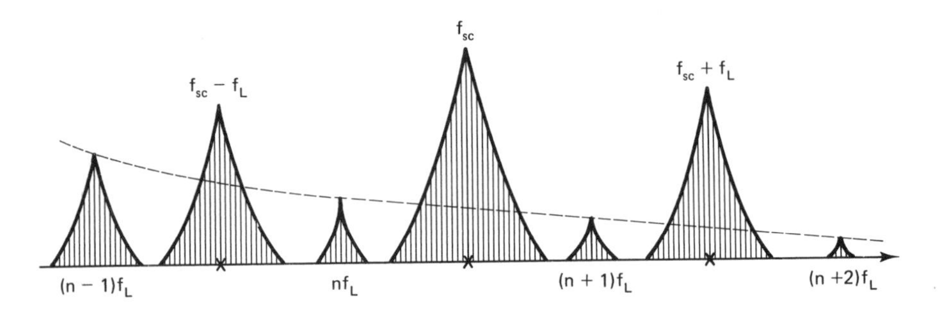

Figure 5.24 Interleaving of brightness and color signals in the frequency spectrum.

The chrominance signal is conveyed by means of a subcarrier. The frequency of this subcarrier was chosen so that its energy would interleave with the energy of the luminance signal. The energy of each of these signals is conveyed by the video carrier. The choice of the subcarrier is the result of the compromise of the two opposite conditions. First, the subcarrier frequency, f_{sc}, should be high enough in the video band that the subcarrier sidebands, when they are limited to a certain bandwidth, do not interfere with reproduction of the luminance signal by a monochrome receiver. For minimum interference, f_{sc} must be as high as possible. Second, the subcarrier frequency should be low enough that sidebands of the chrominance signals are inside the bandwidth of the luminance signal.

As already shown, the energy of the luminance signal is concentrated at frequencies that are whole multiples of the scanning rate. The color signal obviously has the same property. So to get the subcarrier to be in the middle of the two adjacent points of energy concentration, nf_L and $(n + 1)f_L$:

$$f_{sc} = (2n + 1)f_L/2 \tag{5.57}$$

that is, f_{sc} should be an odd multiple of one-half of the line frequency.

Figure 5.25 shows, in principle, overlapping of the luminance and chrominance

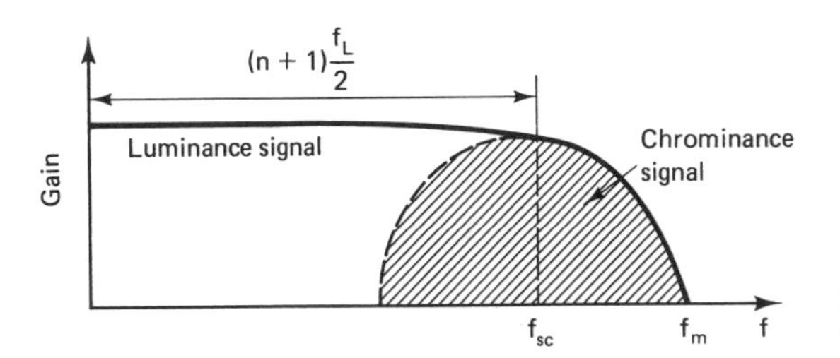

Figure 5.25 Spectra of the composed video signal (baseband).

spectra. After f_{sc} is modulated, the spectrum of the chrominance signal is not necessarily symmetrical around f_{sc}, but it can be.

5.3.2 NTSC System

The basic principles and requirements given up to now are common for all TV systems in color. But methods of subcarrier modulation with color signals are different and will be given separately for the NTSC, PAL, and SECAM systems.

The National Television System Committee (NTSC) is a group formed in the United States in 1940 by the Radio Manufacturers' Association [whose name was subsequently changed to the Radio Electronic, and Television Manufacturers' Association (RETMA)] to formulate standards for monochrome TV broadcast services. In 1950 the NTSC was reformed to investigate proposed color TV systems and recommend a standard system for U.S. broadcast television.

In the monochrome version of the NTSC TV system, a frame contains 525 scan lines, the frame time is $\frac{1}{30}$ s, and each field scan period is $\frac{1}{60}$ s. The choice of the 60-Hz frame rate is motivated by the 60-Hz ac power lines employed in the United States, to avoid bit frequencies between frame and ac frequencies. The video signal linear bandwidth is 4.2 MHz measured to the point where the response falls by 2 dB. The sound carrier is 4.5 MHz higher than the video carrier.

5.3.2.1 Choice of Chrominance Subcarrier Frequency. The choice of a suitable frequency for the color subcarrier represents a compromise between several opposing considerations. The color subcarrier frequency, f_{sc}, was chosen to be an odd multiple of one-half the horizontal line frequency:

$$f_{sc} = (2n - 1)f_L/2 \tag{5.58}$$

in order to get the interlaced spectrum.

The choice of the subcarrier is strongly related to the problem of its visibility. If this signal comes to the CRT, it will appear in the shape of small black-and-white points. Visibility of these points should be as small as possible. Based on the eye property, the visibility of these spots will be reduced if the same picture element (pixel) appears as black in one picture and white in the next. This will be satisfied if the frame period contains odd numbers of one-half of the subcarrier period (one black or white dot corresponds to $T_{sc}/2$):

$$T_F = (2m - 1)T_{sc}/2 \tag{5.59}$$

or

$$f_{sc} = (2m - 1)f_F/2 \tag{5.60}$$

Also, the visibility of the "subcarrier spots" will be reduced if the spots form a chess pattern. This will be obtained if one line period contains an odd number of $T_{sc}/2$, that is

$$T_L = (2n - 1)T_{sc}/2 \tag{5.61}$$

which is equivalent to equation (5.58).

Because $f_L = N_L f_F = 525 f_F$ (N_L is an odd number in all systems), condition (5.61) includes condition (5.59): the product of two odd numbers is odd. With equation (5.58), the optimal f_{sc} is defined.

Thus, in an interlaced system, the phase of the subcarrier alternates in succeeding lines by 180° and four fields are required for picture completion. In addition, the chrominance subcarrier, by definition, becomes zero when no color exists and only shades of gray are to be reproduced via the luminance channel.

The visibility of the dots in a black-and-white picture is small. A notch filter can be used to reduce the f_{sc} component, but the filter is efficient in the large constant-luminance area. The filter influence at the transition states is small because they are connected to the sidebands and are out of the reduced bandwidth of the filter.

The crosstalk of the luminance components changes the color, and the visibility of these changes is small because of the eye persistency. Because of the relation of the f_{sc} and f_L given by equation (5.58), one particular luminance component will, in two adjacent pictures, produce complementary colors which will be greatly compensated by the eye integration. Also, the eye is less sensitive to the color changes than to the luminance signal. The influence of the crosstalk of the luminance depends on the picture content because the luminance component close to the f_{sc} will be detected as a low video frequency, what will boost visibility.

Let's summarize:

1. For minimum interference f_{sc} must be as high as possible.
2. Having specified a bandwidth for the color-difference signal which modulates the subcarrier, f_{sc} must be sufficiently below 4.2 MHz to allow in the upper sideband of the chrominance signal.

In the NTSC standard, f_{sc} is specified, to satisfy both conditions, as

$$f_{sc} = 455 \times f_L/2 \tag{5.62}$$

The frequency $f_L/2$ is obtained by dividing f_{sc}, and the number 455 is suitable because it can be expressed as the product $455 = 13 \times 7 \times 5$, which is suitable for realization of the divider.

For 525 lines per picture and 30 pictures per second, the line frequency is 15,750 Hz, and f_{sc} would be

$$f_{sc} = 3.583125 \text{ MHz} \tag{5.63}$$

This 3.583125-MHz frequency was not adopted for the color transmission standards because of interference from the beat frequency between f_{sc} and the sound carrier, 4.5 MHz. This interference can be made less objectionable if its frequency can somehow be made an odd multiple of one-half of the line-scanning rate, in other words, if the principle of frequency interleaving is applied to it in the same way that it was applied to the choice of color subcarrier.

It was impractical to change the 4.5-MHz frequency, which was already accepted as standard. This made it necessary to select a slightly different frequency for the color subcarrier than the tentative value given; consequently, a slightly different line and field frequency had to be used to retain their frequency relationship to the subcarrier frequency. Using a 15.750-Hz line rate as a basis for determining this multiple, it was found that the 286th harmonic would be at a frequency of 4.5045 MHz. The new line frequency can be computed to have 4.5 MHz as the 286th harmonic.

$$f_L = \frac{4.5 \times 10^6}{286} = 15{,}734.264 \text{ Hz} \tag{5.64}$$

and the FCC standard is $f_L = 15{,}734{,}264 \pm 0.047$ Hz.

The new frame frequency is

$$f_F = f_L/525 = 29.97 \text{ Hz} \tag{5.65}$$

and the new field frequency is 59.94 Hz. Now the subcarrier frequency, f_{sc}, becomes

$$f_{sc} = 455 f_L/2 = 3.579545 \text{ MHz} \tag{5.66}$$

The new scanning frequencies used for color transmission are slightly below the nominal values used in monochrome receivers; however, the changes amount to less than the 1% tolerance allowed and the new frequencies will fulfill the requirements for compatibility in monochrome reception. For the purpose of maintaining close synchronization of color receivers, the tolerance for subcarrier frequency is set at $\pm 0.0003\%$ of about ± 10 Hz, and the rate of change cannot be more than $\frac{1}{10}$ Hz/s.

5.3.2.2 Chrominance Signal: *I* and *Q* Channels. As we saw, the luminance, or monochrome, signal is formed by the addition of specific proportions of the red, green, and blue signals. The red, green, and blue components are tailored in proportion to the standard luminosity curve for the particular values of dominant wavelength represented by the three color primaries chosen for color TV. Therefore, the Y signal has voltage values representative of the brightness sensation of the human eye. The voltage outputs from the three camera views are adjusted to be equal when a scene reference white or natural gray object is being seen, regardless of the color temperature of the scene ambient. However, the colorimetric values have been determined by assuming that the reference white of D_{6500} will always be used as the reproducer. Thus the color values of the reproducer will appear as if the scene had been illuminated with D_{6500} white light.

Signals generating the chrominance information, hue and saturation, are generated by subtracting the luminance signal from the red, green, and blue signals, respectively. In NTSC color standards, the chrominance signals–color difference values are encoded on a common subcarrier in the high-frequency portion of the video domain by simultaneously modulating the phase of the subcarrier to represent the instantaneous hue value, and modulating the amplitude of the subcarrier relative to that of the brightness component to represent color saturation (quadrature modulation). The hue and saturation information can be carried without loss of identity provided that proper timing in a phase scene is maintained between the encoding and subsequent encoding process. This is accomplished by transmitting a reference signal, burst, consisting of eight or nine cycles following each horizontal synchronizing pulse during the blanking interval (Figure 5.26). Quadrature synchronous detection is used to identify the individual color-signal components of the chrominance subcarrier. When individually recombined with the luminance signal in a linear resistive matrix, these color signal components recreate the desired red, green, and blue signals.

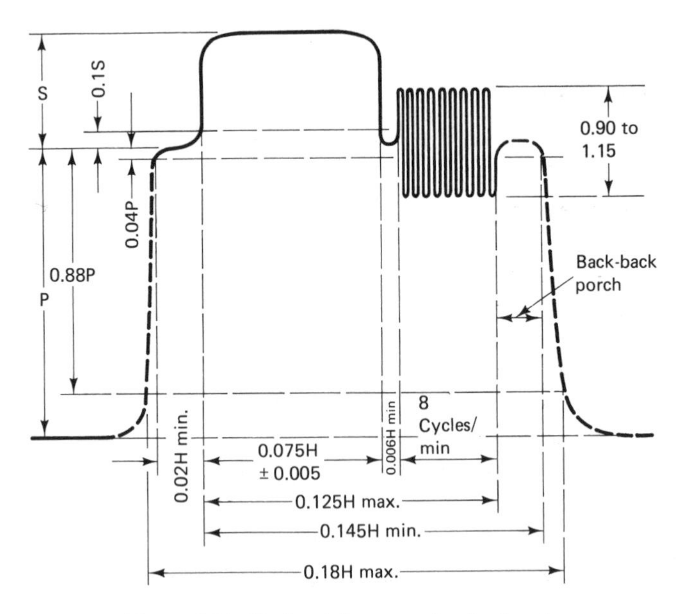

Figure 5.26 Horizontal sync and color burst.

At the transmitter (Figure 5.27) the color-difference signal is initially formed by linear matrix combinations of the red, green, and blue signals. This signal is bandlimited: to about 0.6 MHz in the case of the color-difference signal, designated as Q, containing the green–purple color-axis information; and to about 1.5 MHz in the case of the I color-difference signal, representing the orange-cyan color-axis information. These two signals are then used individually to modulate the color subcarrier in two balanced modulators maintained in phase quadrature

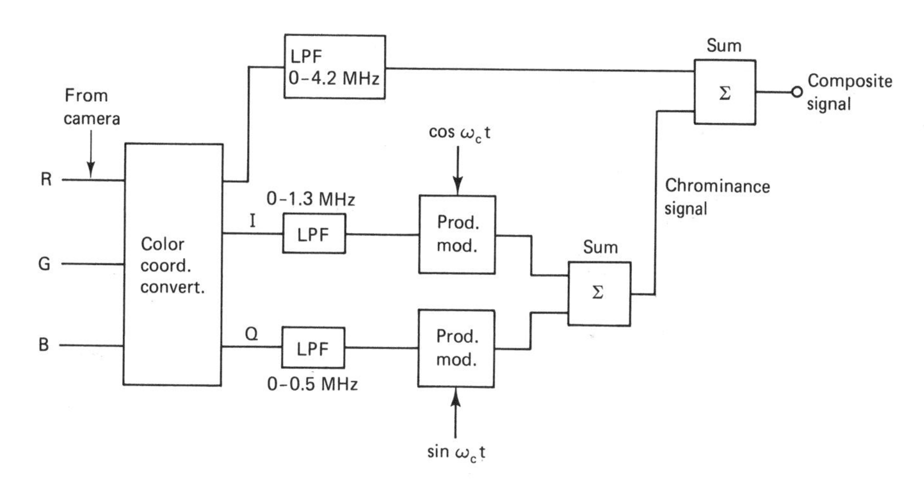

Figure 5.27 Block diagram of the NTSC composite color video signal coding process.

(QAM). The "sum" products are selected and combined linearly to form a composite chromaticity subcarrier whose instantaneous phase represents the hue of the scene at the moment, and whose amplitude—relative to the brightness signal amplitude—is a measure of saturation. This signal is then added to the luminance signal together with the appropriate horizontal and vertical synchronizing and blanking signals including the color-synchronizing burst. The result is the composite color video signal.

It is seen that the picture information in the signal may be divided into three groups (Figure 5.28):

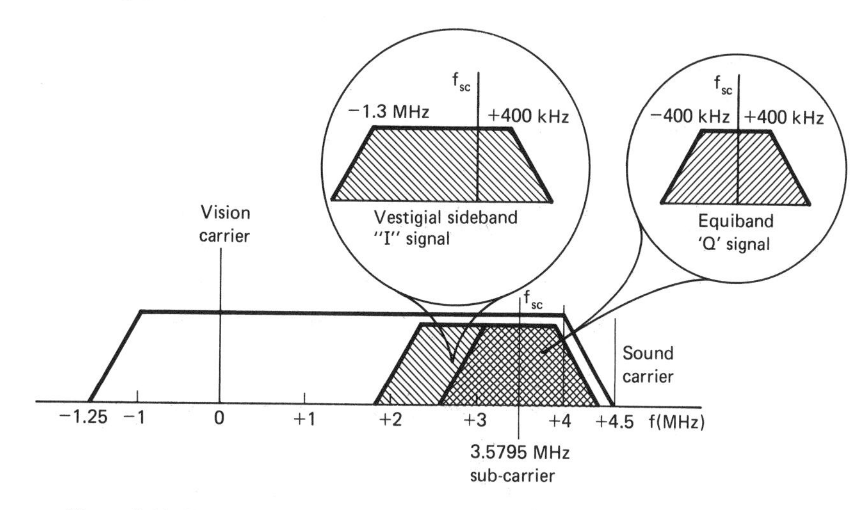

Figure 5.28 Frequency components of the U.S. 525-line NTSC color TV signal—broadcast channel.

1. Chrominance signal Q component, with bandwidth 400 kHz (to the -2 dB point)
2. Chrominance signal I component, with bandwidth 1.3 MHz (to the -2 dB point)
3. Luminance information, with bandwidth 4.2 MHz (to the -2 dB point)

The chrominance signals I and Q are formed by (Figure 5.29)

$$E_I = E_U \cos 33° - E_V \sin 33° \tag{5.67}$$

$$E_Q = E_U \cos 33° + E_V \sin 33° \tag{5.68}$$

where E_U and E_V are obtained by normalizing in order to limit the maximum excursion of the composite color TV signal to the arbitrary value of 1.33 the times the excursion of the monochrome TV signal (more details are given in Section 5.3.2.3):

$$E_U = \frac{E_R - E_Y}{1.44} = 0.877(E_R - E_Y) \tag{5.69}$$

$$E_V = \frac{E_B - E_Y}{2.03} = 0.493(E_B - E_Y) \tag{5.70}$$

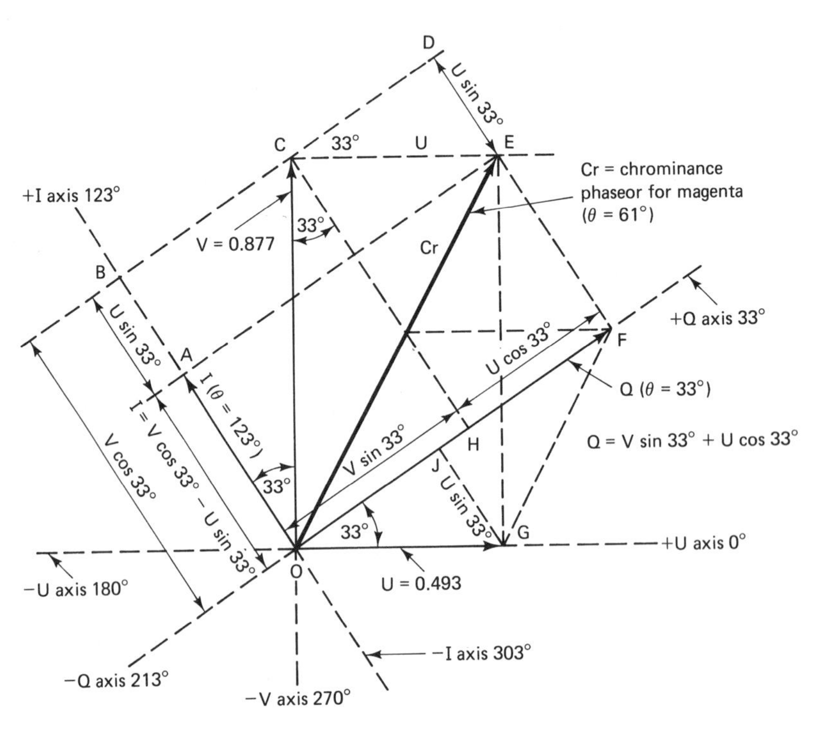

Figure 5.29 Axes of the I and Q signals relative to the U and V axes.

Alternatively, the Y, I, and Q signals can be directly related to the R, G, B camera signals by

$$E_Y = 0.299E_R + 0.587E_G + 0.114E_B \tag{5.71}$$

$$E_I = 0.596E_R - 0.274E_G - 0.322E_B \tag{5.72}$$

$$E_Q = 0.211E_R + 0.523E_G + 0.312E_B \tag{5.73}$$

Inversely, we have

$$E_R = 1E_Y + 0.956E_I + 0.621E_Q \tag{5.74}$$

$$E_G = 1E_Y - 0.272E_I - 0.647E_Q \tag{5.75}$$

$$E_B = 1E_Y - 1.106E_I + 1.703E_Q \tag{5.76}$$

The choice of the I and Q axes (Figure 5.30) is based on the experiment. Better reproduction of color is achieved along the I and Q axes than along the U and V axes. This is particularly true in the reproduction of flesh tones, since they lie along the axes. It was also found that for small areas of color which are well centered in the field of vision, the chromacity diagram degenerates to

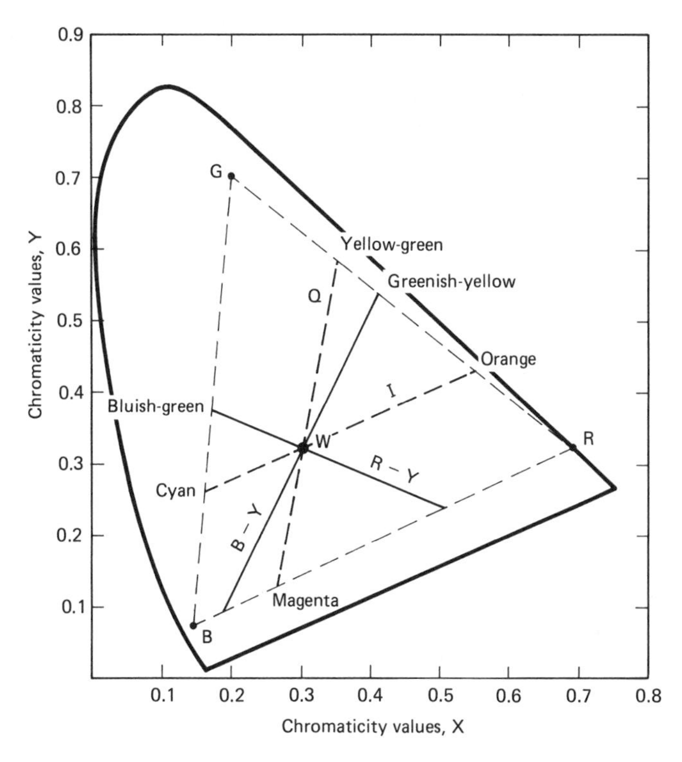

Figure 5.30 Axes of the color-difference signals and the I and Q signals on the color triangle.

a single line. This line is the I-axis; and only two fully saturated colors, orange and cyan, are needed to reproduce colors under these conditions. Actually, the main reason for I- and Q-axis use lies in the need for asymmetrical transmission (in NTSC). Because of this, one axis is placed in the direction of the maximal resolution and the second in the direction of the minimal resolution (for this signal only 0.5 MHz is used).

Figure 5.31 shows the partial block diagram of a color transmitter, with more details than in Figure 5.27. It should be pointed out that signals from the color camera are gamma corrected by passing them through gamma amplifiers. No block for this is shown in Figure 5.31, since gamma correction is also provided in the monochrome transmitter. The output of the other section, where luminance and chrominance signals are combined, is specified by NTSC standards as follows:

$$E_M = E_Y + E_Q \sin(\omega t + 33°) + E_I \cos(\omega t + 33°) \tag{5.77}$$

This can be expressed as

$$E_M = E_Y + E_C \sin(\omega t + 33° + \phi_C) \tag{5.78}$$

where

$$E_C = \sqrt{E_Q^2 + E_I^2} \qquad \tan \phi_C = E_I/E_Q$$

Equation 5.78 clearly shows that the subcarrier is both amplitude and phase modulated. Also, when there is no color information ($E_{R-Y} = E_{B-Y} = 0$), the subcarrier is not included in the color picture signal.

The sync blanking and color burst signals are added to the color picture signal; then the composite color signal is ready for transmission and the vestigial amplitude modulation is used on the vision carrier.

5.3.2.3 Reduction of the Chrominance Signal Amplitude. A block diagram showing the unit used at the studio (transmitter) to form the luminance signal and the color-difference signal is shown in Figure 5.32. As stated earlier, the proportions of the primaries used to form the luminance signal are obtained experimentally, based on the fact that the human eye is most sensitive to green, less sensitive to red, and least sensitive to blue. If three projectors, red, green, and blue, are adjusted so that their light outputs are equal as measured by photoelectric means, white light will be produced when these outputs are superimposed on each other. When they are separated, the green light will appear to the average observer almost twice as bright as the red and from five to six times as bright as the blue. The red light will appear from two to three times as bright as the blue light, which will appear the dimmest.

In Figure 5.32, I and Q signals are formed from the reduced $R-Y$ and $B-Y$ signals, as in the real system. To justify the necessity for these waiting factors and to calculate them, consider the case when the composite video signal, E'_M, is formed without reductions:

$$E'_M = E'_Y + (E'_R - E'_Y) \cos \omega_c t + (E'_B - E'_Y) \sin \omega_c t \tag{5.79}$$

Basically, signal E'_M is a luminent signal with the signal superimposed on the

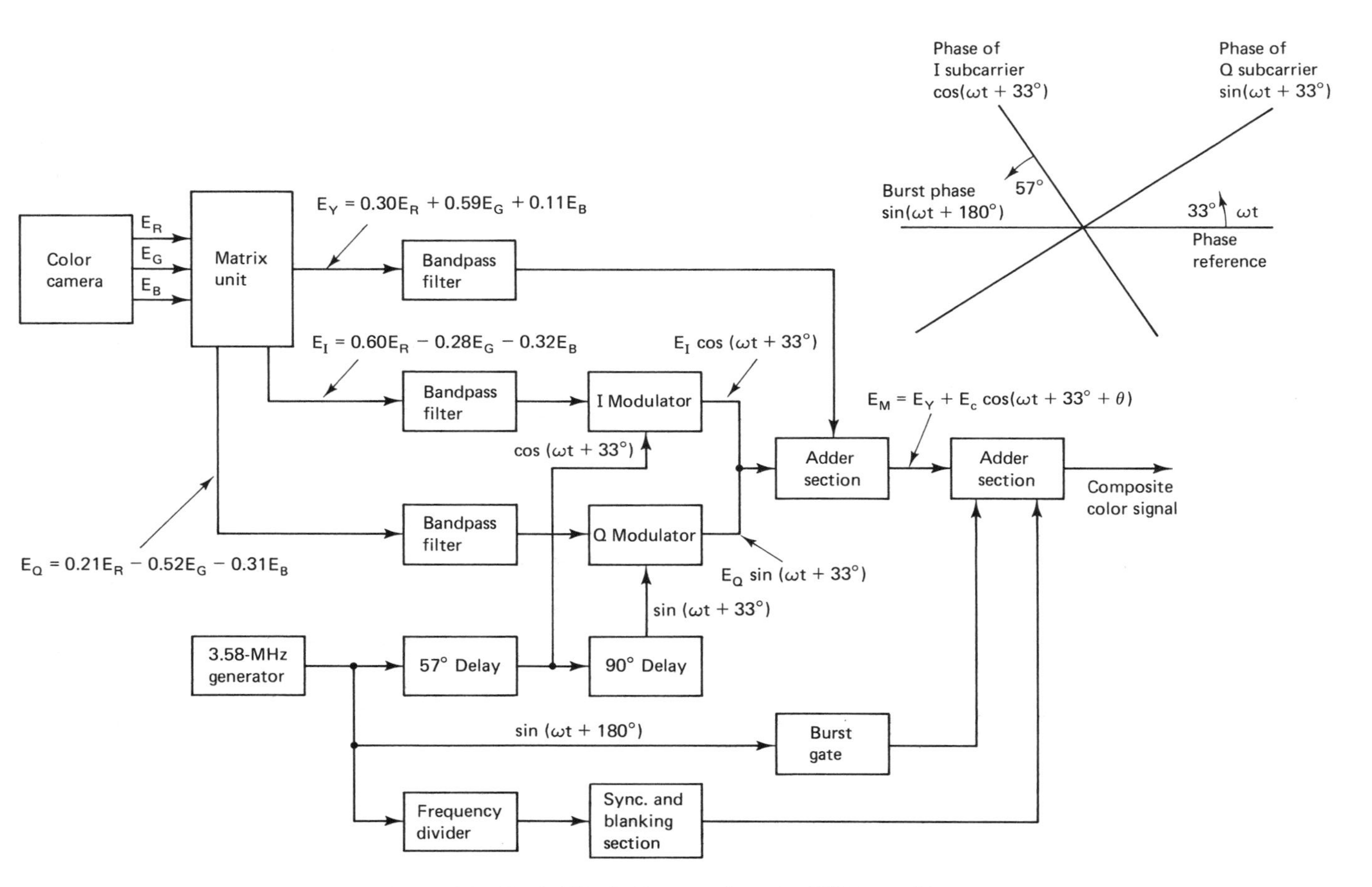

Figure 5.31 Partial block diagram of a color TV transmitter.

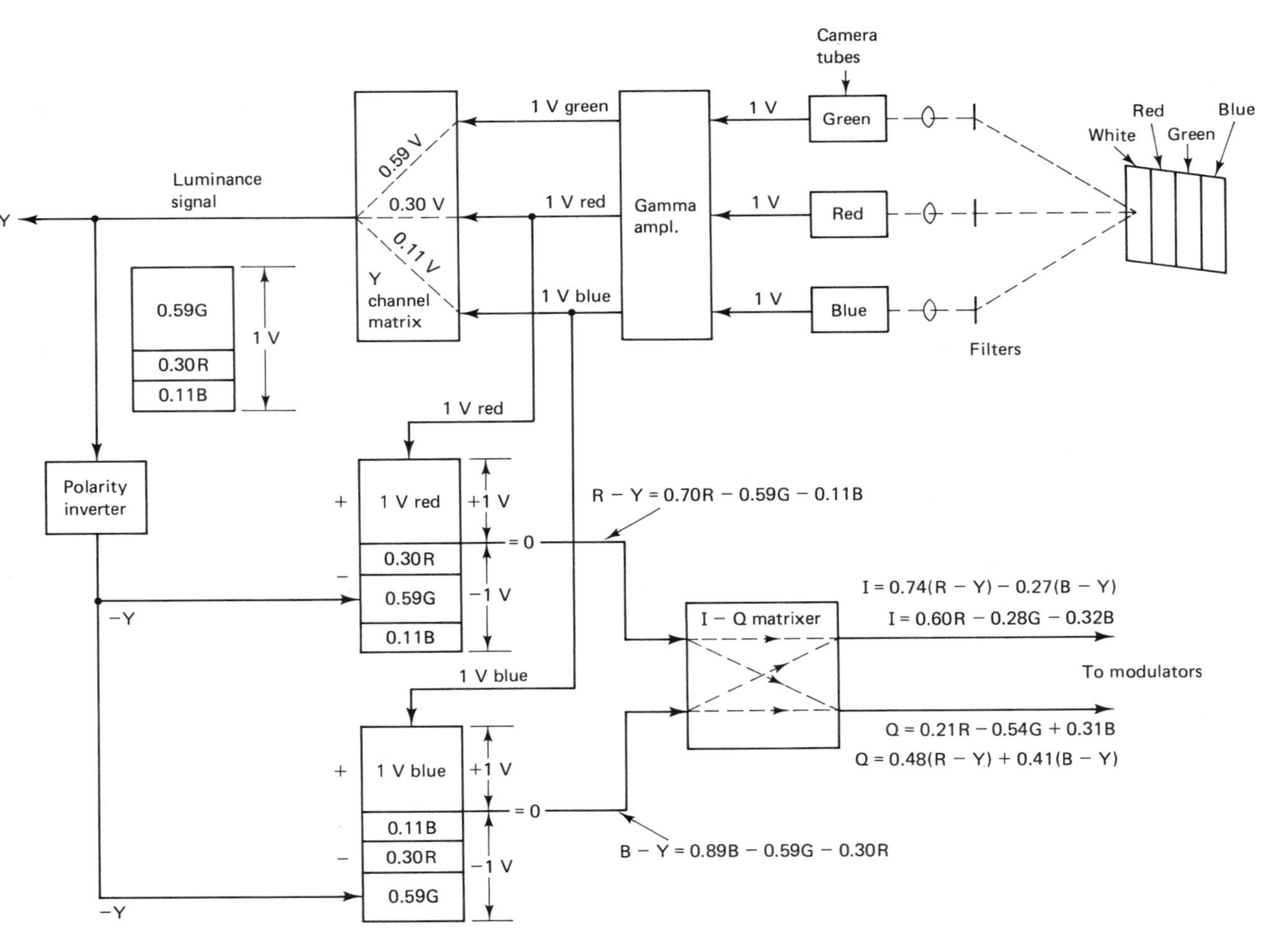

Figure 5.32 Color signal formation.

subcarrier frequency. Thus variations of the composite signal are larger (in amplitude) than corresponding variations of the luminance signal. To state this increase in variations of the luminance (black and white) signal, the variations of the saturated colors should be stated. For this purpose *color-bar signals* are used. A study of color-bar waveforms is a great help toward gaining a clear understanding of the nature of color TV signals.

A standard color-bar waveform produces on a color receiver screen eight vertical bars of uniform width. This includes the three primaries, the three complementaries (these six colors are close to the edges of the color triangle), white, and black. They are arranged in decreasing order of luminance from left to right:

white–yellow–cyan–green–magenta–red–blue–black

The luminance steps produced on a monochrome set are not uniform, and the gray strips on the screen grow progressively darker from left to right.

In practice, color-bar signals are generated entirely electronically, and this signal represents coded information about light and gives rise to the appropriate light output at the receiver. Since electronically generated signals are designed to test the performance of the system, this imposes stringent demands, and in fact the saturation and amplitude of the colors simulated by the generator are seldom reached in normal operation [4, Chap. 3]. However, electronically generated signals are, at present, more frequently used (e.g., for special effects), and a 133% overshoot is becoming a more common phenomenon.

Figure 5.33 shows the color-bar patterns when a color-bar chart is being scanned. The corresponding components of the composite signal, as well as the composite signal, are shown in Figure 5.34 for a one-line interval. It is clear that the particular video waveform shown in Figure 5.34d is unacceptable because it could cause severe overmodulation. To overcome this it is necessary to restrict the peak amplitude of the chrominance signal so that it does not take the vision carrier beyond certain prescribed limits. It is found experimentally that the maximum excursion of the vision signal for short intervals must not exceed 0.33 beyond peak white or black levels. This may be said to represent 33% overmodulation during the active line, and it confines the peak excursion of the chrominance signal to the limits -0.33 and $+1.33$.

Figure 5.34d shows that the highest amplitude of the composite signal in the white direction is for the transmission of yellow, and in the black direction for its complement blue. Overmodulation is the same for both (complementary) colors. The waiting factor obtained for these colors (approximately 0.5) would give optimal amplitudes for them. But from this point of view, the system has two degrees of freedom: signals $R' - Y'$ and $B' - Y'$ can be regulated independently. So two color bars can be chosen to have the optimal amplitude. These must not be colors that are complementary to one another because the equations formed from such a pair are numerically equal and cannot be used. Red and blue can

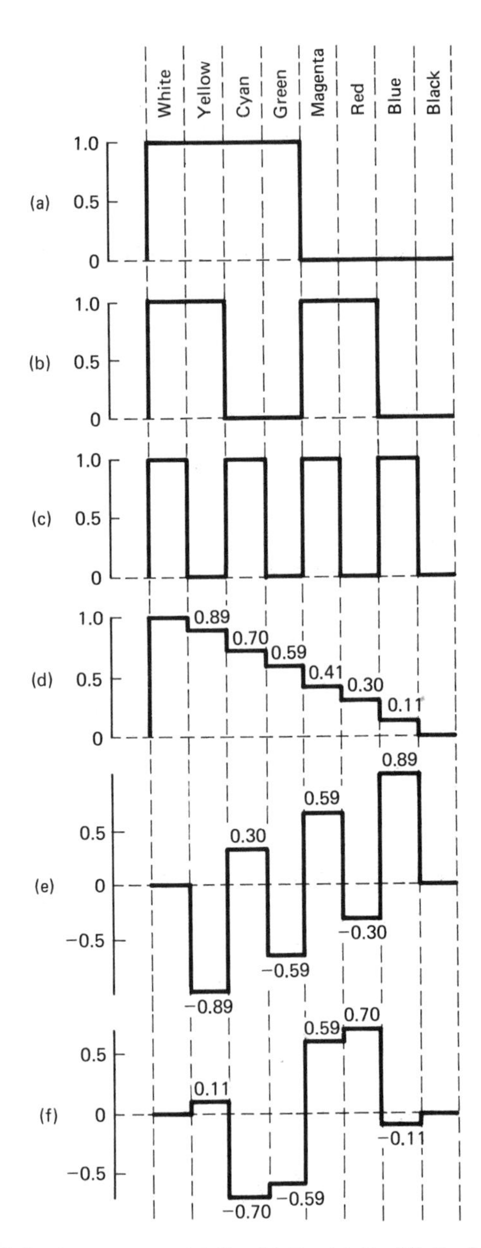

Figure 5.33 Color-bar patterns: (a) E_G; (b) E_R; (c) E_B; (d) E_Y; (e) $E_{B\text{-}Y}$; (f) $E_{R\text{-}Y}$.

be chosen (or yellow and cyan). Then

$$E'_R - E'_Y = a(0.70E'_R - 0.59E'_G - 0.11E'_B) \tag{5.80}$$

$$E'_B - E'_Y = b(-0.30E'_R - 0.59E'_G + 0.89E'_B) \tag{5.81}$$

where a and b are two reduction factors to be determined.

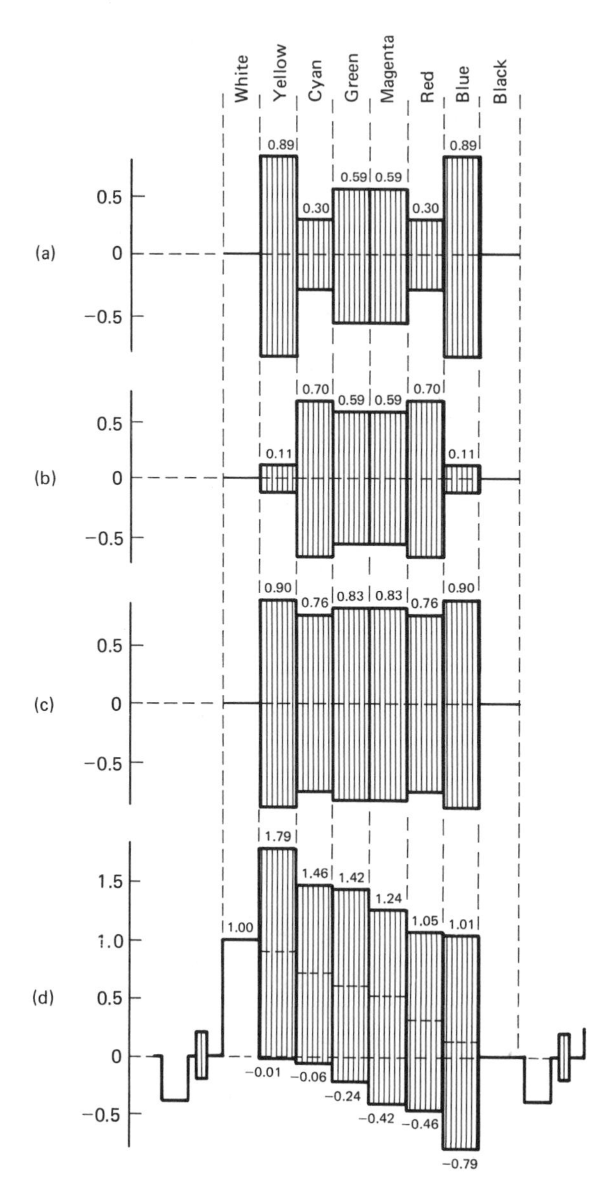

Figure 5.34 Composite color signal pattern for color-bar analysis: (a) E_{B-Y} sin $\omega_c t$; (b) E_{R-Y} cos $\omega_c t$; (c) E_C sin $(\omega_c t + \phi_c)$; (d) E_M.

The maximum permitted amplitudes for the chrominance signal on red and blue are 0.63 (red) and 0.44 (blue). It can be seen that for red, the chrominance signal would be $E'_{R-Y} = 0.70a$ and $E'_{B-Y} = -0.30b$, and for blue, the chrominance signal would be $E_{R-Y} = -0.11a$ and $E_{B-Y} = 0.89b$. The values of a and b can be found by solving two equations:

$$(0.701a)^2 + (0.299b)^2 = (0.632)^2 \tag{5.82}$$

$$(0.115a)^2 + (0.885b)^2 = (0.448)^2 \tag{5.83}$$

what gives $a = 0.877$ and $b = 0.493$ as solutions. In these equations, three decimals are used because that is what is used in practice. A composite video signal with reduced amplitudes for the color bar is shown in Figure 5.35.

Using data from Figure 5.33 and equations for Y', $B' - Y'$, and $R' - Y'$, the Table 5.5 is formed. Based on the information in Table 5.5, the phasor diagram of the chrominance (carrier) signal for the primary and complementary colors is drawn (Figure 5.36). It can be seen that each complementary hue is diametrically opposite to its associated primary, and the amplitudes of these opposite pairs of phasors are equal. This is logical since they are of equal saturation, in this case 100%, and their algebraic sum must be zero since together they make white. Projection from a given phasor to the U and V axes shows the relative amplitudes of the weighted $B' - Y'$ and $R' - Y'$ subcarrier signals, which together form the chrominance signal. The chrominance signal phase angles for the three primaries are: red, 103°; green, 241°; and blue, 347°. Phasors (arrows) for other saturated colors (maximal saturation, maximal normalized amplitude) would be at the solid lines.

The chrominance signal amplitude during color scanning is not necessarily a measure of the saturation of that color. The relative saturation of the color is

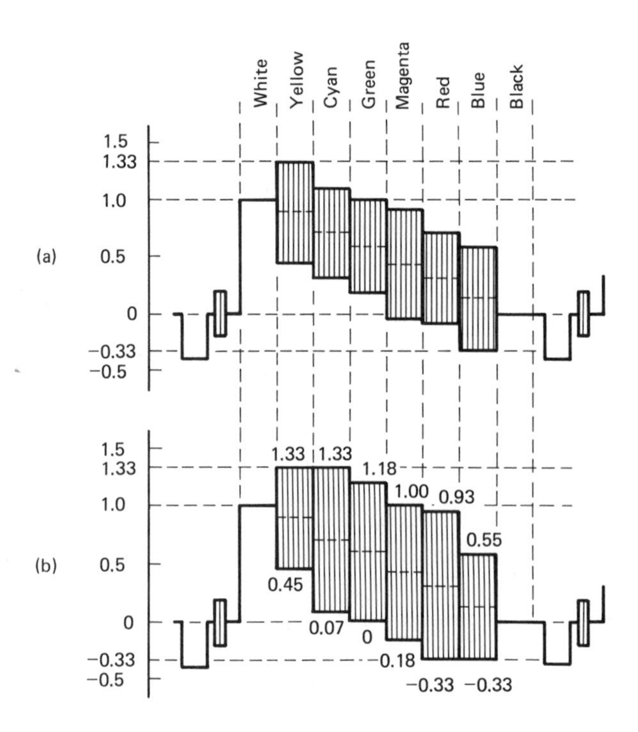

Figure 5.35 Composite video signal after reduction.

TABLE 5.5 SATURATED COLOR-BAR SIGNAL COMPONENTS: 100% SATURATION, 100% AMPLITUDE[a]

Color bar	Y'	$B' - Y'$	$R' - Y'$	$U = 0.493\,(B' - Y')$	$V = 0.877\,(R' - Y')$	*Chrominance amplitude* $= U^2 + V^2$	*Chrominance phase angle (NTSC line)*	*Ratio,* C_h/Y'
White	1.0	0	0	0	0	0	—	—
Yellow	0.89	−0.89	+0.11	−0.4388	0.0965	0.44	167	0.5:1
Cyan	0.7	+0.3	−0.7	+0.1479	−0.6139	0.63	283	0.9:1
Green	0.59	−0.59	−0.59	−0.2909	−0.5174	0.59	241	1:1
Magenta	0.41	+0.59	+0.59	+0.2909	+0.5174	0.59	61	1.44:1
Red	0.3	−0.3	+0.7	−0.1479	+0.6139	0.63	103	2.1:1
Blue	0.11	+0.89	−0.11	+0.4388	−0.0965	0.44	347	4.05:1
Black	0	0	0	0	0	0	—	—

[a] Color-difference signal weighting factors are employed.

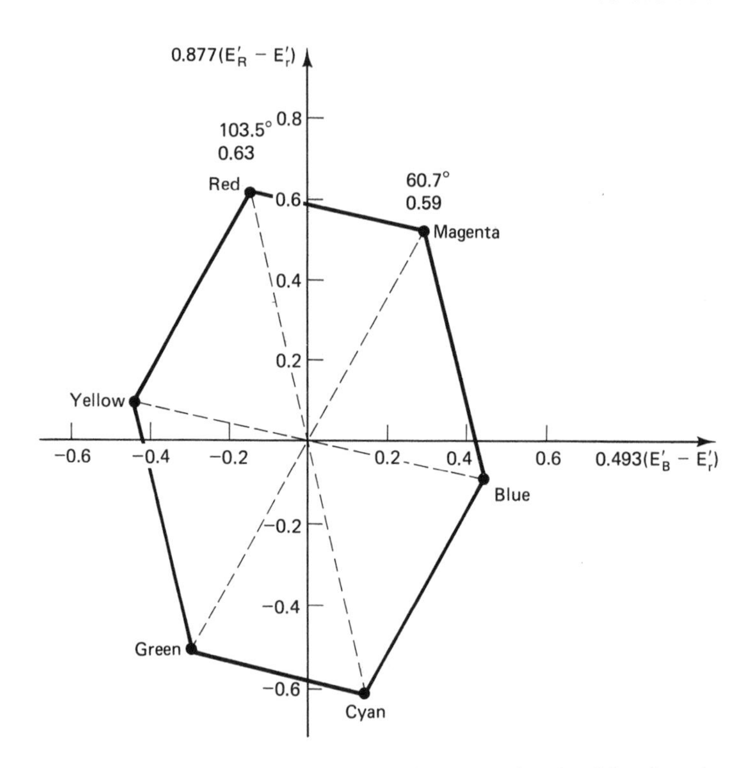

Figure 5.36 Phasor diagram of the chrominance carrier signal for the primary and complementary colors.

conveyed by the ratio between the amplitudes of the chrominance and the luminance signals. The more highly saturated the color is, the higher the ratio becomes. Moreover, the ratio remains fixed for a color with a given saturation regardless of the brightness of the color. The chrominance vector changes in phase whenever there is a change in hue, and it changes in length in accordance with changes in brightness or saturation.

Any color that is less than fully saturated contains white light. Since white light is produced by a combination of equal values of three primaries, a desaturated color may be defined as a mixture of a pure color and its complementary color. A color that has a saturation of 75% is produced when three-fourths of the total light is contributed by the primary color and the other one-fourth is contributed by the complementary color.

Table 5.6 contains details of a 95% saturated, 100% amplitude color-bar signal. The figures in brackets show the phase angles on PAL lines (which will be discussed later). $U = 0.493\ (B' - Y')$, $V = 0.877\ (R' - Y')$. Note the rise in luminance level for the less saturated colors. For example, the introduction of 5% white raises the luminance level from its value of 0.3 on 100% saturated red to 0.48 on 90% saturated red. This would be expected for a "whiter" signal. The phase angles for primary and complementary hues remain the same as for

TABLE 5.6 COLOR-BAR SIGNAL COMPONENTS: 95% SATURATION, 100% AMPLITUDE

Color bar	*R*	*G*	*B*	*R'*	*G'*	*B'*	*Y'*	*B' − Y'*	*R' − Y'*	*U*	*V*	*Chrominance signal*		*Ratio,*
												Amplitude	*Phase angle*	C_h/Y'
White	1.0	1.0	1.0	1.0	1.0	1.0	0	0	0	0	0	0	—	—
Yellow	1.0	1.0	0.05	1.0	1.0	0.25	0.92	−0.67	+0.08	−0.33	+0.07	0.34	167 (193)	0.37
Cyan	0.05	1.0	1.0	0.25	1.0	1.0	0.78	+0.23	−0.53	+0.11	−0.46	0.47	283 (77)	0.6
Green	0.05	1.0	0.05	0.25	1.0	0.25	0.69	−0.44	−0.44	−0.22	−0.39	0.45	241 (119)	0.65
Magenta	1.0	0.05	1.0	1.0	0.25	1.0	0.56	+0.44	+0.44	+0.22	+0.39	0.45	61 (299)	0.8
Red	1.0	0.05	0.05	1.0	0.25	0.25	0.48	−0.23	+0.52	−0.11	+0.46	0.47	103 (257)	0.98
Blue	0.05	0.05	1.0	0.25	0.25	1.0	0.33	+0.67	−0.08	+0.33	−0.07	0.34	347 (13)	1.03
Black	0	0	0	0	0	0	0	0	0	0	0	0	—	—

the 100% saturated, 100% amplitude weighted signal. It can be shown that for an arbitrary color, not primaries and not complementaries, when signals E'_R, E'_G, and E'_B, are not equal to each other, the ratio of the chrominance signals $E'_R - E'_Y$ and $E'_B - E'_Y$ is not constant and is changing when saturation is changing; this means that the phase of the carrier is also changing with saturation.

Note that the very small voltages at the output of the camera are substantially increased by gamma correction. For example, if $E_R = 0.05$ V $= 50$ mV, $E'_R = 0.05^{1/2.2} = 0.25$ V $= 250$ mV. Thus very small amounts of desaturation have a marked effect on all subsequent signal amplitudes. In other words, all (small) errors in the color balance (in the R, G, and B channels) are boosted (in amplitude) after gamma correction.

5.3.2.4 Decoding and Phase Distortion. At the NTSC receiver, the decoding process is essentially the inverse of the encoding function. Figure 5.37 shows a typical receiver functional diagram. The luminance signal in the video domain (baseband) is fed to all three of the reproduction cinescope guns at the appropriate levels to produce a monochrome picture having a D_{6500} color temperature. The chrominance signal is separated by a bandpass filter and impressed on the input of the synchronous demodulators. These demodulators extract the appropriate color-difference signals at the desired phase angle, as determined by a locally generated subcarrier reference signal locked to the incoming "burst signal." As shown in Figure 5.37, the chrominance signal is resolved directly into the weighted $R' - Y'$ and $B' - Y'$ components, that is, into V and U. The synchronous de-

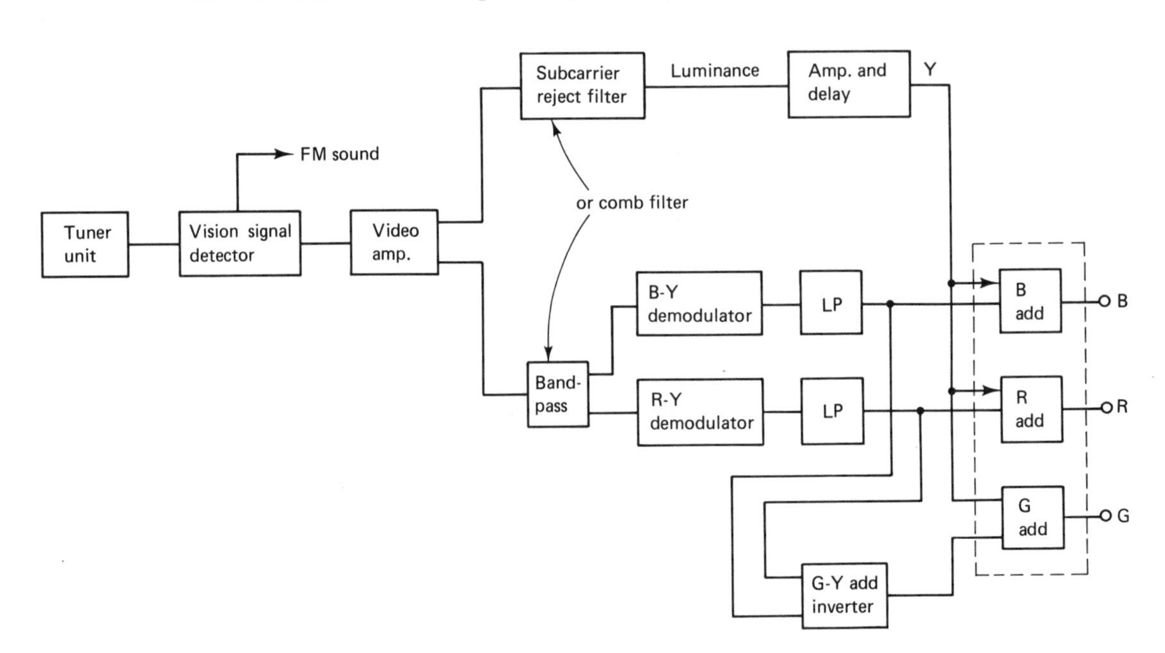

Figure 5.37 Basic NTSC receiver arrangement.

modulators receive subcarriers at 90° and 0°. The U and V components are now treated as equivalent signals having bandwidth equal to that of the Q signal. There are thus fewer finer details in the orange-cyan hues. Such receivers are sometimes referred to as *narrowband chrominance receivers,* whereas those using demodulated I and Q are known as *broadband chrominance receivers*. Because of their relative simplicity, the former are in the majority [4, p. 307], in home TVs. With a technology advancement, the number of comb filters in use is growing.

The overall gain of the chrominance channel determines the reproduced-color saturation (ratio of chrominance to luminance) and the overall phase adjustment of the decoding reference signal provides control of the average hue of the scene reproduced. Delay introduced in the luminance channel compensates for the short delay that is introduced in the narrowband filter within the chrominance amplifier circuit path, so that at the output (of the adders) all signals have the same delay. Also present is a "notch filter" to remove the chrominance signal.

Although the luminance signal is zero during the returning interval, the color synchronizing signal can be visible. It can be shown that reproduced luminance (negative values do not contribute to the luminance) is smallest if the reference signal is shifted 180° from the U-axis. This is why burst is at the 180° axis (Figure 5.29). If this sync signal becomes visible, it will be reproduced as green color with low luminance.

To extract the I and Q color-difference signals from the chrominance signal, it is necessary to demodulate along the I and Q axes: the reference subcarrier inputs to the I and Q synchronous demodulators must have phase equal to that of the suppressed carriers at the transmitter, 123° and 33°, respectively (Figure 5.38). When this is done, the full transmitted information available by matrixing I and Q for the color-difference signals $R' - Y'$, $B' - Y'$, and $G' - Y'$ is recovered:

$$\begin{aligned} R' - Y' &= 0.96I + 0.62Q \\ G' - Y' &= -0.28I - 0.64Q \\ B' - Y' &= -1.1I + 1.7Q \end{aligned} \tag{5.84}$$

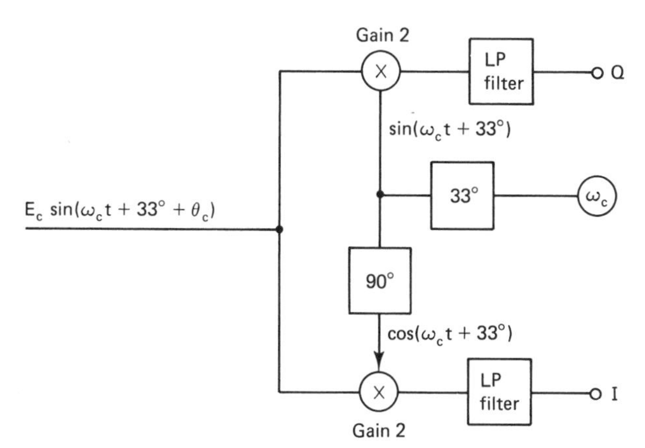

Figure 5.38 I and Q synchronous demodulators.

Consider now the I and Q synchronous demodulator (Figure 5.38). The chrominance signal is

$$E_Q \sin(\omega_c t + 33°) + E_I \cos(\omega_c t + 33°) = E_C \sin(\omega_c t + 33° + \theta) \quad (5.85)$$

where

$$E_Q = E_C \cos\theta \qquad E_I = E_C \sin\theta$$

thus

$$E_C = \sqrt{E_Q^2 + E_I^2} \qquad \tan\theta = E_I/E_Q \quad (5.86)$$

For the sake of simplicity, it is assumed that the gain of the product detectors is 2. The output of the upper product detector, the Q channel, is

$$E_C [\cos\theta - \cos(2\omega_c t + 66° + \theta_s)] \quad (5.87)$$

and the output of the low-pass filter is $E_C \cos\theta = E_Q$; and similarly for the I channel.

This is one possible approach (with a filter for $2\omega_c$); the balance detector, which immediately suppresses the $2\omega_c$ harmonic, can be used, for example.

Suppose now that there is a phase error $\Delta\theta$ between the reference signal and the chrominance signal. The output of the Q channel would be $E_C \cos(\theta + \Delta\theta)$, and the output of the I channel would be $E_C \sin(\theta + \Delta\theta)$. The phase difference between the subcarrier during the active period and the reference signal (burst), θ, defines a hue. If a false change, $\Delta\theta$, occurs at the receiver, the wrong color (hue) will be reproduced. The phase change caused by an amplitude change, either a luminance or a chrominance signal, is called distortion of the differentiated phase. This specific problem forced the development of PAL and SECAM systems. Phase error will exist if phase is changing depending on the luminance signal amplitude. In general, phase changing of $\pm 5°$ will be noticeable, but this visibility is dependent on the picture content. Phase error is introduced if the delay through some circuit in the chain is dependent on the frequency or the signal amplitude. If the amplitude characteristic of some circuits is not even, and the phase is not odd around the subcarrier, undesired FM will be introduced. The relative phase of the reference signal will change instantly but not the mean value. This type of linear distortion will cause crosstalk between chrominance signals, but only on the edge (abrupt change in the signal), in the color areas.

5.3.3 PAL System

The phase alternating line (PAL) system is closely related to the NTSC system. The one component is reversed every other line and simple normalized color difference signals, U and V, with equal-frequency bandwidths are used in place of I and Q. The corresponding chromacity errors tend to be averaged by the viewer; this modulation makes the system less sensitive to subcarrier phase error and minimizes quadrature crosstalk. The receiver is somewhat more complicated, however, and thus more expensive. Throughout Europe the usual pulse system scanning standard is 625 lines at a 50-Hz frame rate with a 2:1 interlace. In South

American countries, which have adopted PAL, the frame rate is 60 Hz, to conform to the local power frequency.

5.3.3.1 Choice of Chrominance Subcarrier Frequency. The principles as to the choice of the chrominance subcarrier frequency are the same for the PAL system as for the NTSC system. In the original NTSC system the chrominance subcarrier was made to fit exactly between two line harmonics, called *half-line offset*. The subcarrier frequency is equal to an odd multiple of half the line frequency; in the United States, $f_{sc} = 455 \times f_L/2$. For the sake of comparison, interleaved spectra of the composite video signal around f_c are shown in Figure 5.39a.

The spectral components of the corresponding signals U and V (corresponding to I and Q in the NTSC system) in the PAL system are not in the same place

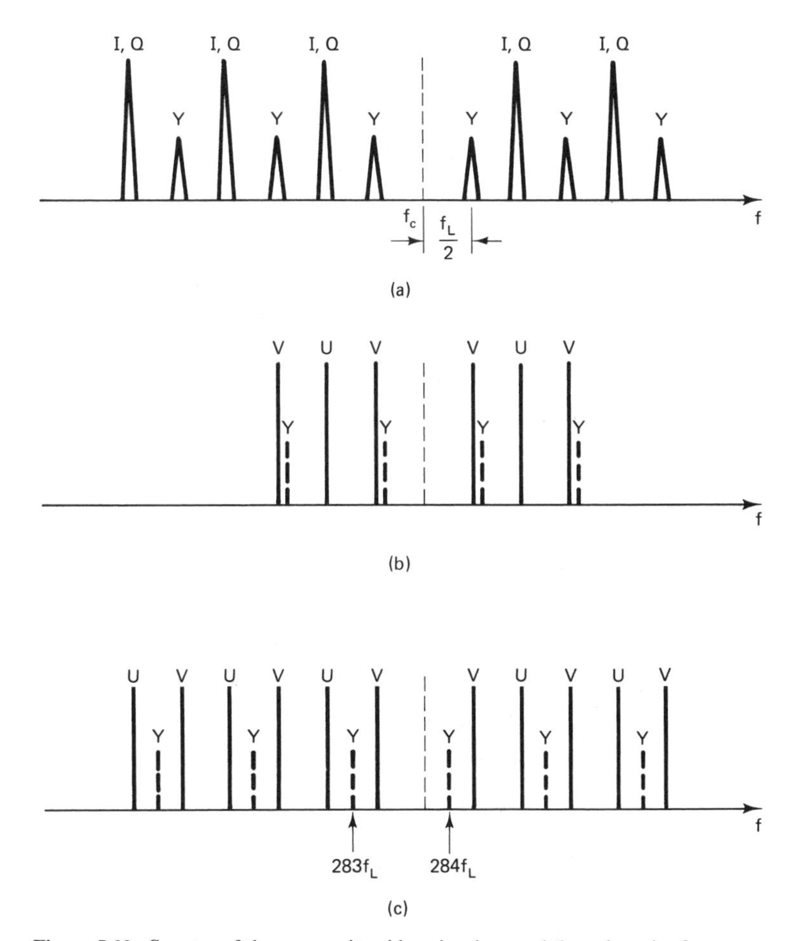

Figure 5.39 Spectra of the composite video signal around the subcarrier frequency: (a) NTSC system; (b) PAL system with half-line offset; (c) PAL system.

on the frequency axis; their basic Fourier components are not identical. By alternating the subcarrier in the V modulator, phase modulation of the subcarrier is introduced. The interleaved spectrum of the composite video signal around f_{sc} is shown in Figure 5.39b for the case in which the principles of "half-line offset" are used. The U components are placed equivalent to the positions of the I and Q components in the NTSC system, while the V and Y components would be on the same place. Thus the visibility of the V components in the luminance signal would be great. Also, the visibility of the dot patterns in the PAL system would not be reduced by half-line offset, because of V-component inversion. To overcome this, *quarter-line offset* is used:

$$f_{sc} = (n - \tfrac{1}{4})f_L \tag{5.88}$$

For optimum results this is modified slightly by the addition of f_F (frame frequency) to create a phase reversal on each successive field, and the actual relationship between f_c and f_L in the PAL system (Figure 5.39c) is

$$f_{sc} = (n - \tfrac{1}{4})f_L + f_F \tag{5.89}$$

The choice of n is related to the synchronizing generator realization. In the 625-line PAL system with $f_L = 15{,}625$ Hz, $f_F = 25$, and $n = 284$,

$$f_{sc} = (284 - \tfrac{1}{4}) \times 15{,}625 + 25 = 4.43361875 \text{ MHz} \tag{5.90}$$

The stability of this frequency according to standards is ± 1 Hz.

It should be mentioned that the dot patterns are of concern only in monochrome receivers working on a color signal, and in color receivers on the colored edges, when $+V$ and $-V$ are not equal (the assumption about PAL is not satisfied!). With the solution taken, the dot pattern will move down the screen diagonally left or right depending on the color content in the picture.

Thus the characteristic frequencies of the PAL system are as follows:

Field (vertical) frequency: $f_V = 50$ Hz
Line (horizontal) frequency: $f_H = 15{,}625$ Hz $= 312.5f_v$
PAL switching frequency: $f_P = 7812.5$ Hz $= 156.25f_v$
Color subcarrier frequency: $f_{sc} = 4{,}433{,}618.75$ Hz $= (1135/4 + 1/625)f_H = 283.75f_H + \tfrac{1}{2}f_v$

The resulting repetition relationship between these four signals is most notable around the field synchronizing period of the composite video signal. From the above-mentioned frequency relationships, it is clear that:

1. The relationship between V and H pulses is repeated every two fields, that is, every 625 lines, or one complete video picture or frame.
2. The relationship between V and P pulses is repeated every four fields, that is, every two successive sets of 625 lines or one complete odd- and even-frame pair. This gives a repetition rate of 12.5 Hz, often referred to as the *PAL four-field sequence*.
3. The repetition structure that results when the color subcarrier is added to

the above-mentioned signal triplet is a little more complex. The formula $f_{sc} = (1135/4 + 1/625)f_H$ shows that one line of the video signal has a number of subcarrier periods, m, where

$$m = 283.75 + \frac{1}{625} = \frac{1135 \times 625 + 4}{4 \times 625} \tag{5.91}$$

A complete frame therefore has $N = 625m$ subcarrier periods, and

$$N = 625m = \frac{1135 \times 625 + 4}{4} \tag{5.92}$$

In order that the subcarrier will have the same phase at the start of any two frames, it is clearly necessary that there be an integral number of periods between these two instants. As the product 1135×625 is not divisible by 4, it is seen that the shortest period that results in this condition is one of four frames. This is called the *eight-field sequence* and it is repeated 6.25 times per second.

5.3.3.2 Basic PAL Coding Process. In PAL simple weighted,

$$E_U = 0.493(E'_B - E'_Y) \qquad \text{and} \qquad E_V = 0.877(E'_R - E'_Y) \tag{5.93}$$

color-difference signals of equal bandwidth are used to modulate the two-quadrature phase, but equal-frequency (f_{sc}) subcarriers are used to form the two components which when added yield the chrominance signal. The composite video signal is formed as

$$E_M = E'_Y + E_U \sin \omega_{sc}t \pm E_V \cos \omega_{sc}t = E'_Y + F_U \pm F_V \tag{5.94}$$

where ω_{sc} is the subcarrier angular frequency. Therefore, in this system, the phase of one chrominance signal component is changing 180° from line to line. Although, in principle, it makes no difference which component of the subcarrier signal changes the phase in the PAL system, the F_V component phase is changing because a chrominance signal is less reduced and its maximal amplitude is less. If the F_U component would change phase, the influence of nonsymmetry in the circuits on the quality of the picture reproduced would be greater.

In lines where the phase of the F_V component is not changed, the coding process is identical in the PAL system to that in the NTSC system, and these lines are called *NTSC lines*. The composite signal for these lines is $E_M = E'_Y + F_U + F_V$. In these lines with inverted F_V component composite signal is $E_M = E'_Y + F_U - F_V$ and these lines are called *PAL lines*.

Figure 5.40 shows a vector diagram of the subcarrier phase and amplitude ratios for 100% saturated colors, and maximal amplitude for odd and even lines. Only primary and complementary colors are shown. A reference signal, burst, for the inverted and noninverted signals is also shown.

A PAL coder is similar to an NTSC coder. The main differences are in the formatting of U and V signals instead of I and Q, and in phase alteration of the subcarrier for the V component. The basic organization of a PAL coder is shown in Figure 5.41. The inputs to the matrix are gamma corrected R', G',

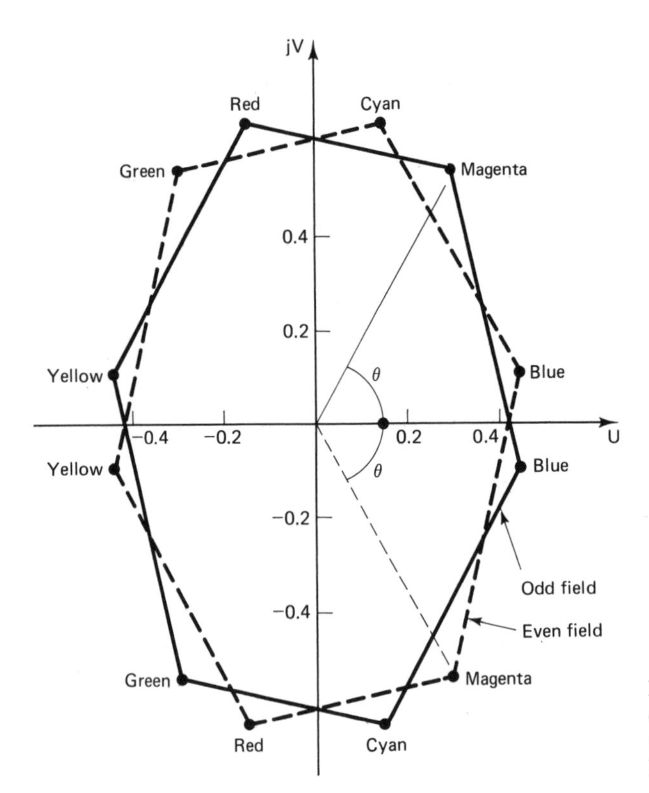

Figure 5.40 Vector diagram of the subcarrier phase and amplitude for 100% saturated colors for the PAL system. Dashed lines: inverted phase (PAL lines).

and B' and outputs Y', $R' - Y'$, and $B' - Y'$. Because the color-difference signals are bandwidth restricted by filters to the range 0 to 1 MHz, they suffer a small delay relative to the Y' signal, which has a broader bandwidth of 5 or 5.5 MHz. To compensate for that, a delay is inserted into the Y' path.

The $R' - Y'$ and $B' - Y'$ signals pass to balanced (product) modulators together with subcarrier inputs at appropriate phases. In the subcarrier path to the V modulator there is a phase-switching circuit which passes to the modulator a subcarrier having a phase of 90° on one line but 270° on alternate lines. Since one switching cycle takes two lines, the square-wave switching signal to the phase switch is of half-line frequency: approximately 7.8 kHz for the 625-line system.

The double-sideband suppressed carrier signals from the modulators are added to yield the QAM chrominance signal. Then the harmonics of the subcarrier frequency being removed from the chrominance signal are combined with the luminance and blanking signals to form a composite video signal.

In the NTSC system, the I and Q signals are formed from the reduced $R' - Y'$ and $B' - Y'$ signals before modulation of the subcarriers. In the PAL system the reduction factors are obtained by adjusting amplification in the $R - Y$ path and in add circuits, which combine the modulator outputs. The weighting

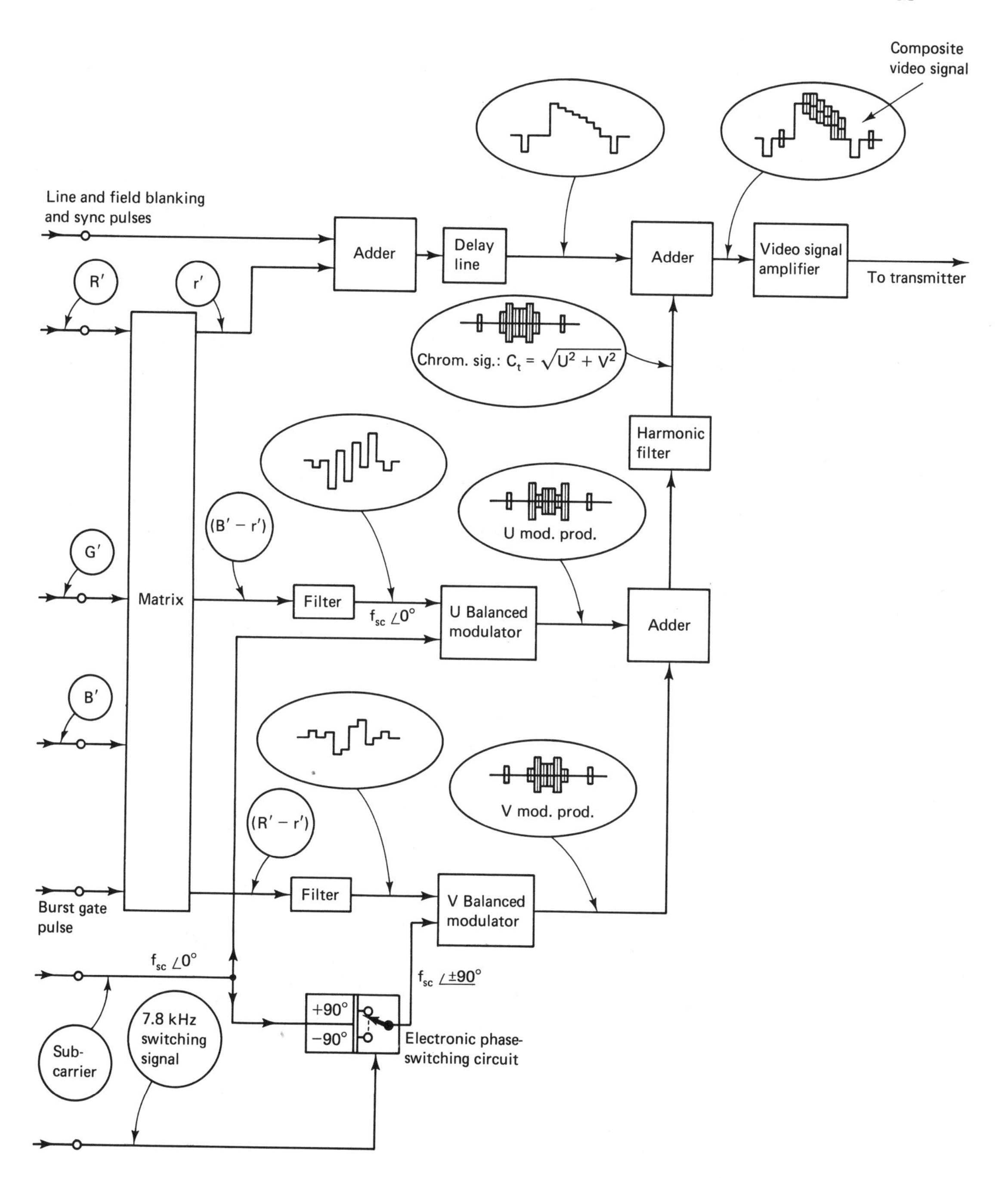

Figure 5.41 Simplified block diagram of the PAL coder.

operations may be carried out in the main "matrix," which will then deliver color difference signals U and V. The end product is the same: the chrominance signal subcarrier frequency and signal wave components have amplitudes in the ratio $U{:}V$.

Since the chrominance signal consists of two suppressed carrier signals, a replacement carrier has to be generated at the receiver. For the purpose of leaking the frequency and phase of the reinserted subcarrier generated by the receiver's "reference oscillator," a burst of subcarrier is transmitted during the back porch in the line blanking period.

This consists of 10 ± 1 cycles. In an NTSC signal the burst phase is 180°. With PAL, however, the burst phase angle is switched on a line-by-line basis from 135° on NTSC lines to 225° on the alternate PAL lines. This is done to allow the receiver to identify NTSC and PAL lines. In the PAL signal, since the burst switches in phase by ±45° line by line (about 180°), its average phase is still 180°, which is the same as that in the original NTSC system. As discussed earlier, this phase angle was chosen to give minimum interference during the returning interval in the color receiver.

In Figure 5.42 the video waveform during the field blanking period is shown. The burst is gated out of the transmitted signal during this interval. To make the burst phase start and stop in identical directions at the beginning and end of each field, in the PAL signal the burst blanking phase is shifted as shown (shifted forward for one line in every field, during four fields, and then returned to the beginning value—all in continuous sequences called *meander* of the PAL blanking), so that it follows a four-field cycle. This is called *Bruh blanking* or *PAL blanking*. If the burst blanking pulse were not staggered in this way but repeated at regular intervals of $312\frac{1}{2}$ lines (about $\frac{1}{50}$ s), the burst phases at the end of each field and at the start of the next would be one direction for fields 1 and 4, but the opposite for fields 2 and 3. Bruh blanking identifies each field in the row of four, but some lines carrying picture information are without information about burst. This does not interfere with the color receiver, but can interfere when exact synchronization with the burst or pulses is required.

In Figure 5.41 one of the inputs to the matrix is labeled the "burst gate pulse." This is a square type of pulse at line frequency, timed to appear during the back-porch period. For its duration the matrix delivers to the two modulators inputs such that the $B' - Y'$ modulator yields a subcarrier burst at 180° while the $R' - Y'$ circuit gives a burst of the same amplitude but having a phase 90° on one line and 270° on the next. The other yields an output which is the vector sum of the two inputs. It delivers a subcarrier frequency and signal wave at 135° on one line and 225° on the next.

5.3.3.3 Basic PAL Decoding Process. In the PAL system the phase of the V modulation produce (F_V) is inverted line by line simply by inverting the polarity of the one modulating component. Reordering of the chrominance data is not so simple. The effect of inverting the V component on PAL lines is to switch

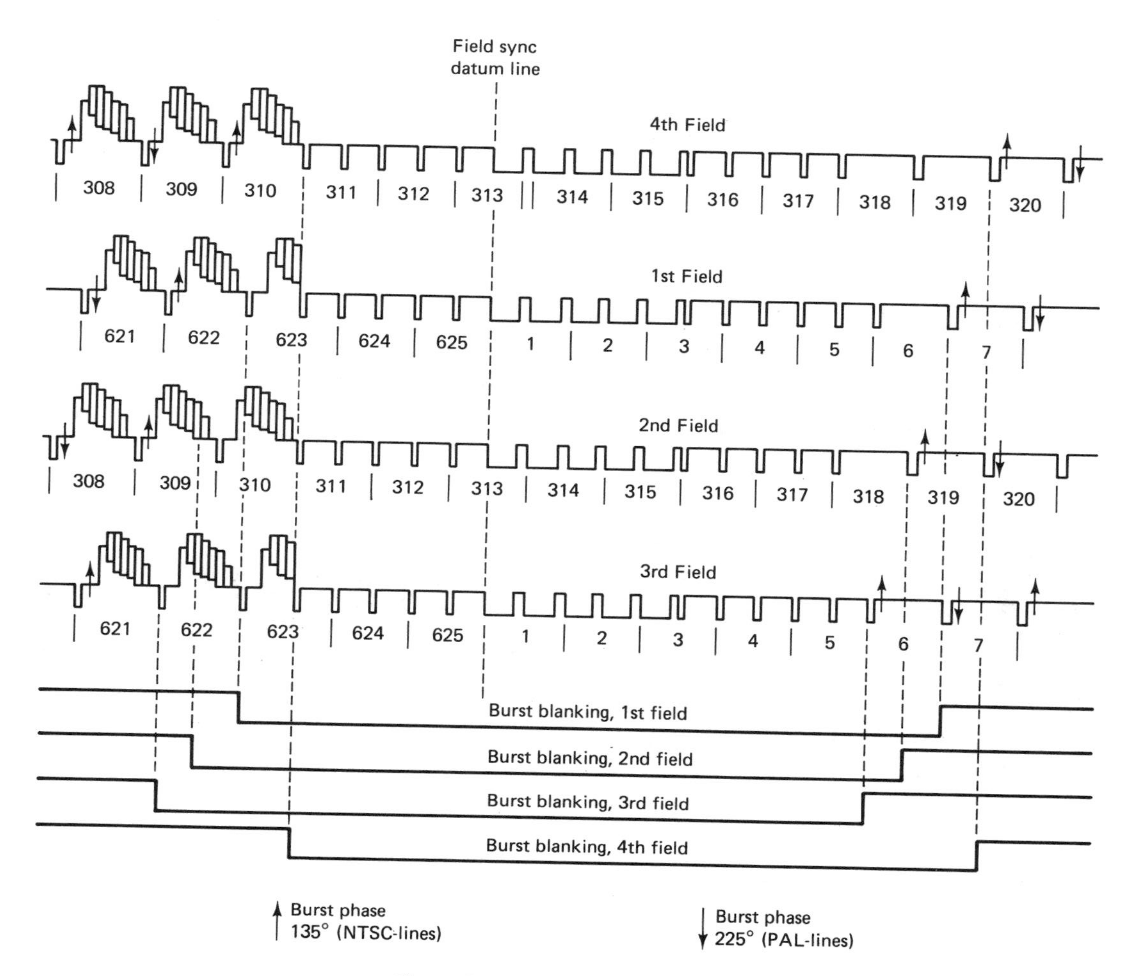

Figure 5.42 PAL burst blanking.

the chrominant phasor to the other side of the x-axis [from the first to the fourth quadrant, etc. (Figure 5.40)]. It must be noted that this is not the same thing as switching the NTSC line phasor by 180°. An NTSC line phasor is the vector sum of the U and V components, and inverting it is equivalent to inverting both the U and the V components, whereas on true PAL lines only the V component is inverted.

In a simple-PAL(PAL-S) receiver, the chrominance signal is fed from the chrominance bandpass amplifier directly to the U and V synchronous demodulators, which then extract the amplitudes of U and V modulation products from the QAM signal.

In a delay-line (PAL-D) receiver, there is an additional circuit between the chrominance amplifier and the demodulators. Thus the F_U and F_V are separated from the QAM chrominance signal, and then these two signals go forward to their respective demodulators.

PAL-S Receiver. This is a satisfactory technique for small phase errors, but beyond a certain order of error (e.g., ±5° or so), the eye is not completely deceived and visible evidence of error conditions becomes apparent to the viewer. One pair of space-sequential lines is complementary in hue to the next pair. But the eye is not uniformly sensitive over the visible spectrum and a difference in relative intensity is noticed. In severe cases a pattern of horizontal bars, sometimes known as *Hanover bars* or *Venetian blinds*, is seen [4, Chap. 9].

In the PAL system the chrominance signal, F_C, is

$$F_C = F_U \pm F_V = U \sin \omega_c t \pm V \cos \omega_c t \tag{5.95}$$

Outputs of the synchronous detectors, after filtering of the second subcarrier harmonic, will be (a gain of 2 is taken for simplicity)

$$2F_C \sin \omega_c t \rightarrow U \tag{5.96}$$

$$2F_C \cos \omega_c t \rightarrow \pm V \tag{5.97}$$

The expression (5.97) can be rewritten as

$$2(\pm V \cos \omega_c t) \cos \omega_c t \rightarrow \pm V \tag{5.98}$$

This means that by synchronous detection the U component can always be obtained, while the V component alternates sign. To get $\pm V$, for every line two methods can be used:

Method 1:

NTSC line: $2(V \cos \omega_c t)(+\cos \omega_c t) = 2(\pm V \cos \omega_c t) \sin(\omega_c t + 90°) \rightarrow +V$
PAL line: $2(-F_V)(-\cos \omega_c t) = 2(-F_V) \sin (\omega_c t - 90) \rightarrow +V$

Method 2:

NTSC line: $2(+V \cos \omega_c t) \cos \omega_c t \rightarrow +V$
PAL line: $2(-V \cos (\omega_c t + 180°) \cos \omega_c t \rightarrow +V$

According to the first version, the reference subcarrier should be shifted $+90°$ for the NTSC lines and $-90°$ for the PAL lines. According to the second version, an alternative method is to leave the phase of the reinserted subcarrier to the V demodulator constant at $+90°$ [$\cos \omega_c t = \sin(\omega_c t + 90°)$] and then to invert the chrominance signal itself on the alternate PAL lines. The circuit arrangements are shown in block diagram form in Figure 5.43.

PAL-D Receiver. The chrominance signal, F_c, in the PAL system can be expressed as

$$F_C = F_U \pm F_V = U \sin \omega_c t \pm V \cos \omega_c t = S \sin(\omega_c t \pm \theta) \qquad (5.99)$$

or

$$F_C = \text{Im}\{Se^{\pm\theta} e^{j\omega_c t}\} = \text{Im}\{Ce^{j\omega_c t}\} \qquad (5.100)$$

where $C = U \pm jV$ is the complex amplitude of the chrominance signal.

NTSC line: $C = U + jV$
PAL line: $C^* = U - jV$

In the PAL system one of the two adjacent lines is the PAL line and the other is the NTSC line. This gives an idea that signals of the two adjacent lines can be combined: one line is C and the other is C^*. Thus after adding it is

$$C + C^* = 2U \qquad (5.101)$$

and after subtracting,

$$C - C^* = j2V \qquad (5.102)$$

or

$$C^* - C = -j2V \qquad (5.103)$$

Figure 5.44 shows the circuit realization based on the previous analysis for extraction of the F_U and F_V components using the delay line. A signal of the direct line is fed from the chrominance bandpass filter output to the adder, the subtractor, and the delay line. Signals from the delay line are fed from the delay-line output to both the adder and the subtractor. The addition of two picture lines yields a signal consisting of U information only: equal in amplitude to twice the amplitude of the chrominance signal's U modulation product. Conversely the subtraction of the two lines produces a signal consisting only of V information, with an amplitude twice that of the V modulation product.

Thus, instead of leaving it to the eye to combine the color of neighboring lines, that is, color change in opposite directions for any differential error signal, corresponding chrominance signals can be combined electronically using delay line and other circuits in the receiver. Each signal F_U and F_V is a combination of the signals of two lines. This causes vertical resolution reduction for the colors. But the U and V components are separated before detection, and because unwanted components are concealed during two lines, crosstalk is avoided.

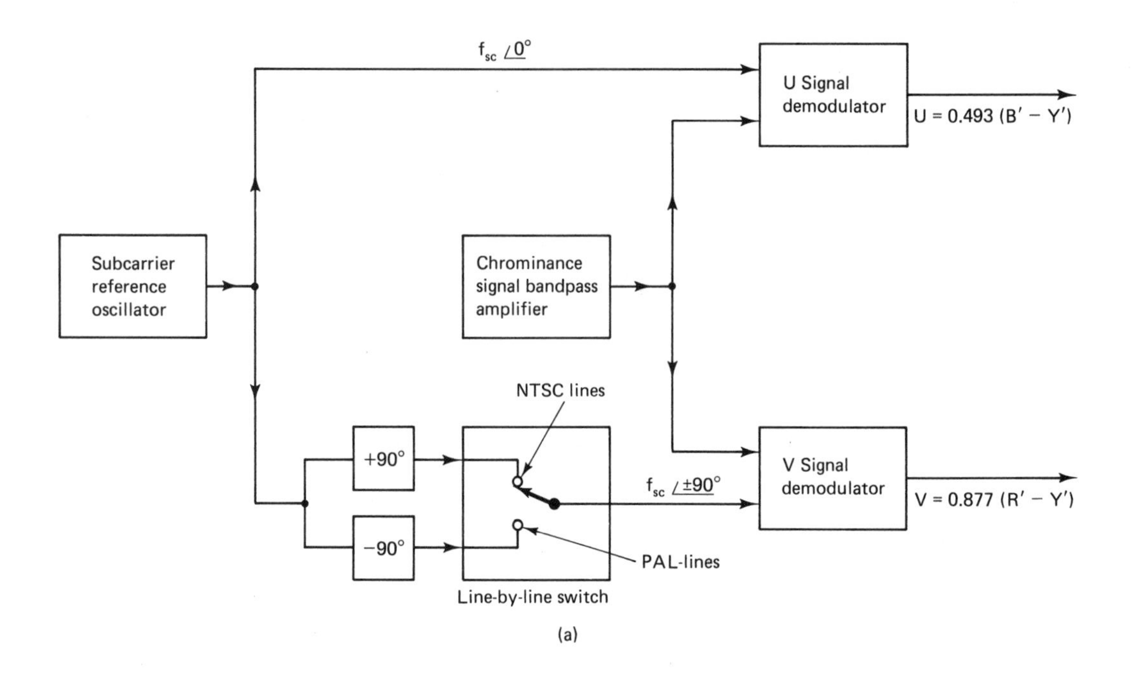

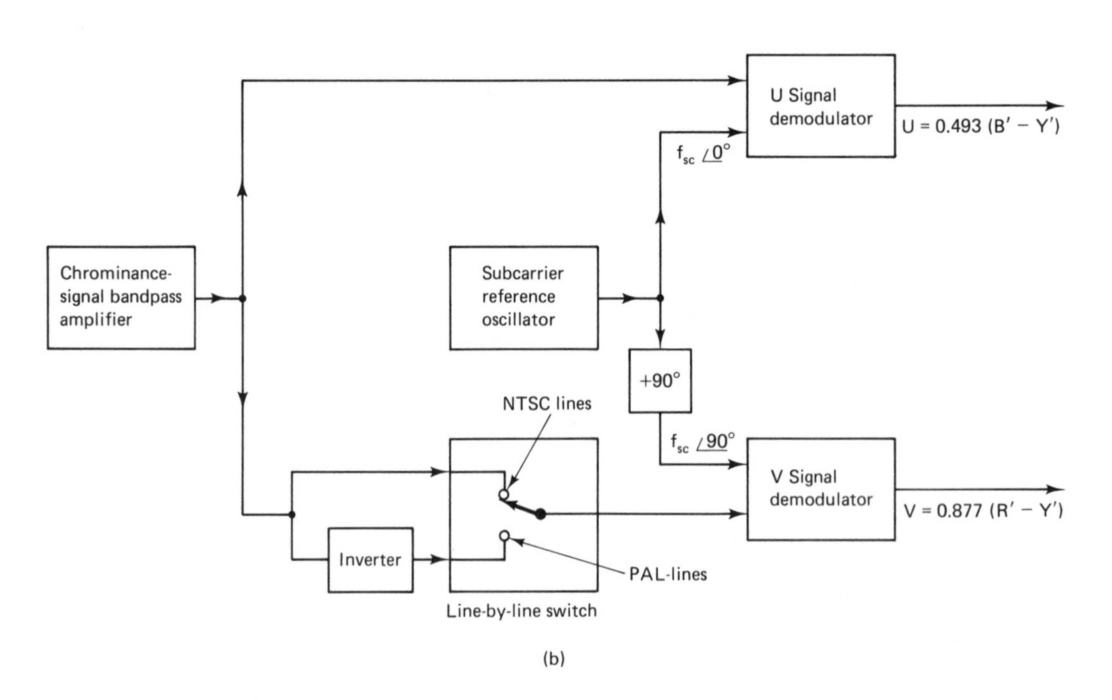

Figure 5.43 Block diagram of PAL-S chrominance signal demodulator: (a) subcarrier inversion; (b) input signal inversion.

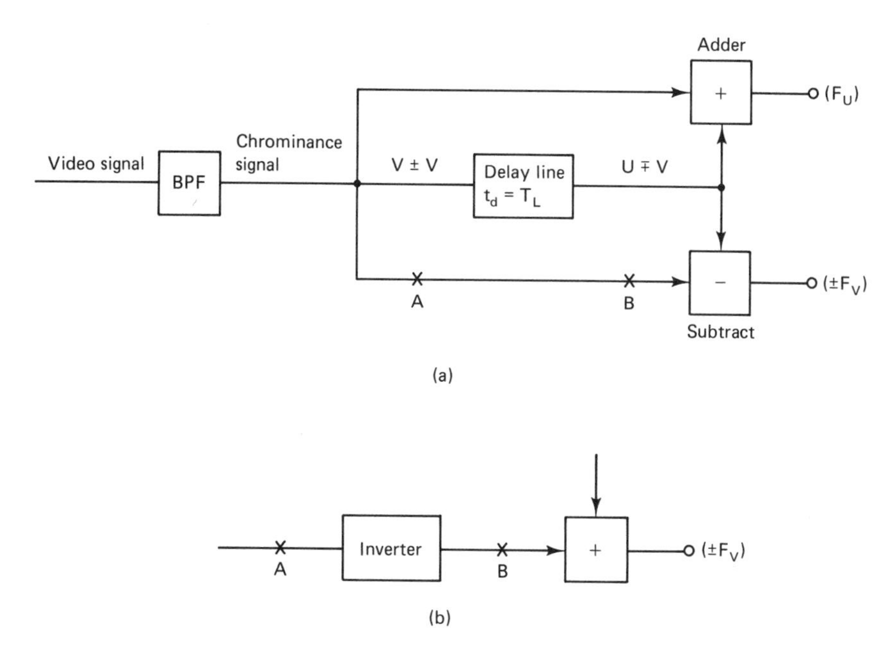

Figure 5.44 Circuit for the extraction of the F_U and F_V components using a delay line: (a) block diagram; (b) subtractor can be substituted by the adder and inverter.

The previous discussion is based on the basic assumption for the PAL system: that color does not change significantly from line to line. When this is not thorough (e.g., horizontal edges), the color (edge) will be blurred.

Two individual signals, F_U and F_V, are fed to their respective demodulators. Each of these must receive a reinserted subcarrier of the correct phase to allow the receiver of the color to recover the color-difference signal waveform. Since it is no longer a QAM signal, there is no longer an inescapable need for a synchronous demodulator capable of dealing with a QAM-type signal. But in practice the synchronous demodulator is used in the PAL-D receiver because with this type of detector there is no difficulty in monitoring an accurate constant "no-color" zero-voltage level, above and below which the sometimes positive, sometimes negative, color-difference signal voltage then varies.

As established earlier when discussing the PAL-S arrangement, there are two ways of taking into account the line-by-line phase reversal of the original $(R' - Y')$ subcarrier. One of these involves a line-by-line switching of the sampling point by 180°. The alternative is to leave the reinserted subcarrier constant in phase at $+90°$ relative to the $+U$ axes and to invert the chrominance signal modulation product with a line-by-line switch. The circuit arrangements for these two methods are shown in block diagram form in Figure 5.45.

Comparing PAL-S and PAL-D block diagrams, the essential differences between the two systems can be pointed out. First, in the PAL-S receiver the

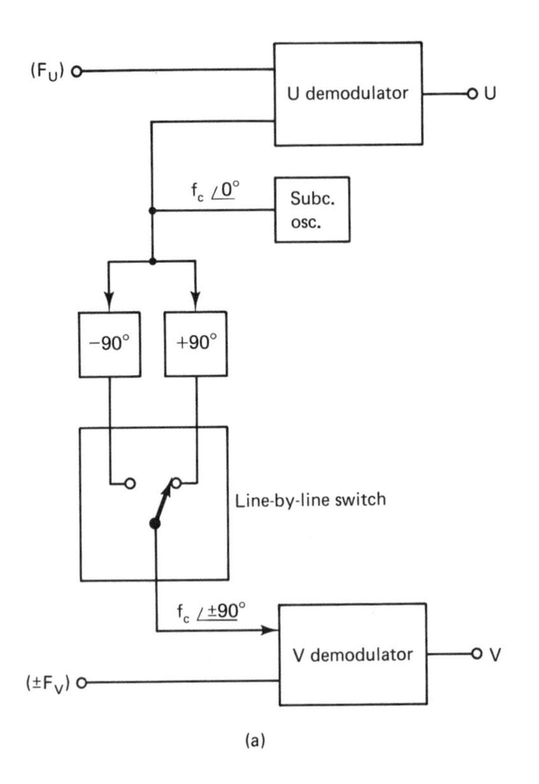

(a)

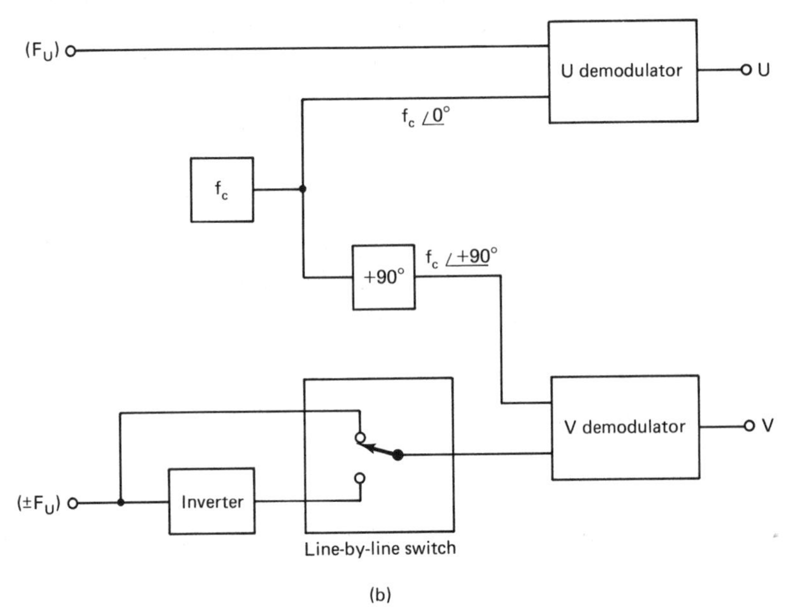

(b)

Figure 5.45 Chrominance signal demodulator in PAL-D receiver.

full QAM chrominance signal goes forward to the synchronous demodulators together with the appropriate reinserted subcarriers. In the PAL-D receiver the chrominance signal is "taken apart" before reaching the demodulators, and each of these then receives only one of the two modulation products, again with the appropriately phased reinserted subcarriers. Both systems finally yield the required color difference voltages. Second, these two systems have different maximum vertical resolution for color details. In the PAL-D system, where pairs of lines are of the same hue, the vertical resolution of color detail is reduced to one-half that of the PAL-S value. But this point of comparison does not represent a disadvantage of the PAL-D system because the bandwidths of the color signals, E_U and E_V, are reduced compared to monochrome bandwidth and the available vertical resolution of color detail in the PAL-D system is therefore perfectly adequate to deal with the bandwidth-restricted chrominance signal. The luminance resolution in both receivers is equal to that of normal monochrome receivers. However, there is a relative performance under phase-error conditions in which the systems differ.

5.3.3.4 Influence of the Phase Error and Asymmetry of the Sideband Components. If the frequency-response characteristic of the circuit for video signal transmission is not symmetrical around the subcarrier frequency, the upper and lower sideband components will no longer be equal, and the delay times will differ. Because of this asymmetrical transmission, the amplitude of the subcarrier and its delay will change with the change of modulating frequency. Further, because of the phase distortion, the projections on the normal axes will not be correct; in the NTSC system this will cause crosstalk and distortion in the reproduced color.

When the phase error is introduced, the chrominance signal for two adjacent lines in the PAL system will be:

$$\textit{NTSC line: } F = E_C \sin(\omega_c t + \phi + \theta)$$
$$\textit{PAL line: } F_C^* = E_C \sin(\omega_c t + \phi - \theta)$$

It can be noted that it does not matter whether the distortion angle ϕ is introduced by differential phase error or by reference subcarrier unstability.

In Figure 5.46, phasor presentation of the effect of the phase distortion of the PAL signal is shown. Parts (a) and (b) show the original error-free chrominance phasors for the NTSC and PAL lines. For reference only, the positions of the red and blue colors are indicated, according to Figure 5.40. In parts (c) and (d), the NTSC and the PAL lines are redrawn bearing the same (ϕ) phase error. It can be noted that whereas on the NTSC line the received V signal is greater (in this example) than it should be (the U is smaller), the reverse is true (always) on the PAL line. In a PAL-S type of receiver, the effect of this will be that on a given receiver tube any pair of time-sequential scanning lines on the raster will show complementary errors. In this example, the NTSC line will be too red and the PAL line, too blue. The eye averages this out optically and the viewer

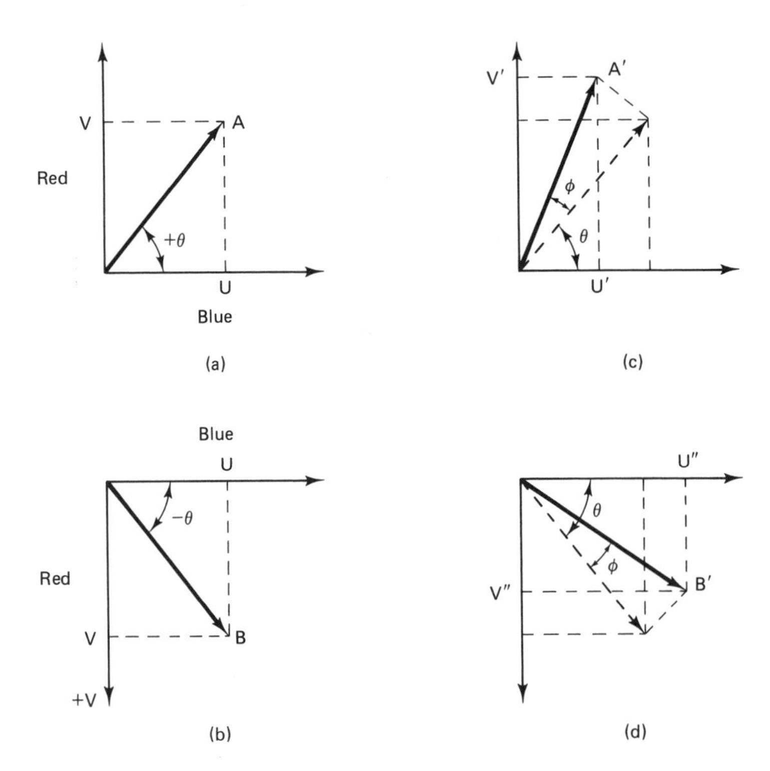

Figure 5.46 Phasor representation of the effect of phase distortion on a PAL signal: (a) nondistorted NTSC; (b) PAL line; (c) distorted NTSC; (d) PAL line.

appears to see the correct hue. This ability to negate the effect on the reproduced hue of the phase distortion is a basic advantage of the PAL system over the original NTSC system.

From Figures 5.46a and b it is clear that inverting the PAL phasor, which implies shifting the signal wave by 180°, places the phasor in the second quandrant. However, the V component of this inverted phasor is equal to the component of the NTSC-line phasor, since both components lie along the axis.

The same phase-distorted chrominance signal is represented in Figure 5.47. Part (a) shows the PAL-line phasor. The appropriate amplitude of the U component (U'') is extracted after this is passed, with its phase unchanged, to the U demodulator. This is so because the formation of the transmitted PAL line did not involve any change in the U component; only the V component is inverted. The phasor B'' represents the chrominance signal fed to the V demodulator on the PAL line and it is obtained after being inverted; the demodulation method shown in Figure 5.43b is supposed. From this inverted phasor the V demodulator then extracts the positive color-difference signal voltage of amplitude defined by the V'' vectorial component of B'' shown at Figure 5.47a. The recovered U and V amplitudes U'' and V'' are again portrayed in Figure 5.47b, and they are components of the

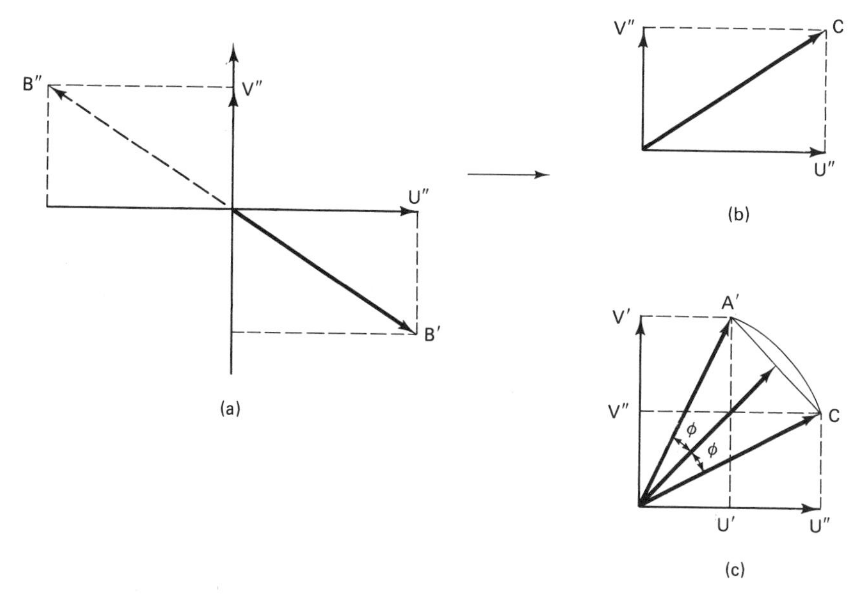

Figure 5.47 Effect of the phase distortion of the PAL signal demodulated in the PAL-S receiver: (a) V component recovered from the inverted PAL line; (b) equivalent NTSC-system phasor; (c) combined phasor diagrams.

phasor *C*. The imaginary phasor *C* is the equivalent chrominance phasor, which, when treated with a normal NTSC-line demodulator, would yield the same *U* and *V* amplitudes—and thus create the same hue—as the PAL-line (part a) yields in the PAL-S receiver [4, Chap. 9]. In Figure 5.47c the two successive equivalent NTSC lines are now portrayed on one phasor diagram, and the third "average" phasor between these two is drawn. It can be seen that having two hue errors on successive lines gives rise to hues that together seem to create the optical illusion of the correct hue. This follows from the fact that the color from successive lines is represented by phasors equally spaced on opposite sides of the true phasor direction. The average phasor can be seen to be shorter in length than the original error-free phasor, and this implies a small loss of saturation; the distance between the card and the arc is a measure of the loss of saturation. Clearly, the larger the phase error, the greater the loss of saturation.

Although two lines occur in time sequences, the ending process is due to the eye property if it is giving a single response to two adjacent but different pieces of color information, not to the electrical summation of the two phasors. This point leads directly to the concept of designing a receiver in which two successive chrominance signal lines are added electrically (PAL-D). In delay-line (PAL-D) receivers, the time-sequential lines are electronically averaged. Instead of two successive raster lines being different, although complementary hues, both will now be of the original correct hue.

The phasor diagrams in Figure 5.48 illustrate the functioning of the PAL-

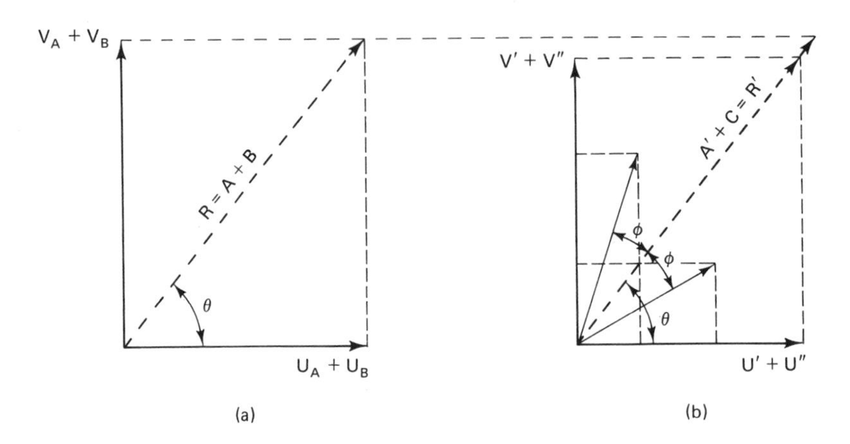

Figure 5.48 Phase error canceling in the PAL-D receiver: (a) effective summation of two successive lines; (b) effective summation of error-bearing NTSC and PAL lines.

D receiver. The diagram in part (a) is obtained based on the diagram in Figures 5.46a and b, and it shows the two separated modulation products as two phasors, one of the amplitude $U_A + U_B$ lying along the $+U$ axis, and the other of the amplitude $V_A + V_B$ along the $+V$ axis. The equivalent, but nonexistent single chrominance signal phasor that would bear these two quadrature components in an NTSC signal is shown dashed as R.

In part (b) the same signal is shown bearing a phase distortion of $\phi°$. This diagram is obtained based on Figures 5.46c and d or Figure 5.47c, which is equivalent. The net U and V components are both smaller than they would be if no phase distortion were present. However, these still bear the same ratio to one another and hence despite the chrominance signal phase error, the outputs from the U and V synchronous demodulators are still in the correct ratio and the hue reproduced is correct [4, Chap. 9]. The only effect of the phase error is a small loss of saturation due to the diminished amplitudes. This can be demonstrated either by a scale drawing or by simple trigonometry.

In the PAL-D receiver, the CRT is activated by information from the effective sum of one NTSC and one PAL line, and is never activated by information from single lines. Thus the resultant phasor R' (Figure 5.48b) truly represents the single equivalent NTSC-type chrominance phasor which would give rise to the hue actually reproduced on the screen. The eye is no longer called on to integrate different hues. But the actual displayed degrees of saturation R and R' are the effective equivalent amplitudes of the chrominance signal under normal and phase-error conditions, respectively, and can be arrived at only by computing the net R', G', and B' inputs to the tube and then raising these to a power appropriate to the tube's gamma value. Calculations of the actual estimated visual effects of the saturation loss that accompanies phase errors in the PAL-D system show that these are negligible for the levels of phase distortion likely to be encountered in practice, [4, Chap. 9].

The loss of saturation can be compensated at the receiver by employing the principle of *chrominance lock* [4]. In this technique the reference oscillator is controlled by the synchronizing signal derived from the chrominance itself. The initial subcarrier phase of the start of each raster line is set by the back-porch burst; this also overcame a possible 180° phase ambiguity. Along the raster line, if the chrominance signal should show a phase distortion, the synchronizing signal derived from it imparts the same phase shift to the generated subcarrier and the effect of the distortion is canceled.

5.3.3.5 Video Signal Bandwidth. Figure 5.49 shows the bandwidth of the normal 625-line vision signal. There is compatibility with the corresponding monochrome signal. The chrominance signal extraction introduces the horizontal resolution loss. For example, a notch filter eliminates some spatial patterns from the reproduced image. The main difference in comparing this to the basic NTSC system is that the U and V channels are both symmetrical around f_{sc}, whereas the I and Q channels have not been.

5.3.4 SECAM System

The Sequential Color à Mémoire (SECAM) system [7–8] was developed in order to overcome the phase-error sensitivity of the original NTSC system. During the development process, many variations of this system have been used, called by different names such as SECAM I, II, and III, but in practice the improved version, SECAM III, is simply called SECAM. In the SECAM system during each line scan the luminance signal Y is transmitted together with one of the

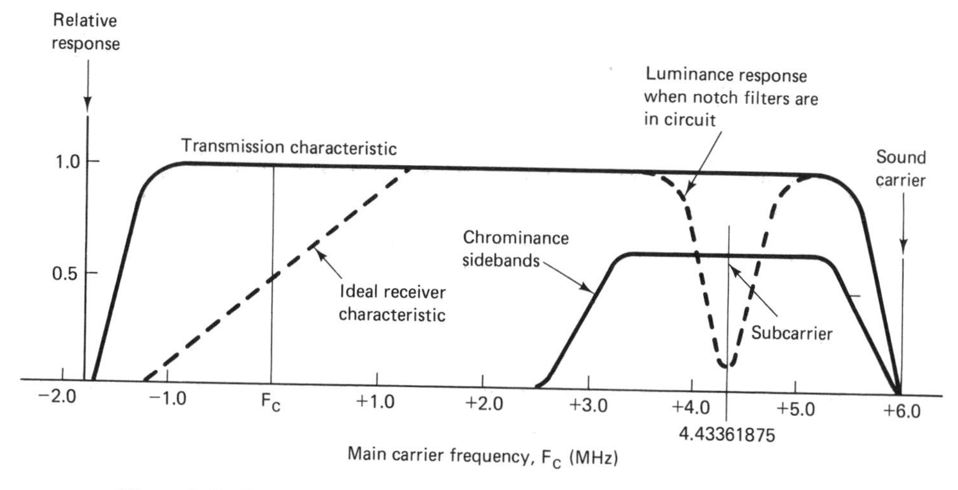

Figure 5.49 Frequency spectra of the video signal in the 625-line PAL system.

chrominance signals $R - Y$, or $B - Y$ on an alternate-line basis. At the decoder the alternate-line chrominance signals are stored and repeated at each line display. The chrominance signal to be transmitted is modulated onto a subcarrier added to the luminance signal in an interleaved fashion similar to the NTSC modulation process, but frequency modulation is used instead of amplitude and phase modulation. The French SECAM System transmits 625 lines of a 50-Hz frame rate with 2:1 line interlacing.

5.3.4.1 Coding and Decoding Processes. A simplified block diagram of the SECAM coder is shown in Figure 5.50. In the matrix circuits the luminance signal, $Y' = 0.3R' + 0.59G' + 0.11B'$, and reduced difference signals, D'_R and D'_B, are formed, where

$$D'_R = k_1(E'_R - E'_Y) \tag{5.104}$$

$$D'_B = k_2(E'_B - E'_Y) \tag{5.105}$$

The constants k_1 and k_2 differ from the corresponding constants in the PAL system: $k_1 = -1.9$ and $k_2 = 1.5$. The delay line in the luminance channel compensates for delay introduced in the chrominance channel.

The normalized color-difference signals D'_R and D'_B go through the video preemphasis and limiting circuits (preemphasis is about 9 dB at 800 kHz). Then they frequency modulate separate subcarriers f_{CR} and f_{CB}.

An electronic switch that works at half-line frequency proceeds to alternate F_{DR} and F_{DB} signals to the input of the phase-reversing circuits. The phase-reversing-switch alternates the phase of the subcarrier in every field and in every three lines.

The RF preemphasis, also referred to as *mise en forme*, is a filter that reduces the amplitude of the subcarrier (peak to peak) by about 23% compared

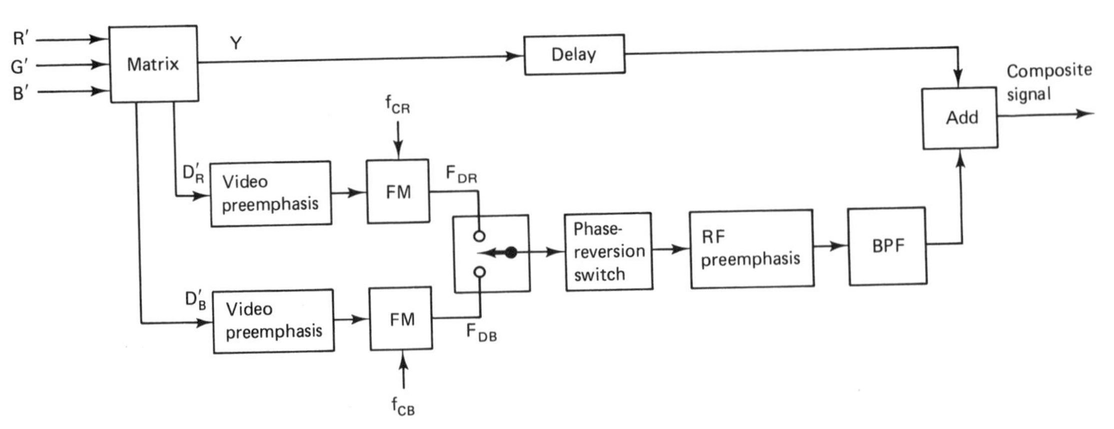

Figure 5.50 SECAM coder.

with the modulating white level. The purpose is to improve a signal-to-noise ratio of the chrominance signal.

The bandpass filter (BPF) has a bandwidth of approximately ±1.5 MHz around the subcarrier. A composite video signal is obtained after mixing the luminance and chrominance signals. Sync signals are not indicated in Figure 5.50.

A simplified block diagram of the SECAM decoder is shown in Figure 5.51. A notch filter eliminates the subcarrier frequency in the luminance channel; otherwise, the subcarrier would cause color desaturation by making the brightness higher, and a dot pattern would be produced. The delay circuit introduces about a 0.8-μs delay in the luminance channel to equalize the delay in the chrominance channel because of the limited bandwidth.

The bandpass filter separates the frequency-modulated subcarrier. The RF deemphasis filter has inverse characteristic to the preemphasis filter in the coder. It is also referred to as a *remise en forme* or *bell filter*.

In the SECAM system the subcarrier transmits one chrominance signal at a time. On the other hand, both signals should be supplied simultaneously to the matrix circuits. Thus a memory element has to be used in this system. One signal is taken directly from the present line, and the second from the previous line. This process, as in the PAL system, uses the color-resolving property of the eye and it is assumed that distribution of the chrominance characteristics along two adjacent lines is approximately equal.

The SECAM delay line has a delay of 64 μs ±170 ns. This tolerance is much larger than the ±3 ns in the PAL system because the exact color registration

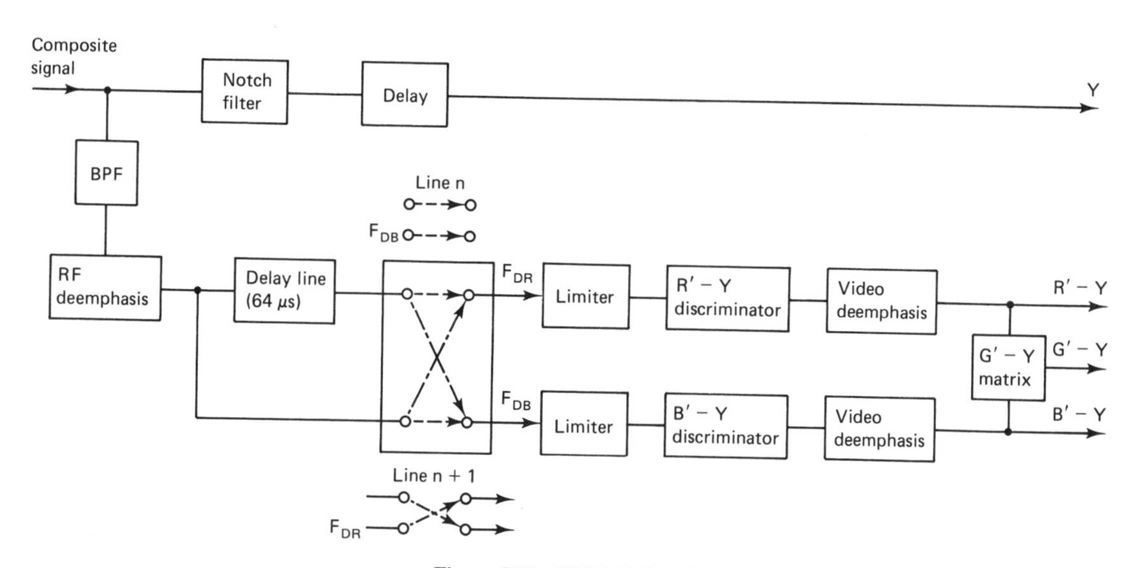

Figure 5.51 SECAM decoder.

is important and there is no condition for the exact number of half-periods of the subcarrier.

When working properly, the double-pole electronic switch sends F_{DR}, to the D'_R, $R' - Y'$, discriminator, and F_{DB} to the D'_B or $B' - Y'$ discriminator through corresponding limiters. For example, for the first line, when F_{DB} is at the delay line input, the switch should be in the horizontal position (line n); for the next line (line $n + 1$), when F_{DR} is at the delay line input, the switch should be in the cross-section position. After video deemphasis, the $R' - Y'$ and $B' - Y'$ signals are combined to give a $G' - Y'$ signal:

$$G' - Y' = -0.51(R' - Y) - 0.19(B' - Y) \quad (5.106)$$

All three color difference signals can then be combined with the luminance signal, Y, to give color signals R', G', and B' for the color tube.

5.3.4.2 System Parameters. The SECAM system is not sensitive to the differential phase error or to the crosstalk, but because FM is used, there are some difficulties in getting good compatibility and in avoiding dot-pattern visibility in black-and-white receivers.

Through the constants k_1 and k_2 the chrominance signal amplitude is normalized in order to get more efficient use of the frequency bandwidth. In the SECAM system, the chrominance signal is normalized to 1 for specified colors, when gamma-corrected voltages E'_P, E'_G, and E'_B are 75% of the maximal amplitude. The specified saturated colors are red and its complement for the color difference signal $E'_R - E'_Y$, and blue or yellow for the $E'_B - E'_Y$ signal.

For gamma = 2.2, to correct amplitudes of 0.75-V signals, the camera output should be $0.72^{2.2} = 0.53$ V. Now, for saturated red color

$$E'_R - E'_Y = 0.75 - 0.75 \times 0.3 = 0.525 \quad (5.107)$$

and for saturated blue,

$$E'_B - E'_Y = 0.75 - 0.75 \times 0.11 = 0.67 \quad (5.108)$$

Therefore, the normalizing factors for the SECAM systems are

$$k_1 = -1/0.525 = -1.8 \quad \text{and} \quad k_2 = 1/0.67 = 1.5 \quad (5.109)$$

so

$$D'_R = -1.9(E'_R - E'_Y) \quad (5.110)$$

and

$$D'_B = 1.5(E'_B - E'_Y) \quad (5.111)$$

The negative sine for the chrominance signal D_R is related to the modulation method of the subcarrier frequency; the frequency deviation should be opposite for the F_{DR} and F_{DB} signals.

In the SECAM system it is found that the compatibility is higher if the subcarrier frequency has a different value for the D_R' and D_B' signals. The difference is $10f_L$:

$$f_{CR} = 282 \times f_L = 4.40625 \text{ MHz} \tag{5.112}$$

$$f_{CB} = 272 \times f_L = 4.25000 \text{ MHz} \tag{5.113}$$

for the 625-line system. The corresponding unity deviations are

$$\Delta f_{CR} = 280 \pm 28 \text{ kHz} \tag{5.114}$$

$$\Delta f_{CB} = 230 \pm 23 \text{ kHz} \tag{5.115}$$

The maximal frequency deviation is limited to +350 kHz when $D_R' = 1.25$ and to −506 kHz for $D_R' = -1.79$; and for D_B' the maximal frequency deviation is +506 kHz for $D_B = 2.18$ and −350 kHz for $D_B' = -1.52$. In the SECAM system, because FM is used, preemphasis and deemphasis are used to get a better S/N ratio.

The electronic switches in the receiver are synchronized by the special indent signal. This signal is made of the nine synchronizing pulses; each of them lasts for one line interval. The indent signal is transmitted once in every field, during vertical blanking intervals. The first of nine pulses starts 14 lines before the beginning of the picture information.

REFERENCES

1. D. G. Fink (Ed.-in-Chief) and A. A. McKenzie (Assist. Ed.), *Electronics Engineers' Handbook,* McGraw-Hill, New York, 1975.
2. G. M. Glasford, *Fundamentals of Television Engineering,* McGraw-Hill, New York, 1955.
3. P. Mertz and F. Gray, "A Theory of Scanning and Its Relation to the Characteristics of the Transmitted Signal in Telegraphy and Television," *Bell Syst. Tech. J.,* Vol. 13, July 1934, pp. 464–515.
4. G. H. Hutson, *Color Television Theory—PAL System Principles and Receiver Circuitry,* McGraw-Hill, London, 1971.
5. H. V. Sims, *Principles of PAL-Color Television,* Iliffes, Butterworth, Woburn, Mass., 1970.
6. H. de France, "Le Système de télévision en couleurs séquentiel-simultane," *L'Onde Electr.,* Vol. 38, 1958.
7. H. de France, "The Sequential-Simultaneous-Color Television System," *Acta Electron.,* Vol. 2, 1957–58, pp. 392–397.
8. F. C. McLean, "Worldwide Color Television Standards," *IEEE Spectrum,* Vol. 3, No. 6, 1966, pp. 59–60.
9. R. Bracewell, *The Fourier Transform and Its Applications,* McGraw-Hill, New York, 1965.
10. N. S. Piskunov: "Diferentsigolnoe i Integralniye i schisleniyo." Fizichko-Mat-Ematichesuoy Literaturi" Moskow, 1963.

APPENDIX 5A: VERTICAL INTERVAL IN THE NTSC STANDARD

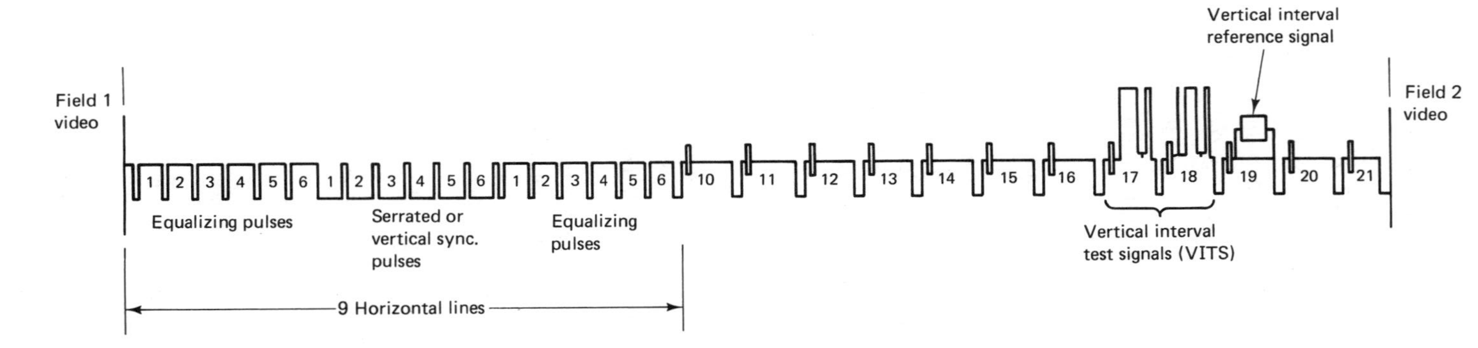

Figure 5A.1 Entire vertical interval (21 blanked horizontal lines).

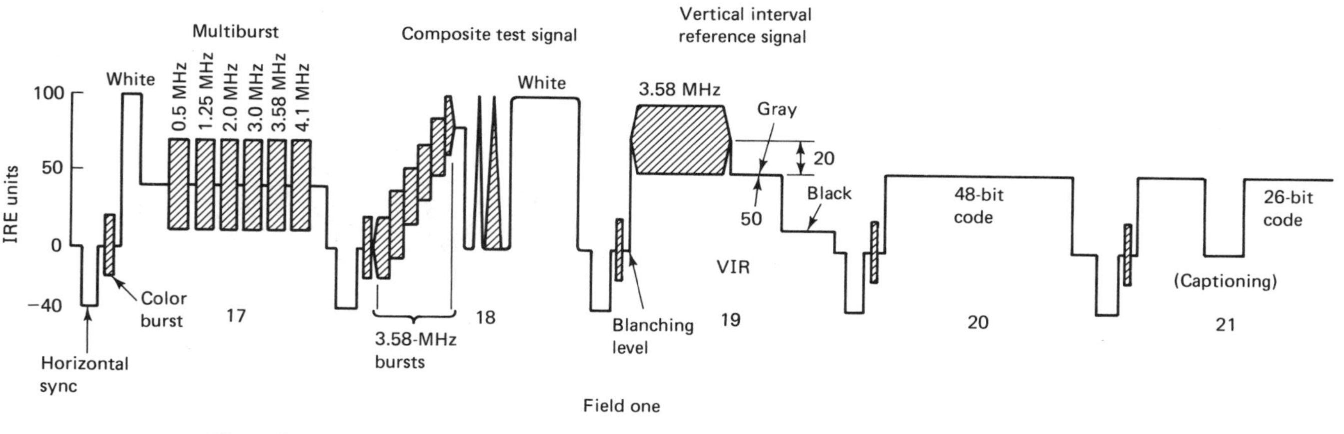

Figure 5A.2 Vertical interval test signal (VITS): lines 17–21 of the vertical blanking pulses.

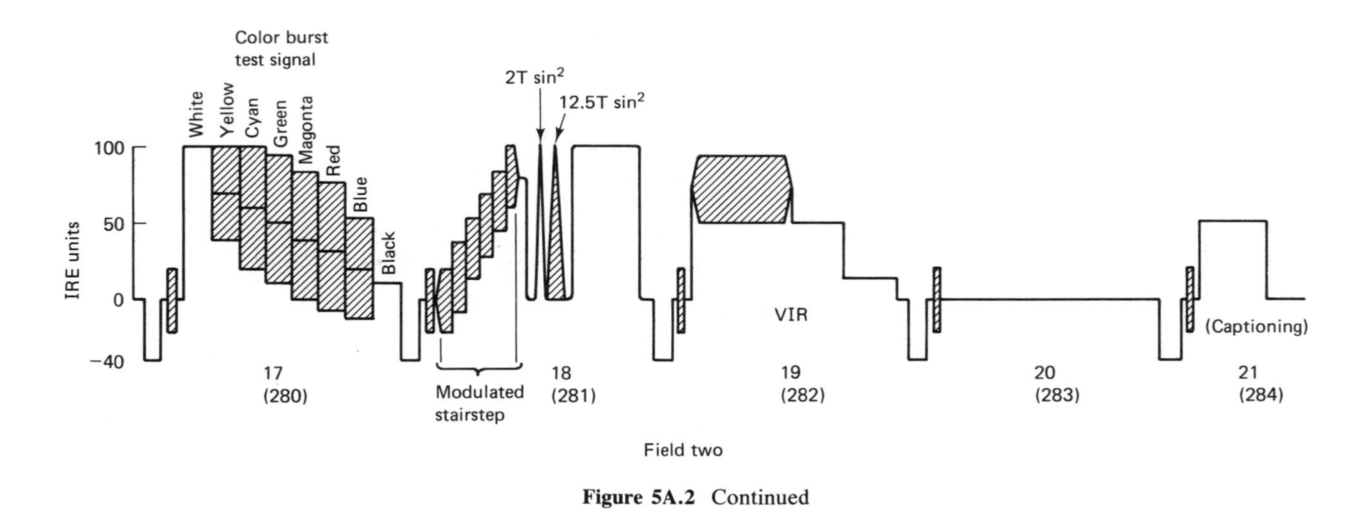

Figure 5A.2 Continued

Entire vertical interval has twenty-one horizontal lines. Vertical Interval Test Signals (VITS) may occur during lines 17 to 21 of each field and may differ from one field to the next:

Line 17, first field: multiburst signal
Line 17, second field: color test signal
Line 18, both fields: composite test signal
Line 19, both fields: vertical interval reference (VIR) signal
Line 20 is being developed as a network source identification signal using a 48-bit digital signal.
Line 21 is being used by some PBS (public broadcast service) stations as well as some other TV stations for digital programming of captions to aid the deaf in understanding TV program content. In the receiver, a decoder is used to convert the digital signals into the proper form to display captions on the screen.
Lines 10 to 18 are sometimes used for a teletext.

APPENDIX 5B: AMPLITUDE SPECTRA

For the rectangular pulse, $s(t) = E\Pi(t)$, where $\Pi(t)$ is

$$\prod(t) = \begin{cases} 0, & |t| > \frac{\tau}{2} \\ \frac{1}{2}, & t = \frac{\tau}{2} \\ 1, & |t| < \frac{\tau}{2} \end{cases} \tag{5B.1}$$

The spectrum is

$$F = A \int_{-\tau/2}^{\tau/2} \prod(t) e^{-j\omega t}\, dt = E\tau \frac{\sin \pi f \tau}{\pi f \tau} = E\tau \operatorname{sinc}(\pi f \tau) \tag{5B.2}$$

where $\omega = 2\pi f$. This is shown in Figure 5B.1a. The cosine square pulse (dashed line) and its spectral amplitude density are shown for comparison. The spectral amplitude density of two rectangular pulses is shown in Figure 5B.1b. In general, for

$$s(t) = \sum_n E_n \prod(t - t_n) \tag{5B.3}$$

$$F(f) = \sum E_n \tau_n e^{-j2\pi f \tau_n} \operatorname{sinc}(\pi f \tau_n) \tag{5B.4}$$

where E_n and τ_n are the amplitude and duration of the nth pulse, and t_n is its displacement [9, p. 128].

For the periodic signal, the spectrum is discrete:

$$F = \tfrac{1}{2} C_n = a_n^2 + b_n^2 \tag{5B.5}$$

$$\tan \theta = -b_n / a_n \tag{5B.6}$$

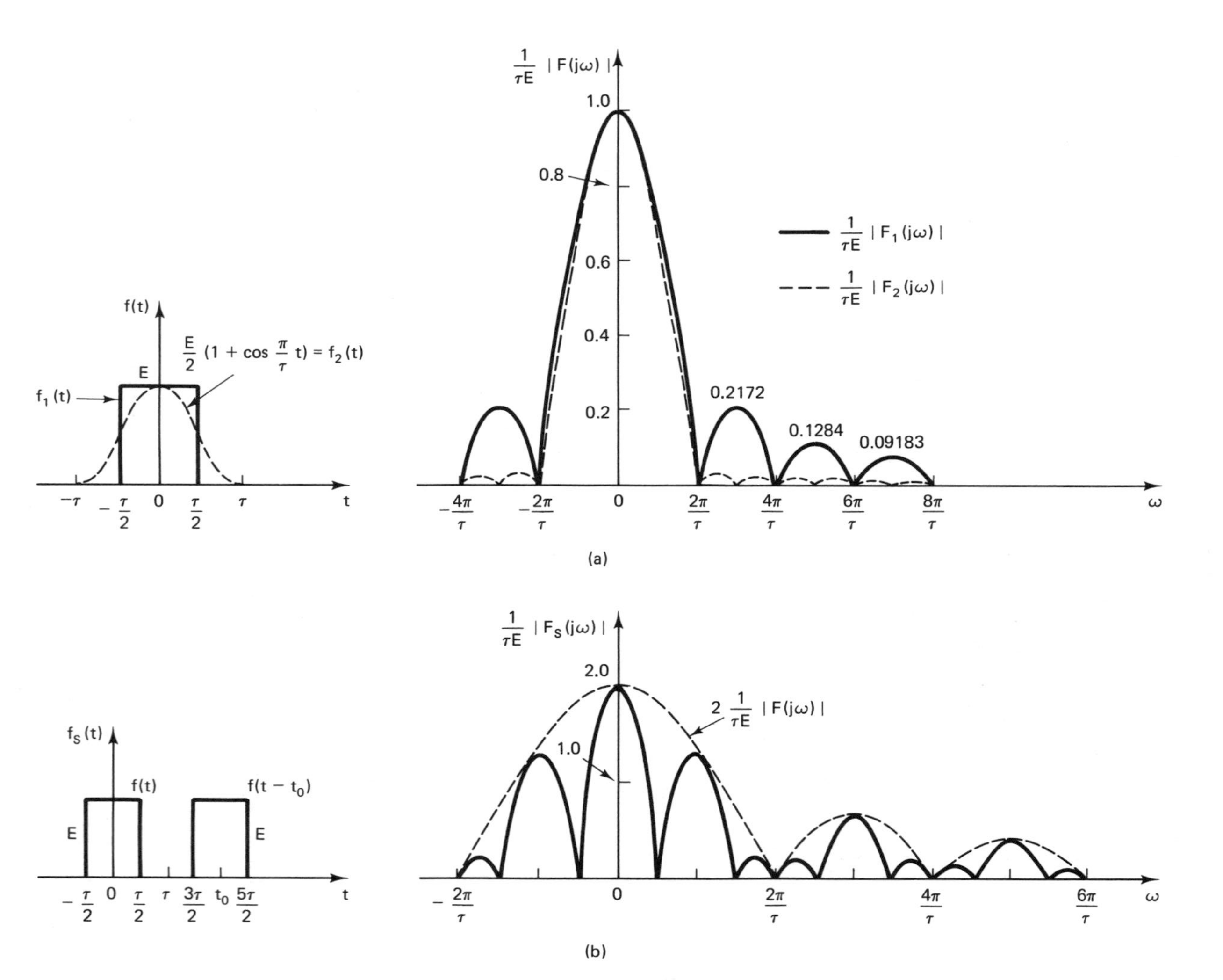

Figure 5B.1 Spectral amplitude density of (a) rectangular and cosine square (dashed line) pulses, and (b) two rectangular pulses.

where for the period $2l$ [10, Chap. XVII],

$$a_0 = (1/l) \int_{-1}^{l} f(x)\, dx \tag{5B.7}$$

$$a_n = (1/l)) \int_{-1}^{1} f(x) \cos (k\pi/l)\, x\, dx \tag{5B.8}$$

$$b_n = (1/l) \int_{-1}^{1} f(x) \sin k\pi/l)\, x\, dx \tag{5B.9}$$

and

$$f(x) = (a_0/2) + \sum_{n=1}^{\infty} a_n \cos (k\pi/l)\, x + b_n \sin (k\pi/l)\, x \tag{5B.10}$$

This is the Fourier series.

For rectangular pulses $s(t)$ (Figure 5B.2a) with duration τ, period T, and amplitude A,

$$a_0 = (2/T) \int_{-T/2}^{T/2} s(t)\, dt = (2/T) \int_{-\tau/2}^{\tau/2} A\, dt = 2A\, (\tau/T) \tag{5B.11}$$

$$a_n = (2/T) \int_{-T/2}^{T/2} s(t) \cos (2k\pi/T)\, dt = (2/T) \int_{-T/2}^{T/2} A \cos \tag{5B.12}$$

$$(2k\pi/T)\, dt = 2A \operatorname{sinc} (n\omega_0\tau/2)$$

$$b_n = (2/T) \int_{-T/2}^{T/2} s(t) \sin (2k\pi/T)\, dt = 0 \tag{5B.13}$$

where $\omega_0 = 2\pi f_0 = 2\pi/T$. Thus

$$s(t) = E\, (\tau/T) + 2E \sum_{k=1}^{\infty} \operatorname{sinc} [n\omega_0\, (\tau/2)] \cos n\omega_0 t \tag{5B.14}$$

The physical meaning of this expression is that the function $s(t)$ is based on the harmonic (Fourier) analysis, decomposed into some of the cosine components. The amplitude of each spectral component (harmonic) is given by the Fourier coefficients.

Figure 5B.2b shows the distribution of the Fourier coefficient or discrete amplitude spectrum. The envelope of the spectrum is given by sinc $(n\omega_0\tau/2)$; the zero crossings of the envelope are defined by pulse durations, τ, and they are k/τ, $k = 1, 2, 3, \ldots$. Spacing between adjacent components is $f_0 = 1/T$; that is, it is defined by period T. The number of spectral lines up to the first zero is an integer of the duty cycle T/τ. In this particular case, it is assumed that $T = 5\tau$.

Sometimes it is convenient for mathematical manipulation to express $s(t)$ in terms of components with negative frequencies. For rectangular pulses

$$s(t) = E\frac{\tau}{T} + E \sum_{\substack{k=-\infty \\ k\neq 0}}^{\infty} \operatorname{sinc} (n\omega_0\tau/2) \cos n\omega_0 t \tag{5B.15}$$

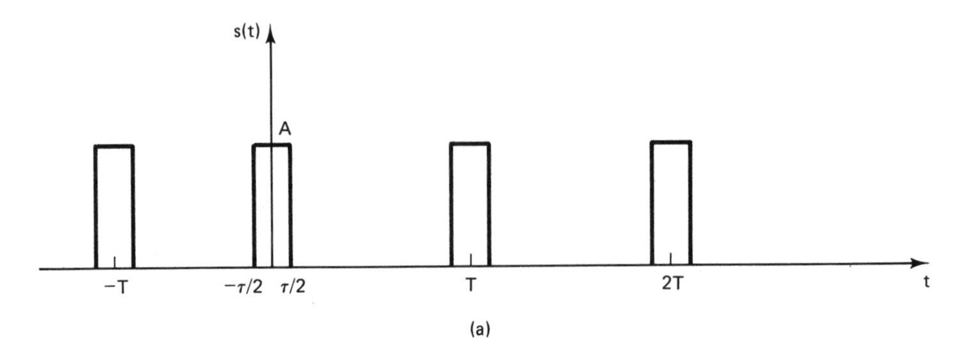

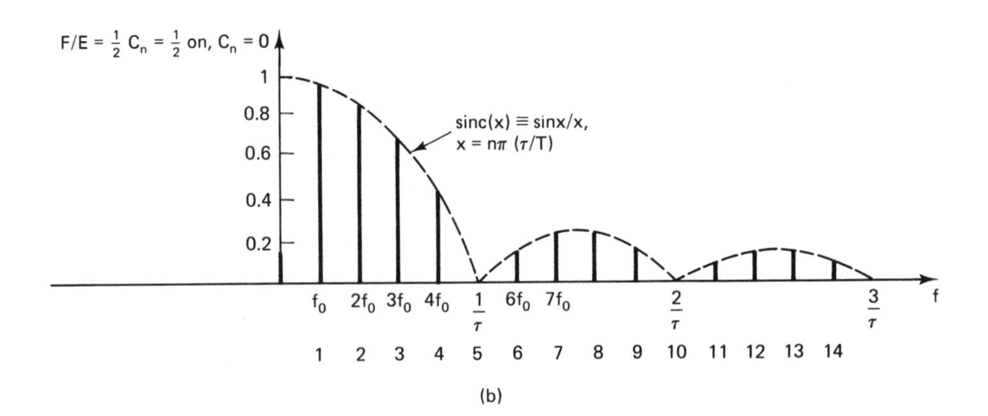

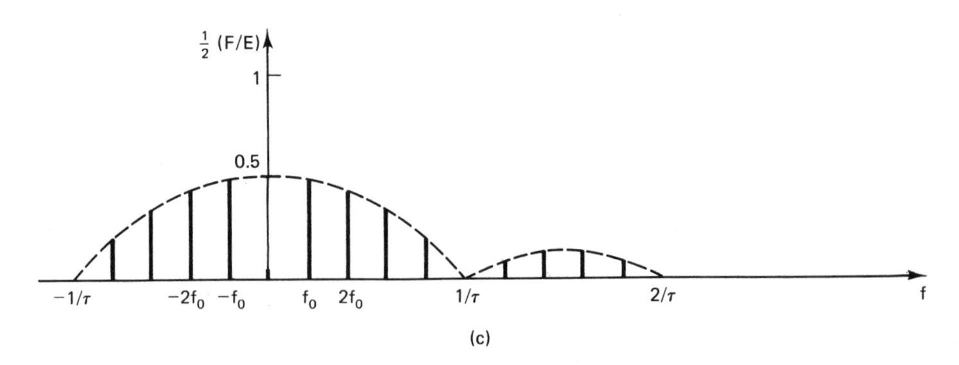

Figure 5B.2 Amplitude spectra of the periodical rectangular pulses: (a) pulse train; (b) positive spectra; (c) double-sided spectra.

Corresponding (discrete) amplitude spectral distribution is shown in Figure 5B.2c.

Finally, an example of amplitude keying is shown in Figure 5B.3, where a carrier with frequency f_c and amplitude A_c is keyed by a pulse train with period T_p, pulse duration τ_p, and amplitude A_p. The amplitudes of individual spectral lines are shown in Figure 5B.3c. It can be seen that the number of spectral lines

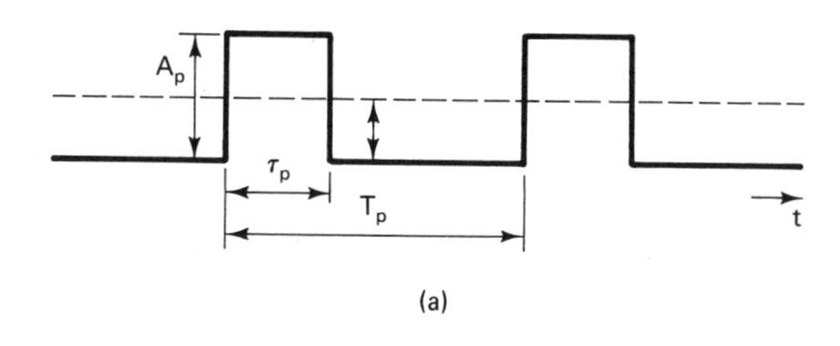

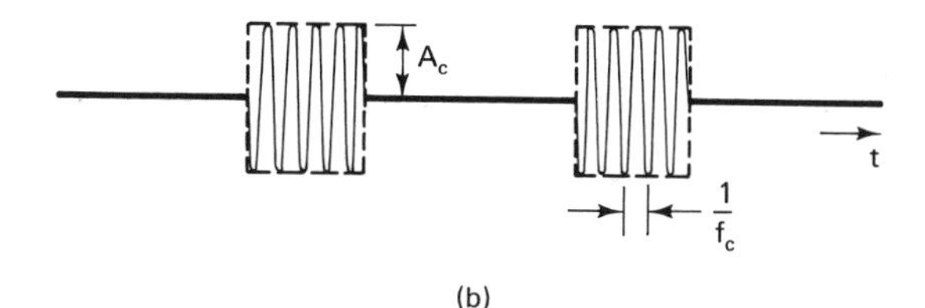

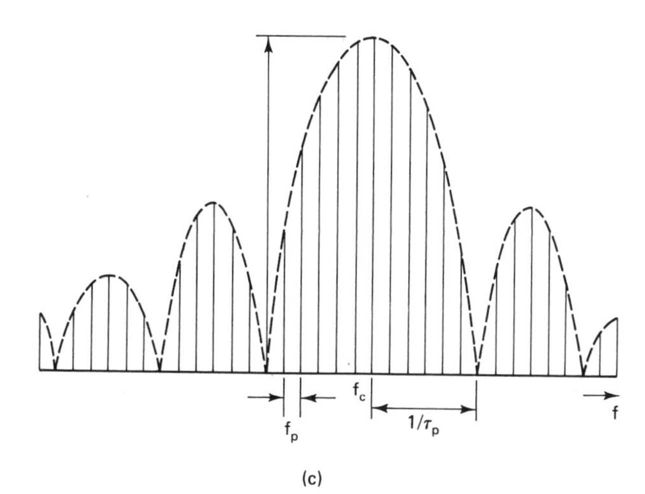

Figure 5B.3 Amplitude keying; (a) keying pulse train; (b) keyed carrier; (c) frequency spectrum of part (b).

up to the first zero is an integer of the duty cycle T_p/τ_p; the distance between two neighboring spectral lines corresponds to pulse repetition frequency f_p and the zeros are $f_c \pm n/\tau_p$.

Based on the convolution theorem [9, p. 108], the spectrum in Figure 5B.3c can easily be obtained. If $s_p(t)$ describes the pulse train and the carrier signal is

$$s_c(t) = A_c \cos \omega_c \tag{5B.16}$$

then a keyed signal $s_k(t)$ is

$$s_k(t) = s_p(t)s_c(t) \tag{5B.17}$$

That is, the spectrum of the keyed signal, F_k, is given by convolution of the carrier (F_c) and pulse train (F_p) spectra.

six

TV Signal Recording

6.1 INTRODUCTION

In this chapter a description is given of the use of the videodisc as an analog TV channel. To avoid any possible confusion, it is necessary to specify clearly a number of assumed facts about the system before going into any great detail. Unless otherwise specified, the following will be assumed.

As a normal video (TV) signal, a standard broadcast composite color signal will be considered with the exception of the audio portion. A *standard signal* means a signal tailored according to one of three worldwide TV standards, NTSC, PAL, or SECAM, and it does not matter if it is intended for use in broadcast or closed-circuit TV installations. The term *normal signal* does not, for example, refer to a signal tailored for extended play based on compression techniques or bandwidth reduction even if the case is covered by the Laservision (LV) standard. As is customary in television techniques, both time- and frequency-sharing principles are used. In the time domain, sync and active video intervals are interleaved; during sync intervals the reference subcarrier signal (''burst'') is placed in the specified positions. During active video intervals, based on frequency sharing, four (or five) signals are recorded simultaneously: luminance (brightness), two color signals, and one stereo (or two mono) sound signals. Identification signals also can be made during sync intervals. The number of fields per second is either 50 or 60. Vertical resolution is usually 525 or 625 lines per frame (the number of active lines is about 20% less), and the horizontal resolution is between 380 and 420 picture elements per active line.

There is a single track on each side of the disc in which all the information is stored for the reproduction of a color television program with two channels of sound (or one stereo) and one video channel. Although it is possible, at least in principle, to use systems with more than one information channel in parallel, this is not commonly used in videodisc systems.

In principle, both constant angular velocity (CAV) and constant linear velocity (CLV) can be used for TV signal recording. The CAV can be considered as "normal" because of its relative simplicity compared to the CLV system (which will be considered in reference to extended play). If a recorded TV signal has 25 frames per second, the disc should rotate at 1500 rpm; or 1800 rpm for 30 frames per second.

The (TV) signal is recorded from the inside radius, R_i (or R_{min} or R_1), to the outside radius, R_o (or R_{max} or R_2). If R_{max} is fixed, there is no particular reason to record from inside to outside radius; however, during development various R_{max} radii were envisioned and starting at the inside radius enabled the start of play to remain the same regardless of R_{max}. This is illustrated in Figure 6.1, where an approximate arrangement of the lines in one frame is also shown. One TV frame is stored on one 360° revolution of the spiral information track.

Figure 6.2 shows more details of the videodisc layout for the CAV disc.

Apart from video and audio information, the disc contains a number of

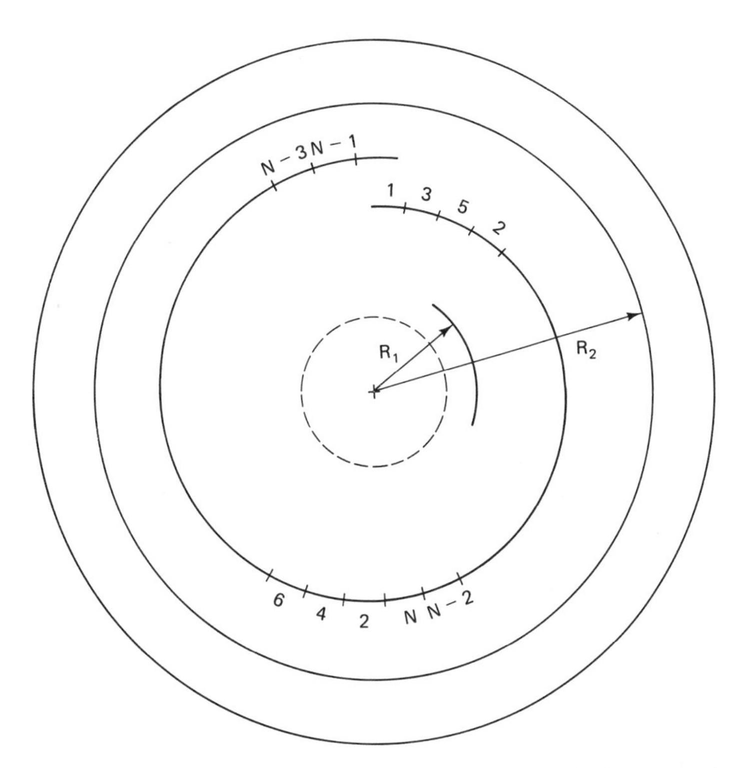

Figure 6.1 Approximate arrangement of the lines in one TV frame stored on the disc surface.

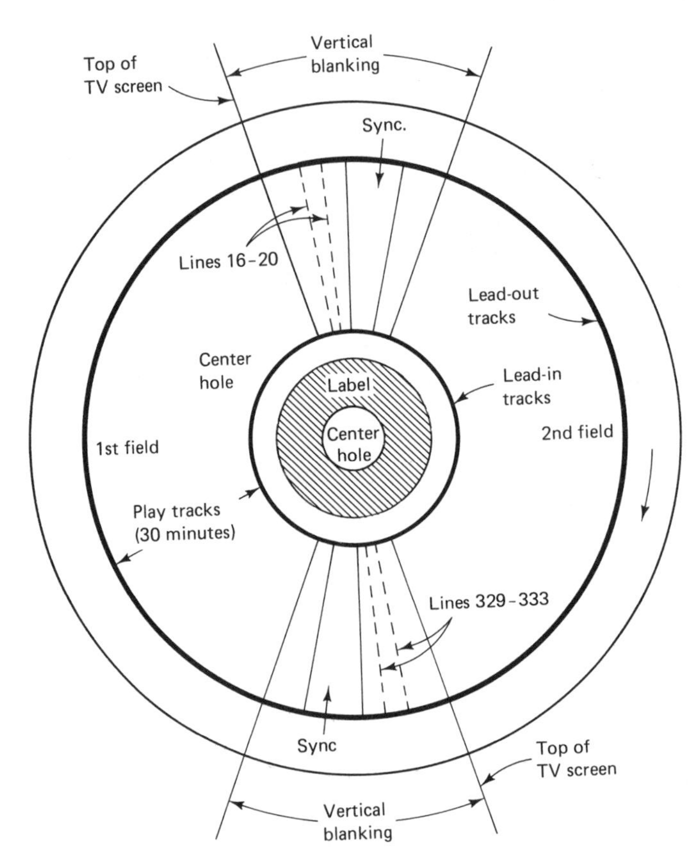

Figure 6.2 Disc layout for the CAV disc.

signals inserted in the nonvisible lines during the blanking periods. There are, for example, VIR (vertical interval reference) and VIT (vertical interval test) signals for test purposes (lines 19, 20, 332, and 333 in 625/25 systems); also a digital address signal, for various purposes (lines 16, 17, 18, 329, 330, and 331 in 625/25 systems). The address signals have the following functions:

Lead-in tracks. Some minimal number of tracks (e.g., 1000) prior to the start of the actual program contain a start code which directs the reading spot to the beginning of the program on the videodisc at high speed (e.g., nine times normal).

Lead-out tracks. A certain minimal number of tracks (e.g., 600) immediately following the end of the program contain an end code which directs the reading spot back to the start at very high speed (e.g., 75 times normal). A muting circuit in the player suppresses video and audio signals during this period.

Whether the videodisc is conductive or reflective is not generally essential for the principal elements of this chapter. If there is a significant difference, it will be pointed out; otherwise, the specific system examples describe the reflective optical disc. More specifically, a Laservision (LV) standard, in use since 1978, is assumed.

6.2 SELECTION OF OPTIMAL MODULATION TECHNIQUES

The disc record/playback process restricts the possible modulation techniques to a carrier system with binary characteristics. That is, the nonlinearity of the master recording process, which may be partly or wholly photographic, limits the choice of possible encoding techniques, and a two-level signal recording is the most attractive solution. The information can then be present only in spatial variations between successive transitions from one level to the other. A system of coding is therefore necessary if four or five signals are recorded simultaneously on a single information track in which all the information is stored for the reproduction of a color TV program [1–9].

Although the videodisc is well suited for digital modulation techniques (PCM, DPCM, or DM), the TV signal is not. At least, the circuitry needed for the digital coding of the TV signal was more complex and more expensive at the time of the system development. Therefore, taking into account the equipment complexity and the (videodisc) channel property, a modulation technique should be selected from analog pulse techniques possessing the properties of a binary system.

Figure 6.3 shows several typical spectra of analog pulse modulations. For a given baseband with a maximum frequency f_m, the typical spectrum of the modulated signal with a carrier pulse frequency f_0 will contain multiples of the carrier f_0, nf_0, each accompanied by two sidebands of bandwidth f_m. Only the first sidebands are shown, although more can exist, and the amplitudes of the spectral components decrease with frequency. In general, two easily distinguishable cases are possible: the spectrum of the modulated signal contains (Figure 6.3b) or does not contain (Figure 6.3c) baseband components. In the first group are, for example, pulse position modulation (PPM) and pulse FM with constant pulse duration. The second group contains pulse FM with a constant duty cycle (see Section 4.6). Figure 6.3d shows a special case of the second group when the duty cycle is 50%. No even harmonics of the pulse repetition frequency ($2kf_0$) are present.

The choice between these two spectra is obvious. In the case shown in Figure 6.3c, the band between 0 and $f_0 - f_m$ is not occupied—which is good for at least two reasons:

1. This bandwidth can be used for other signals (audio subcarriers, for example).
2. The possible distortion caused by the baseband influence is lower in this case.

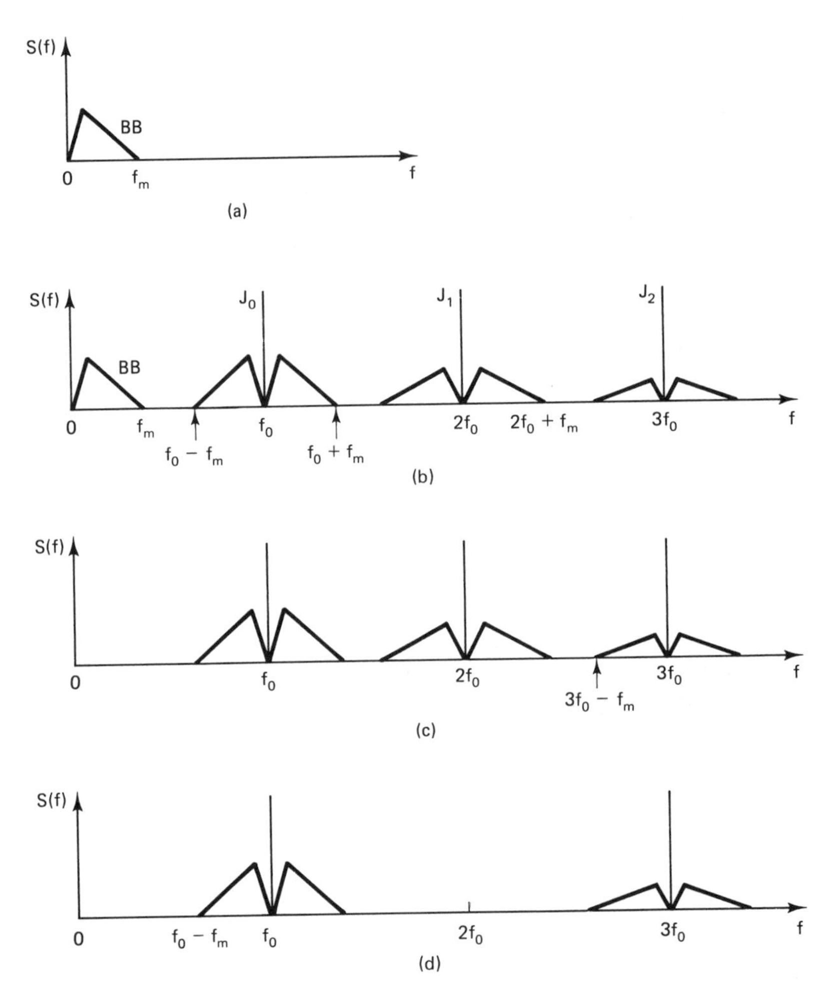

Figure 6.3 Typical spectra of analog pulse modulations: (a) baseband spectrum; (b) spectrum that contains a baseband; (c) and (d) no baseband in the modulated spectra.

This means that pulse frequency modulation best fits the channel requirements along with the economics required for this class of equipment. Because of its characteristics, FM is a most suitable technique. The realization of circuits is much simpler than that for pulse code modulation and FM accommodates a much larger variation in signal amplitude than with amplitude modulation.

Figure 6.4 shows an example of FM signal conversion for a videodisc. After the continuous-wave FM signal is passed through the limiter, the pulse FM signal is obtained. This signal is then recorded on the disc.

In general, the spectrum of the FM signal shows energy in numerous sidebands extending well away from the carrier. Fortunately, in many cases, including the

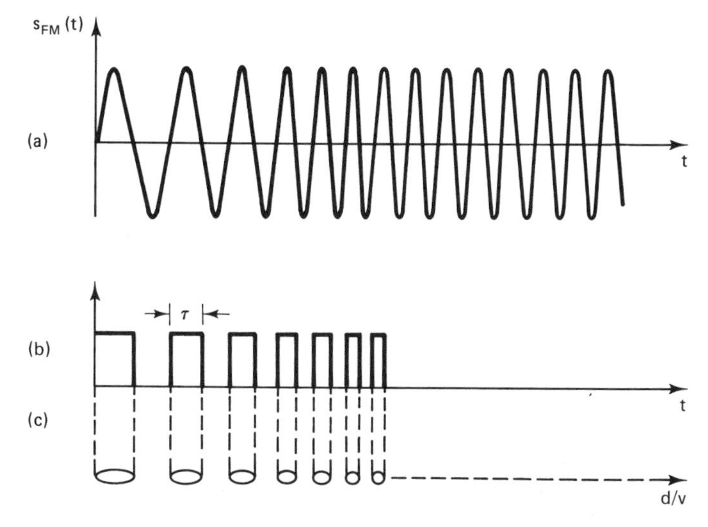

Figure 6.4 FM signal conversion to videodisc: (a) FM signal; (b) limiter output (zero-crossing detector); (c) videodisc layout.

videodisc system, the majority of the sideband energy is concentrated in a band close to the carrier. In most FM systems this band is well separated from the spectrum of the modulating signal (baseband); that is, the bandwidth of the modulating signal is small compared with the carrier frequency.

The videodisc system employs FM having the following two properties, which differ from usual systems:

1. The carrier frequency f_0 is closer to the highest modulating frequency.
2. The modulation index $(\Delta f / f_m)$ is lower than in most other FM systems.

The problem now is to select the carrier frequency so as to minimize destructive interference products, which are unavoidable, and also the products that result from second harmonic distortion.

From the expression for pulse FM with a constant duty cycle obtained in Section 4.6, it can be seen that for $\tau = T/2$ (50% duty cycle), the second harmonic of the carrier, J_1 in Figure 6.3, is zero (with all its sidebands). Also, the overall low-pass characteristic of the videodisc system, limited mainly by the MTF of the lens in the reading process, practically eliminates the higher harmonics of the subcarrier, nf_0, although they will be recovered after the signal passes through the hard limiter. So basically, continuous-wave FM can be analyzed to obtain the desired solution.

Using preemphasis in the modulation process to reduce the effect of triangular noise in the demodulated signal, higher-frequency deviations will happen for higher frequencies present in the video signal. However, this fact is of less concern at the moment. It will be seen that the main interest is in the behavior of the color subcarrier through the system.

It is convenient to think of a full-field color picture and its associated video signal as a constant luminance level with a constant amplitude and phase sinusoidal subcarrier of frequency f_{sc} added to it:

$$g(t) = a + b\cos(\omega_{sc}t + \theta) \tag{6.1}$$

The instantaneous angular frequency of the FM signal is then

$$\omega_i(t) = \omega_c + ma + mb\cos(\omega_{sc}t + \theta) = \omega_0 + \Delta\Omega\cos(\omega_{sc}t + \theta) \tag{6.2}$$

where ω_c is the main carrier, $\omega_0 = \omega_c + ma$ is the instantaneous frequency for the monochrome TV signal, and ω_0 is between ω_b and ω_w, which correspond to black and white levels, respectively.

The amplitude of the nth-order chroma sidebands is given by the Bessel function term $J_n(k\beta)$, where β is the modulation index, defined as $\beta = \Delta\Omega/\omega_{sc}$, and k is the preemphasis factor. The preemphasis affects only the amplitudes of the sidebands, not their position on the frequency axes.

Suppose now that the energy contained in the carrier (J_0) and low-order sideband components (J_1, J_2) is enough for FM signal recovery. Figure 6.5 shows one example of this. For reference only, in Figure 6.5a the spectrum is shown for a very high carrier. In Figure 6.5b the spectrum is shown for the case when f_0 is greater than $2f_c$: all components are in the natural positions. Finally, for

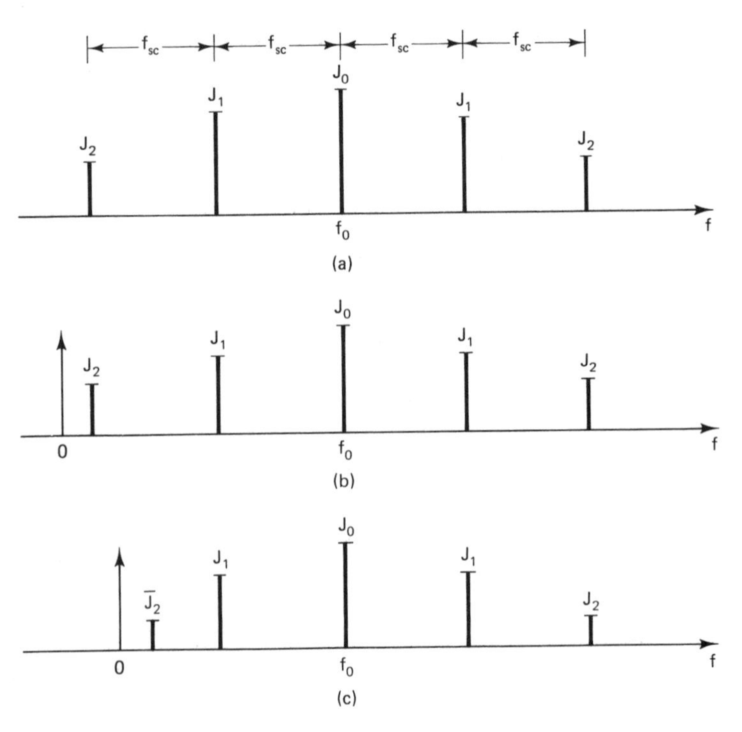

Figure 6.5 Spectrum of the FM signal: (a) $f_0 >> f_{sc}$; (b) $f_0 > 2f_{sc}$; (c) $f_{sc} < f_0 < 2f_{sc}$.

the case $f_{sc} < f_0 < 2f_{sc}$ (Figure 6.5c) the second-order lower sideband component is folded back from negative frequencies. The folded sidebands will produce an unwanted output from the demodulator at $2f_0 - 2f_{sc}$ frequency.

Consider now a monochrome picture with a gray level varying sinusoidally around a constant level; the video signal is then

$$g(t) = a + A_m \cos \omega_m t \tag{6.3}$$

Figure 6.5 can again be used to illustrate the FM signal spectrum. In all three systems (NTSC, PAL, SECAM) the color subcarrier frequency, f_{sc}, is lower than the maximum frequency of the luminance signal, $f_{m,\max}$. But fortunately, the energy contained in the higher frequencies is low so that a maximum angular frequency deviation, $\Delta\Omega$, for these signals is small (even including the preemphasis). Practically, the corresponding modulation index is small, and the first-order sidebands and carrier are enough to recover the FM signal. Thus the lower component is at frequency $f_0 - f_{m,\max} = f_0 - B$, where B is the TV signal bandwidth.

Further, in the detection process, the baseband signal has to be recovered. Thus, to avoid distortions caused by spectra overlapping, the relationship should be

$$B \leq f_0 - B \tag{6.4}$$

or

$$f_0 \geq 2B \tag{6.5}$$

This is one of the conditions for the FM carrier; the f_0 is $f_{0,\min}$ (usually, $f_0 = f_b$). Another condition can possibly be obtained from the distortion caused by the beat frequency between the FM and audio carriers, for example.

6.3 SIGNAL PROCESSING

In the videodisc system, all the information sufficient for reproduction of the color TV picture and the corresponding sound signals is contained in the pit pattern on the disc track. Therefore, adequate processing of the signal is needed prior to mastering and after reading.

6.3.1 Basic Block Diagram

Figure 6.6 contains a simplified block diagram illustrating preprocessing, that is, preparation of the signal for mastering on the videodisc. All three signals (one video, two audio) are passed through preemphasis circuits. For the video signal, the time constants are, for example, 50 ns and 125 ns, while the time constant for the audio channels is 75 μs.

The two audio carriers are summed with the FM carrier after limiting. The output signal is used to modulate the intensity of the laser beam passing through an electro-optical or acoustic-optical modulator in the master recording machine. In the summing block, basically, a frequency-sharing process is performed. For some purposes a pilot signal with frequency f_p may be added.

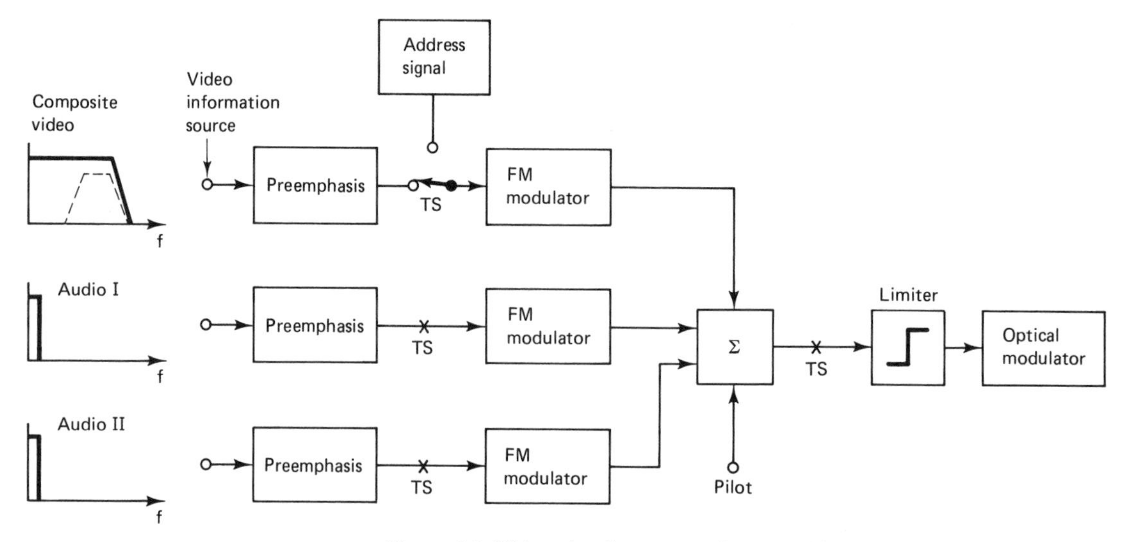

Figure 6.6 Video signal preprocessing (encoding).

A picture number code, whereby each picture of a program can be identified and displayed on the TV screen, is desired. This is always present in CAV reflective discs in each field of a full frame and may be succeeded in the next field by a stop code which would cause normal play to be interrupted and obtain a still picture. Basically, the insertion of picture numbers is a time-sharing (TS) process. As an address signal the picture number is placed in a special position in each frame. The time sharing can also be performed at other points on the diagram, marked as TS points in Figure 6.6.

Figure 6.7 presents a block diagram of the electronic part of the system for playback (player). Basically, this system should ensure that the signal read from the disc should first be formed, then the video signal and audio signals extracted in their natural bands. This information can then be played to a VHF (RF) modulator on a free channel (e.g., channels three or four for the NTSC system). This signal can then be connected to the antenna input of any standard TV receiver. The block diagram contains the basic blocks necessary to illustrate this.

A reflected light returning from the disc falls on a PIN photodiode and its output is amplified by a wide-band, low-noise preamplifier located near the photodiode. The signal is corrected to compensate for the modulation transfer function (MTF) of the reading lens. The MTF can change due to the difference in spatial frequency of the inner and outer radius and also due to the defocusing of the read spot. The compensation for changes in MTF is automatic and can be, for example, controlled by maintaining the ratio of the carrier to the first-order sidebands of the color burst. Obviously, for the capacitive disc, the PIN diode and MTF compensation are substituted with corresponding circuits.

After MTF correction the signal is limited and applied to the corresponding

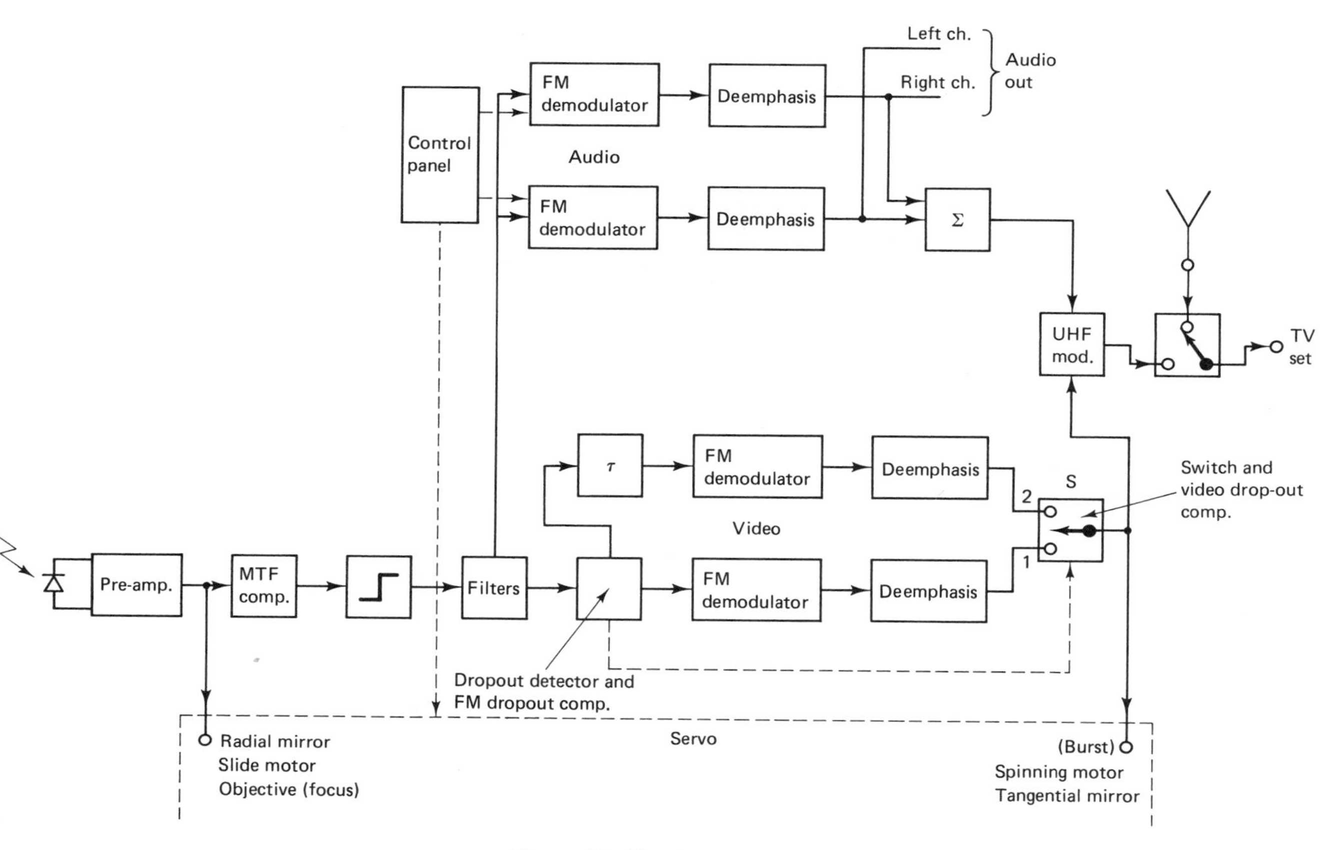

Figure 6.7 Signal decoding.

filters. A low-pass filter separates the two audio channels, which are further separated by two bandpass filters.

A high-pass filter separates the video information, and this signal is applied to the FM dropout compensator. Due to the nature of the record and reproduce process, dropouts tend to cause a missed half-cycle of the FM carrier. This requires a different compensation technique than for systems such as magnetic tape, where the duration of the dropout is generally a large portion of a horizontal line. If not compensated, the dropout would produce a very distracting spot in the TV image. To reduce the viewer's awareness of dropouts, the playback electronics includes an FM dropout compensator which detects the missing half-cycle and synthesizes a signal to replace it. Although it contains no information, the synthetic pulse greatly reduces the visibility of the dropout. This entire section, including the multiplying type of discriminator, is implemented with digital techniques for cost savings and easy alignment. When dropouts with a duration of several carrier cycles occur, the compensation is most easily done in the video domain. This is accomplished by passing the video signal through a zero-order hold which operates normally in the sample mode. When a multiple-cycle dropout is detected, the circuit is switched to the hold mode. This replaces the video with the luminance signal from the previous line for an (arbitrary) fixed duration.

The separated FM signal which carries the video information is then demodulated and deemphasis is applied to compensate for the preemphasis employed before recording, to achieve a better S/N ratio and a more uniform frequency response.

To generate an RF television signal the audio subcarrier and video signal (separated by filters) are each applied as one input to a pair of balanced modulators. The other pair of inputs is derived from an oscillator tuned to the carrier frequency of the unused TV channel selected. The outputs of the modulators are summed to form the RF signal for application to the antenna terminals of a TV receiver. A switching arrangement linked to the player power switch connects the normal TV antenna to the TV receiver when the player is not in use.

6.3.2 Encoding

The video signal here is expressed in the main as a frequency-modulated carrier which together with the audio carrier(s) is integrated into the videodisc frequency spectrum. FM video is also combined in a time-sharing process with the addressing and synchronizing signals.

6.3.2.1 Video Frequency Modulation. The selection of optimal modulation techniques for TV signal recording on a videodisc has been discussed in Section 6.2. The pulse FM with constant duty cycle (50%) was suggested as the best solution. The constant duty cycle was chosen to avoid the baseband of the modulating signal (the TV signal, in this case) in the spectrum of the modulated signal. The 50% duty cycle was chosen to avoid the second harmonic of the pulse repetition frequency and its sidebands; thus more energy is contained around the first harmonic.

Figure 6.8 illustrates a possible instantaneous frequency deviation during the line interval of an FM signal. For the sake of simplicity only the luminance (monochrome TV) signal is illustrated. For reference, the scale for signal amplitude expressed in IRE units is shown (see also Figure 5.4). By convention, 100 IRE units correspond to the amplitude between the blanking level and the white level. The burst signal is also shown, because some standards recommend that burst always be present in both monochrome and color video signals.

Characteristic frequencies are marked:

f_w: white level (peak white)
f_b: blanking level
f_s: sync tip
f'_b: black level

It will be assumed that blanking and black levels are equal; that is, $f_b = f'_b$.

Sometimes, frequencies between f_b and f_w are referred to as the luminance carrier frequencies, f_0; the period of the pit pattern is determined by f_0. One possible choice of these frequencies is: $f_s = 7.6$ MHz, $f_b = 8.1$ MHz, and $f_w = 9.3$ MHz. The difference $\Delta F = f_w - f_s$ is the maximum frequency deviation (1.7 MHz, in the numerical example).

In Figure 6.8, it is shown that the carrier is deviated higher by the white level (a higher level of the modulating video signal) and lower by the sync tips. This is referred to as *positive modulation*. A case also exists for using the inverse system (sync tips high), referred to as *negative modulation*. Positive modulation is preferable in a videodisc system because dropouts have a lesser influence.

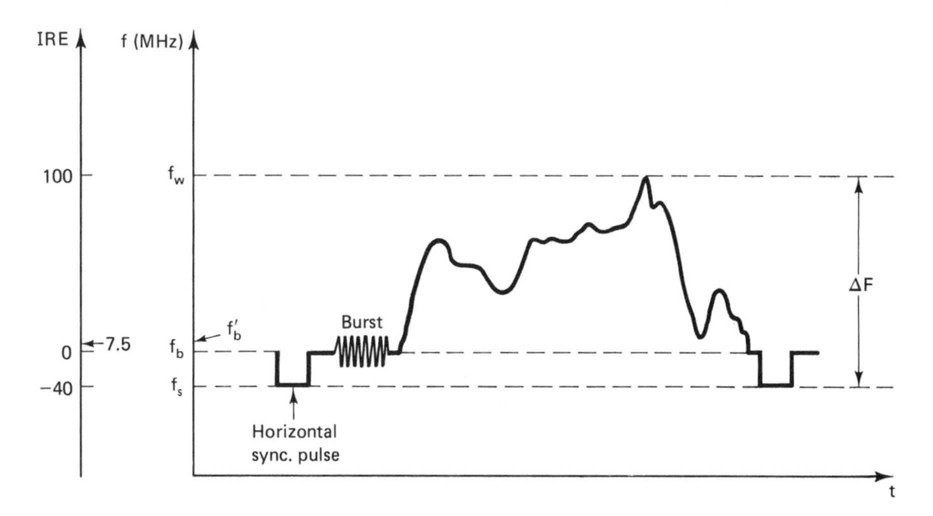

Figure 6.8 FM signal deviation (instantaneous frequency) during a line interval.

The requirements for preemphasis have already been discussed in Chap. 4. The circuits should be simple, passive, and easy to specify. The preemphasis circuit is fed from a low-impedance source and feeds a high-impedance load. The normalized transfer function, $F(\omega)$, can be specified by two time constants, τ_1 and τ_2:

$$F(\omega) = \frac{1 + j\omega\tau_1}{1 + j\omega\tau_2} \tag{6.6}$$

The condition for boosting high frequencies is $\tau_1 > \tau_2$. The 3-db frequencies are $f_1 = 1/(2\pi\tau_1)$ and $f_2 = 1/(2\pi\tau_2)$.

The asymptotes are:

For low frequencies, $\omega \rightarrow 0$: $F(\omega) = 1$
For high frequencies, $\omega \rightarrow \infty$: $F(\omega) = \tau_1/\tau_2$

The high frequencies are therefore boosted by a factor of τ_1/τ_2.

The amplitude characteristics are shown in Figure 6.9 for two cases: (a) $\tau_1 = 125$ ns, $\tau_2 = 50$ ns, and (b) $\tau_1 = 320$ ns, $\tau_2 = 120$ ns.

6.3.2.2 Audio Frequency Modulation and Audio/Video Combination (Frequency Sharing). If the disc contains two independent audio channels, this offers possibilities of:

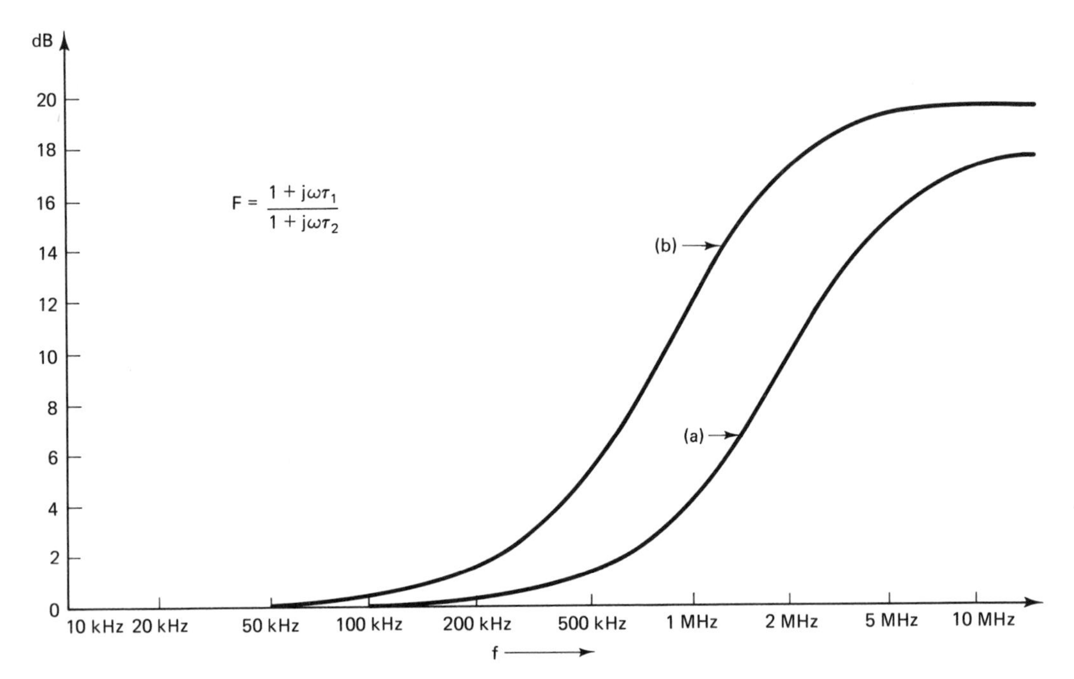

Figure 6.9 Preemphasis amplitude characteristics: (a) $T_1 = 125$ ns, $T_2 = 50$ ns; (b) $T_1 = 320$ ns, $T_2 = 120$ ns.

1. Stereophonic sound
2. Monophonic sound: two independent audio programs
3. Monophonic sound: one audio program on both channels
4. Use of one or both channels for control or cueing information

In principle, the audio information can be added to the video signal before video FM is performed, but there are reasons not to use this method. The FM triangular noise property would make the audio channel noisy. The MTF would introduce further distortion because the video carrier, as already discussed, should be as high as possible, whereas considerations of inexpensive optics and a long depth of focus dictate as low a cutoff frequency for the lens as is acceptable. Thus the audio information should be added to the video (FM) information based on a frequency-sharing principle.

The signal waveforms in Figure 6.10 illustrate the principle of adding an additional audio carrier signal, frequency f_a, into an already frequency-modulated signal of frequency f_0. Figure 6.10a presents a video FM signal, and Figure 6.10b an audio carrier signal. Figure 6.10c shows the signal obtained by summing the two prior signals. It may be noted that the shifting of zero crossings by passing this signal through the limiter depends on the signal in Figure 6.10b. The signal in Figure 6.10d is obtained after limiting the output from Figure 6.10c, and contains the spectral components $f_0, f_a, f_0 \pm kf_a, f_0 \pm m(f_0 - f_a)$; $k, m = 0, 1, 2, \ldots$. As will be seen, the most important undesired component is $2f_0$ −

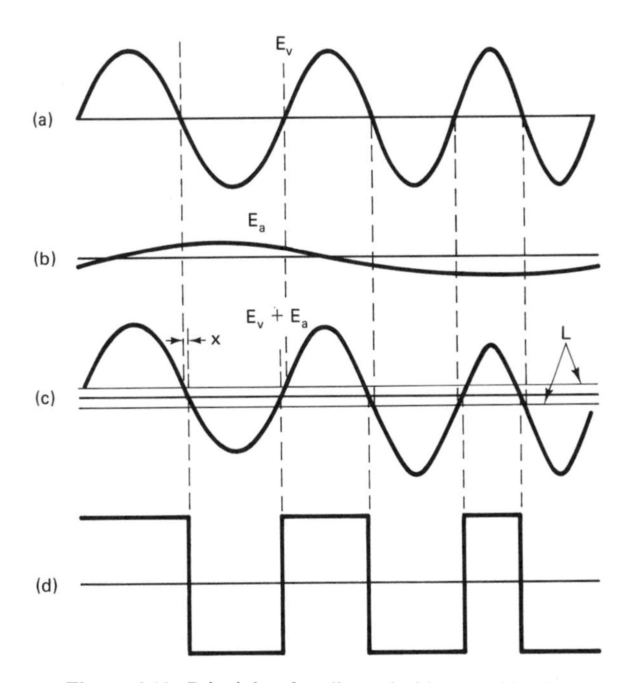

Figure 6.10 Principle of audio and video combination.

f_a, because this lies in the luminance signal band and cannot be removed by filtering [10].

Practically, the audio component is added using duty-cycle modulation. As shown in Section 4.6, the pulse FM signal contains baseband components. As an example, consider the signals carrying video and audio information:

$$s_0(t) = A_0 \cos \omega_0 t \tag{6.7}$$

$$s_a(t) = a \cos \omega_a(t), \qquad a \ll A \tag{6.8}$$

As a result of hard limiting the sum of these signals, the spectral components of interest are shown as the solid lines in Figure 6.11. The components $f_0 \pm 2f_a$ are in fact due to amplitude modulation of the video carrier and can be removed by bandpass filtering and limiting if they are not in the bandwidth of the luminance signal.

The processes in the videodisc system are such that occasionally the average duty cycle of the modulated signal at the output will have an offset:

$$s(t) = s_0(t) + s_a(t) + \alpha \tag{6.9}$$

This is equivalent to adding a dc value to the sum of the signals. For small values of α the average duty cycle will be $0.5 + \alpha/\pi$ rather than 0.5. In this case, the spectral components would also contain the dashed lines in Figure 6.11.

In the previous examples, for the sake of simplicity, only one signal carrying audio information is considered. With two audio carrier signals with frequencies f_{a1} and f_{a2}, the situation differs in the sense that the beat frequency $(f_{a2} - f_{a1})$ can cause many mirror components, which will be considered as signal distortion.

The baseband audio signals cannot be added directly to the video FM signal. Two main reasons for this are: (1) we need to add more than one audio channel, and (2) at very low frequencies (close to dc) the channel has more noise. So, audio subcarriers should be used. Because of the properties already discussed,

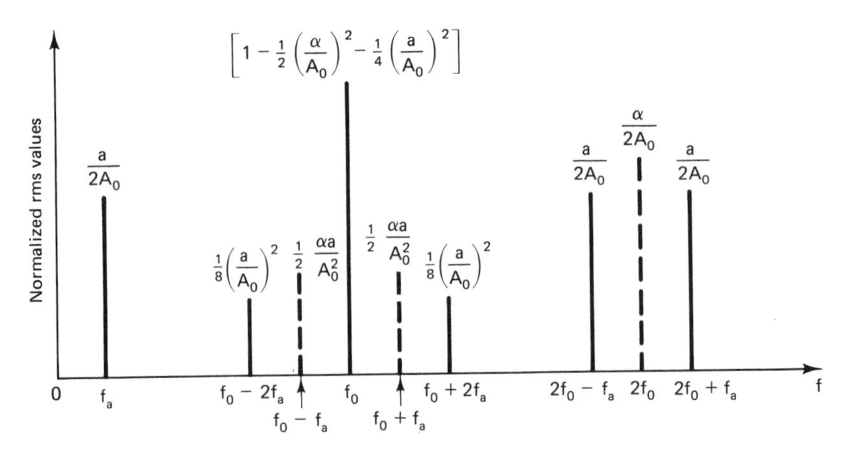

Figure 6.11 Spectrum of duty-cycle modulation.

FM is a good choice (a deviation of ± 100 kHz for 100% modulation would be sufficient). Both audio subcarriers should have the same polarity of modulation. The audio signals, prior to FM, should have a preemphasis of +6 dB/octave (with, e.g., a time constant $\tau = 75\ \mu$s for a 20-Hz to 20-kHz audio bandwidth). The level of the audio subcarriers in the recorded frequency spectrum should be low enough with respect to the unmodulated main carrier (e.g., -26 dB) so that no Moiré (interference) components are seen in the reproduced picture.

The placement of the audio subcarriers also requires careful consideration. Very low frequencies are undesirable because large luminance components in baseband video may cause interference in the audio channel(s) (see Section 6.5.4) (duty-cycle offset should be expected with the videodisc system). In addition, there is more disc noise at low frequencies. The upper limit is $f_{0\min} - f_m$, where $f_{0\min} = f_b$ (Figure 6.8) and f_m is the maximum spectral component in the luminance signal. This means that the audio subcarriers in the videodisc system (f_{a1}, f_{a2}) should be selected inside the video baseband in order to avoid a color subcarrier (f_{sc}) influence. Also, to lower the visibility of the beat components, these components should be interleaved with the video spectral components. For the NTSC standard, one possible choice is $f_b = 8.1$ MHz (Figure 6.8) and:

Channel I: $f_{a1} = 146.25 f_H = 2.301136$ MHz (nominal)
Channel II: $f_{a2} = 178.75 f_H = 2.812499$ MHz (nomimal)

The offset of $\frac{1}{4} f_H$ and $\frac{3}{4} f_H$ is chosen to minimize audio interference due to the $f_{a2} - f_{a1}$ beat frequency.

Pilot Signal. An additional requirement in videodisc recording can be to record a pilot signal (tone). To prevent the pilot tone (frequency f_p) from interfering with the video and audio FM signals, it must be placed either above or below the FM band. It has already been shown that it is important to place the FM band as high as possible, so it is advantageous to place the pilot below the FM signal. If a pilot is added to the FM signal it must be ensured that crosstalk between the signals is minimal. One possible choice for the previous example is $f_p \simeq 400$ kHz, although the disc/laser (recording) noise can be significant at this frequency.

The physical process for a pilot signal is similar to that for audio signals. Frequency sharing, that is, superposition in the frequency domain, should be used.

Originally, the pilot signal in early videodisc systems was added for the time-base error correction. It is shown both experimentally and theoretically [Ref. 27, App. 3.4], that in the presence of noise, the burst technique performed better than the pilot. Because the idea of continuous pilot signal may be, as found lately, attractive in some applications for interactive video-simulators (flight); for example, let us consider, in short, offset and no offset criteria for the f_p.

There are two basic problems: visibility (interference) of the pilot signal and focus error after jump. After FM detection a "residue" component of the pilot signal is in the video bandwidth. If visible, the visibility is less if there is $\frac{1}{2}f_H$ frequency offset in the (baseband) detected video signal. Also some bit components can be located in the video baseband. The amplitude of the pilot signal should be made small enough (-26 dB in reference to the unmodulated video carrier) at the recording process, so that there is no visible interference, and f_p location is irrelevant.

On the other hand, the phase of the reference signal is important any time, including the time after a jump. For that reason it is convenient if $f_p = nf_H$, so that there is no phase change from line to line.

6.3.2.3 Vertical Interval Control and Address Signals (Time Sharing). The code signals on a videodisc provide special information which can be utilized by the player to control special functions and provide picture frame or time information. For reference only, let us recall some CCIT recommendations for the 60-Hz/525-line NTSC system. It is recommended that the video signal contain a vertical interval reference (VIR) signal on lines 19 and 282 (Figure 5.A.2). The VIR signal will not be present in a monochrome video signal. It is also recommended that the video signal contain a composite test signal on line 20 and a combination test signal on line 283 (Figure 6.12). These are international test signals (ITS).

In the video signal, lines 10 to 18 and 273 to 281 are reserved for address and data signals. The lines not specified have a video content set at the blanking level and are reserved for future applications.

Basically, there are two formats for optical videodiscs: CAV and CLV. The CAV format has the following types of codes:

1. Lead-in
2. Lead-out
3. Picture numbers
4. Picture stop
5. Chapter numbers

On CLV format, the codes are:

1. Lead-in
2. Lead-out
3. Program time code
4. Chapter number
5. CLV code

Sometimes it is necessary to uniquely identify one field in two with NTSC or one in four with PAL and SECAM; thus a method of unique addresses for each frame or field on the videodisc is needed. This can be accomplished by recording a sequential digital code on the cue track with a number unique to that frame. This is usually inserted as a 24-bit digital word (Philips code, or 40 Bit-MCA code) in selected video lines during the vertical interval (Fig. 6.13). In

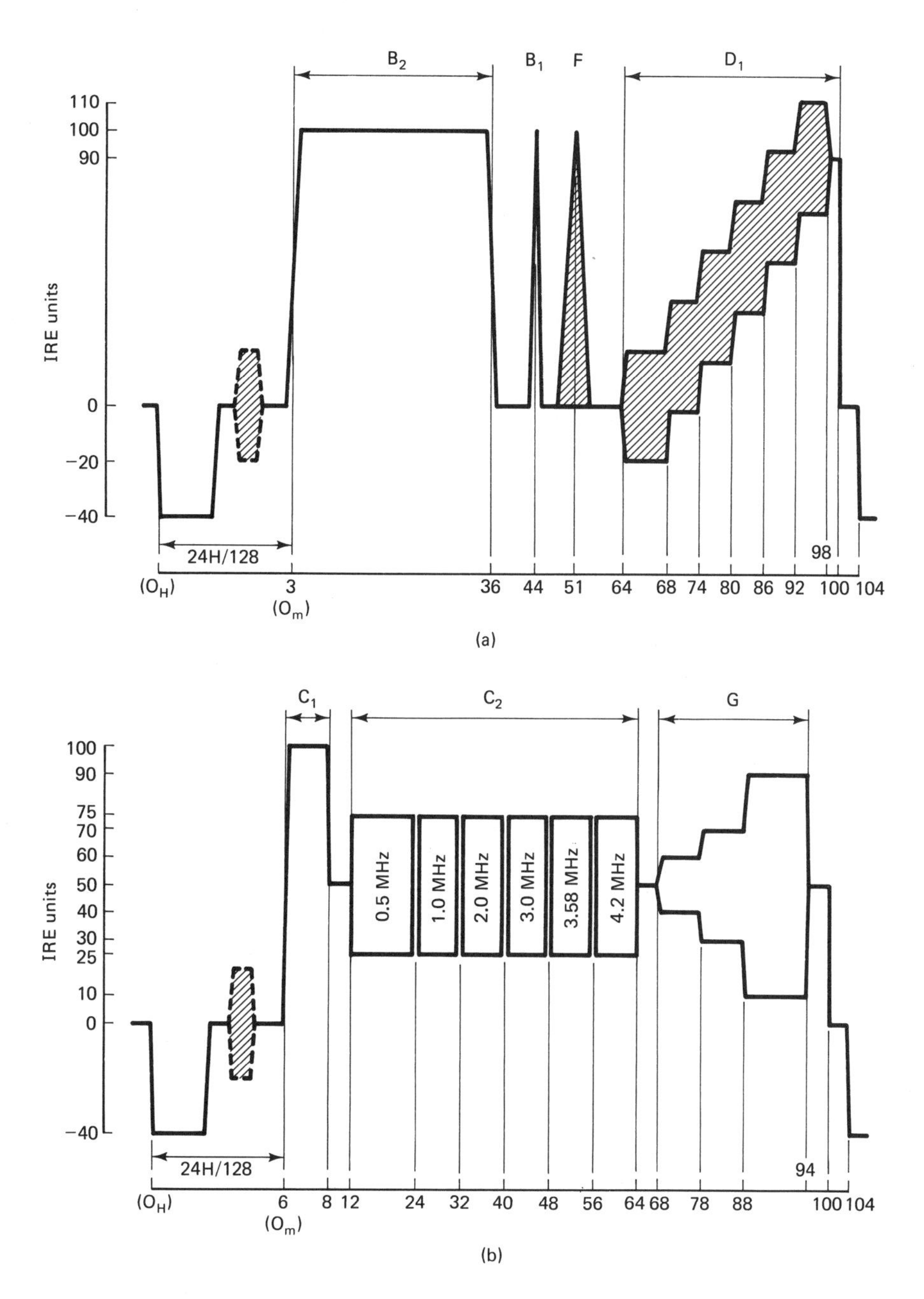

Figure 6.12 Test signals: (a) composite; (b) combinational.

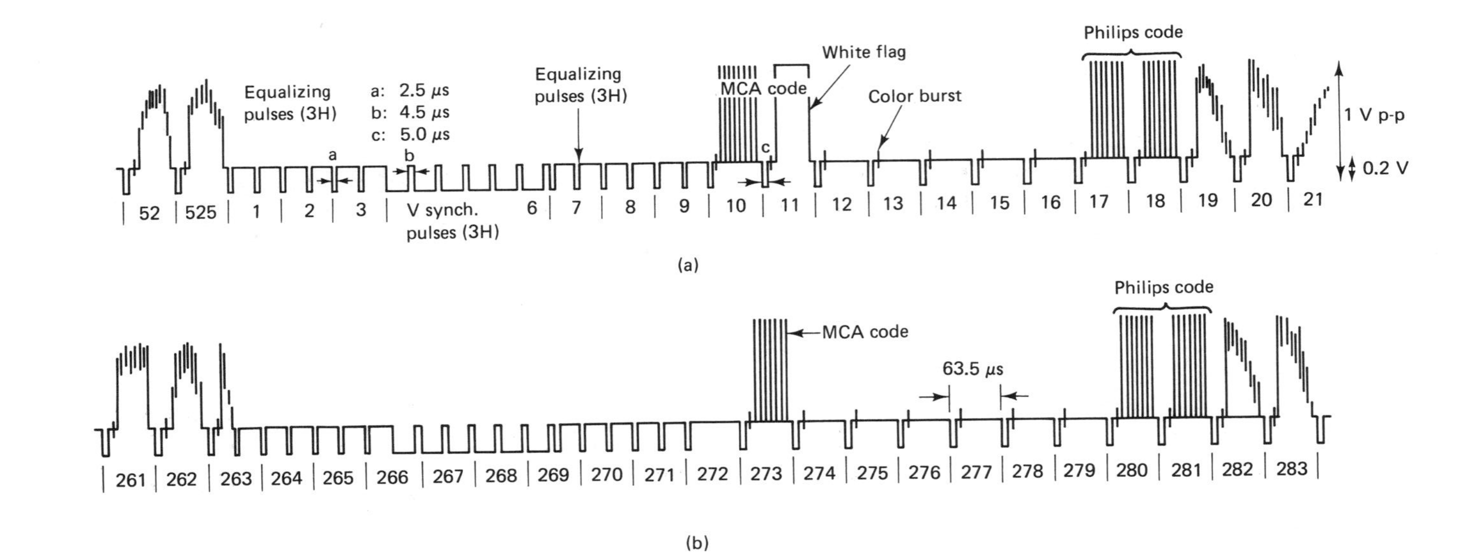

Figure 6.13 Vertical interval signal.

this case, each digital word is subdivided into six groups of 4 bits and each group can be any hexadecimal word. The first group of 4 bits is the key and starts with a logic 1. Each bit cell has a fixed duration and a digital level between 0 and 100 IRE.

The lead-in code indicates where the program starts: for example, a 24-bit lead-in code with the hexadecimal value 88FFFF inserted into lines 17, 18, 280, and 281 (60-Hz/525-line NTSC) during at least 900 tracks prior to the active program start.

The lead-out code immediately follows the active program for at least 600 tracks in duration. A 24-bit lead-out code with, for example, a hexadecimal value of 80EEEE can be inserted (for 60-Hz/525-line NTSC) in lines 17, 18, 280, and 281.

The picture numbers are always present on CAV discs. They are unique and in normal ascending sequence. In the video source material (regular TV signal) the 24-bit code is inserted in lines 280 and 281 (NTSC standard). For film source material, if there are two or more fields scanned from the same photographic picture or if there are two fields that are made equal by electronic processing, the 24-bit code can be inserted in lines 17 and 18 or in lines 280 and 281, depending on which field is the first field of the picture.

On CAV discs, the picture stop code enables the playback equipment to switch automatically to a still-picture mode or from normal speed to slow motion. The 24-bit picture stop code (with the hexadecimal value 82CFFF, for example) can be inserted in lines 16 and 17 or 279 and 280 (NTSC) of the film immediately following the field in which the 24-bit picture number was inserted, to enable stopping on the selected picture. On CLV discs there are no picture stops. During the early years of development, the fifth bit of the picture number code was used to indicate (provide for) a picture stop.

Chapter numbers indicate part of the program as chapters and are optional. They are unique and in a normal count sequence starting with 1 at the start of active program material. The 24-bit coded chapter numbers, if present, can be inserted in lines 18 and 281 of those fields that do not have picture numbers inserted on CAV discs. On CLV discs these are inserted in line 281.

Each chapter number starts with a stop bit (the first bit after the key bit) at a zero logic value for 400 tracks, followed by at least 400 tracks with the stop bit at a logic 1 value until the next chapter number starts. The zero-value stop bit is inserted to disable the search action of the player. The hexadecimal value, for example, may be $8X_1X_2DDD$, where X_1 and X_2 are the chapter numbers. The maximum number in this example is therefore 79.

The program time code is always present on CLV discs during the active program and indicates the running time expressed in hours and minutes. The 24-bit time code with, for example, a hexadecimal value of $FX_1DDX_2X_3$ can be inserted in lines 17 and 18, where X_1 indicates hours, and X_2 and X_3 indicate minutes.

The CLV code is always present during the active program on a CLV disc.

It is intended to disable certain functions of the player. The 24-bit CLV code with a hexadecimal value of, say, 87FFFF would be inserted in line 280.

Recording Codes. Many different wave shapes for encoding digital information are being used, each with its own application and limitations [11–26]. If clock and data are derived from the read waveform, transitions must occur frequently enough to provide synchronization pulses for the clock. On the other hand, consecutive transitions must be far enough apart to limit the interference to an acceptable level for reliable detection. For this reason, the binary data are coded in binary sequences that correspond to waveforms in which the maximum and minimum distances between consecutive transitions are constrained by prescribed coding rules. The most common waveforms are shown in Figure 6.14. A digital word 00010101110 is used to compare the various codes. Corresponding codes are defined in Table 6.1. The codes can be classified as: Nonreturn-to-zero (NRZ)-a) b) and c, Return-to-zero (RZ)-d) e) f) g), Double frequency-h) i) j), and Double density codes-k) l).

NRZ codes are not run-length-limited codes. Large strings of 0's or 1's could result in no transitions, giving the signal a very low frequency component. In consequence, the "eye" pattern can become completely closed, rendering simple binary threshold detection impossible.

The return-to-bias (RB) system is more efficient than RZ but still requires high resolution. It also requires a reference at the bit rate because if a long series of zeros are being retrieved, the number of missing pulses must be determined.

Ratio code is similar to RZ in the sense that it is a self-clocking code. Compared with RB, the price for self-clocking is a lower signal-to-noise ratio.

In the class of double-frequency (line) coding, redundancy is introduced by using a signaling rate equal to twice the input binary data rate. The coded (line) signal is dc free and contains a large number of transitions from which timing information can be recovered. Double-frequency coding (and its variants) has many names, such as biphase, split-phase, Manchester codes, phase encoding, frequency-shift keying, frequency modulation (FM), conditioned diphase, or two-level AMI class II coding.

In biphase-level (or split-phase, phase-encoding, Manchester II + 180°) (Figure 6.14h): A transition is made during a clock time only if it is required to give the correct phase transition for the following bit; this means that a transition always occurs in the middle of each bit interval and also between identical bits. In biphase-mark (or Manchester I) (Figure 6.14i), the direction of the transition is immaterial; therefore, the phase or polarity of the signal is not important. The decoding of the signal is more complex because the clock edges and the bit information edges have to be separated.

The biphase-space (or Manchester I + 180°) code (Figure 6.14j) is the complement of biphase-mark.

Double-density codes are known by two other names: delay modulation and modified frequency modulation (MFM). They are basically phase shift codes.

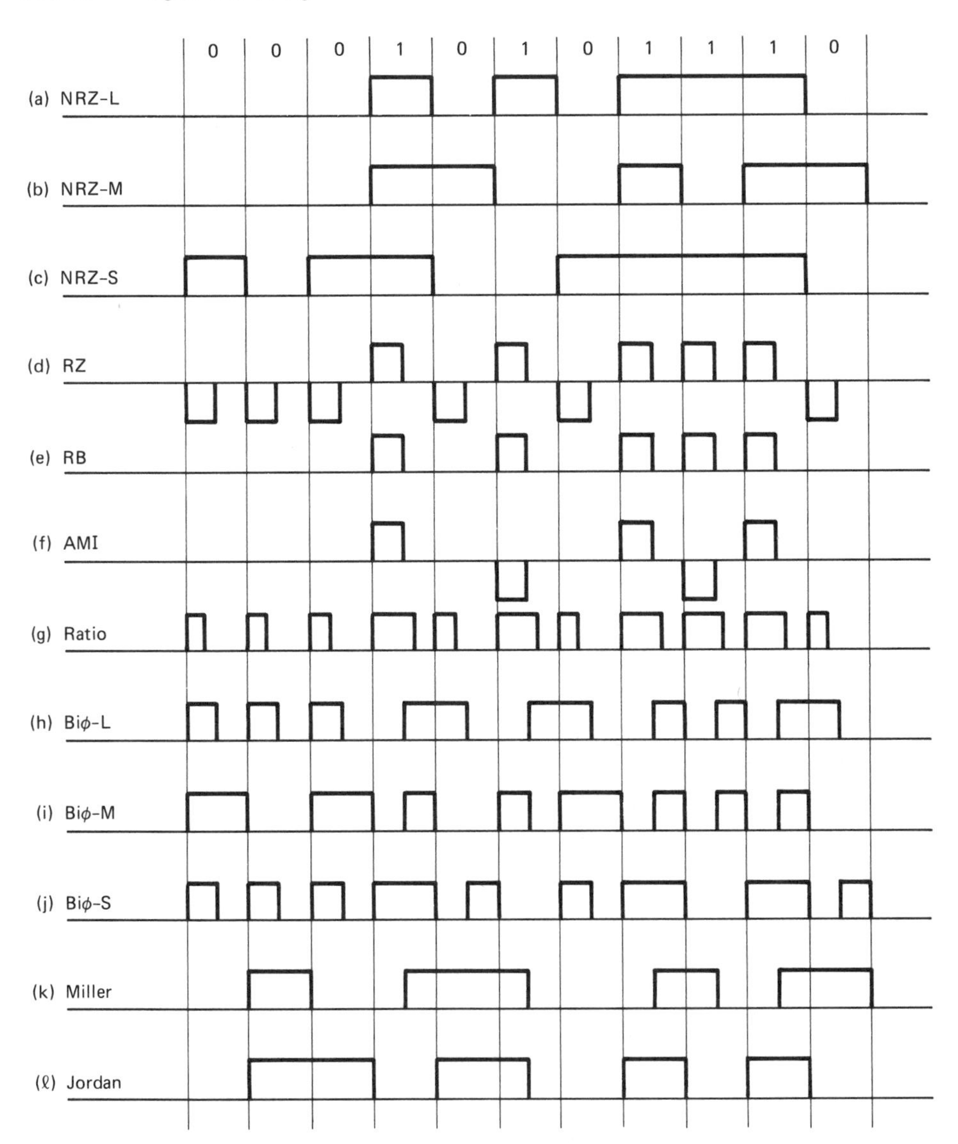

Figure 6.14 Code waveforms.

The bandwidth requirements of these codes are slightly greater than for NRZ. The price of this is the double frequency clock, a 3-dB loss in the signal-to-noise ratio, and the requirements for the data pattern to establish a correct phase of the clock. The encoders for double-density codes are relatively simple, while the decoder requires a slightly more sophisticated system. The synchronizing data patterns are: "101" for the Miller code and "000" for Jordan code.

In Figure 6.15, the spectral power density curves for NRZ-L, biphase-L

TABLE 6.1 DEFINITION OF DIGITAL SIGNAL ENCODING FORMATS

Code Group	*Names and Algorithms*
NRZ	a) Nonreturn to Zero-Level (NRZ-L) 1 = High level (level L_1) 0 = Low level (level L_0) b) Nonreturn to Zero-Mark (NRZM) 1 = a change in level 0 = no change in level c) Nonreturn to Zero-Space (NRZ-S) 1 = no change in level 0 = a change in level
RZ	d) Return to Zero (RZ) 1 = positive pulse (a half-bit wide) 0 = negative pulse (a half-bit wide) e) Return to Bias (RB) 1 = pulse in first half of bit interval 0 = no pulse f) Bipolar or Alternate-Mark-Inversion (AMI) 1 = pulse in first half of bit interval, alternating polarity from pulse to pulse 0 = no pulse g) Ratio Code 1 = pulse wider than a half-bit interval 0 = pulse narrower than a half-bit interval
Biphage (double frequency)	h) Biphase-Level (Manchester) 1 = transition from high to low in middle of bit interval 0 = transition from low to high in middle of bit interval i) Biphase-Mark Always a transition at beginning of bit interval 1 = transition in middle of bit interval 0 = no transition in middle of bit interval j) Biphase-Space Always a transition at beginning of interval 1 = no transition in middle of interval 0 = transition in middle of interval
Double density	k) Miller Code 1 = transition in middle of interval 0 = no transition if followed by 1; transition at end of interval if followed by 0 l) Jordan Code 1 = transition(s) in the middle or at both ends of the bit interval 0 = transition at the beginning or at the end of bit interval

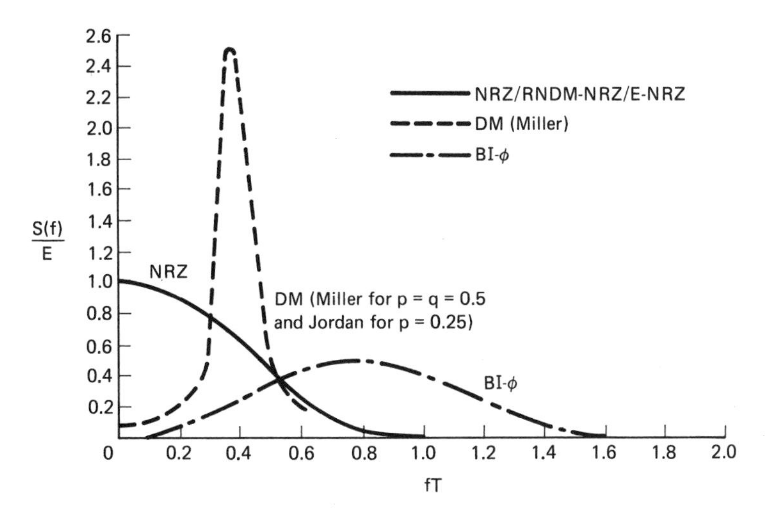

Figure 6.15 Plots of the power spectral density distributions for various codes. T = bit period.

and delay codes are compared. The normalized frequency scale, 1/T, where T equals the bit period in seconds, is used. Note that the curves differ significantly.

In Figure 6.16 an idealized form of the (video) signal containing binary data is shown. Four bits (1001) are shown in the biphase-L code. Each bit cell is 2 μs long with a digital level between 0 and 100 IRE. This is the Philips code: a 24-bit digital word; in the MCA code 40-bit digital words are used in the biphase-M code.

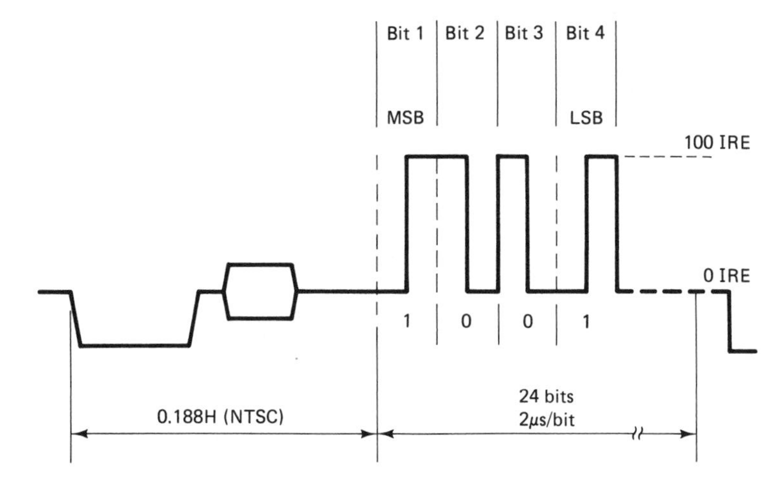

Figure 6.16 Bit-cell length and digital level.

6.3.3 Frequency Spectrum of the Recorded Signal

The nature of the input signals, as well as the encoding process and nonlinearities in the channel, make the frequency spectrum of the recorded signal very complex. The information carried by each individual input spectral component is widely spread over the frequency domain after the encoding operation. Also, the interaction of the input signals (e.g., audio and video) produces the beat components between them, and the way the signal system deals with them is also complex. What follows is therefore only a simplified presentation of the idealized system. First consider the case of video encoding by FM (higher harmonics of the pulse instantaneous frequency, kf_0, for $k = 2, 3, \ldots$ in Figure 6.3, and their sidebands, are disregarded). The frequency spectrum of the FM modulated video signal is shown in Figure 6.17. Also, the video modulating signal for the constant color picture is depicted in the time and frequency domain. The corresponding instantaneous carrier is marked as f_0 (dashed line). The chroma sidebands are shown in Figure 6.17 as dashed lines:

J_1: first-order chroma sideband, $f_0 \pm f_{sc}$
J_2: second-order chroma sideband, $f_0 \pm 2f_{sc}$
J_3: third-order chroma sideband, $f_0 \pm 3f_{sc}$

The lower third-order chroma sideband, $3f_{sc} - f_0$, is a foldover component in this example; for $f_0 = 2.5f_{sc}$ the lower third- and second-order chroma cannot be distinguished in frequency.

In general, the main carrier deviates from f_b and f_w, corresponding to the black and white luminance levels, respectively. The instantaneous frequency for the sync pulse level is f_s. When the level of the luminance signal changes from the black to the white level, the instantaneous frequency (f_0) would change from f_b to f_w; this is marked with an arrow associated with J_0. The corresponding chroma sidebands will change as shown with arrows on J_1, J_2, and J_3. Notice the inversion of direction for the lower third-order chroma sideband (J_3).

The numerical example in Figure 6.17 is for $f_b = 8.1$ MHz, $f_w = 9.3$ MHz, $f_{sc} = 3.6$ MHz, and $f_s = 7.6$ MHz. The boundary frequencies are (NTSC standard) for $k = 1, 2, 3$:

$f_b - kf_{sc}$: 4.5 MHz, 0.9 MHz, 2.7 MHz
$f_b + kf_{sc}$: 11.7 MHz, 15.3 MHz, 18.9 MHz
$f_w - kf_{sc}$: 5.7 MHz, 2.1 MHz, 1.5 MHz
$f_w + kf_{sc}$: 12.9 MHz, 16.5 MHz, 20.1 MHz

or:

J_0: 8.1 to 9.3 MHz
J_1: lower sideband, 4.5 to 5.7 MHz; upper sideband, 11.7 to 12.9 MHz
J_2: lower sideband, 0.9 to 2.1 MHz; upper sideband, 15.3 to 16.5 MHz
J_3: lower sideband, 1.5 to 2.7 MHz; upper sideband, 18.9 to 20.1 MHz

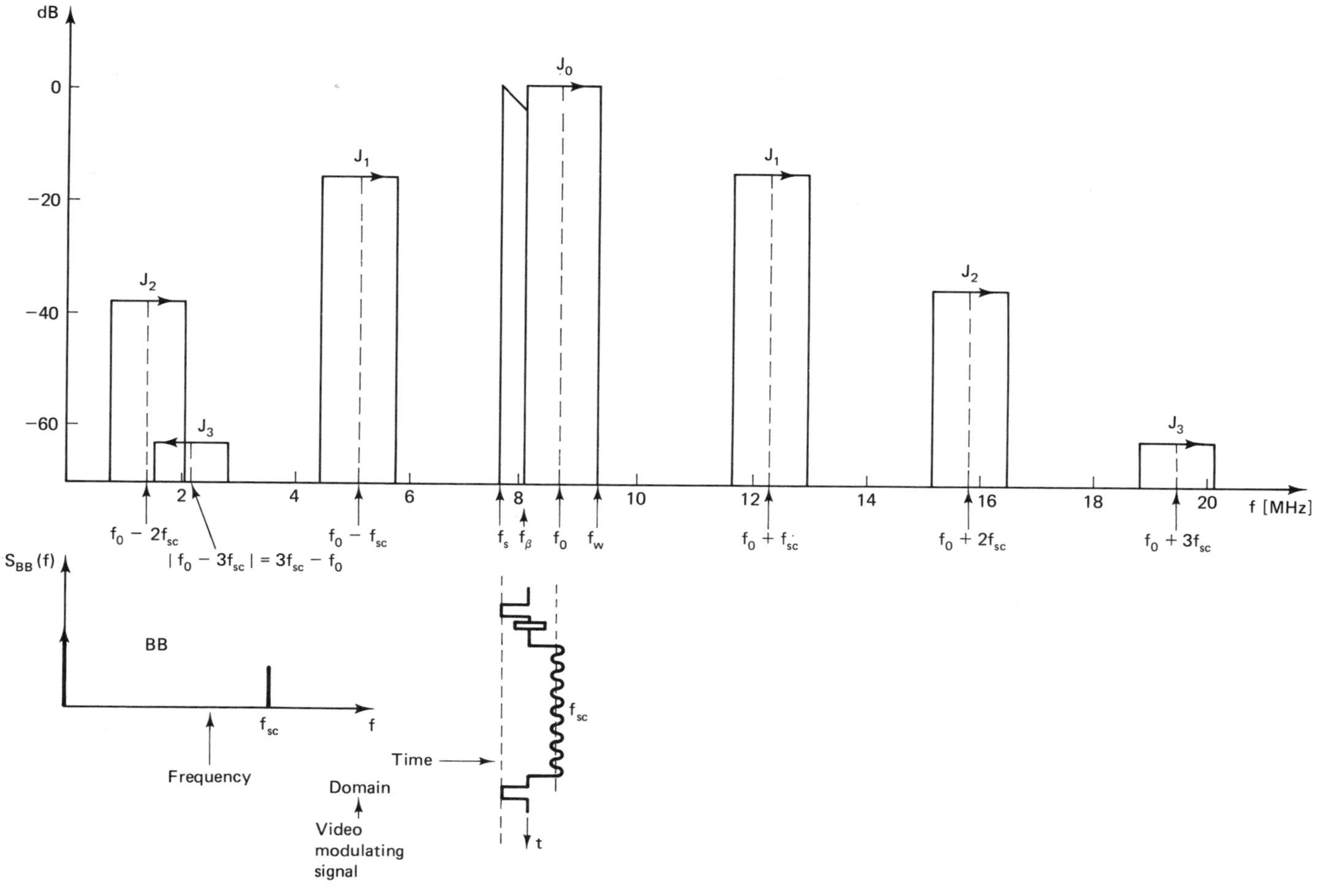

Figure 6.17 Frequency spectrum of the FM modulated video signal.

Figure 6.18a shows, for an idealized case, the envelope of the spectrum in the channel. Only the first-order sidebands of the video signal are marked. The typical envelope of the video signal spectrum (BB) is shown for reference. The dashed line refers to the monochrome video signal. It is assumed that spectral components in the baseband video higher than the maximum frequency, f_m, are filtered out.

To reduce the intermodulation components, the pilot (if it exists) and the audio carriers, f_{a1} and f_{a2}, are chosen about 26 dB below that of the unmodulated main carrier. Because of the effects of the limiter on the spectrum (which generates the new spectral components, i.e., spreads the signal energy), these amplitudes should be 6 dB higher before limiting: 20 dB below the unmodulated carrier.

Based on previous discussions, the frequency occupation in the channel is shown in Figure 6.18b. This is a simplified version and no distortion or beat components are indicated. Normally, the pilot is not used. Lately, in that frequency range, the CD audio channel-digital audio is placed.

6.3.4 Decoding (Demodulation)

The basic block diagram of the electronic part of the player is presented in Figure 6.7. It shows that the audio subcarriers and the video FM signals are separated by filters. First consider the case of audio decoding. For the audio encoding, duty-cycle modulation is used. From the theoretical considerations in Chapter 4 (Section 4.6) it follows that the modulating signal can be recovered by a low-pass filter without harmonic distortion. This is also illustrated in Figures 6.3b and 6.11. The distortion can come from the lower sidebands of harmonics of the pulse repetition frequency and video baseband and sideband components, which fall into the demodulating filter's passband. After a low-pass filter separates the two audio channels (and pilot signal, if it exists), they are further separated by bandpass filters. For audio signal demodulation, standard FM demodulators can be used.

A high-pass filter (Figure 6.7) separates the video (FM) information. In general, the disadvantage of using more conventional frequency discriminators, such as the Foster-Seeley or ratio detector, on videodisc FM signals is the problem of linearity over such a wide deviation. Such circuits provide excellent performance at center frequencies of 10.7 MHz with deviations of ± 75 kHz in the audio modulating frequency range. With deviations of ± 1 MHz, it is difficult to obtain linearity, and at modulating frequencies comparable to the center frequency, it is difficult to separate the RF from the video.

For continuous-wave (CW) FM, the modulating signal $g(t)$ can be demodulated by the product FM demodulator with the delay line (Figure 6.19). If the input FM signal is

$$s(t) = A_c \cos\left(\omega_c t + \mathrm{m}\int_0^t g(t)\,dt\right) \tag{6.10}$$

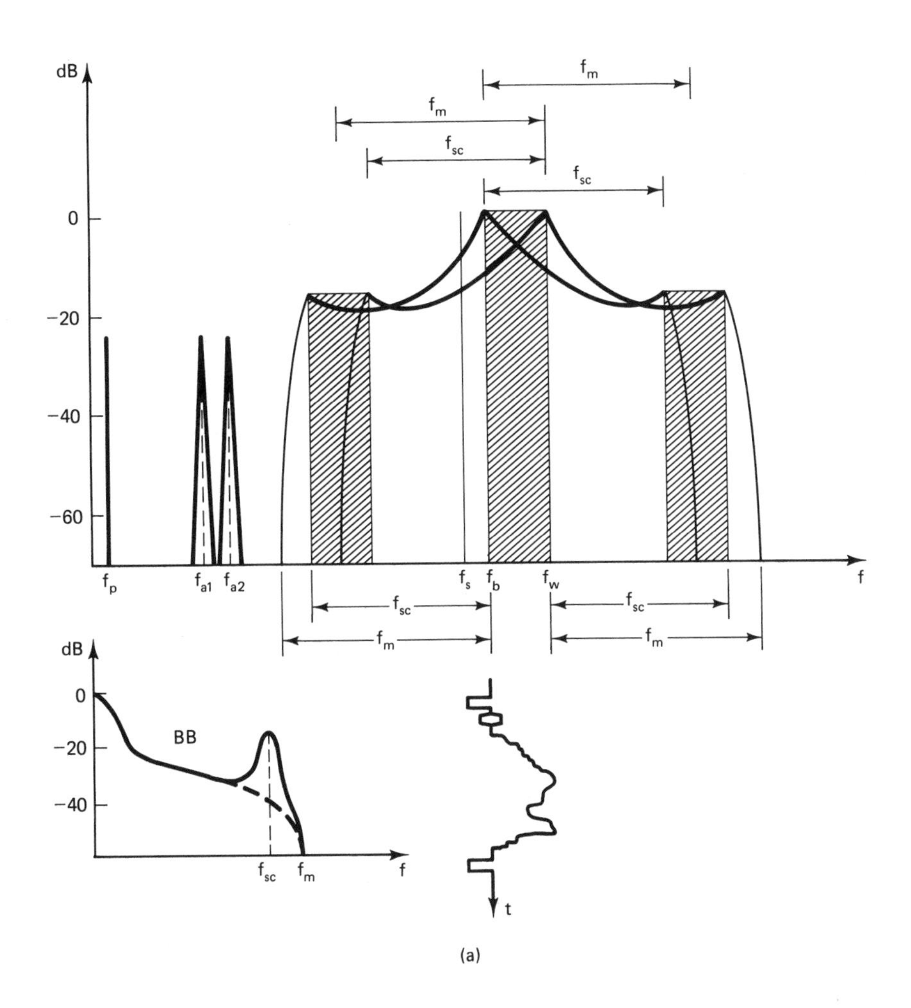

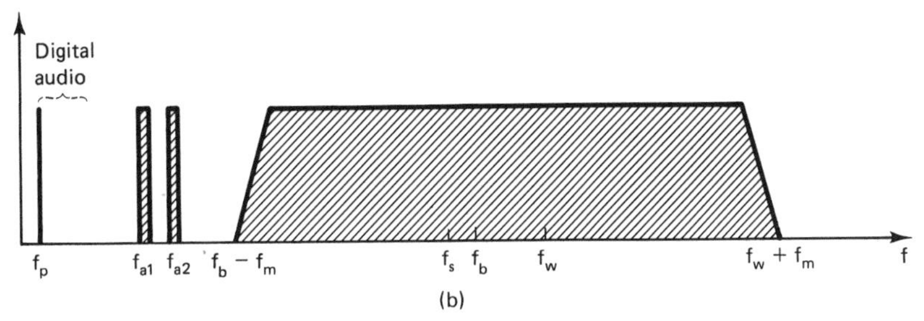

Figure 6.18 (a) Envelope of the spectra in the channel, idealized case; (b) frequency occupation in the channel.

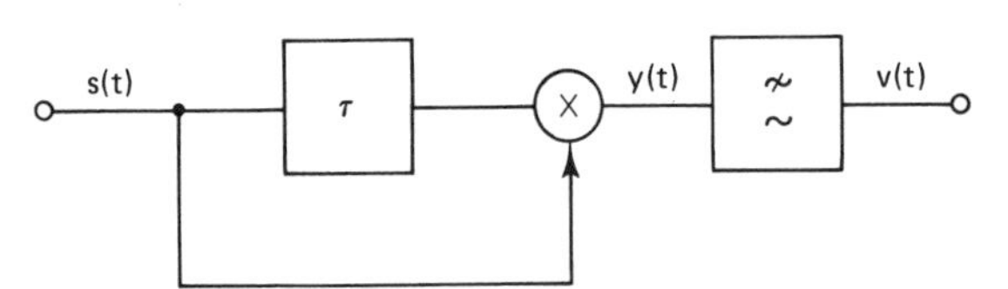

Figure 6.19 FM demodulator with a delay line.

the output of the multiplier is

$$y(t) = s(t)s(t - \tau) \tag{6.11}$$

After passing through the low-pass filter, which eliminates the second harmonic, the output signal is

$$v(t) = (A^2/2) \cos \left[\omega_c \tau + m \int_{t-\tau}^{t} g(t)\, dt \right] \tag{6.12}$$

For $\omega_c \tau = \pi/2$, using the approximation $\sin x \simeq x$, it follows that

$$v(t) \approx kg(t) \tag{6.13}$$

where k is a constant. Thus the modulating signal, $g(t)$, can be recovered in this manner. But it is better (e.g., because of the AM influence) and easier to perform FM demodulation by pulse signal processing.

The "pulse count" type of discriminator is most suitable. In the pulse FM with constant duty cycle (this type of FM is used for video encoding) the low-pass filter cannot be used for demodulation because the dc component of the pulse spectrum does not have a sideband of the modulating frequency baseband (Chapter 4 and Figure 6.3c). But the pulse FM with constant duty cycle can easily be transferred to the duty-cycle modulation. The signal part of the complete demodulator is shown in Figure 6.20.

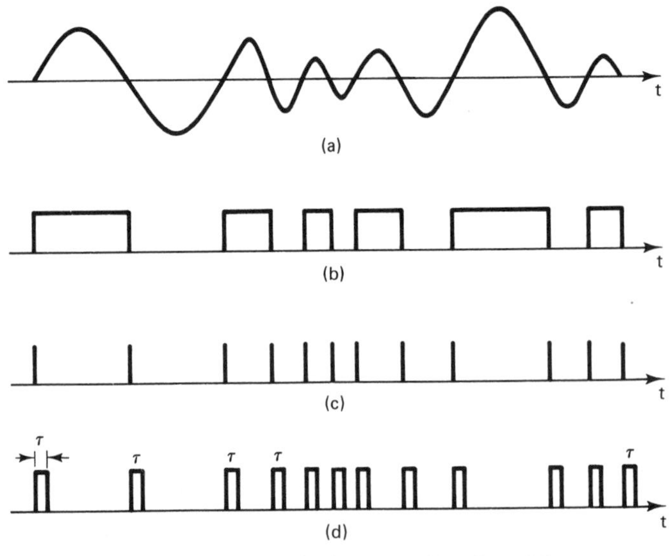

Figure 6.20 Signal part of the complete demodulator.

Because of the MTF and other factors, the FM video is amplitude modulated (Figure 6.20a). This can be removed by limiting or zero-crossing detection (Figure 6.20b). In the case of limiting, several stages may be necessary to remove the AM. At least 50 dB of limiting is required to provide an output square-wave signal whose edges are unaffected by input amplitude changes. After the edge detection, the signal obtained (Figure 6.20c) can be used to start a one-shot circuit which generates pulses with duration τ (Figure 6.20d); thus duty-cycle modulation is obtained. The integrated low-pass filter can be used to produce a signal whose amplitude is proportional to the input frequency.

It should be noted that the pulse repetition frequency is doubled because both positive- and negative-going edges are used. This leads to better filtering (i.e., fewer unwanted components will fall into video baseband). If, instead of the one-shot, some other pulse-forming circuit is used, care must be taken to ensure that the pulse amplitude of the pulses corresponding to the positive- and negative-going edges are the same. Failure to do this would result in a residual carrier component.

6.4 SYSTEM PARAMETERS

The basic assumptions for the foregoing considerations of the videodisc system have been:

1. The picture quality should be high—professional-grade resolution is necessary; the signal in the channel should be specified according to an existing standard (NTSC, PAL, or SECAM).
2. The unique features of optical videodisc systems, including the still and search modes should be retained.

Besides the signal system parameters discussed above, other system parameters are of importance. The main parameters of the disc are given in Table 6.2. The main characteristics of a player with an optical pickup are given in Table 6.3. It is assumed that the maximum luminance level of the video signal will not exceed 110 IRE units and the maximum chroma saturation will not exceed 100%.

The signal parameters discussed above cannot be independently chosen and must rely on outside constraints, the most important of which are the cutoff frequency of a reflective-type optical stylus, minimum recording wavelength (ν_{min}), and effects of defects. There are, of course, many other constraints.

The first of these is the cutoff frequency of a reading lens, given by [27, Chap. 5]

$$f_{co} = 2\pi R n \nu_{co} = 2\pi R n (2\text{NA}/\lambda) \tag{6.14}$$

where f_{co} = temporal cutoff frequency

ν_{co} = spatial cutoff frequency of the reading lens with the numerical aperture NA

TABLE 6.2 MAIN PARAMETERS OF THE OPTICAL DISC

Characteristic	*Requirements*
Diameters of disc (two versions)	300 and 200 mm
Diameter of center hole	35 mm
Thickness of double-sided disc	2.7 mm
Sense of rotation (seen from objective)	Anticlockwise
Speed of rotation	
CAV	1500 rpm (PAL/SECAM); 1800 rpm (NTSC)
CLV	1500–570 rpm approx. (PAL/SECAM); 1800–680 rpm approx. (NTSC)
Rate of rotation	
CAV	1 rotation/TV frame
CLV	Between 10 and 11.4 m/s; never to exceed an angular velocity equivalent to one rotation per TV frame
Starting diameter lead-in tracks	<107 mm
Starting diameter program area	>110 mm
Maximum diameter program area	12-in. disc, 290 mm; 8-in. disc, 192 mm
Minimum number of lead-out tracks	600
Track pitch	1.5–1.8 μm
Refractive index	1.5
Reflectivity	75–85%

λ = wavelength of the light (laser)
n = disc rotation speed
R = radius of the track to be read

The choice of 0.4 for the numerical aperture of the reading lens, together with the 0.633-μm wavelength of the laser light, leads to a spatial frequency cutoff of 1264 cycles/mm or about 16.67 MHz at the innermost (70-mm radius) track of a disc spinning at 1800 rpm. With the same numerical values for NA, λ, and radius (R_i), for the PAL/SECAM with $n = 25$, the temporal cutoff frequency is 13.9 MHz. For a higher NA (e.g., between 0.45 and 0.65), f_{co} would be higher.

The next critical constraint to be considered is the minimum recording wavelength Γ on the disc. This parameter should desirably be as large as possible. Longer wavelengths are easier to record, easier to replicate, and easier to resolve with the optical pickup. Defects are inevitable in the mass production of discs. The system must be able to handle these defects gracefully. For this reason, a large Γ_{min} is also important, since a defect will cover a smaller number of pits.

For a CAV disc (constant rpm, n = const.), the playing time is

$$T_p = (R_o - R_i)/(nq) \tag{6.15}$$

where q is the track pitch (tracks/mm).

TABLE 6.3 MAIN PARAMETERS OF THE PLAYER

Characteristic	*Requirements*
Disc readout	Optical
Light source	He-He laser/1 mW, 632.8 nm (for example)
Objective lens	20 × NA 0.40 or 0.45
RF modulator	UHF channels 31–43 (PAL); VHF channels 3 or 4 (NTSC)
RF output	2.5 mV/75 Ω
Video output	2 V (p-p), 1 V (p-p) terminated
Video frequency range	5 MHz (−6 dB)(PAL); 4.2 MHz (−6 dB) (NTSC)
Video signal-to-noise ratio	>37 dB (PAL), >40 dB (NTSC)
Time-base stability	<10 ns
Maximum time base error at:	
Diameter 110 mm	10 μs p-p of 30-Hz roll-off with 12 dB/octave
Diameter 290 mm	4 μs p-p of 30-Hz roll-off with 12 dB/octave
Shift between two adjacent tracks	±25 ns
Audio output	1 V rms/1500 Ω(PAL), 220 mV rms/1500 Ω(NTSC)
Audio channels	2
Audio frequency range	40 Hz–20 kHz
Audio distortion	<1%
Audio signal-to-noise ratio	60 dB (PAL); 55 dB (NTSC)
Crosstalk between audio channels	< −55 dB
Playing modes	Normal forward and reverse still pictures forward and reverse, slow motion forward and reverse fast forward
Program access	Maximum search time for 12-in. disc 3 s
Picture and chapter	Five-digit picture number (CAV)
Identification	Two-digit chapter number (CAV + CLV)
Remote control	Optional
Main voltage	220 V, 50 Hz (PAL); 120 V, 60 Hz (NTSC)

The minimum tangential (linear) velocity is

$$v_{\min} = 2\pi n R_i \tag{6.16}$$

If $f_{\max}$ is the highest temporal frequency, then

$$\Gamma_{\min} = v_{\min}/f_{\max} \tag{6.17}$$

The track pitch can be expressed as

$$q = \frac{2\pi(R_o - R_i)R_i}{T_p v_{\min}} \tag{6.18}$$

For fixed T_p and $\Gamma_{\min}$, the track pitch will have a maximum value when $R_i = R_o/2$.

Alternatively, $\Gamma_{\min}$ can be expressed as

$$\Gamma_{\min} = \frac{2\pi(R_o - R_i)R_i}{T_p f_{\max} q} \tag{6.19}$$

For fixed q, T_p, and f_{max}, Γ_{min} is a maximum when $R_i = R_o/2$. Therefore, conditions of both large b and large Γ_{min} are optimized for $R_i = R_o/2$ under the foregoing constraints. This optimum, however, is broad, and small changes about $R_i = R_o/2$ are inconsequential.

The frequency deviation, Δf, and modulation index, β, of the video FM are important signal parameters. The maximum sinusoidal deviation that can occur for a video signal is

$$s(t) = [(S_w - S_b)/2](1 + \cos \omega t), \qquad \omega \leq \omega_m \tag{6.20}$$

where S_w and S_b are peak white and black levels, respectively. This signal has an average midgray level and a frequency component with a peak value for the peak white to black level. In monochrome signals such a situation is rare for higher frequencies. In color signals where the chrominance information is transmitted on a high-frequency subcarrier, large amplitudes of high frequency are more common.

Assuming the worst case of 100% amplitude, 100% saturated color bars, the maximun peak-to-peak subcarrier amplitude is always less than 0.9 of the peak-to-peak excursion from sync tip to peak white. On this basis the maximum peak-to-peak deviation and the resultant modulation index due to the color subcarrier is:

Peak to peak: $\Delta f = \Delta f_{pp} = (f_w - f_s)0.9$ (without preemphasis)
Peak: $\Delta f = \frac{1}{2}(f_w - f_s)0.9$

When preemphasis of the video signal is performed prior to modulation, this is modulated by the amount of high-frequency boost applied (Figure 6.9):

$$\Delta f = \tfrac{1}{2}(f_w - f_s)0.09 \times \tau_1/\tau_2 \tag{6.21}$$

The modulation index β_{sc} is

$$\beta_{sc} = \Delta f/f_{sc} \tag{6.22}$$

As an example, let us take $f_s = 7.6$ MHz, $f_b = 8.1$ MHz, and $f_w = 9.3$ MHz; then

$$\Delta f = \frac{9.3 - 7.6}{2} 0.9 \times \frac{125}{50} = 1.9125 \tag{6.23}$$

For the 525-line/60-Hz NTSC system ($f_{sc} = 3.58$ MHz),

$$\beta_{sc} = 0.5342 \tag{6.24}$$

For the 625-line/50-Hz PAL system ($f_{sc} = 4.43$ MHz),

$$\beta_{sc} = 0.4317 \tag{6.25}$$

These are maximum values and show, with reference to the Bessel function (Chapter 4), that most of the energy is in the first-order sidebands. Thus the second (higher)-order sidebands can normally be ignored and their loss poses negligible distortion.

6.5 DISTORTION IN FM SIGNALS

In the videodisc system the FM signal takes the form of all three domains: electrical, optical, and mechanical. In each of these domains, as is shown in Ref. 27, there are many sources of impairments. The system should be designed to tolerate a practical amount of impairments and still guarantee a good-quality TV picture.

The most important distortions in FM signals will be discussed next.

6.5.1 Nonflat Frequency Response

In the videodisc system, the overall FM frequency response is not flat, due to at least the lens frequency characteristic (MTF). This will cause amplitude modulation, which is of little consequence since it can be practically removed by limiting. Also, this will result in angular modulation, which is a more important factor: the amplitude of the demodulated signal will be reduced and the reduction can be frequency dependent.

For a quantitative analysis let us take a one-tone FM signal: For $\beta \leq 0.2$,

$$s(t) = S_c \cos [\omega_0 t + \beta \cos (\omega_m t - \pi/2]$$

$$\cong S_c \cos \omega_0 t + \frac{1}{2}\beta S_c \cos (\omega_0 + \omega_m)t - \frac{1}{2}\beta S_c \cos (\omega_0 - \omega_m)t$$

After passing through the system with transfer characteristics as shown in Figure 6.21a, all three components will be changed (multiplied) by different factors. If the normalization is done for the carrier (ω_0) component, the output signal would be

$$s^*(t) = S_c \cos \omega_0 t + aS_c(\beta/2) \cos (\omega_0 + \omega_m)t - bS_c(\beta/2) \cos (\omega_0 - \omega_m)t \quad (6.26)$$

or

$$s^*(t) = S(t) \cos (\omega_0 t + \theta) \quad (6.27)$$

where

$$S(t) \simeq S_c[1 - (\beta/2)(b - a) \cos \omega_m t]$$

and

$$\tan \theta(t) = \frac{(a - b)(\beta/2) \sin \omega_m t}{1 - (a - b)(\beta/2) \cos \omega_m t}$$

If the influence of amplitude modulation is neglected, the frequency-demodulated signal would be

$$g^*(t) = k_d(d\theta/dt) = G_0 + G_1 \cos \omega_m t + G_2 \cos 2\omega_m t + G_3 \cos 3\omega_m t \quad (6.28)$$

where:

$$G_1 = [\omega_m(a + b)(\beta/2][1 - (\beta^2/4)(3a^2 + 3b^2 - 5ab)]$$

$$G_2 = [\omega_m(a + b)(\beta/2)][(b - a)(\beta/2)(1 - (\beta^2/2)\, ab)]$$

$$G_3 = \omega_m(a + b)(\beta/2)[(\beta^2/4)\, ab]$$

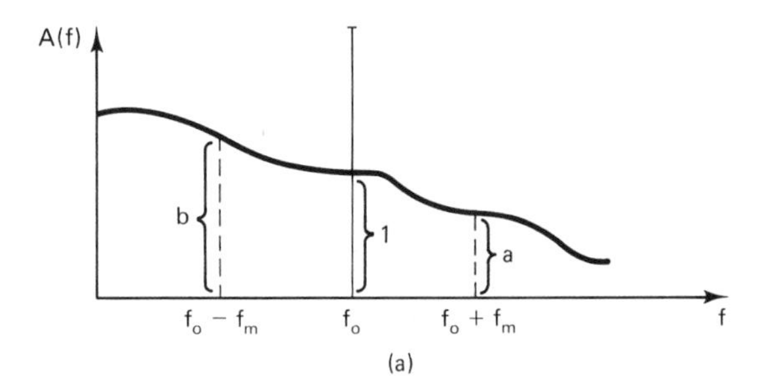

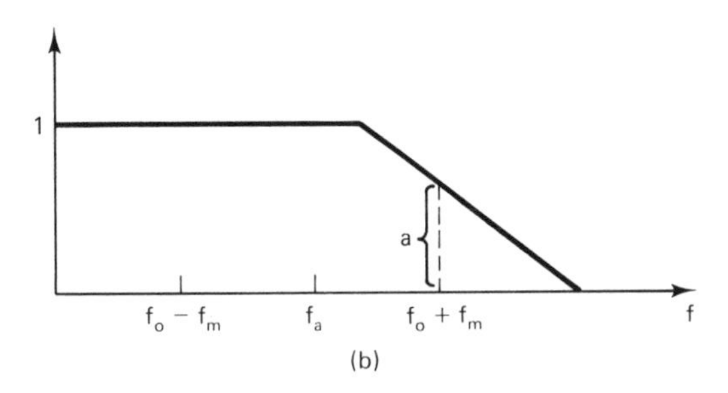

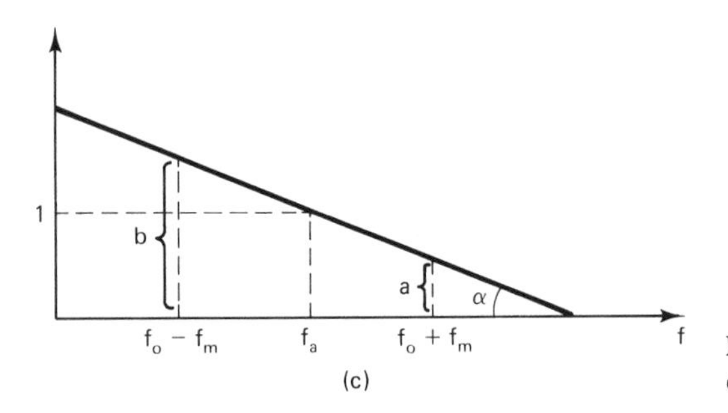

Figure 6.21 Effect of a nonlinear frequency response.

The coefficient of the second harmonic distortion is:

$$d_2 = \frac{G_2}{G_1} = \frac{(\beta/2)(b - a)[1 - (\beta^2/2)\, ab]}{1 - (\beta^2/4)(3a^2 + 3b^2 - 5ab)} = d_2(a, b, \beta) \tag{6.29}$$

If a modulated signal is passed through a network with a frequency response as shown in Figure 6.21b, the upper sideband ($f_0 + f_m$) is attenuated with respect to the center frequency (f_0) and the lower sideband ($f_0 - f_m$): $b = 1$, $a < 1$. For this case,

$$d_2 = d_2(a, \beta) = \frac{(\beta/2)(1 - a)[1 - (\beta/2)\, 2a]}{1 - (\beta^2/4)(3a^2 - 5a + 3)} \tag{6.30}$$

Thus for higher frequencies d_2 would be large, but for higher f_m, the component $2f_m$ will be filtered out.

If the response is linear, as shown in Figure 6.21c, $f_0 + f_m$ is reduced relative to f_0, and $f_0 - f_m$ is proportionally increased; thus the total projection on the y-axis remains constant.

For this case,

$$d_2 = d_2(k, f_m, \beta) = \frac{|k| f_m[1 - (\beta^2/2)(1 - k^2 f_m^2]}{1 - b^2/4(1 + 11k^2 f_{m)}^2} \tag{6.31}$$

where $k = \tan \alpha$ (1/Hz).

The discussion referring to d_2 is valid when the second harmonic is not filtered out. In practice, this is valid for low-frequency luminance signals (lower than $B_L/2$, where B_L is the luminance bandwidth). The largest distortion influence is manifested by the color subcarrier, f_{sc}, although the component $2f_{sc}$ is filtered out. The main point here is that the amplitude of the subcarrier is a function of a, b, and $\beta[G_1(f_{sc}) = G_1(a, b, \beta)]$. See Appendix 6A for a more detailed discussion. It is not obvious which kind of frequency response would introduce severe distortion: comparisons should be performed under particular noise conditions. Initially, it would seem that it is important to have a flat response. It has been found, however, that some improvement in the signal-to-noise ratio and unwanted beat interference can be achieved by having a linear fall off in response. If the response is linear, as shown in Fig. 6.21c, $f_o + f_m$ is reduced relative to f_o, and $f_o - f_m$ is proportionately increased. The net result is that although amplitude modulation occurs, β remains constant. The advantage gained in using such a response is that less of the upper sidebands are used and as these tend to be noisier, owing to compensation of high frequency loss and the presence of unwanted beat components (harmonic distortion), an improvement in overall signal/noise ratio and Moiré is made.

6.5.2 Nonlinear Phase Response

A linear phase response or constant group delay is required by the system in order to preserve the modulation index β. To equalize the group delay distortion (of the playback low-pass filter), the video group delay should be predistorted. This involves complex equalization, and it is performed before recording.

An example of video group delay predistortion (Figure 6.22) for the NTSC (525-line/60-Hz) signals is as follows:

f (MHz)	t_d (ns)
0.5	0 reference
2	−15 ± 15
3	−45 ± 15
3.58	−80 ± 15
4	−135 ± 30
4.2	−200 ± 50

For the other systems, diagrams similar to that in Figure 6.22 can be defined.

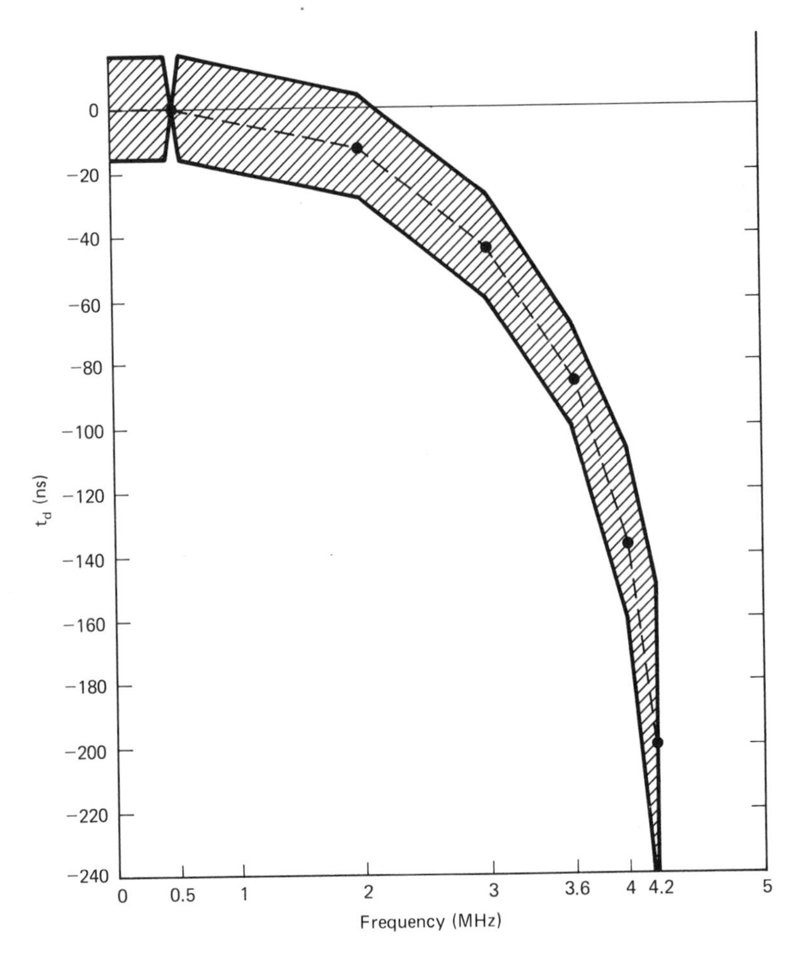

Figure 6.22 Group delay predistortion.

6.5.3 Random Noise

The possible sources of the noise in the videodisc system and its model have been discussed in Refs. 27 and 28, while the FM signal (continuous wave) with white Gaussian noise has been analyzed in Section 4.4. It is shown in Refs. 27 and 28 that the random noise in the videodisc systems can be, in the first approximation, assumed to be Gaussian white noise, although it is significantly tailored by the MTF of the reading lens. For input (passband) white Gaussian noise, the spectral density of the output (of the FM demodulator) noise is a square function of the frequency (Figure 4.7). The amplitude of the noise in the demodulated signal is thus proportional to the frequency (triangular noise spectrum). The signal-to-noise ratio for the particular spectral components depends on their positions on the frequency scale. For the given bandwidth Δf, the output noise gets larger when this bandwidth is moved toward higher frequencies. To compensate for this, the preemphasis/deemphasis technique is used. But boosting (around f_{sc}) to much higher frequencies would cause too high a value of β. For high video frequencies this would produce energy in the higher-order sidebands. So a compromise should be reached.

For video, the signal-to-noise ratio for the continuous random noise is defined as the ratio expressed in decibels of the nominal peak-to-peak (p-p) amplitude of the picture luminance signal to the rms amplitude of the noise measured under the following conditions:

1. The noise is passed through a specified passband filter to delineate the effective frequency range and also, where appropriate, through a specified weighting network or equivalent.
2. The measurement is made with an instrument having, in terms of power, an effective time constant or an integrating time of 1 s.

The unweighted video noise is measured in the nominal frequency range (e.g., 7.5 kHz to 5.5 MHz for the PAL system). But the eye sensitivity to the noise is not equal at all frequencies: the sensitivity is higher for low-frequency components and decreases toward higher frequencies. For subjective measurements of video noise, so-called *weighted noise* is measured. A special weighting network is used; the frequency response is shown in Figure 6.23 for the PAL system (the bandwidth is 7.5 kHz to 5 MHz). The curve in Figure 6.23 is for random noise in the luminance channel and defined for the network that has a time constant of 200 ns, giving a weighting effect of 6.5 dB for flat random noise and 12.3 dB for triangular random noise.

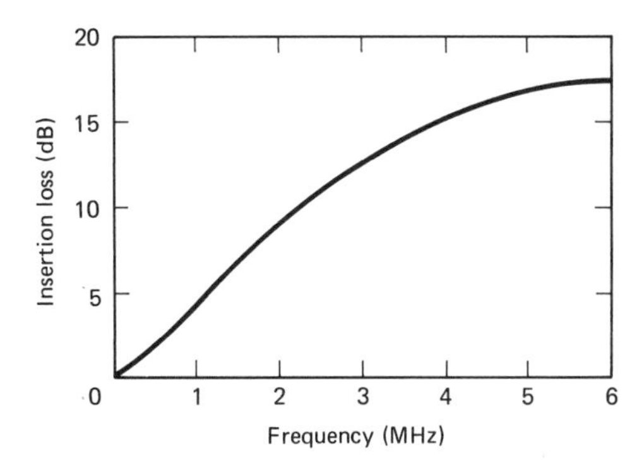

Figure 6.23 Characteristic of weighting network for random noise in the luminance channel (CCIR Rec. 451-2).

Figure 6.24 shows the characteristic of the weighting network for random noise in the chrominance channels for the nominal frequency range of 3.5 to 5.5 MHz (PAL). For each subcarrier sideband, the filter provides a weighting effect which is approximately equal to that of the luminance weighting network in the 0- to 1-MHz band.

For audio noise, noise measurements are usually expressed with reference to a 1-kHz tone at a level of 0 dB. An audio weighting network, as defined in CCIR Recommendation 468-1, with an attenuation characteristic that follows the curve reproduced in Figure 6.25 can be used. The input to the channel to be measured should be terminated.

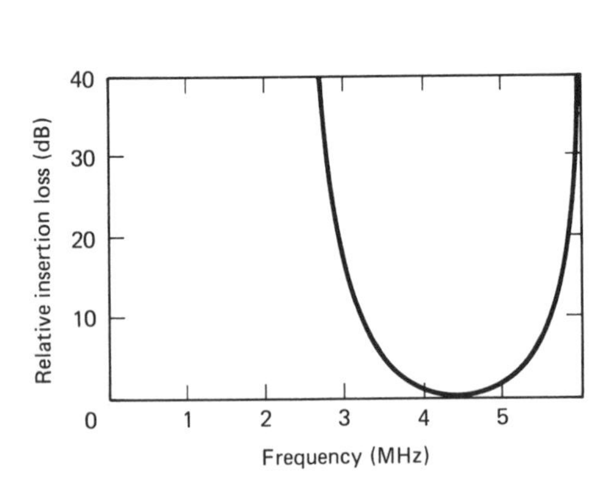

Figure 6.24 Characteristic of passband filter and weighting network for random noise in the chrominance channel (CCIR Rec. 451-2).

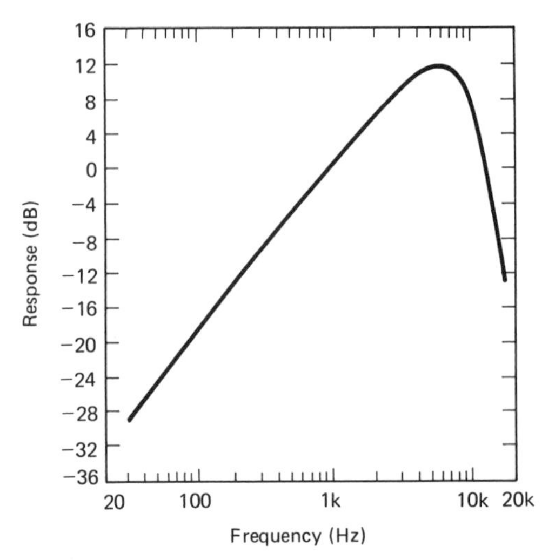

Figure 6.25 Characteristic of the audio noise weighting network (CCIR Rec. 468-1).

In Ref. 27 the dropout phenomenon is explained, including statistics for the videodisc system; the compensational principle is shown in Figure 6.7, and more discussion is given in Section 6.7.

6.5.4 Moiré Patterns

The Moiré components are those obtained upon detection which do not appear in the original video signal. In general, all the frequency-domain information of a bandlimited function is contained in the interval $(0, f_m)$, or for the double-sided presentation in the interval $(-f_m, f_m)$. When this interval is corrupted by unwanted components, the phenomenon is frequently referred to as *aliasing* and the corresponding effect is called *aliasing error*. In the field of optics, aliasing errors are called *Moiré effects* or *Moiré patterns*.

The total FM signal, $s_{\text{FMT}}(t)$, is

$$s_{\text{FMT}}(t) = s_{\text{FMI}}(t) + a_{\text{FM}}(t) \tag{6.32}$$

where $s_{\text{FMI}}(t)$ is an ideal FM signal and $a_{\text{FM}}(t)$ represents the aliasing-error artifact.

In the video-signal domain (after FM demodulation and low-pass filtering) the total video signal is

$$s_T(t) = s_I(t) + a(t) \tag{6.33}$$

where $s_I(t)$ is an ideal video signal.

The unwanted components in the FM domain are created, mainly, by folded sidebands, baseband signals, harmonic distortion and (usually the worst interference) intermodulation products. The visibility of the Moiré components depends on their amplitudes, and also on their frequency position: inclined or interleaved with multiples of f_H.

6.5.4.1 Folded Sidebands. For a low modulation index, energy in higher-order chroma sidebands is low and their loss does not cause noticeable distortion in the picture reproduced. But higher-order lower sidebands can be folded back ($\bar{J}_2$, Figure 6.5c, and $\bar{J}_3$, Figure 6.17), and, after demodulation, can be in the passband; that is, they appear as spurious frequencies equal to the difference between the unwanted sideband position and the center frequency.

For $f_0 = 8.1$ MHz (black level) and for the 625-line PAL system, the third-order lower sideband is $f_0 - 3f_{sc} = 5.19$ MHz. After demodulation, this is $8.1 - 5.19 = 2.91$ MHz; or, directly, $2f_0 - 3f_{sc} = 16.2 - 13.29 = 2.91$ MHz.

For the 525-line NTSC system and $f_0 = 8.1$ MHz also, the third-order lower sideband after demodulation is $2f_0 - 3f_{sc} = 5.46$ MHz (out of band). The fourth-order lower sideband after demodulation is $2f_0 - 4f_{sc} = 16.2 - 14.32 = 1.88$ MHz.

A low-frequency Moiré pattern is more evident if a picture contains periodic structures and if unwanted components are just right for the Moiré patterns to occur. In practice, one does not often see pictures with strong periodic components. Cases that do arise include pictures of a plowed field, streets in high-altitude photographs or urban areas, ocean wave patterns, and wind patterns in sand. Also, those frequencies in the original (picture) video signal that are outside the bandwidth and irretrievably lost in reproduction lead to a loss of resolution and other distortions. These remarks are valid, in general, for the Moiré patterns.

6.5.4.2 Harmonic Distortion. In the videodisc system, an FM signal is passed through nonlinear devices and harmonics of the fundamental are produced. When a 50% duty cycle is used, even harmonic distortion can be minimized by good design. But odd harmonic distortion, predominantly the third, is present in the system. As shown in Chapter 4, the harmonics appear as separate FM signals with their own sidebands; the harmonics deviate if the center frequency deviates. The third harmonic is outside the fundamental passband, but the lower sidebands of the third harmonic can appear within the passband of the system and become an aliasing component. The problem, basically, results from the fact that the modulation index increases by the order of the harmonic [29]:

$$\beta_n = n\beta \tag{6.34}$$

where β and β_n are the modulation index for the fundamental and the nth harmonics, respectively. As β increases the number of significant sidebands increase (Chapter 4): the higher modulation index increases the energy in the higher-order sidebands of the harmonics.

In Figure 6.26, Moiré components due to third harmonic distortion are shown for an 8.1-MHz carrier. Asterisks mark the worst in-band components. When one considers this kind of distortion, it is convenient to remember that for a modulation index $\beta \leq 0.2$, the significant components of the FM signal are the carrier and first-order sidebands, and for $\beta \leq 0.9$, the significant components are carrier and first- and second-order sidebands.

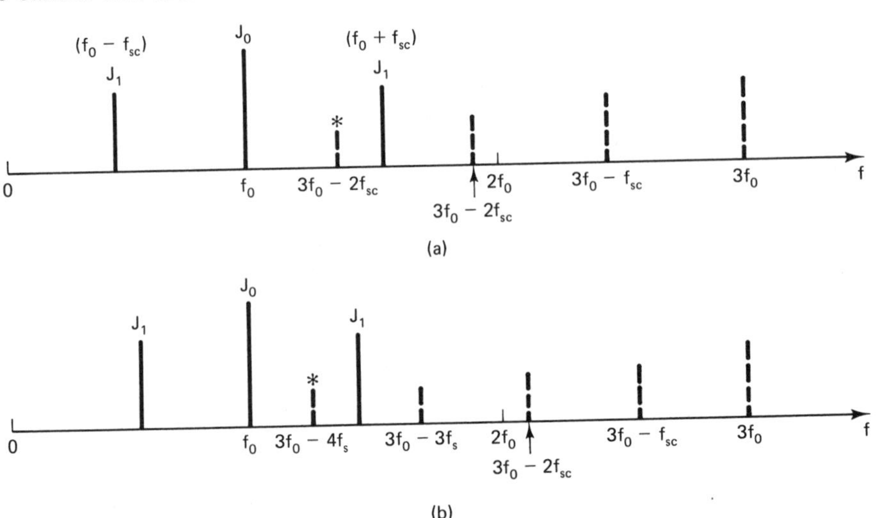

Figure 6.26 Moiré components due to third harmonic distortion for (a) the PAL 625/50, and (b) the NTSC 525/60 systems.

Moiré components can also arise whenever baseband components arise (caused by, for example, nonlinearities and fixed-length asymmetries in the video-disc system). For $\beta \leq 0.2$, the FM signal can be approximated as

$$s(t) = A_0 J_0(\beta) \cos \omega_0 t + A_0 J_1(\beta) \cos \left[\left(\omega_0 - \omega_m\right) t - \frac{\pi}{2}\right] \tag{6.35}$$

$$+ A_0 J_1(\beta) \cos [(\omega_0 + \omega_m) t + \pi/2$$

where f_m is the frequency of the modulating signal (video). Approximate values for the Bessel functions are $J_0(\beta) \approx 1$ and $J_1(\beta) \simeq \frac{1}{2}$. The corresponding phasor, R, is (Figure 6.27a)

$$R = A_0 + \tfrac{1}{2}\beta A_0 \exp[-j(\omega_m t - \pi/2)] + \tfrac{1}{2}\beta A_0 \exp[j(\omega_m t + \pi/2)] \tag{6.36}$$

Figure 6.27b illustrates the phasor diagram when the baseband signal

$$s_b(t) = a \sin (\omega_m t + \phi) \tag{6.37}$$

is added. Using an analysis as in Chapter 4 (relating to Figure 4.7) it can be

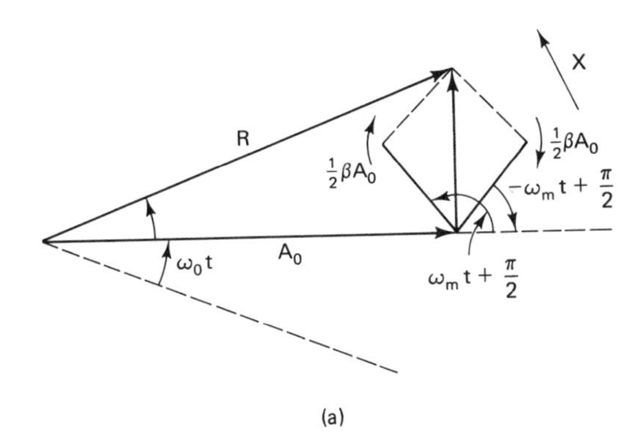

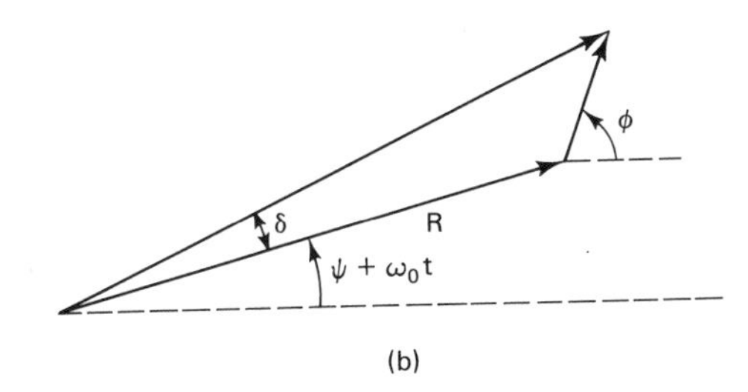

Figure 6.27 Phasor representation of the tone modulated signal: (a) FM signal $\beta \leqslant 0.2$; (b) FM and baseband signal.

shown that the phase error of the FM signal (δ, Figure 6.27b) is

$$\delta \approx a/R \cos [\psi + (\omega_0 - \omega_m)t - \phi] \tag{6.38}$$

The discriminator output is

$$d/dt\,(\psi + \delta) = \frac{d\delta}{dt} + \frac{d\psi}{dt}\left(1 + \frac{\partial \delta}{\partial \psi}\right) \tag{6.39}$$

In equation (6.39) only the dominant terms are shown. The pictorial presentation is shown in Figure 6.28 for the case $f_0 = 8.1$ MHz, $f_m = 4.5$ MHz, $\beta = 0.2$, and $a/A_0 = 0.1$.

Because of the nonlinearities in the system, many intermodulation products and beat frequencies are generated in the FM signal. After demodulating, some of them appear as Moiré components. The most significant originate from the audio carriers (f_{a1} and f_{a2}) added to the video FM and then processed in a nonlinear manner. These components are positioned at:

$$kf_0 \pm nf_{a1} \pm mf_{a2}, \qquad k = 1, 2, 3, \ldots; \quad n, m = 0, 1, 2, \ldots$$

The worst in-band components are

$$f_{a1}, \quad f_{a2}, \quad f_{a2} - f_{a1}$$

and they are obtained after demodulating components:

$$f_0 \pm f_{a1}, \quad f_0 \pm f_{a2}, \quad f_0 \pm (f_{a2} - f_{a1})$$

Of course, other products are generated, but in general, their amplitude will be sufficiently small that they may be neglected. Some are eliminated by

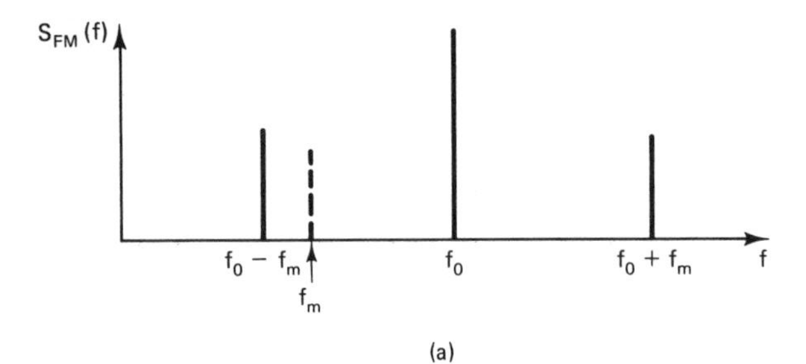

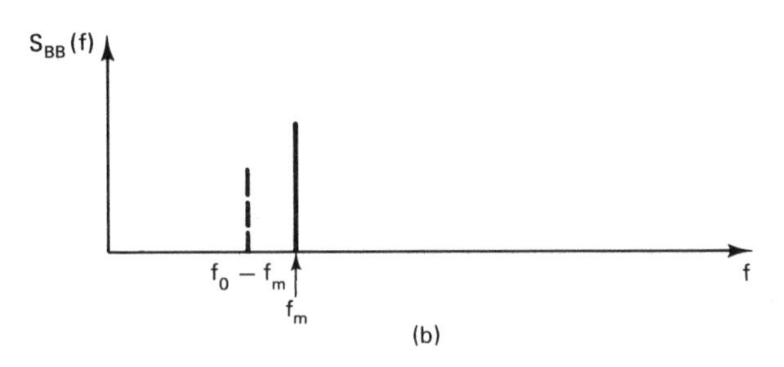

Figure 6.28 Moiré component caused by the baseband: (a) FM spectrum with the baseband; (b) demodulated signal.

passing the FM signal through a limiter before demodulation. This happens whenever the phase relationship between every three components, f_0 (or kf_0, in general), $f_0 - f_x$, and $f_0 + f_x$, is such as to amplitude modulate f_0 as a carrier with a signal of frequency f_x. On the other hand, the products cannot be removed by limiting when the phase relationship between every three components, f_0 and $f_0 \pm f_x$, is such as to angular modulate f_0 as a carrier with a signal of frequency f_x.

6.5.5 Asymmetry

The foregoing deformations are characteristic for all FM systems. The deformation due to asymmetry is characterized for the videodisc systems in Figure 6.29. If a series of square pulses with period T is mastered on the disc in the symmetric case, the same signal will be obtained by the reading process from the disc.

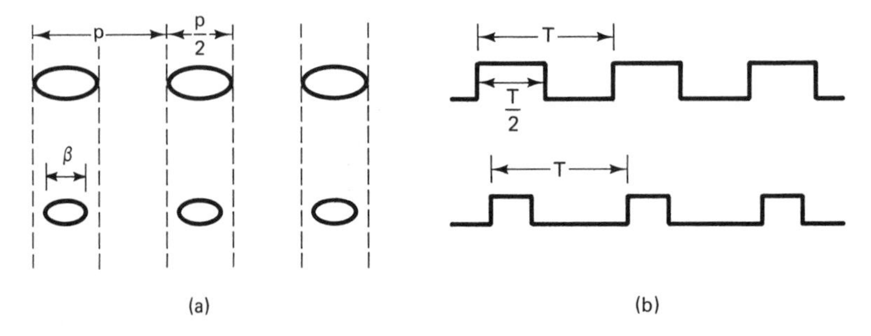

Figure 6.29 Symmetrical and asymmetrical signals: (a) on the disc surface; (b) in the time domain after optical readout.

However, if during mastering it happens that the pit length is not equal to the distance between pits, a signal containing additional components will be obtained upon reading.

When three sinusoidal signals of frequency f_0, f_{a1}, and f_{a2} with different amplitudes are added and then processed in a nonlinear manner, frequency components are created with amplitudes that are strongly dependent on the symmetry of this process.

The asymmetrical signal is shown in Figure 6.29 in both the space domain (disc surface) and the time domain. The asymmetry (AS) can be defined as

$$\text{AS} = \frac{(T - \tau) - \tau}{T} \times 100\% = |1 - 2\theta| \times 100\%, \qquad \theta = \tau/T \tag{6.40}$$

where θ is a duty cycle of the distorted (with asymmetry) signal.

The asymmetry is caused by nonlinearities in the videodisc system. In mastering a videodisc there is a chain of at least three major nonlinear operations:

1. The electronic hard limiter as the last stage in the coding scheme for the frequency modulator.
2. The electro-optical (EO) modulator acting as a soft limiter having a characteristic corresponding to a half sine wave. The operating point of the EO modulator is the same as the dc level to the input of the electronic limiter.
3. The characteristic of the photoresist, its exposure and development. Here, the over- or underexposure and the over- or underdevelopment of the photoresist afterwards take the form of a dc-level shift.

The recording process also introduces some bandwidth restriction due to the spatial cutoff frequency given by the numerical aperture of the recording objective. The net result is a signal on the disc which is read at playback through another objective having a lower numerical aperture and thus introducing more bandwidth reduction.

One way to represent asymmetry is to consider the nonbandwidth-restricted output square wave (of the limiter) (Figure 6.30). Then positions of the frequency components introduced (by asymmetry) can be estimated. In the case of bandwidth restriction, asymmetry can be regarded as the enhancement of components located previously. The bandwidth restriction after the limiter only affects the amplitude, and eventually the phase of the asymmetry components, but does not change their positions.

There is no asymmetry in a so-called 50-50 duty cycle; this is shown in Figure 6.30a for reference only. Basically, two kinds of asymmetry are introduced in the videodisc system. In proportional asymmetry (Figure 6.30b), the process is equivalent to a dc-level shift of the signal at the input to the limiter. This has already been discussed relating to Figure 6.11. Suppose now that signal $s(t) = A \sin \omega t$ is passed through the hard limiter or zero-crossing detector. The output signal would be a square-wave pulse train with a 50% duty cycle, that is, a symmetrical signal. If there is a small dc-level shift, a, the zero crossings are

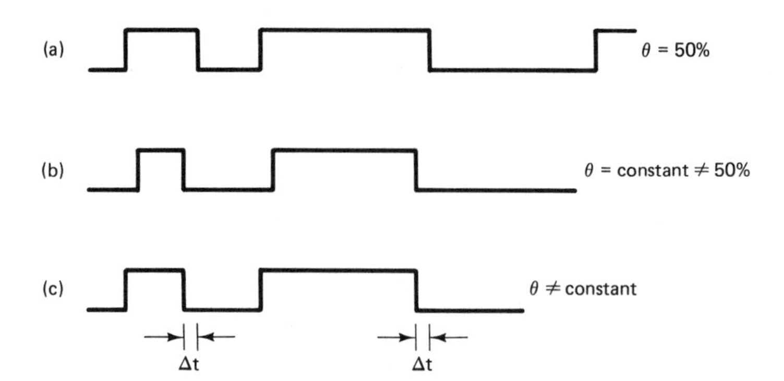

Figure 6.30 Three types of asymmetry: (a) symmetrical signal (for reference); (b) proportional asymmetry; (c) fixed-length asymmetry.

defined by

$$A \sin \omega t = a, \qquad a \ll A \tag{6.41}$$

The zero crossings are obtained for

$$\omega t = (2n + 1)\,\pi - \alpha_0 \tag{6.42}$$

$$\omega t = 2n\pi + \alpha_0 \tag{6.43}$$

where $\sin \alpha_0 = a/A$.

Then the pulse duration is

$$\tau = \frac{\pi - 2\alpha_0}{\omega} \tag{6.44}$$

and the duty cycle is

$$\theta = \frac{\tau}{T} = \frac{1}{2} - \frac{\alpha_0}{\pi} = \frac{1}{2} - \frac{a/A}{2} = \text{const.} \tag{6.45}$$

where the period $T = 2\pi/\omega$ and $\alpha_0 \approx a/A$ for $a \ll A(\sin \alpha_0 \approx \alpha_0)$.

In fixed-length asymmetry (Figure 6.30c) the pulse duration is changed for the fixed time duration, $\pm\Delta t$, independently of the pulse duration, that is, independently of the frequency. For the simple case considered, $s(t) = A \sin \omega t$, the duty cycle is

$$\theta = \frac{T/2 - \Delta t}{T} = \frac{1}{2} - \frac{\Delta t}{T} = \theta(\omega) \tag{6.46}$$

Of course, $+\Delta t$ could be inserted in place of $-\Delta t$.

From the analysis in Section 4.6, the main disadvantages of the two types of asymmetry are:

1. Proportional asymmetry (θ = const. $\neq$ 50%) introduces a second harmonic of the carrier ($2f_0$); this component deviates (with double deviation) when the main carrier (f_0) deviates.
2. Fixed-length asymmetry ($\theta \neq$ const.) introduces a baseband signal; for a

constant color picture the subcarrier f_{sc} will be introduced (in the video FM), and for a picture with changing gray levels, corresponding baseband components will appear.

In the videodisc system, three carriers are usually present: f_0, f_{a1}, and f_{a2}. When asymmetry is present during processing of these signals even-frequency components will appear in the spectral area around all even harmonics of the main carrier, and odd-frequency components around all odd harmonics of the main carrier at levels that depend on the asymmetry introduced. Additional unwanted audio components will in general, appear, in the spectral range [5]:

$$2nf_0 \pm 2mf_{a1}, \quad 2nf_0 \pm 2mf_{a2}$$

and

$$(2n + 1)f_0 \pm (2m + 1)f_a, \quad f_a = f_{a1}, \quad f_a = f_{a2}$$

where n, m = 0, 1, 2.

Because f_0 changes with the luminance level, the first effect of the presence of asymmetry components will be a Moiré whose frequency is luminance-level dependent in the demodulated luminance signal. The second effect can occur when the distance of the asymmetry components of the frequencies carrying the color information becomes so small as to fall within the bandwidth of the color demodulating circuits of the receiver. In this case a low-frequency Moiré that varies with the luminance level will appear in the color-difference signal.

Finally, we should mention that the proportional type of asymmetry is more likely to be generated by the electronic limiter and electro-optical modulator, whereas fixed-length asymmetry is expected to be predominate in the development portion-recording steps.

There are a number of properties of the signal system that counteract the interferences:

1. Not all of the frequencies occur at the same time.
2. Components that are on only one side of the video or audio carriers are reduced 6 dB in the limiting process.
3. Deemphasis in the player reduces the interference significantly.
4. The visibility of some interferences is low even though their measured values may be significant (e.g., interference in the picture has reduced visibility if the interfering frequencies are interleaved with multiples of f_H).

Within the frequency tolerance of the videodisc modulation system, it is possible that the luminance component has nearly the same frequency as the audio carrier. In this case there will appear a disturbing low-frequency beat. To avoid this, the modulation carrier should be properly chosen. In Table 6.4 it is illustrated that it is better to take f_b = 8.1 MHz than f_b = 8 MHz for the same other conditions: f_{a1} = 2.3 MHz, f_{a2} = 2.81 MHz, and f_{sc} = 3.58 MHz (NTSC). For f_b = 8.1 MHz, there is still enough space between the J_2 (chroma) component and the first audio carrier.

TABLE 6.4 CHOICE OF THE FM CARRIER (f_0) AND ASYMMETRY COMPONENT $f_0 - (f_{a1} + f_{a2})$

HF component	*Level* (*dB*)	f_s	f_b	f_w	$f(f_{a1})$[a]	f_s	f_b	f_w	$f(f_{a1})$
f_0	0	7.5	8	9.2	—	7.6	8.1	9.3	—
$f_0 - (f_{a1} + f_{a2})$	−52	2.35	9.2	4.05	10	2.45	2.95	4.15	110
J_2 (chroma)	−38	—	0.84	2.04	300	—	0.94	2.14	200

[a] $f(f_{a1})$, Distance to audio carrier f_{a1}.

6.5.6 AM-to-FM Conversion

When a message is transmitted by an FM signal, any amplitude change is unwanted. There are many mechanisms that cause unwanted AM on the FM signal in the videodisc systems. For example, the FM signal becomes amplitude modulated whenever an outside electrical perturbation is added. Or nonlinear characteristics of the phase-delay parts of videodisc systems are also manifested as parasite AM on the FM signal. Therefore, before the FM signal is demodulated, it is passed through a limiter to remove the AM. Its efficiency is measured by the degree of compression, C:

$$C = \frac{(dA_i)/A_i}{dA_o/A_o} = \frac{d(\ln A_i)}{d(\ln A_o)} \tag{6.47}$$

where dA_i is the input signal amplitude change, and dA_o is the corresponding change in the output signal. If

$$A_o = \begin{cases} G_1 A_i, & A_i < A_1 \\ G_1 A_1 + G_2(A_i - A_1), & A_i > A_1 \end{cases} \tag{6.48}$$

then for $A_i > A_1$, $C = G_1/G_2$, where G_1 and G_2 are corresponding slopes. Obviously, the best situation is when $G_1 = 0$. It can be shown that in that case, the amplitude of the first harmonic of the output, A_{o1}, is

$$A_o \leq A_{o1} < (4/\pi)A_o. \tag{6.49}$$

In other words, these are limits for the output first harmonic when the input signal is changing from A_o to ∞.

If the tank (LC) circuit is included with the limiter, the parasite AM of FM signals will be converted to angular modulation, also the parasite. The degree of conversion is proportional to the frequency deviation (Δf).

6.5.7 Crosstalk

In videodisc systems, crosstalk is usually between tracks. When the reading beam is centered on one of these tracks, the detector signal contains components originating from neighboring tracks. The condition that these crosstalk components not overly disturb the signal from the central track determines the distance between the tracks and thus the information density in the radial direction [7, pp. 2029–2036].

In Figure 6.31 the crosstalk is plotted as a function of defocusing (the NA is fixed). In practice the crosstalk should be kept below −35 dB (referenced to the main signal). The upper curve in the figure is for a reading beam exactly in focus. The crosstalk between tracks grows rapidly with defocusing. This is understandable because more energy goes into the outer rings of the diffraction pattern. The behavior of crosstalk as a function of defocusing is somewhat different for rectangular and circular apertures. For a circular aperture the crosstalk depends not only on the distance between the tracks but also on the (frequency) signal recorded on adjacent tracks.

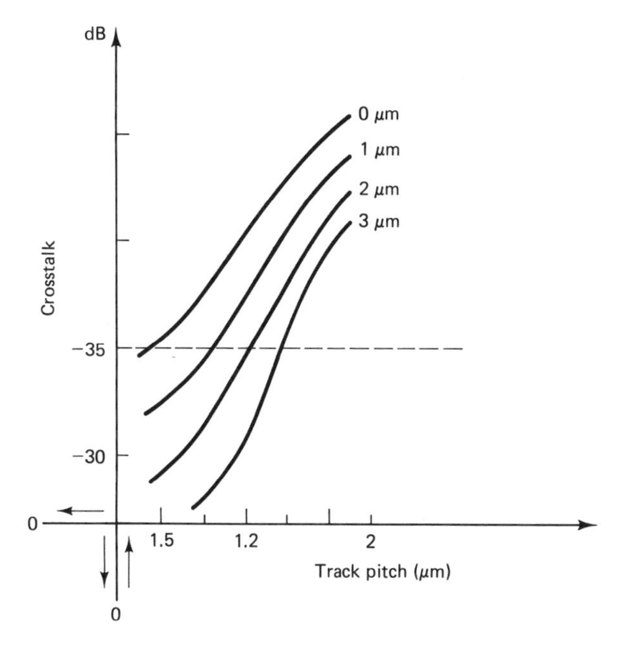

Figure 6.31 Crosstalk as a function of track pitch and defocusing.

The crosstalk is also greatly influenced by the tracking error. This is illustrated in Figure 6.32 for circular apertures. The parameter r in Figure 6.32 is

$$r = \frac{\text{track (pit) width}}{\text{track pitch}} \tag{6.50}$$

The curves in Figure 6.32 are valid when the reading beam is exactly in focus. The signal-to-noise ratio decreases rapidly with defocusing.

6.5.8 Distortion Due to Limited Bandwidth

Suppose that we have an ideal low-pass system. The transfer function of this system is given by

$$H(j\omega) = A(\omega)e^{-j\theta(\omega)} \tag{6.51}$$

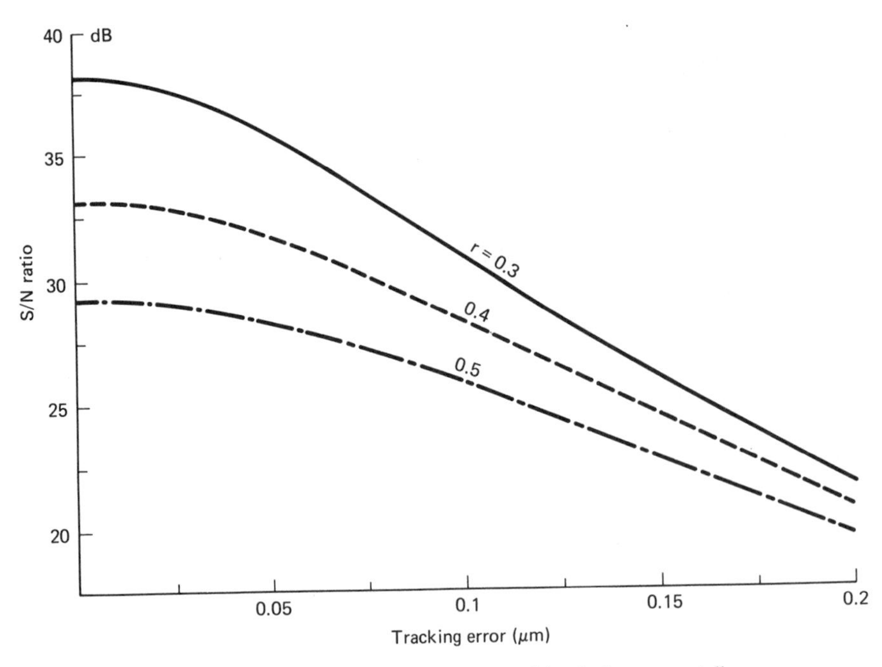

Figure 6.32 Effect of slit aperture (at objective) on crosstalk.

where

$$A(\omega) = \begin{cases} A = \text{const.}, & |\omega| < \omega_N \\ 0, & \text{elsewhere} \end{cases}$$

$$\theta(\omega) = \omega t_0$$

Suppose also that the input signal, $x(t)$, is a single rectangular pulse. Its Fourier transform is

$$X(j\omega) = \tau E \frac{\sin(\omega\tau/2)}{\omega\tau/2} \tag{6.52}$$

where τ is the pulse duration and E its amplitude.

The Fourier transform, Y, of the output signal, $y(t)$, is given by

$$Y(j\omega) = H(j\omega)X(j\omega) \tag{6.53}$$

The output signal is then

$$y(t) = \tfrac{1}{2}\pi \int_{-\infty}^{\infty} Y(j\omega)e^{j\omega t}\, d\omega \tag{6.54}$$

After some mathematical transformation it is found that

$$y(t) = (AE/\tau)\,\{S_i[\omega_N(t - t_0 + \tau/2)] - S_i[\omega_N(t - t_0 - \tau/2)]\}$$

where

$$S_i(x) = \int_0^x \frac{\sin x}{x}\, dx \tag{6.55}$$

is the sine integral of x.

In Figure 6.33, the output waveforms are shown for three bandwidths $B = f_N$: $B\tau \ll 1$, $B\tau = 1$, and $B\tau \gg 1$. The influence of the system bandwidth on the output waveform is obvious and instructive, although the case is idealized and not physically realizable.

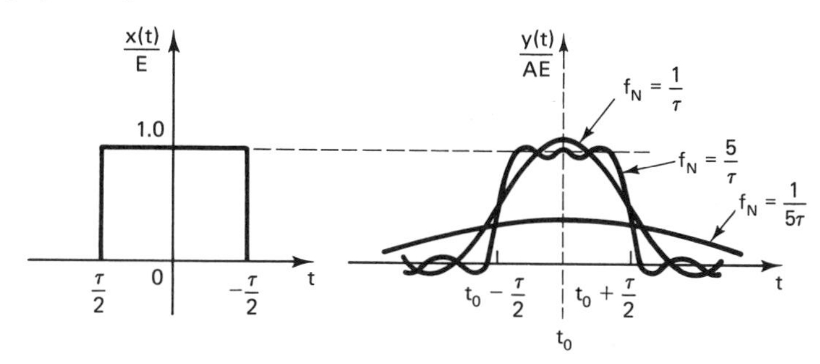

Figure 6.33 Output of the low-pass system. The corner frequency f_N is the parameter.

Previous results can be used to estimate the resolution of the system (low-pass). This is illustrated in Figure 6.34. When the input signal contains two rectangular pulses, the output is

$$\begin{aligned} y(t) = (A/\pi) - [S_i[\omega_N(t - t_1 - t_0)] \\ - S_i[\omega_N(t - t_2 - t_0)] \\ + S_i[\omega_N(t - t_3 - t_0)] - S_i[\omega_N(t - t_4 - t_0)]] \end{aligned} \tag{6.56}$$

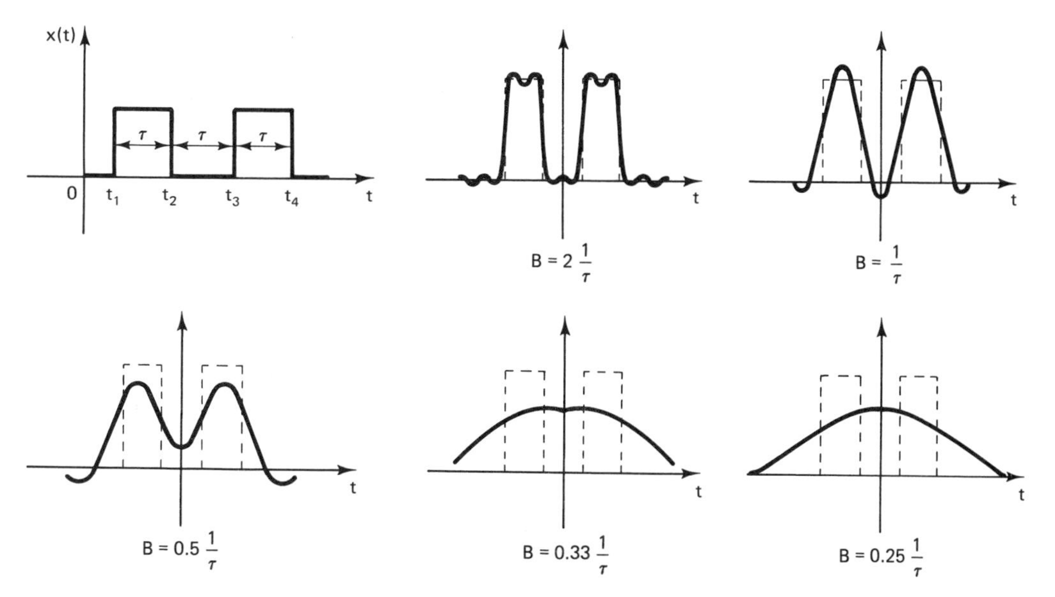

Figure 6.34 Outputs of the low-pass system for two pulses as an input signal. The bandwidth $B = f_N$ is a parameter.

The output signals for various values of system bandwidth $B = f_N$ and duration of pulses $\tau = t_2 - t_1 = t_3 - t_2 = t_4 - t_3$ are plotted in Figure 6.34. If we consider $f_0 = 1/T_0$ as a pulse repetition frequency and $\tau = T_0/2$, the corresponding bandwidths in Figure 6.34 are $4f_0$, $2f_0$, f_0, $0.66f_0$, and $0.5f_0$.

With $B = f_0$, pulses can be recovered by an ideal limiter. But an offset of the limiter would have a larger influence, for example, for $B = f_0$ than for $B = 2f_0$.

6.6 TIME-BASE-ERROR CORRECTION

In a videodisc recorder the electrical waveform is recorded on a disc in a manner such that the original time scale of the waveform is transformed. The recording process allocates to each unit of time in the waveform a corresponding unit of space on the disc. On reproduction these recorded elements are converted back into units of time. During the process of translating the waveform on and off the disc, errors in timing may occur which affect the geometry of the reproduced picture and/or its stability with respect to a stable reference source.

It has been demonstrated that the unprocessed demodulated video can have timing errors because the linear speed of the track, as seen by the objective, is not constant, or in general, differs from the corresponding recording speed. The causes of this difficulty are to be found in deviations of the disc, in unavoidable tolerance on the centering of the disc on the turntable, and in tracking servo instability.

The eccentricity of the track is the main cause of time errors located at the rotational speed of the disc. Since the player/disc combination allows for a maximum eccentricity, E, of typically 100 μm, the resulting peak time error

$$\Delta t = E/\omega R \tag{6.57}$$

and for rotating speed $f = 25$ Hz ($\omega = 2\pi f$) and radius $R = 70$ mm is

$$\Delta t = 9.1\ \mu\text{s} \tag{6.58}$$

A maximum time error of 10 ns can be permitted to ensure a satisfactory performance in combination with any TV receiver. Thus a reduction of approximately 59.2 dB is required. Higher-frequency time errors occur, but it has been found that their statistical amplitude decreases at a rate of approximately 12 dB per octave.

The time-base error causes two problems—poor color reproduction and horizontal sync instability—since errors of this size and rate will not normally be corrected by the color or horizontal locking circuits of a TV receiver due to bandwidth limitation in these circuits. A shift of a few degrees in the chroma can cause objectionable variations in hue. Therefore, these chroma phase errors must be reduced at least 60 to 70 dB, or more. Horizontal sync instabilities can cause objectionable picture motion, but these tend to be tracked to some extent by the receiver and 20 to 30 dB correction is sufficient.

To minimize time errors it is first necessary to keep the rotational speed of the disc as accurate as possible. For this a separate servo is used. It is not possible, however, to obtain an acceptable reduction in time error frequency at a rotational speed of 25 to 30 rps (Hz) and above. For a further reduction a second pivoting mirror, scanning the track tangentially, is used. The mirror is controlled by a separate time-base-error correction servo.

6.6.1 Time-Base-Error Correction Techniques

The inevitable distortion of the reproduced image from time-base errors may be reduced below subjective recognition by the servo correction. This can be achieved in a variety of ways and several techniques have been developed, each with their own limitations and cost-effectiveness. All systems operate in a manner similar to adjusting the delay of the video signal to cancel out any timing discrepancy between the disc video and a stable reference. Care must be taken to ensure that the calibration of the detector in terms of V/μsec error exactly matches that of the delay line in microseconds of delay per volt of correction. Failure to do this results in under- or overcorrection.

The system for time-base-error correction can be connected in an open- or closed-loop configuration. In the open-loop system the timing of the input video is compared to a stable reference and a correction voltage or digital signal is derived, the magnitude of this voltage being a function of the timing difference between the two signals. A better system is the closed-loop one, where the delay is adjusted until the output is correctly timed.

Although several techniques have been developed for the correction of timing errors [30, 31], four will be discussed here to illustrate the range of options available:

1. EVDL-type time-base corrector
2. Binary quantized delay line, type TBEC
3. Heterodyne color-recovery system
4. Digital time-base-error correction

6.6.1.1 EVDL-Type Time-Base-Error Correction. The principal element of this system is the electronically variable delay line, or EVDL (Figure 6.35). To obtain a time-base stability acceptable for broadcasting standards, that is, less than 10 ns peak to peak, three stages of correction are cascaded.

The first is a closed-loop tangential servo, which eliminates much of the error by moving the reading beam tangentially in a manner that maintains substantially constant relative velocity between the reading beam and the recorded information on the disc. In this electromechanical servo system the timing of vertical and horizontal sync from the playback video is compared with the reference vertical and horizontal sync. The timing error is fed back to the tangential transducer.

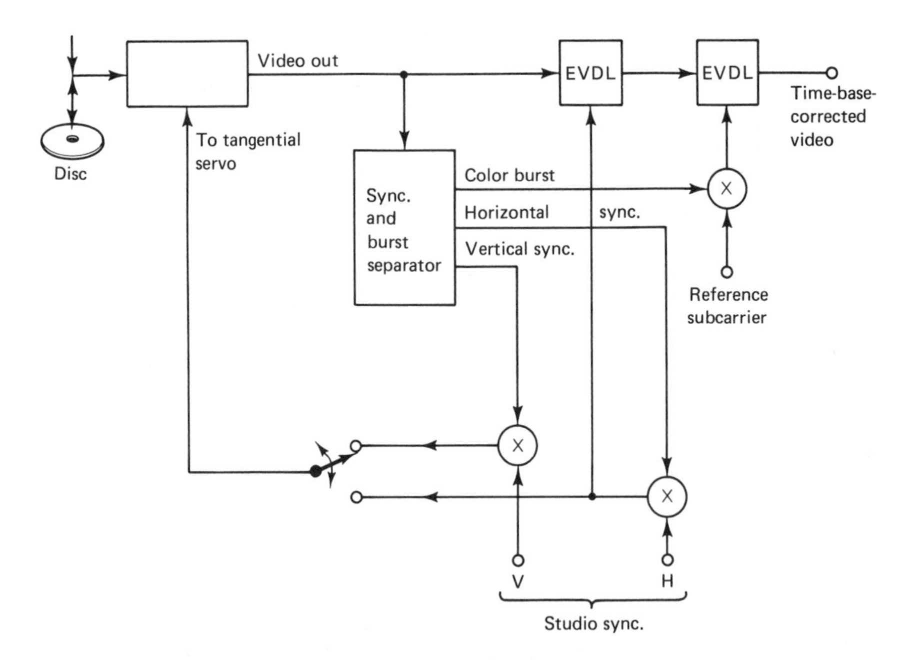

Figure 6.35 EVDL-type time-base corrector.

In the second stage the timing of the horizontal sync from the playback video is compared with the reference horizontal sync and the timing-error voltage is fed to the EVDL.

In the third stage the phase of the color burst from the playback video is compared with the phase of the reference burst. The timing error thus produced is fed to another EVDL. After the third stage, the timing error is normally less than 10 ns peak to peak.

Of course, the basic block diagram in Figure 6.35 can be modified. The EVDL can be in the closed loops, and instead of the burst the pilot signal can be used; and so on.

An EVDL can be realized as a lumped constant delay line with all its values/capacitors fixed, and a delay time

$$\tau = \sqrt{LC} \qquad \text{seconds per section} \tag{6.59}$$

A variable delay can be made if the capacitive elements are composed of back-biased varactor diodes. By changing the bias a change in capacitance will cause a change in delay. In this way the delay line length can be varied as much as $\pm 20\%$ of its nominal value without substantially degrading its transmission characteristics.

There is, however, a small degradation since the frequency response will change as the delay is adjusted. Each section acts as a low-pass filter with a cutoff frequency dependent on the value of capacitance. Tracking equalization is therefore necessary with the change in response depending on the amount of the delay.

The capacitance of the varactor diode is

$$C_V \simeq 1/\sqrt[N]{V} = V^{-1/N} \tag{6.60}$$

where V is the bias voltage and N is between 2 and 3.

The delay of the line is

$$\tau \sim \sqrt{C_V} \tag{6.61}$$

Thus

$$\tau \sim V^{-(1/2)N} \tag{6.62}$$

If $N = 2$,

$$\tau \approx V^{-1/4} = 1/\sqrt[4]{V} \tag{6.63}$$

The delay of the line is proportional to the reciprocal of the fourth root of the voltage. It must be ensured that the correction voltage follows this law.

The EVDL can also be made of charge-coupled devices (CCDs) (Figure 6.36). The delay introduced by CCD elements depends on the frequency of the

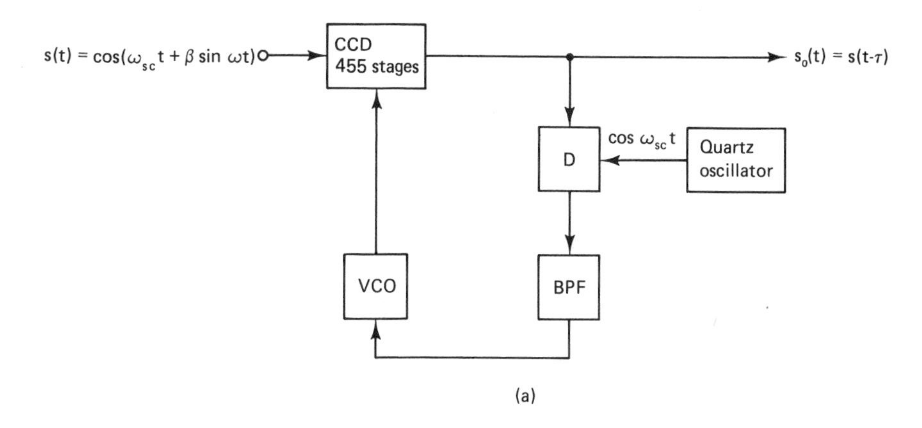

(a)

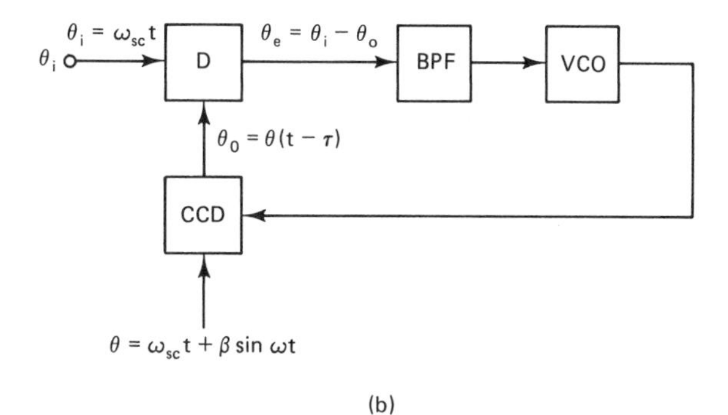

(b)

Figure 6.36 Time-base correction by means of the CCD elements.

voltage-control oscillator (VCO). The oscillation frequency of the VCO depends on the control voltage, consisting of the error voltage at the discriminator (D) output, which compares the signal phase from the quartz oscillator, and from the CCD block output. Figure 6.36b presents the system of Figure 6.36a but in a form suitable for analysis.

If we denote by H the closed-loop gain of the system, then:

Signal from the oscillator: $\theta_i = \omega_{sc}t$
Signal from the video disc: $\theta = \omega_{sc}t + \beta \sin \omega t$
Delayed signal, CCD block output: $\theta_0 = \theta(t - \tau)$
Error signal: $\theta_e = \theta_i - \theta_0$

Because

$$\theta_0 = H\theta_e \tag{6.64}$$

it follows that

$$\theta_0 = \frac{H}{1 + H}\theta_i \tag{6.65}$$

For instance, if the VCO has the change of 2 MHz/V and the CCD block delay of 2.3 ns/MHz, these two blocks together give a delay of 4.6 ns/V.

6.6.1.2 Binary Quantized Delay Line. A lumped constant delay line could be constructed with fixed components and with output taps along the line at discrete intervals. The video that is applied at the input could be delayed by any quantized amount by selecting the appropriate output. The delay is not continuously variable and therefore may not be exactly the correct amount. For very long delays the number of outputs required on a simple delay line becomes prohibitive. It is also difficult to design a lumped constant delay line that does not cause group delay problems. The longer the delay, the greater the problem.

Fixed-length delay lines are generally arranged in geometrical series (to solve both these problems) having, for example, lengths of 32, 16, 8, 4, 2, 1, $\frac{1}{2}$, $\frac{1}{4}$, and $\frac{1}{8}$ μs (Figure 6.37). The delay time required for compensation of time bias error is achieved by the insertion of the appropriate delay lines, the total length of which compensates for the time-base error. The length of the shortest delay line represents the smallest maximum increment of the time bias instability still remaining after correction. To obtain a finite level of correction, an EVDL TBC is sometimes used following a switched delay-line series.

Special attention should be paid to the switching problem. When simple electronic toggle switches are used, the system suffers from very serious disadvantages. There are "dead" intervals after the switching actions during which, partially, nothing would happen and, partially, no signal is followed by correct video. For example, nothing would happen for 32 μs, after which there is 16 μs of nonsignal. With double-pole, double-throw switches the total delay can be adjusted only in increments of the smallest delay, in this case $\frac{1}{8}$ μs.

If quartz delay lines are used, attention should be paid to their very poor

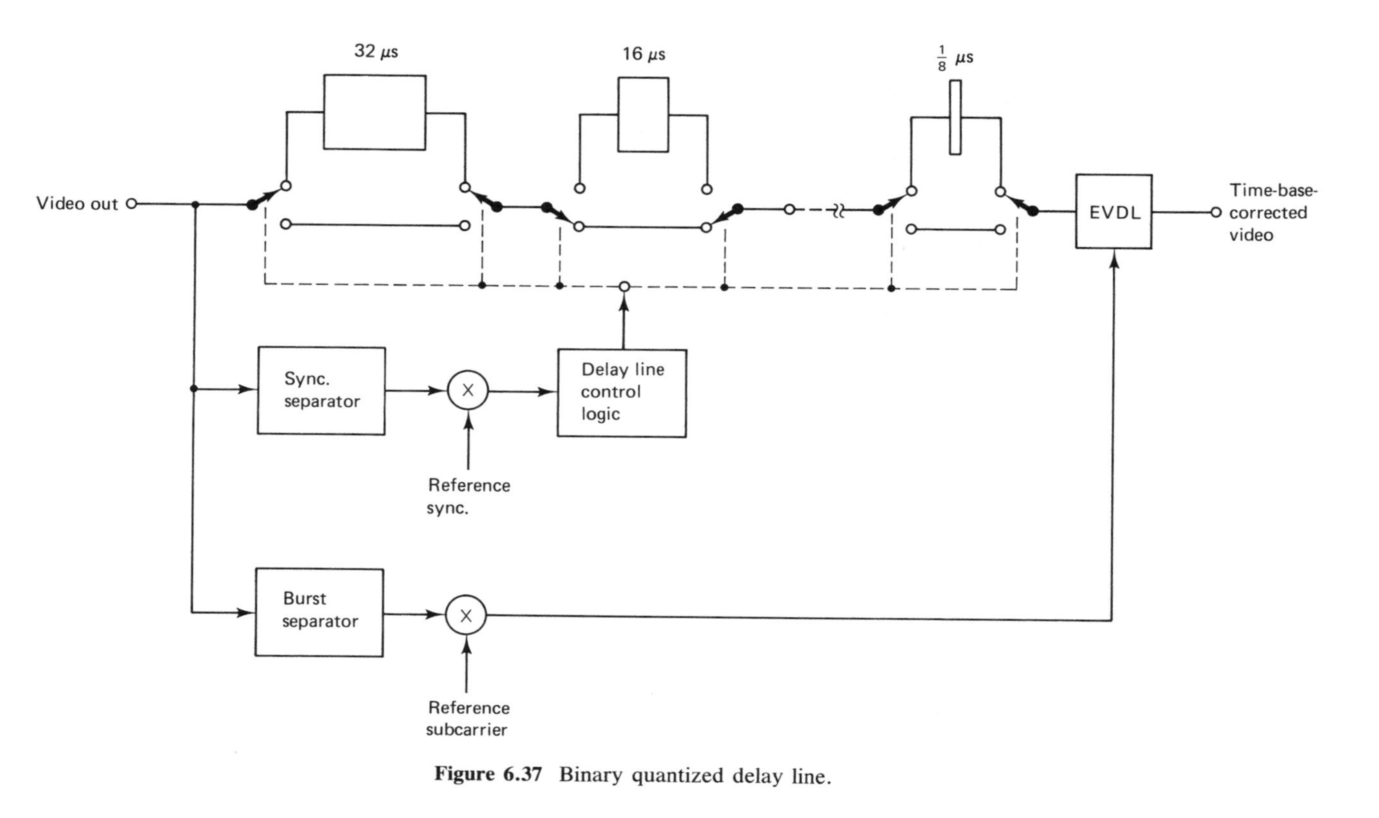

Figure 6.37 Binary quantized delay line.

low-frequency response. This makes them unsuitable for video signals without some form of modulation. If FM modulation is used, the problem of gain stability is alleviated.

6.6.1.3 Heterodyne Color Recovery System. This is an example of a low-cost color recovery technique. The principle of a heterodyne color recovery system is shown in block diagram form in Figure 6.38. The aim of the technique is to remove the time-base jitter, shown as Δt in Figure 6.38, from the chrominance signal so that the video could be displayed on an internally synchronized picture monitor with sufficient stability. Removal of time-base jitter from the luminance signal is not intended because the time constant of the horizontal lock circuit of a television receiver is capable of following the time-base jitter, thus producing a stable raster [30]. Heterodyne color operation is accomplished as follows:

1. The chrominance is separated from the composite video signal.
2. The continuous color subcarrier, containing Δt, is generated from the video signal.

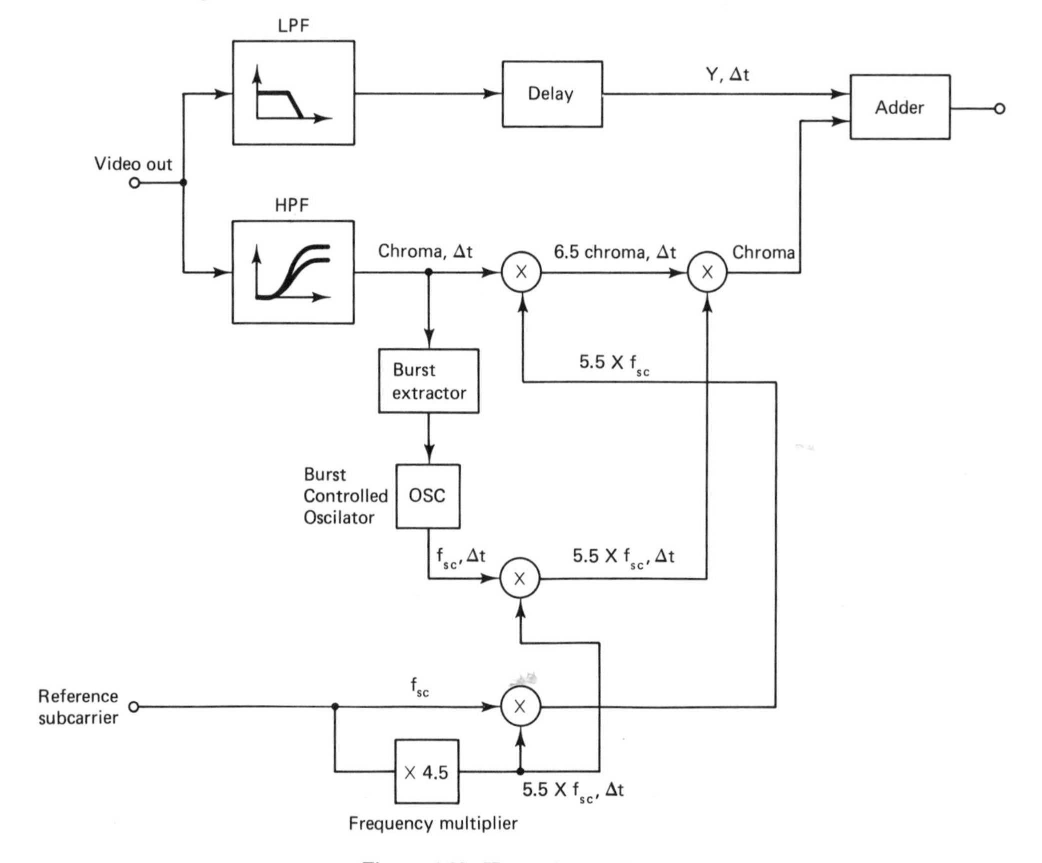

Figure 6.38 Heterodyne color recovery system.

3. The frequency of the chroma signal (f_{sc}) is up-converted by heterodyning it with the local (stable) subcarrier. Then the chroma signal will have a carrier frequency nf_{sc} and will still contain Δt.
4. The video-generated color subcarrier is up-converted to $(n - 1)f_{sc}$. This signal also contains Δt.

By mixing the two signals nf_{sc} $(+ \Delta t)$ and $[(n - 1)f_{sc}/(+ \Delta t)]$, a stable chroma can be produced.

6.6.1.4 Digital Time-Base-Error Correction. In lumped constant delay lines, there is a problem of phase and frequency response, with the difficulty of tracking equalization as the delay is adjusted. Switchable binary quantized delay lines require extremely precise equalization for each delay path if frequency, amplitude, and group delay response changes are to be avoided. Accordingly, these devices are expensive, particularly when it is considered that they are followed by an analog delay for time and velocity errors. But, in practice, when corrected video is fed into the color monitor or receiver, a subjectively stable picture is produced. However, when the criterion used to judge their usefulness in broadcast applications is time-base stability with respect to an absolute reference, their stability is rather poor. Using digital methods and techniques prior problems can be solved. Figure 6.39 is a basic block diagram of a digital TBC.

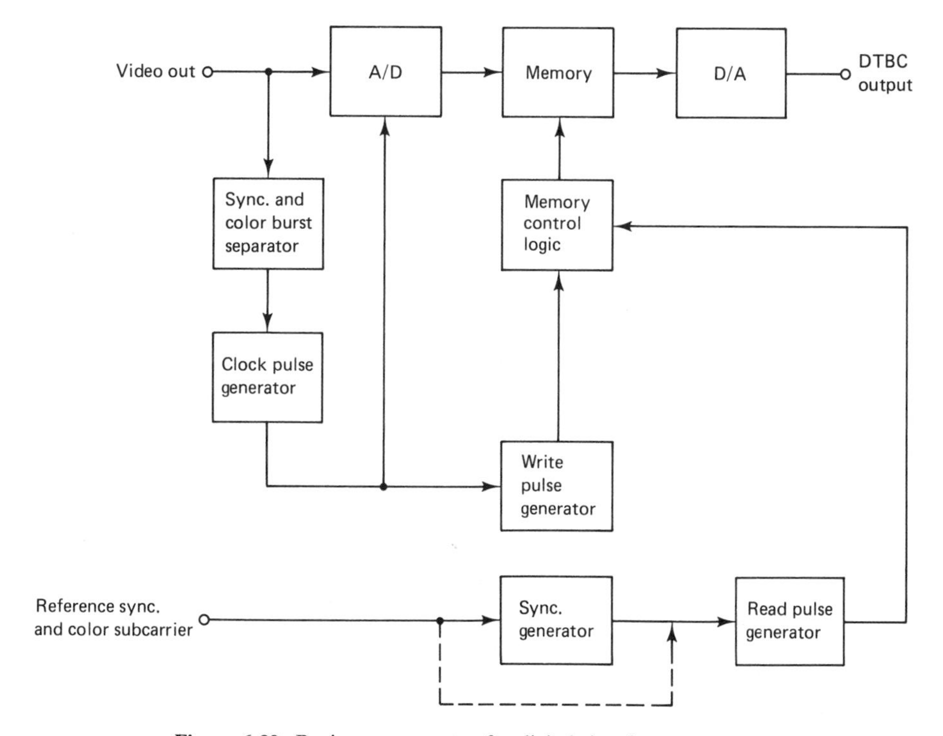

Figure 6.39 Basic components of a digital time-base corrector.

Digitization of video is done by an analog-to-digital (A/D) converter driven by clock pulses locked to the incoming video. A commonly used clock frequency is three times the color subcarrier (10.74 MHz for NTSC, 13.29 MHz for PAL), or four times the color subcarrier (14.32 MHz for NTSC, 17.72 MHz for PAL). Lately, a 13.5 MHz sampling rate is frequently used for both NTSC and PAL. The bits per sample are either 8 or 9. Memory is either in shift registers or in RAM (random access memory).

Since the incoming video signal with its time-base error is written into memory by clock pulses that are synchronous with the video, the digitized video is stored in memory in an organized manner. The stored digitized video is read by a clock pulse locked to the studio sync. In this way one obtains video without time-base error as well as concrete synchronization with the studio sync.

Additional advantages of digital time-base correction (DTBC) are that velocity error correction is accomplished almost automatically, and an extra video delay is readily available for dropout compensation.

The correlation of video SNR of a digital video processing system, such as a DTBC, and its sampling and bit rate, are of particular interest to potential users and designers. Detailed analyses are shown in several publications and are beyond the scope of this discussion, but an approximation is given.

The signal-to-noise ratio is

$$S/N(\text{dB}) = 20 \log (V_S/V_N) + 18 \text{ dB}$$

where V_S = video amplitude (peak to peak)

V_N = noise amplitude (peak to peak)

The 18 dB is a factor used to convert the peak noise to an rms value.

For $V_N = V_S/2^n$, where n is the number of bits, $S/N(\text{dB}) = 6n + 18$. Of course, some corrections can be introduced. Other forms of distortion are normally a result of quantization errors.

6.6.2 Color Correction: Color Burst versus Pilot Signal

In the NTSC and PAL systems the color information is contained in the phase of the subcarrier, making these systems very sensitive to spurious phase modulation, due to possible deviations from the required conditions for reading. SECAM, on the other hand, uses a line sequential modulated system which is more tolerant to the videodisc timing instability. Stability requirements for the SECAM system are no more stringent than for monochrome systems. For the PAL system the phase error should be limited to $\pm 5°$, and for the NTSC system to $\pm 4°$, so that the error visibility will be negligible. The corresponding servo systems have been discussed in Chapter 1.

In these servo systems a comparison between the signal recorded on the videodisc with the quartz-oscillator-generated frequency signal at the same frequency as the disc signal is made. The error signal is, in fact, a correction signal in these servo systems.

In the case of magnetic mastering of the video signal and in the case of the videodisc, the following possibilities of mastering the control signal were experimentally investigated:

1. Color burst signal of frequency f_{sc}(f_{sc} = 3.58 MHz for NTSC and f_{sc} = 4.43 MHz for PAL)
2. Control pilot signal (frequency $f_p \simeq$ 400 kHz)

It was experimentally found for both magnetic tapes and videodiscs that the first method yields much better results, although no obvious reason was observed. The following question could be raised. How large should the amplitude of the pilot signal, S_p be (superimposed over the FM signal) to obtain the same visual result as with the burst signal? The pilot signal amplitude cannot be too large, because it lies in the baseband of the video signal, which could cause a Moiré effect to be noticed in the reproduced picture. The following expression can be obtained analytically [27, Chap. 3]:

$$(S/N)_p(\text{dB}) = (S/N)_{obr}(\text{dB}) - 3\text{ dB} + 20 \log f_{sc}/f_p + 20 \log \beta_{dp} \qquad (6.66)$$

where $(S/N)_p$ = signal-to-noise ratio of the pilot signal at recording
$(S/N)_{obr}$ = signal-to-noise ratio of the burst signal having the same subjective effects as the respective pilot signal
β_{dp} = $(f_p/f_d)(\Delta R/R)$; f_d is disc rotation per second
R = instantaneous radius of the disc
ΔR = eccentricity

Figure 6.40 illustrates the foregoing statements for the values f_p = 400 kHz, f_{sc} = 3.6 MHz (NTSC), and $(S/N)_{obr}$ = 25 dB. Then

$$(S/N)_p = 40\text{ dB for } \beta_{dp} = 4.4 \text{ (outside radius)} \qquad (6.67)$$

$$(S/N)_p = 34\text{ dB for } \beta_{dp} = 8.8 \text{ (inside radius)} \qquad (6.68)$$

It should be kept in mind that the noise at f_p is about 2 to 4 dB greater than the noise at f_0, which means that S_p should be larger than the value shown.

It was experimentally found that the amplitude of the pilot signal should be about 25 to 30 dB lower than that of the carrier (f_0) frequency for the Moiré component to be unobservable in the produced picture. This means that until techniques ensure S/N = 70 dB at the carrier frequency, the system with the burst signal will yield better results with respect to color correction because of the influence of time-base errors.

Figure 6.41 shows the error signals when the pilot and burst signal were used with an open feedback loop, while Figure 6.42 represents the signal with a closed feedback loop and with circuit reinforcement of about 60 dB. The lower flow in Figure 6.42b is relevant to the pilot signal and the others to the burst signal. Figure 6.42a shows the error signals for a burst packet. Namely, the error signal is detected only when the burst signal exists. It could be noticed that the error is largest for the first burst period; then it reduces tending to zero. Figure 6.42b presents the error signal for the two subcarrier (burst) periods, while Figure

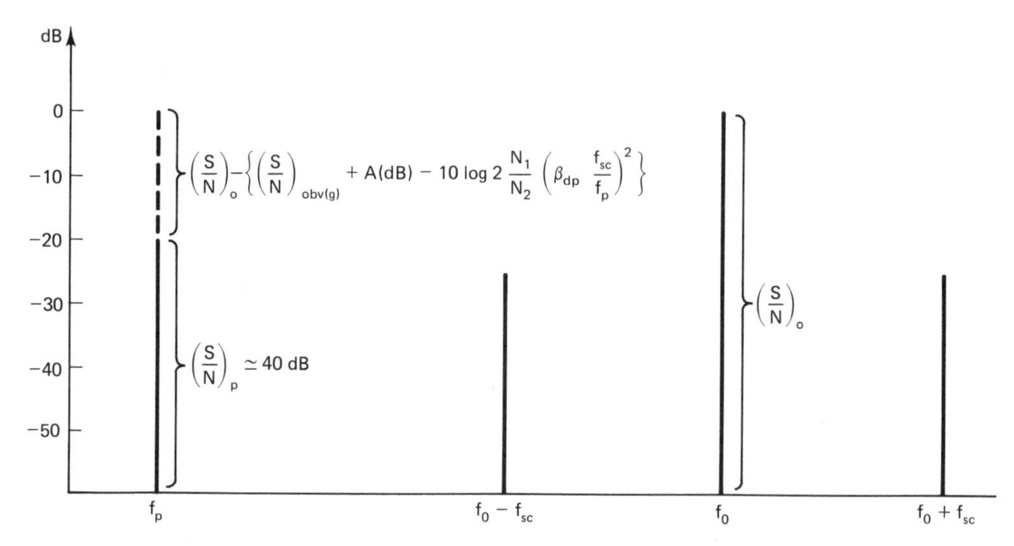

Figure 6.40 Corresponding S/N's.

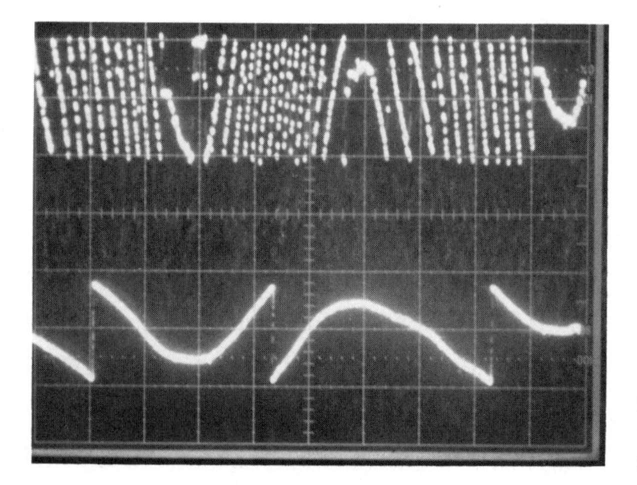

Figure 6.41 Signal outputs of respective detectors when the reference signal is a pilot signal and burst.

6.42c shows the spectrum of the error signal when the burst signal is used and when the servo system is closed. The components of 30 Hz, 60 Hz, and 90 Hz should be noted as well as the fact that the noise is higher at low frequencies (it behaves like $1/f$ noise).

6.7 DROPOUT COMPENSATION

The random dropouts in the videodisc system cause a loss of FM. The loss in FM causes a burst of noise or impaired video signal. At best this produces a subjective annoyance and at worst can cause a false sync or elimination of a sync edge or color burst or an incorrect program code. Dropouts in the FM

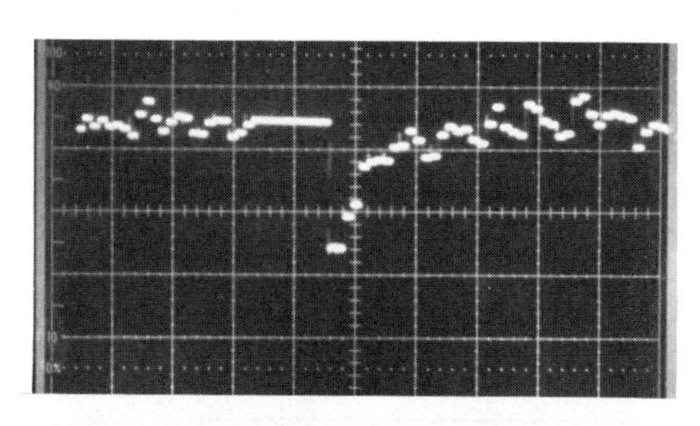

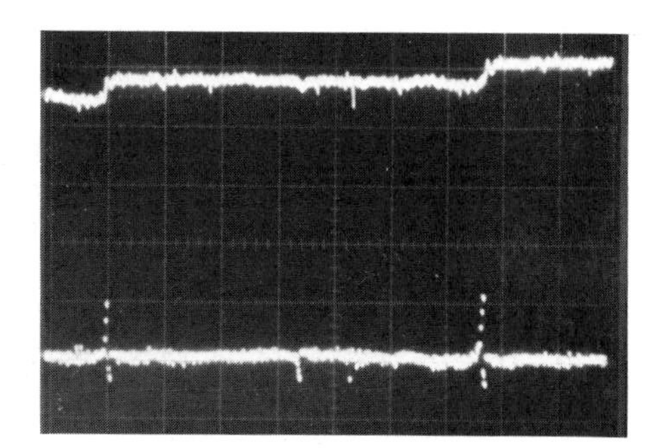

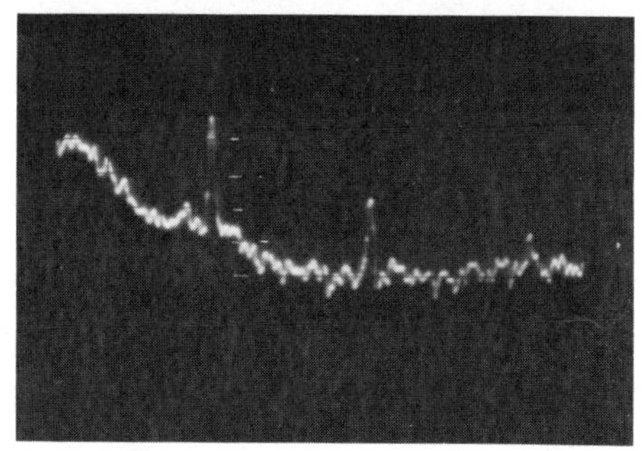

Figure 6.42 Error signal of the closed loop.

signal picked up by the photodiode are caused by the presence of dust, scratches, and other irregularities on the surface of the videodisc due to various inhomogeneities of the material and imperfection of the technological procedures used when the disc stamper and the disc are made. Subsequently, this leads to distortion in the video output signal. The two following cases may occur:

1. False pits on the videodisc appear because the existing pits are either deformed or because holes appear between the real pits.
2. The pits are not mastered into the videodisc, or several pits are joined into one large pit.

When reading, in the first case, false pulses occur, which correspond to an increase in the instantaneous pulse frequency: this is also called a "drop-in."

Because frequency modulation is used, as presented in Figure 6.43, the instantaneous frequency could be enhanced or diminished due to the dropout and is marked by 1 or 2, respectively.

The dropout marked by 2, when sufficiently long, could be taken as a sync pulse so that particular measures must be undertaken, for example, substitution of this line (at playback) by the previous line.

To minimally reduce the visibility of these dropouts, the compensation requires two separate functions: detection of the dropout and substitution during the dropped-out signal. Figure 6.7 presents the compensation principle. While dropouts are not detected, switch S is in position 1 and the system operates normally. However, when loss of the FM signal is detected, switch S is switched

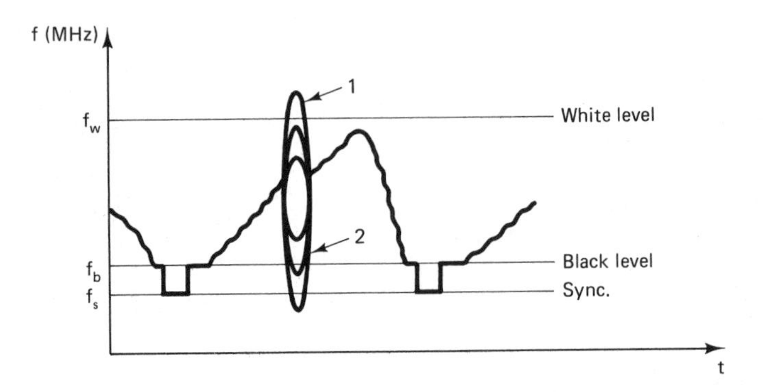

Figure 6.43 Distortion in the reproduced signal due to dropouts.

to position 2, so that a portion of the previous line is substituted. This procedure considerably decreases the visual sensation of the dropout. Instead of the FM signal, the video signal can be delayed using, for example, the CCD delay lines.

The corresponding time chart is outlined in Figure 6.44. The signal "in focus" indicates when the system is out of focus. The dropout signal (DOS) is indicated in the last line; the DOSF signal indicates when the system is out of focus and a dropout is present.

6.8 FILM PICTURE RECORDING ON THE DISC

Standard films, 35 mm and 16 mm, have a speed of 24 frames per second. Therefore, conversion to 25 pictures per second for PAL and to 30 pictures per second for NTSC is necessary.

For PAL this is accomplished simply enough by scanning the film at a rate of 25 pictures per second, thus compressing the original in time by 4%. An analogous procedure for NTSC would be unacceptable due to the large difference in speed that would result. A solution has been found by scanning every other film picture two or three times, resulting in five fields for two successive film pictures. In this way 60 fields per second are obtained. This method is commonly called *3-2 pulldown*. This is illustrated in Figure 6.45. Film picture 1 (Figure 6.45a) is scanned twice, and two fields, 1′ and 1″, are recorded on the disc (Figure 6.45b). The next picture, number 2, is scanned three times and three fields, 2′, 2″, and 2‴, are recorded on the disc. Picture 3 will be scanned twice; and so on.

Before mastering program material from the film it should be transferred from a flying spot scanner to tape (the master tape). This is a premastering process. Premastering is the process during which the basic program material is processed in such a way that it results in a master tape containing a composite video signal plus all other information as required for mastering. In principle, any type of program material, be it tape or film, can be used as a basis for

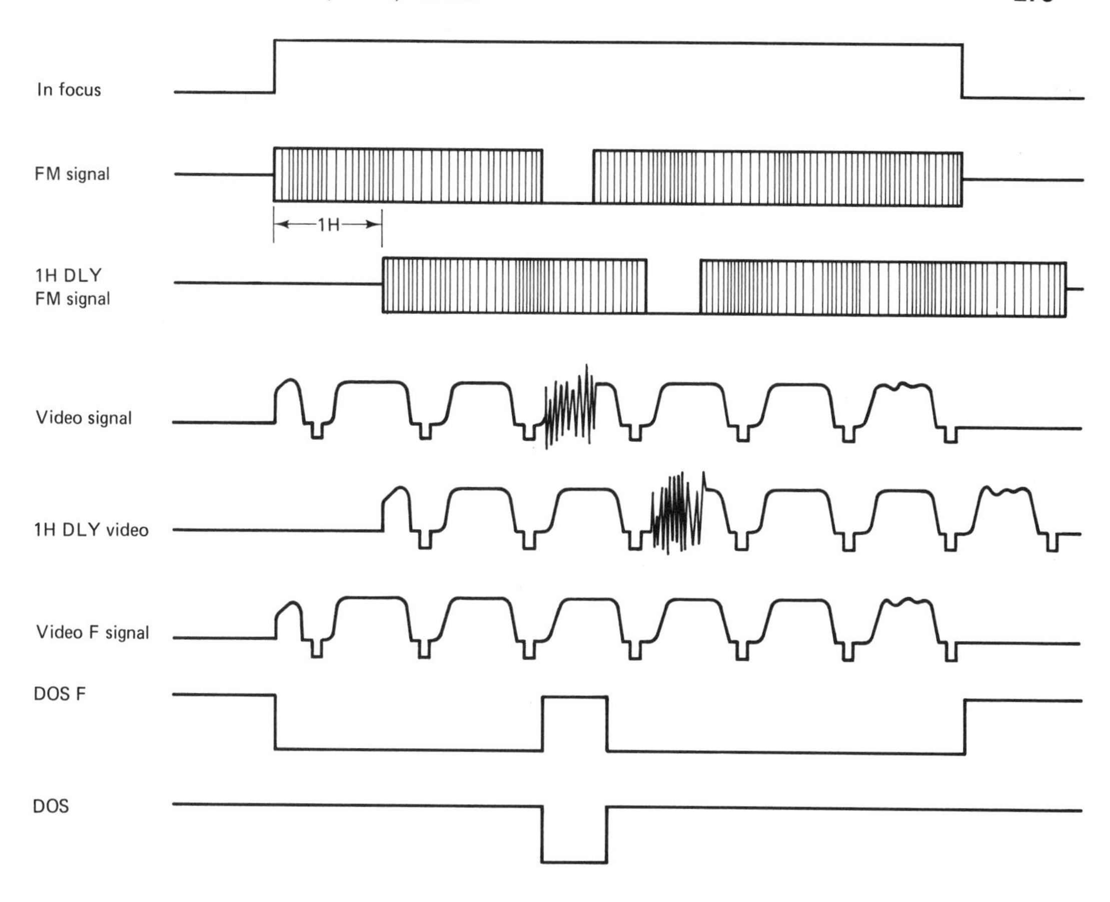

Figure 6.44 Time chart.

videodiscs. The optical sound track of film is recorded during the scanning process on the tape. Synchronizing with the film scanner is required.

6.9 FRAME (PICTURE) NUMBER

To allow automatic searching to a particular frame of the program material on disc, a method of unique addresses for each frame recorded is needed. As has been shown in Section 6.3.2, this can be done by recording a sequential digital code on the disc with a number unique to that frame.

For video source material the frame number is inserted in the second field (lines 280 and 281 for the NTSC 60-Hz/525-line system). For the film source material the frame number appears in one field or the other, depending on the film frame, but never both fields in the same frame.

Using the white flag in Figure 6.46, it is shown where the frame number changes field, for the film source material, if the video signal is tailored according

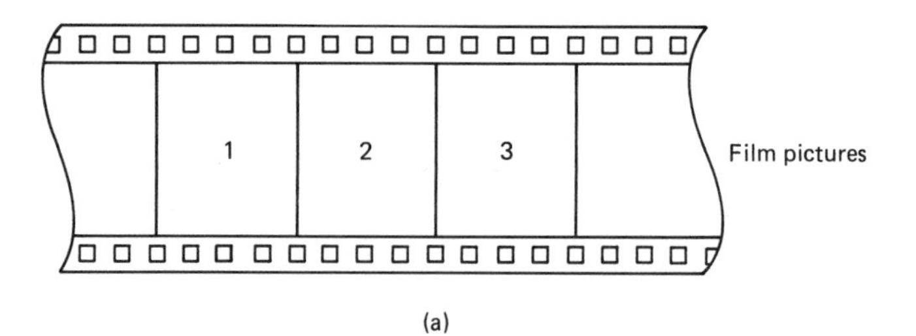

(a)

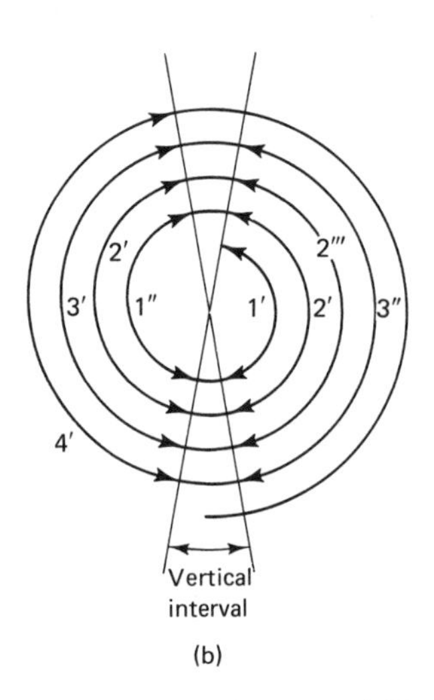

(b)

Figure 6.45 Relation between (a) film pictures and (b) video frames on the disc.

to the NTSC 525-line/60-Hz system. White flags are marked on both film and master tape (or videodisc). For both, white flag shows the end of the picture (frame). Also, for the CAV disc, for each revolution two fields are recorded (Figure 6.45b). Finally, the frame (picture) number is always inserted in the last field of each picture. Notice that the time scale in Figure 6.46 is common for film and tape (disc).

6.10 SPECIAL EFFECTS AND PLAYING MODES

Due to the fact that in the videodisc system the reading is performed optically—without physical contact between the disc and the reading "head"—some specific applications are possible: for example, randomly accessing frames within the disc (this may be done very rapidly). A particularly interesting feature of the videodisc system is the possibility of several playing modes, such as:

Fast forward
Slow motion

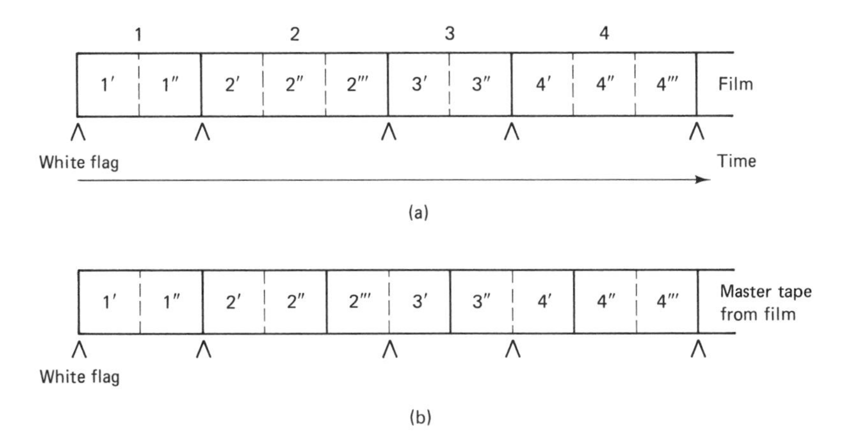

Figure 6.46 White flag appearance: (a) on the film; (b) on the master tape (and disc).

Freeze frame (single frame hookup)
Picture presentation similar to projector systems
Reverse
Time loop forward
Time loop backward

These possibilities exist only with CAV discs, which are therefore eminently suitable for instructive programs.

The foregoing effects originate from the possibility of moving the light beam rapidly from one track to another during the playback period and reading of the frame (ordinal) number of the recorded picture. Figure 6.47 shows the way in which the various playing modes are realized.

Twice per revolution, during field playback, an opportunity exists to jump from one track to another. A reverse jump after every revolution yields a stationary picture (Figure 6.47a). A reverse jump after every half revolution produces reverse motion at normal speed (Figure 6.47b). Jumping one track after every two revolutions yields half-speed forward. A forward jump every half-revolution results in three times normal speed forward (Figure 6.47c). Double-speed forward is obtained by skipping one track after each revolution. Two reverse jumps after every half revolution results in three times normal speed (Figure 6.47d).

These different types of operations can be realized because the disc rotation speed (25 or 30 rps) agrees with the frame frequency of TV (25 or 30 frames per second). Since a TV image consists of two sequential half-frames, this breakdown is also used by the videodisc. Thus for two half-frames there are also two synchronization signals that consequently are always at the same place on the track (i.e., diametrically opposite each other), blanked on each track spiral. During these times the TV image is made dark, and hence within narrow sectors the scanning point of light can jump from one track to another without being

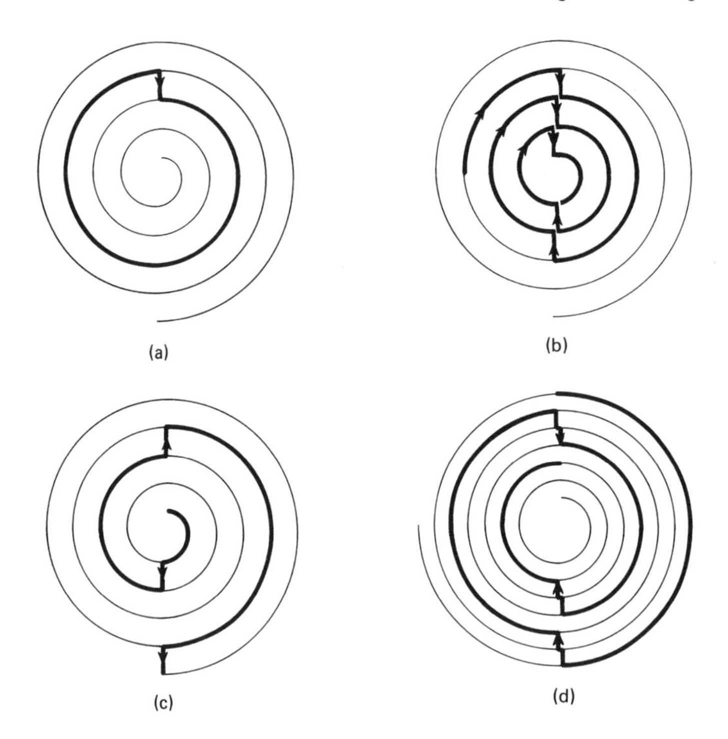

Figure 6.47 Playing modes of the videodisc system: (a) freeze frame: reverse jump after every revolution; (b) reverse motion normal speed: reverse jump after every half-revolution; (c) three times normal speed forward: forward jump every half-revolution; (d) three times normal speed reverse: two reverse jumps after every half revolution.

visible in the playback TV image (picture). To enable the beam to jump from one track to another during this period, use is made of the fact that the open control loop of the radial servo has a low-frequency response, so that it behaves like a ballistic galvanometer. The beam jump is thus effected by opening the control loop and applying an accelerating current pulse followed by an equally large retarding pulse through the coil of the radial tracking mirror and then closing the loop gain.

Figure 6.48 shows the principle of the circuit [32]. Switch S_1 is normally closed if the laser is "on" and the objective is in focus so that the radial servo, as described before, acts under the influence of the difference signal. To achieve track jumping a pulse P_2 is applied, the duration of which determines the period during which S_2, and as a consequence S_1, is open. The timing of pulse P_1 in relation to P_2 determines the duration of the jump: forward or backward. The combination of two pulses in the track changes circuit results in pulse P_{1A}. Under the influence of pulses P_1 and P_{1A}, switches S_3 and S_4 are temporarily closed, resulting in a voltage at the mirror as illustrated. In the same figure, the angular velocity of the mirror and its position are quoted as a function of time.

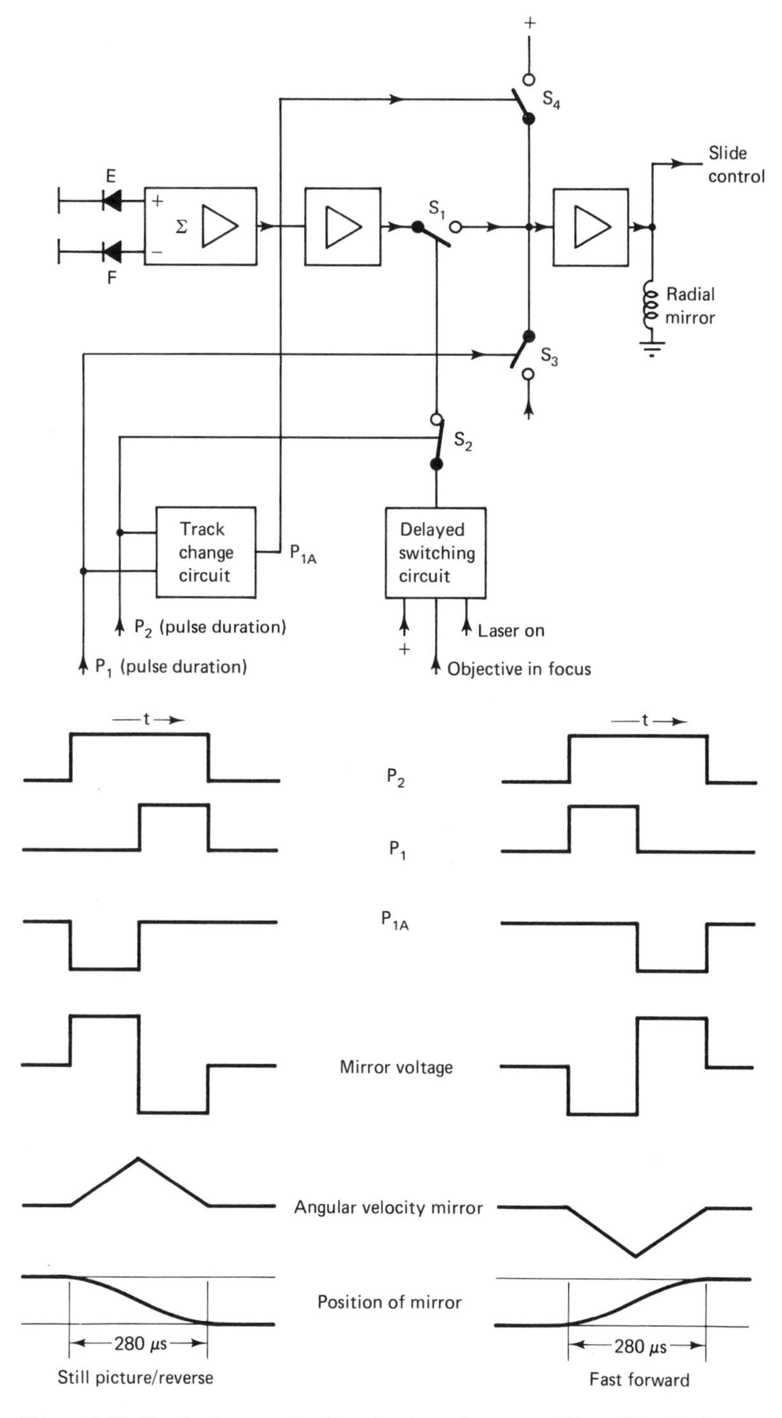

Figure 6.48 Track change circuit and schematic presentation of track change pulses and resulting jump of mirror.

Starting from the situation "freeze frame" it is obvious that by suppressing or adding a pulse sequence regularly as a function of time, slow-motion forward or reverse can be obtained. By varying the frequency with which this is done by means of a variable resistor, a continuous variation of the slow-motion speed is possible.

6.11 OPERATING CONTROLS

To take advantage of the versatility of the system, the videodisc player should be equipped with a number of operating controls for a variety of functions. Apart from the latch for releasing the lid, the other controls are of the pushbutton type and are arranged in a logical sequence at the front panel [32].

Besides forward and reverse play at normal speed, the control system controls other main features of the videodisc system:

1. Immediate access to any program part (frame).
2. Freeze frame (still picture)—the ability to repetitively stop the playing of the disc at any one of the selected picture frames on the TV screen as long as desired. Next ("step forward") and preceding ("step reverse") picture features are included.
3. Slow motion—usually separate pushbuttons for forward and reverse action combination, with a sliding potentiometer to vary the speed between normal (25 or 30 frames per second) and, say, 4 seconds per picture.
4. Search, fast forward, and reverse—the ability to access any selected frame, passage, or segment of the disc in a few seconds.
5. Frame crawl—the ability to view each frame of the disc separately, picture by picture, as a series of stills in a desired, timed sequence similar to a regular slide projector.
6. Coding possibilities for automatic actions such as access, start, and stop.
7. Individual identification of each picture by means of visual display—picture number and chapter number can be made to appear on the screen ("indexing").
8. Two discrete channels for stereophonic reproduction or, for example, two languages: for each sound channel there is one button that can be operated separately to enable or disable either of the two sound channels or both.

Some other capabilities can be included, as, for example, interchangeable wireless or remote control, multiplayer or external synchronization with subcarrier and composite sync, control by external computer, and so on. Also, a great many automatic functions are incorporated in the videodisc player to ensure a flawless performance of the apparatus under all circumstances in order to make it foolproof.

For many features, operations are under microprocessor control. One of these operations is the search for an arbitrary picture. For example, let the ordinal number of the picture N_1 be given on the panel, and the present position (number) is N_0. The sign of the difference $N_1 - N_0$, sign $(N_1 - N_0)$, defines the direction of search: forward (+) or reverse (−). The magnitude of the difference $N_1 - N_0$ defines the speed. If N_1 is greater than N_0, then, for example:

If $N_1 - N_0 > K_2$, three times normal speed forward.
If $K_1 < N_1 - N_0 < K_2$, double speed forward.
If $N_1 - N_0 < K_1$, normal speed.

Naturally, other algorithms are also possible to achieve this.

6.12 EVALUATION OF PICTURE (COLOR) RENDERING IN THE VIDEODISC SYSTEM

There has recently been considerable interest in the objective evaluation of the picture, especially the color reproduction properties of the complete (color) imaging system. Much progress has been made toward the development of quantitative measures of image fidelity and intelligibility [33, Chap. 7]. However, those measures that have been developed are not perfect; counterexample images can often be generated that possess a high quality rating but that are subjectively poor in quality, or vice versa. The key to the formulation of improved image quality measures is, no doubt, better understanding of the human visual system.

The effects of viewing conditions should be considered in the design as well as in the evaluation of a color reproduction system [31]. Also, any valid color rendering index for an imaging system should correlate with the subjective color quality.

The CIE color rendering index (CRI) of a light source [34–36] might be useful for the intended evaluation of the videodisc system. The CRI was intended originally and primarily only as a means of characterization of a light source—an illuminant description—by means of the colorimetric shifts produced in the CIE test colors. It was not intended as a means of color description of objects having a similar color apppearance, but using different colorants and measured under the same illuminant.

The extension of the CRI concept, which is based on colorimetric matching, assumes that the highest subjective color quality will be obtained when the CIE tristimulus values of the reproduction are equal to those of the original colors. In the proposal, the same illuminant is assumed for both the reproduction and the original.

To compute the proposed CRI one would first photograph, under the light of a specific illuminant, a set of standard test colors such as those specified for the determination of the CIE illuminant color rendering index. Next the CIE tristimulus values X, Y, Z would be measured for both the original colors and their reproductions using the same illuminant. These values would then be transformed to the CIE 1964 $U^*V^*W^*$ uniform color space [37] and the colorimetric difference E between the original and reproduced colors would be computed using the 1964 CIE color difference formula

$$E = \sqrt{(U_0 - U_R)^2 + (V_0 - V_R)^2 + (W_0 - W_R)^2} \qquad (6.69)$$

where U_0, V_0, and W_0 are $U^*V^*W^*$ coordinates of the original color, and U_R, V_R, and W_R are $U^*V^*W^*$ coordinates of the color reproduced. Finally, a CRI would be computed for each test color using equation (6.80) and the average of these CRI values would be used as a measure of a system color quality:

$$\text{CRI} = 100 - 4.6\Delta E \qquad (6.70)$$

The results would be a number ranging up to 100, which would represent a perfect (colorimetric) color reproduction. It is usually assumed that a CRI number of 75 to 80 is intended to represent a "passing" or acceptable level of color quality. The basic assumption underlying this proposed CRI is that the highest subjective color quality will be obtained with the color reproduction that colorimetrically matches the original colors. This assumption is wrong for the videodisc application.

The CIE system is a physical measurement using a set of spectral responses which are linear transformations of average human visual spectral responses. A CIE specification will tell us if two colors will match colorimetrically when viewed under identical conditions. However, the CIE specification tells us nothing about the appearance of colors and will fail to predict matching when the two colors are viewed under different conditions.

In general, development of a quantitative measure for evaluation of color rendering is much more difficult than the development of such a measure for monochrome images not only because of the increased dimensionality, but also because of the larger number of perceptual phenomena that must be satisfied. The color rendering index should quantitatively measure just noticeable color differences on a point-by-point basis, and also be consistent with spatial effects such as color Mach bands and color adaptation.

The list of other major factors that affect the relationship between visual response and the color stimulus includes color of illuminant, level of illuminant, size of the image, and luminance of the area surrounding the image relative to that of the image. The effects of a dark surrounding and small image size on the relationship between brightness is a visual response [38] and luminance (a psychophysical measurement of the amount of light) has been extensively studied. It has been shown [39] that the gamma of an image viewed in a dark surrounding needs to be greater than unity to achieve correct relative brightness reproduction. Several functional relationships have been published to describe the change in brightness response that occurs in an image with a dark surrounding. One example [34] is:

$$B = a_1 L^{1/3} + a_2 \qquad (6.71)$$

where B is the brightness, L the luminance, and a_1 and a_2 are constants. The effect of the dark surrounding is to alter the value of a_2 in this equation as a function of average scene luminance. A television engineer will recognize a_2 as a dc- or black-level control. The description of the effect is very much like the automatic flare correctors in modern television cameras. But, one wonders if the dark surrounding effect may just be the result of a sort of a psychological flare corrector. Similarly, a television engineer will recognize a_1 as a gain or exposure control. At any rate, this effect is a good example of how our visual response to a constant physical stimulus can change with the content of our visual field.

Recently, it has been shown [40] that the only significant change required when there is too much appearance of an image viewed with a dark surrounding is a change in luminance contrast as indicated by equation (6.81). Assuming the same illuminant, the chromacity coordinates in the image should approximately match those in the original.

Presently, the most common and most reliable judgment of color rendering in the videodisc systems, and judgment of output image quality, is subjective rating by human observers. In some instances, untrained, "nonexpert" observers are utilized so that the judgment represents image quality as perceived by an average viewer. Tests are also conducted with trained "expert" observers who are experienced in processing image and system distortions and are allegedly better able to provide a critical judgment of image quality. Presumably, expert viewers have acquired the ability to notice small-scale image degradations that a nonexpert viewer might overlook. Probably, the greatest difference between "expert" and "nonexpert" is in the dropout noticeability.

REFERENCES

1. Three papers on Philips VLP, *Philips Tech. Rev.,* Vol. 33, No. 7, 1973, pp. 177–193.
2. Special Issue, *J. SMPTE,* Vol. 83, No. 7, 1974.
3. L. Michelson, J. Winslow, and K. D. Broadbant, "Use of the Laser in Home Video Disc System," *Ann. N.Y. Acad. Sci.,* Vol. 267, Jan. 1976, pp. 477–481.
4. K. D. Broadbant, "A Review of the MCA DiscoVision System," *115th SMPTE Tech. Conf. Equip. Exhib.,* Los Angeles, Apr. 26, 1976.
5. P. W. Bogels, "System Coding Parameters Mechanics, and Electromechanics of the Reflective Video Disc Player," *IEEE Trans. Consum. Electron.,* Vol. CE-22, No. 4, 1976, p. 309.
6. Special Issue on Video Discs, *RCA Rev.,* Vol. 39, No. 1, 1978.
7. Five papers on video long-play systems, *Appl. Opt.,* Vol. 17, No. 13, 1978.

8. Special Issue on Video Disc Optics, *RCA Rev.*, Vol. 39, No. 3, 1978.
9. *MCA Video-Disc Player Model PR-7820: Service Manual,* MCA DiscoVision Inc., 1979.
10. J. H. Wessels and W. Von Den Bussche, "Method of Recovering a Video Signal," U.S. Patent 3,893,163, July 1, 1975.
11. C. E. Johnson, Jr., "Digital Recording in Low Cost Transporters," *Digit. Des.*, June 1977, pp. 38–48.
12. G. King, "Cassette, Cartridge and Diskette Drives," *Digit. Des.*, June 1977, pp. 50–66, 88–90.
13. A. M. Patel, "New Method for Magnetic Encoding Combines Advantages of All the Techniques," *Comput. Des.*, Aug. 1976, pp. 85–91.
14. B. R. Jarrett, "Could You Design a High Speed Manchester-Code Demodulator?" *EDN,* Aug. 20, 1974, pp. 75–80.
15. W. E. Bentley and S. G. Varsos, "Squeeze Mode Data onto Mag Tap," *Electron. Des.*, Vol. 21, Oct. 11, 1975, pp. 76–78.
16. R. C. Franchini and D. L. Wartner, "A Method of High Density Recording on Flexible Magnetic Disc," *Comput. Des.*, Oct. 1976, pp. 106–109.
17. P. S. Sidhy, "Group-Coded Recording Reliably Doubles Diskette Capacity," *Comput. Des.*, Dec. 1976, pp. 84–87.
18. N. O. Duc and B. M. Smith, "Line Coding for Digital Data Transmission," *ATR,* Vol. 11, No. 2, 1977, pp. 14–27.
19. A. L. Knoll, "Spectrum Analysis of Digital Magnetic Recording Waveforms," *IEEE Trans. Electron. Comput.*, Vol. EC-16, No. 6, 1967, pp. 732–743.
20. M. Hecht and A. Guida, "Delay Modulation," *Proc. IEEE,* Vol. 57, July 1969, pp. 1314–1316; see also correction in *Proc. IEEE,* Vol. 58, Jan. 1970, p. 182.
21. J. C. Mallinson and J. W. Miller, "Optimal Codes for Digital Magnetic Recording," *Radio Electron. Eng.*, Vol. 47, No. 4, 1977, pp. 172–176.
22. A. M. Patel, "Zero-Modulation Encoding in Magnetic Recording," *IBM J. Res. Dev.*, Vol. 19, No. 4, 1975, pp. 336–378.
23. G. V. Jacoby, "A New Look Ahead Code for Increased Data Density," *IEEE Trans. Magn.*, Vol. MAG-13, No. 5, 1977, pp. 1202–1204.
24. J. Isailović, "Method and Means for Encoding and Decoding Digital Data," U.S. Patent 4,204,199, 1980.
25. J. Isailović, "A New Code for Digital Data Recording/Transmitting," Publication of Electrical Engineering Faculty, University of Belgrade, Series: *Electron. Telecommun., Autom. 136–141,* 1980, pp. 41–51.
26. J. Isailović, "Realization of a Decoder for Two MFM Codes," *Electron. Lett.*, Vol. 17, No. 3, 1981, pp. 117–119.
27. J. Isailović, *Videodiscs and Optical Memory Systems,* Prentice-Hall, Englewood Cliffs, N.J., 1985.
28. J. Isailović, "Channel Characterization of the Optical Videodisc," *Int. J. Electron.*, Vol. 54, No. 1, 1983, pp. 1–20.

29. P. F. Panter, *Modulation, Noise and Spectral Analysis,* McGraw-Hill, New York, 1965, Chap. 7.
30. K. Sadashige, "Overview of Time Base Correction Techniques and Their Applications," *SMPTE J.*, Vol. 85, Oct. 1976, pp. 787–791.
31. J. F. Robinson, *Videotape Recording: Theory and Practice,* Focal Press, London, 1978.
32. "The Philips and MCA Optical Video Disc System," Internal Issue, 1976.
33. W. K. Pratt, *Digital Image Processing,* Wiley, New York, 1978.
34. L. F. DeMarsh, "Color Rendering in a Television," *IEEE Trans. Consum. Electron.*, Vol. CE-23, No. 2, 1977, p. 149.
35. *CIE Publ. 13,* "Method of Measuring and Specifying Color Rendering Properties of Light Sources," 1st ed., 1965.
36. *CIE Publ. 13.2,* "Method of Measuring and Specifying Color Rendering by Light Sources," 1974.
37. J. I. Kaufman (ed.) and J. F. Christensen (assoc. ed.), *IES Lighting Handbook,* 5th ed., Illuminating Engineering Society, New York, 1972, pp. 1–6, 5–2, 5–18 through 5–21.
38. G. Wyszecki, "Proposal for a New Color Difference Formula," *J. Opt. Soc. Am.*, Vol. 53, 1963, p. 1318.
39. C. J. Bartleson and E. J. Breneman, "Brightness Reproduction in the Photographic Process," *Photogr. Sci. Eng.*, Vol. 11, 1967, pp. 254–262.
40. E. J. Breneman, "The Effect of Surrounding on Perceived Saturation," *J. Opt. Soc. Am.*, Vol. 67, May 1977, p. 657.

APPENDIX 6.A DISTORTION IN FM SIGNALS DUE TO NON-FLAT FREQUENCY RESPONSE

An FM signal is tolerant of amplitude variations but it is important that the modulation index (β) is preserved.

For the one tone narrow-band FM, the modulated signal (Eq. 4.37) is:

$$s(t) = S_c \cos \omega_o t + \tfrac{1}{2}\beta S_c \cos (\omega_o - \omega_m)t - \tfrac{1}{2}\beta S_c \cos (\omega_o - \omega_m)t$$

Because

$$\cos \alpha = \mathrm{Re}\{e^{j\alpha}\}$$

then

$$s(t) = S_c \mathrm{Re}\{V\}$$

where

$$V = e^{j\omega_0 t} (1 + \tfrac{1}{2}\beta e^{j\omega_m t} - \tfrac{1}{2}\beta e^{-j\omega_m t})$$

Corresponding phasor presentation, but not in proportion, is shown in Figure 6.A.1a. Similar drawings are obtained for the equation 6.26 and b = 1, a = 0.5 (Figure 6A.1b) and b = $\tfrac{3}{2}$, a = $\tfrac{1}{2}$ (Figure 6A.1c). It can be seen that in

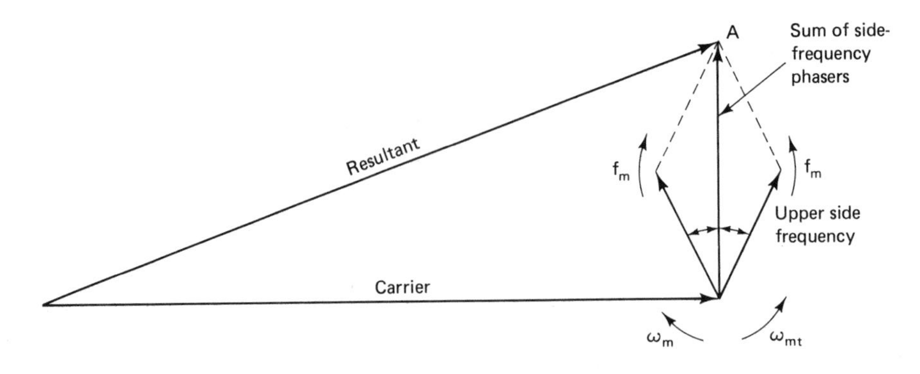

(a)

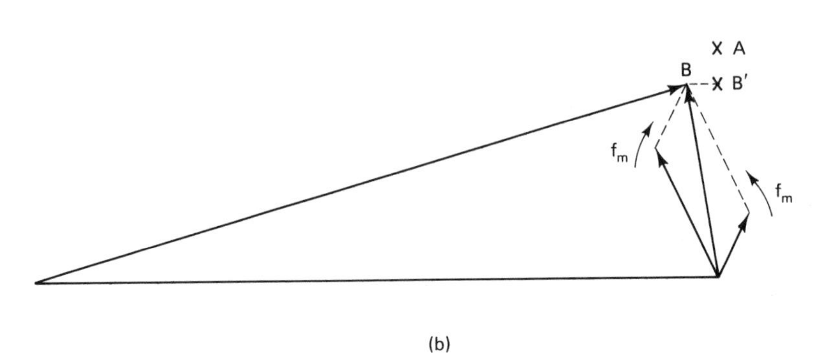

(b)

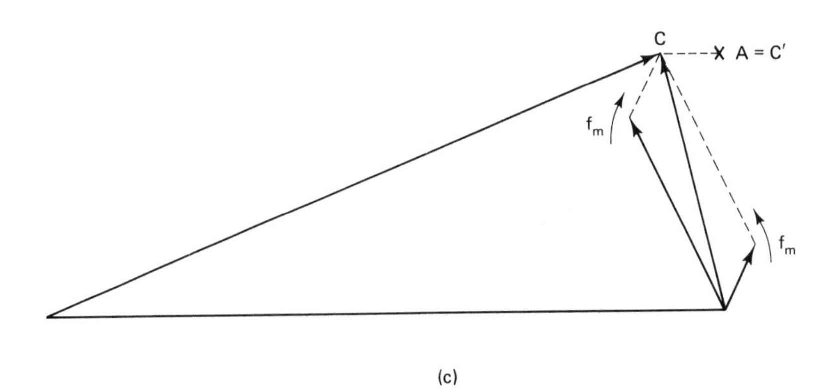

(c)

Figure 6A.1

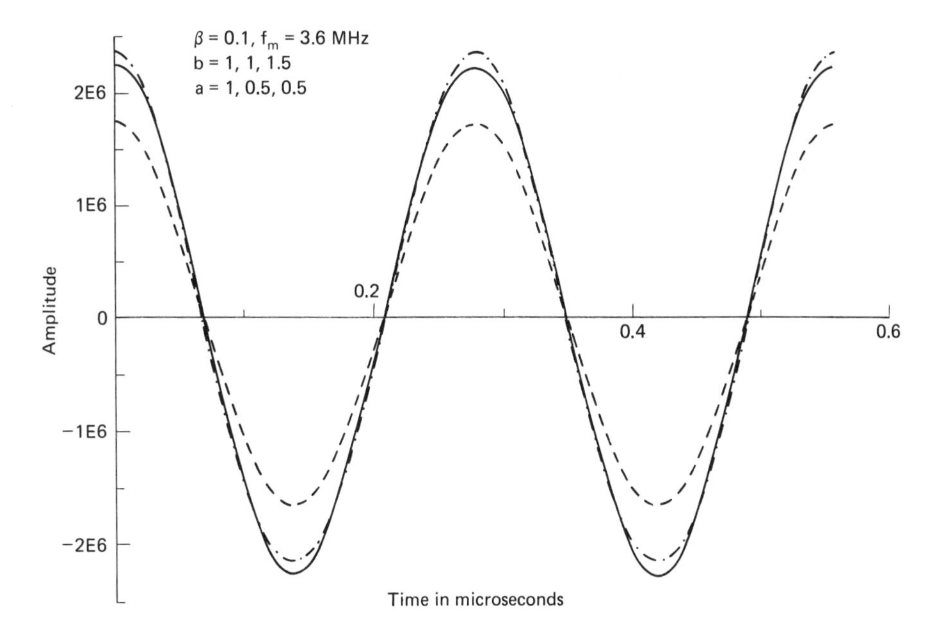

Figure 6A.2

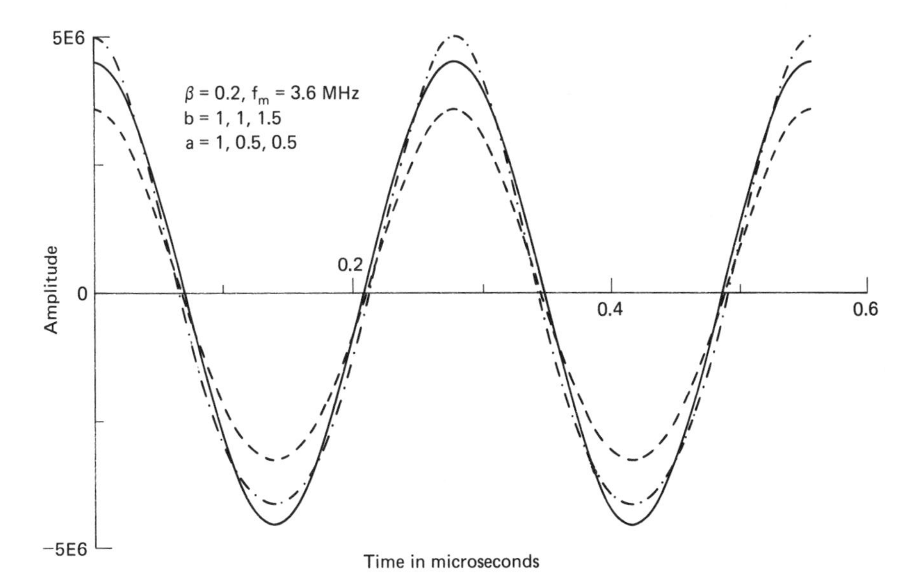

Figure 6A.3

the first and last case the projection of the sum of side-frequency phasors on the y-axis is the same.

Equivalent time domain presentation is shown in Figure 6A.2 for $\beta = 0.1$, and Figure 6A.3 for $\beta = 0.2$; in both cases $f_m = f_{sc} = 3.6$ MHz and the carrier frequency is in the range of the LV carrier for the NTSC standard. Three signals in each figure are for the systems with the frequency responses:

flat
as in Figure 6.21b; $b = 1$, $a = \frac{1}{2}$
as in Figure 6.21c; $b = \frac{3}{2}$, $a = \frac{1}{2}$

It can be seen that for large β there is large variation in the zero crossings.

seven

Extended Play

7.1 INTRODUCTION

Many television pictures contain far more detail than is perhaps necessary. Even in extended views lasting half a minute or more, people do not examine all parts of a picture; surely they must examine even less when the picture changes more rapidly. Human beings are severely limited in their ability to process information, particularly details, but human perception is not a passive process. The proper visual diet can be prescribed only after we know more about psychophysics. For example, in presenting pictures electronically, we must satisfy the biophysical requirements of the visual system. Also, we must realize that viewers' eyes are in service to their minds and that interpreting pictures involves many mental operations.

By information reduction, extended play can be obtained. In a broad sense, extended play means optimization of·videodisc systems. Play-time extension of the (optical) videodisc is especially attractive from the commercial viewpoint: feature films and similar prolonged programs are expected to be the most popular product envisioned for future consumer videodisc players. Preliminary analysis shows that the mean duration of films is about 90 minutes, with a standard deviation of about 20 minutes. This means that a 1-hour-per-side disc (2 hours per disc) would be desirable.

There are many ways to extend playing times. But basically three distinct groups are:

1. Increasing disc information capacity
2. More efficient use of the disc surface
3. Signal processing

These three groups are all mutually compatible and can be combined.

Design considerations for extended play discs should include, among other things:

Picture quality
Special features of (optical) videodisc systems [1]
Complexity involved in accommodating the audio
Player modifications (complexity)

7.2 INCREASING DISC INFORMATION CAPACITY

The simplest way to increase disc capacity, that is, to increase maximal number of recorded pits, would be to increase the disc radius. For example, for the CAV disc, in order to double the capacity the radius should be increased $\sqrt{2}$ times; for $R_o = 14$ cm, the new radius would be $R'_o \simeq 20$ cm. But this is related to technical problems and problems of disc size standardization. Also, this does not include optimization techniques of the videodisc system.

The information capacity of the disc is limited by various factors, including uncertainty in the phase reference of the information, the nonhomogeneous spectrum of the information, the crosstalk, and the binary instead of analog character of laser beam recording. The recording process may also restrict the information density, depending on the properties of the recording material. For materials without an energy threshold, the spatial frequency spectrum of the recorded information is described in terms of the modulation transfer function of the objective lens in the system. Ablative materials show a distinct energy threshold for recording [2]. Infinitesimally small holes have to be recorded by use of the very top of the light (energy) distribution in the spot. Such a system is sensitive to changes in laser power, the sensitivity of the material, and the focus setting. The energy considerations limit miminum hole size. For optimum recording, hole size and spot size have to be matched. Recording with minimum hole sizes near the half-power width of the spot gives a good compromise between noise and information density.

Consider, for example, metal film as a recording medium. A lower limit of the thermal energy required to form a hole of diameter d in thin metallic film is set by the energy required to heat a cylinder of diameter d in the metal adiabatically to the melting point. Thermal diffusion losses in the substrate materials tend to increase the minimum energy by a factor on the order of 5. Radial heat loss into the surrounding parts of the metallic layer are less important for exposure frames smaller than a few hundred nanoseconds, provided that the material is a semimetal or a small-gap semiconductor (B_l, Te) rather than a highly conductive true metal. In addition, there will be reflection and transmission losses of a factor on the order of 3 (almost avoided in a multilayer structure). As a result, 3×10^{-10} J per pit will present a lower limit for ablation of pits in single metallic layers [2].

Inherent in these considerations is the assumption that a hole is formed once a small area in a metallic layer has attained a temperature above the melting point. Material parameters, such as the surface energies of the substrate and of the metal, as well as of the interface, determine the size of the smallest holes. The hole can be formed only where sufficient material is molten. For example, for Te-based materials, hole sizes have been made less than 0.6 μm in diameter.

Now, if we accept a fixed outside disc diameter R_o and a maximal recording resolution, the aim of optimization of the disc system can be set to recording rather than reading, which puts a limit on the information capacity of the disc.

The probes (optical or capacitive) used to read the recorded information have reading resolution limitations. The spatial variation on the disc is translated to a signal strength variation in the probe which decreases in amplitude for small recorded wavelengths. It is customary to record these variations as a function of the spatial frequency, or reciprocal wavelength, of the probe.

In Figure 7.1, the pickup transfer functions, for optical (lens MTF, NA = 0.45) and capacitative probes, are shown. From these two curves the reading resolutions can be compared but the recording capacities (resolutions) cannot. The disc track pitch should also be considered. This is indicated in Figure 7.1.

The smallest resolvable element for the optical readout is

$$l_c = 1/\nu_c = \lambda/2\text{NA} \tag{7.1}$$

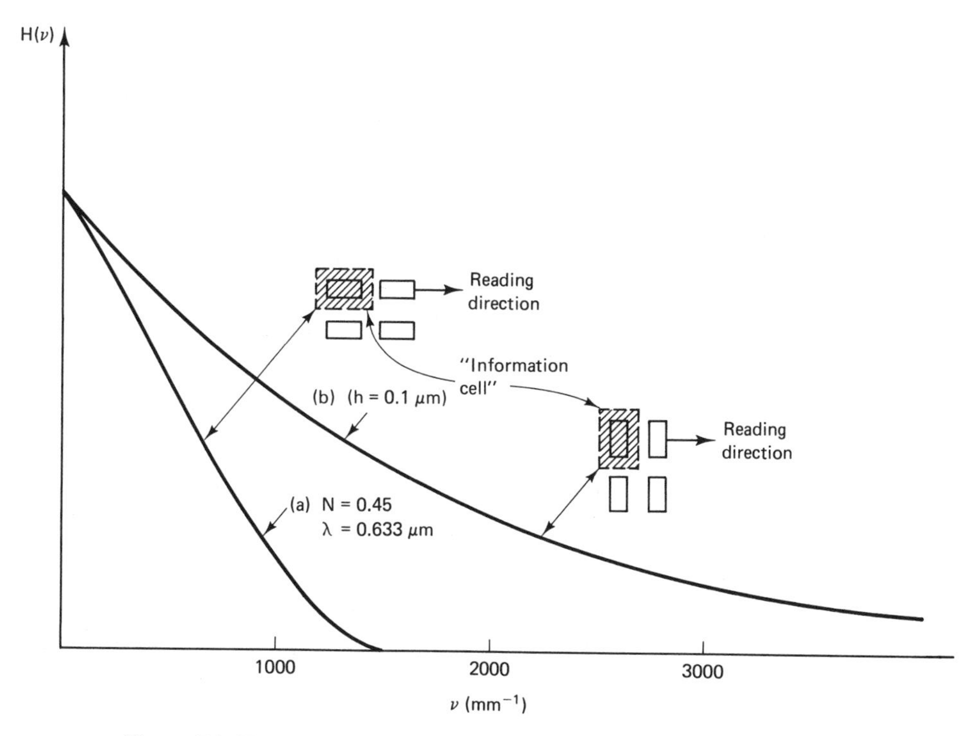

Figure 7.1 Frequency response (reading resolution) of the (a) optical and (b) capacitive probes.

where ν_c is the cutoff spatial frequency of the lens. High resolution (small l_c, high ν_c) can be obtained by:

1. Lowering the laser light wavelength (λ)
2. Taking lenses with higher NA

Besides technical limitations, the first solution automatically means a high price for the laser. For mass production, this is a drawback of the system. The second solution must be compromised with the depth (D) of focus:

$$D = \frac{\lambda}{(\text{NA})^2} \tag{7.2}$$

It is desired for the depth of focus to be as high as possible. The relationship

$$D\nu_c^2 = 4/\lambda \tag{7.3}$$

shows how the depth of focus drops rapidly with increasing cutoff frequency and how the maximum useful cutoff frequency falls as focus error increases.

For example, to get the smallest resolvable element $l_c = 0.6\ \mu\text{m}$, with $\text{NA} = 0.6$, the light source with $\lambda = 480$ nm is needed. Corresponding focus servo should be able to work with the depth of focus $D = \frac{4}{3}\ \mu\text{m}$. This would be sufficient for 1 hour of play per disc side.

The recording disc capacity can be increased by decreasing track pitch. The upper portion of Figure 7.2 shows a diffraction pattern of a circular aperture obtained in the vicinity of the focus spot [3]. This pattern comprises a central spot and alternately dark and bright concentric rings. The distribution of intensity as a function of the distance to the center (reckoned as a number of times $f\lambda/a$, where a is the radius of the aperture of the objective, f is the focal distance, and λ the wavelength of the reading light) with respect to the maximum intensity of the center of the central spot is represented in the lower portion of Figure 7.2. The maximum illumination in the first bright ring is very distinctly less than the illumination of the center (on the order of 1.75%).

If it is assumed, which is usually the case, that the optical noise is rather low when the illumination received by the neighboring sections is lower than 1% of the illumination received by the read section, the track pitch is determined as a function of the numerical aperture of the reading objective. Thus a reading objective having a numerical aperture equal to 0.45, when the reading radiation has a wavelength of 0.63 μm, results in a satisfactory reading provided that the pitch of the track is not less than 1.6 μm. This means that the adjacent tracks of the read track are located outside the first bright ring of the diffraction pattern (in Figure 7.2 the pitch of the track is equal to 1.6 μm). Such a positioning causes the exterior sections to receive no more than 1% of the illumination received by the read section, which is satisfactory.

Figure 7.3 is similar to Figure 7.2 but shows the diffraction pattern of an aperture in the form of a slit. The distance x from a point on the disc to the track center is, as before, reckoned in numbers of times of $f\lambda/a$, where f is the focal distance of the pupil, a is half the width of the slit, and λ the reading wavelength. The maximum intensity at the center of the first bright fringe is

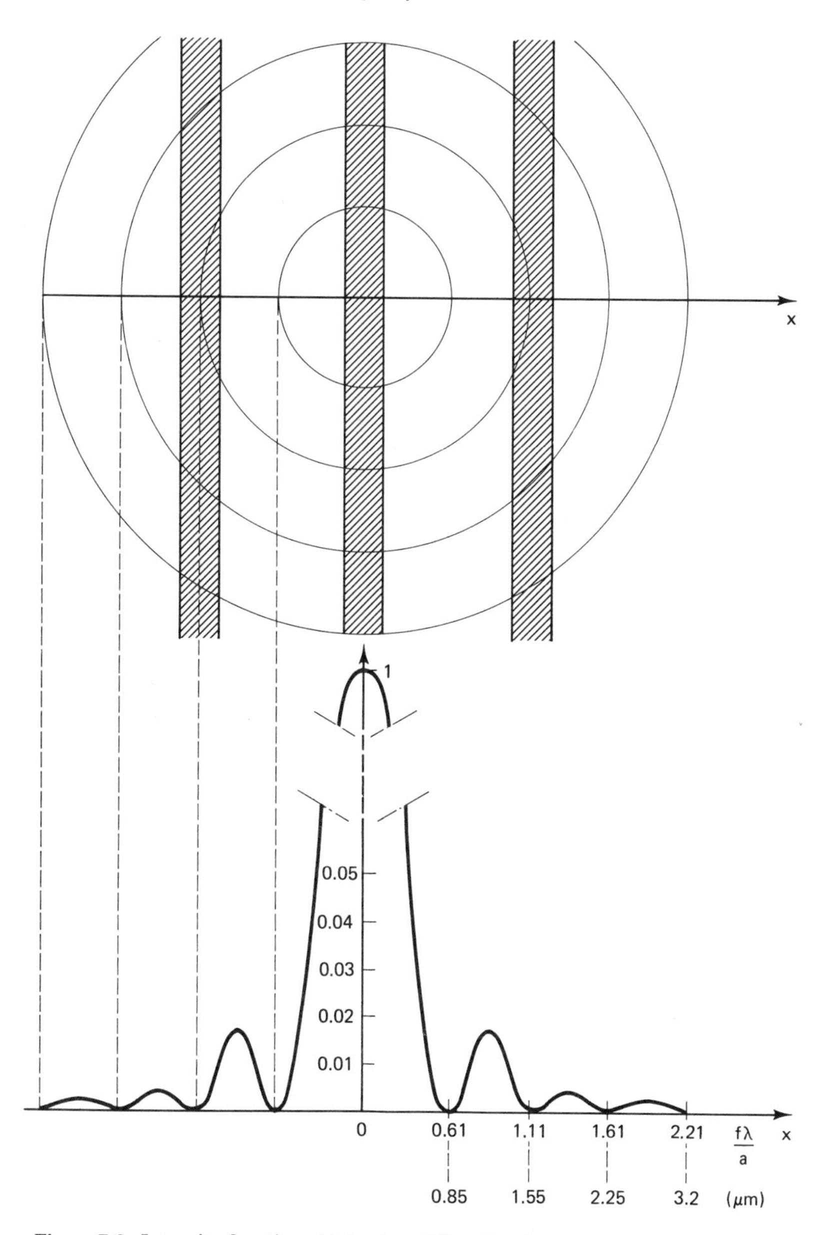

Figure 7.2 Intensity function obtained on diffraction through a circular aperture.

markedly higher than the intensity at the center of the first bright ring in the case of the circular pupil. However, as the fringes are rectilinear, the hole of the track section in the vicinity of the read section may be placed in minimum illumination; the illumination in the first bright fringe is therefore unlikely to notably disturb, upon diffraction, the radiation diffracted by the element illumi-

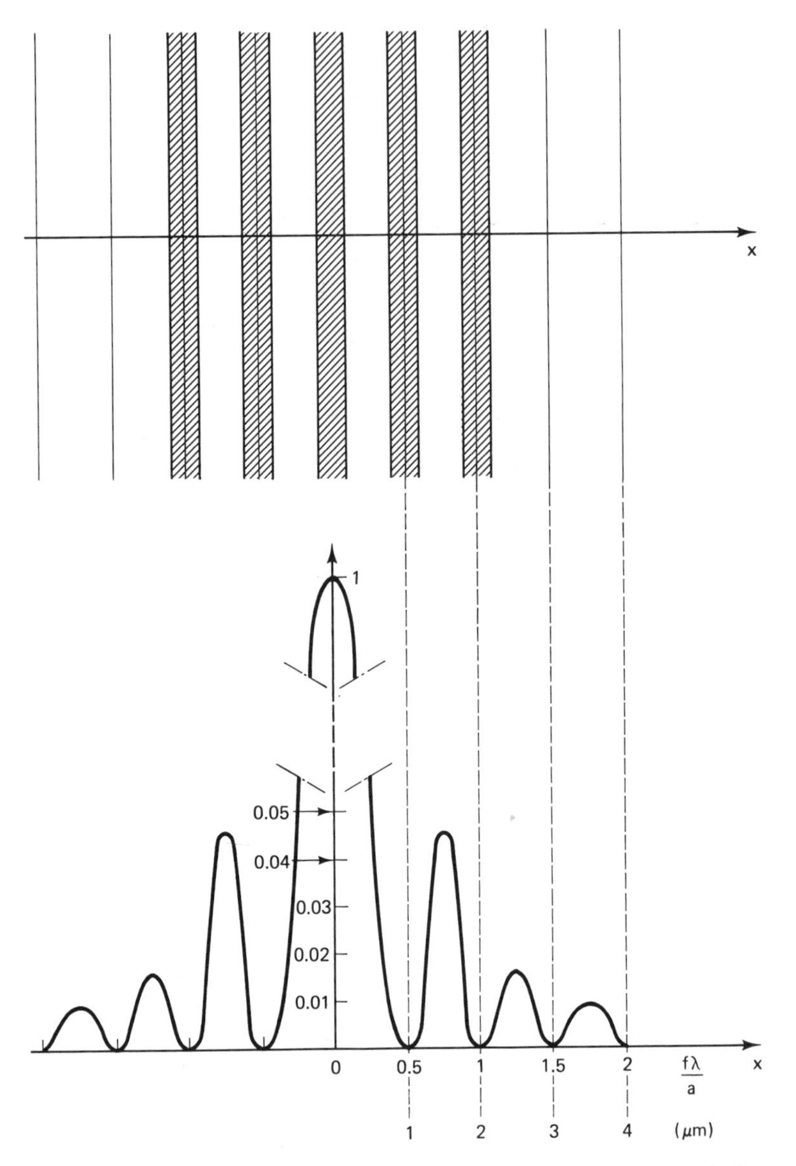

Figure 7.3 Intensity function obtained on diffraction through an aperture in the form of a slit.

nated by the central fringe. The width of the slit must therefore be adapted to the pitch of the track to be read. The even spacing of the fringes also causes the following track sections to coincide with the dark fringes.

If it is considered that this slit is obtained by a partial shutting off of a microscope objective having an aperture 0.45, similar to that employed in the example above, the semiwidth a of the slit is equal to $a/\sqrt{2}$, if the width of the bright fringe is equal to 2 μm, the pitch of the track may be chosen to be equal to 1 μm, so that the middles of the track sections in the neighborhood of the read section coincide with the first illumination minimum. Although the illumination rapidly increases on each side of this minimum, such a disposition, when the track has a width of 0.4 μm, does not result in an illumination of the neighboring sections exceeding 1% of the illumination received by the central section. Even if the decentering of the disc with respect to the diffraction pattern reaches an amplitude of 0.1 μm, the optical noise remains low.

Optical devices for reading, using an aperture in the form of a slit [3], are shown in Figures 7.4 and 7.5. For the sake of simplicity, instead of reflection,

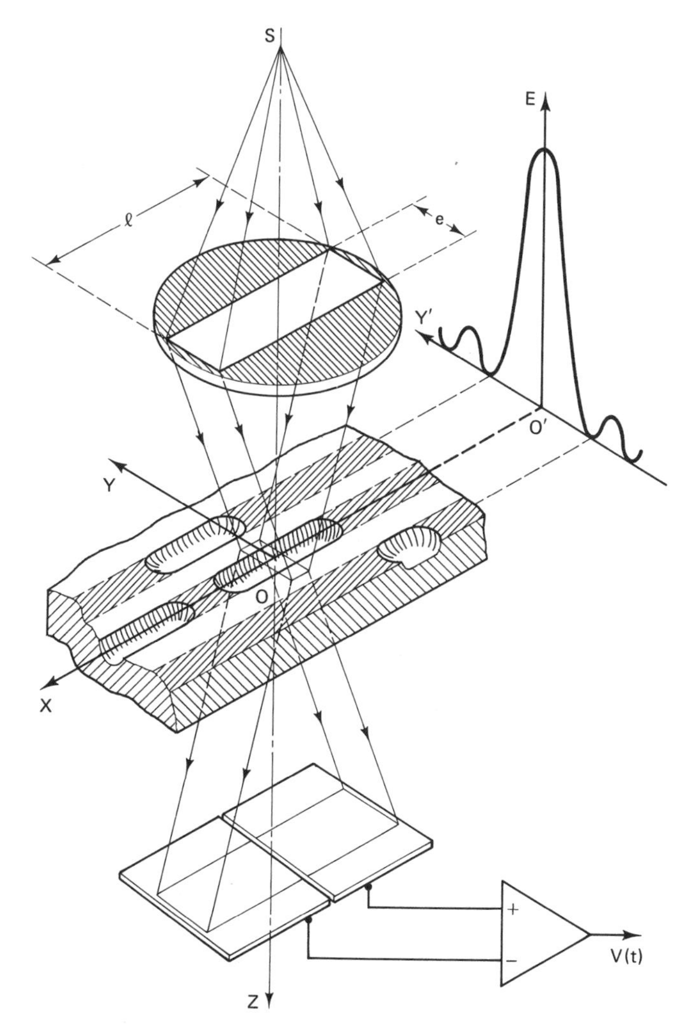

Figure 7.4 Schematic of a device for recording using an aperture in the form of a slit.

transmission is considered. The dimensions e and l of the pupil are such that the projected diffraction pattern has dark fringes which coincide with the middles of the adjacent tracks. The radiation diffracted by the read track is received by the differential detector (two photodetector cells). The output signal $V(t)$ is characteristic of the recorded information.

In Figure 7.4, a monochromatic radiation source is located in S. An objective forms at 0 the image of the source S. The opaque stop has a rectangular window and limits the radiation beam transmitted to the disc.

In Figure 7.5, the source of radiation delivers a parallel beam. The projection objective has an aperture of rectangular shape ($e \times l$). To adopt the input beam to this aperture, the projection device comprises an anamorphic device comprising a divergent and a convergent cylindrical lens. The projection lens is a spherical

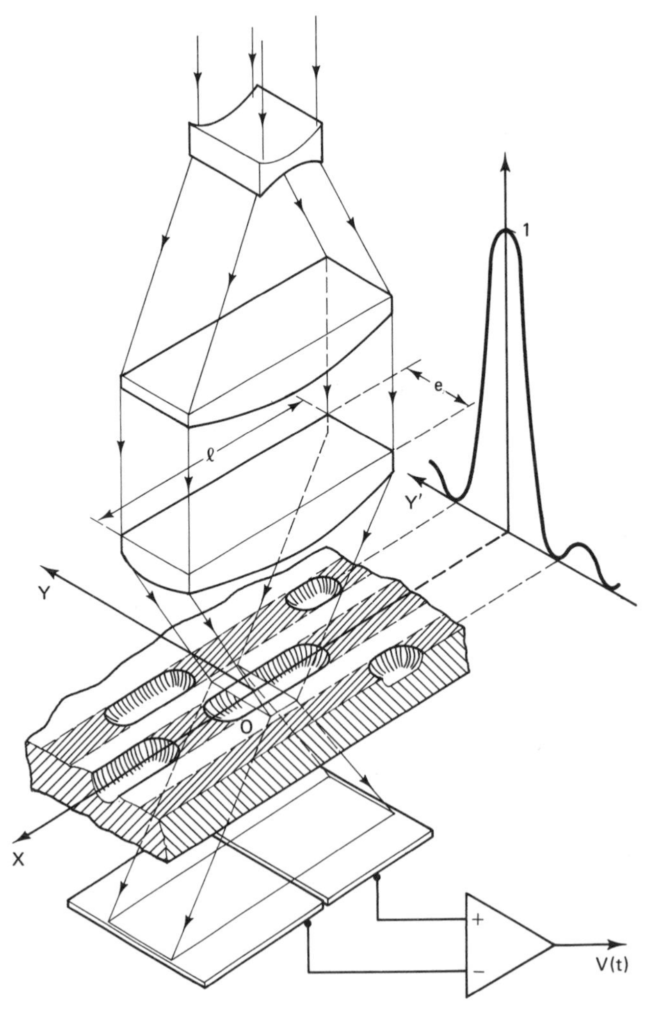

Figure 7.5 Second schematic of a device for reading.

lens which projects onto the disc surface a diffraction pattern similar to that previously described, the aperture being rectangular.

In this way, a program duration of about 50 minutes, instead of 30 minutes when the pitch is equal to 1.6 μm, can be recorded. It is also possible to record the same amount of information on a disc having a pitch of 1.6 μm in the same area. The benefit will then be with respect to the reduction of the speed of rotation of the disc, which permits the utilization of slower, and therefore cheaper, position control devices.

The main problem with the previous method is that it is overidealized. In practice, positioning of the first diffraction null over the neighboring track gives an optical stylus without radial modulation, and hence poor tracking characteristics [4].

A method that yields a sufficiently low crosstalk level in the case of track spacings as small as 0.5 λ/NA is suggested in Ref. 5. The method is based on the fact that the optical depth of the pits influences the phase difference between the light diffracted in the forward track direction and the light diffracted in the backward track direction. By modulating the optical depth of the pits from track to track, a split detector that captures separately the forward and backward scattered light can discriminate between the signals from neighboring tracks when those signals are in phase opposition. But, at the lower spatial frequencies, the separation between forward and backward scattered light on the split detector becomes less and less pronounced.

Another way to reduce pitch is to alternate tracks with pit depths of $\lambda/4$ and $\lambda/8$ and, correspondingly, to use single and split detectors. Of course, crosstalk should be kept low.

Multiple information planes with, for example, a $\lambda/2$ slot top plane and a $\lambda/4$ slot bottom reflective plane, can be the subject of future work. Compared with the present systems, it can be expected that more complex focus and tracking, more assembly steps, and lower yield will result.

For a given size of disc, the information capacity can be increased by an effective increase in the disc surface. To effectively increase the encoding area of the disc, we could accordian-pleat the surface tangentially and read optically from two viewing angles. Similar ideas are applied in the grooved capacitive system, where "V" shaped grooves are used to enhance a pick-up capacitance. But, if the capacitive system primary region for "V" shaped grooves is not pickup sensitive, the tracking simplicity dictates the grooves' shaping.

7.3 MORE EFFICIENT USE OF THE DISC SURFACE

After fixing the physical limitations (reading and recording resolutions):

1. The minimum recording wavelength on the disc.
2. The cutoff frequency of, for example, the reflection-type optical stylus, given by

$$f_{co} = 2\pi nR\,(2\text{NA}/\lambda) \tag{7.4}$$

where NA is the lens numerical aperture value, λ the light wavelength, n the disc rotation speed, and R the radius of the track to be read.

3. Track-to-track pitch.

The next approach to the extended play would be to get more efficient use of the disc surface.

One way to increase the use of the disc surface would be to store video and audio signals only; sinc pulses could then be regenerated electronically by the player. The increase in playing time (i.e., in efficiency) is proportional to the percentage of the time being occupied by the sync pulses, according to the corresponding standard. Conventional television standards allow almost 18% of the horizontal line time and about 6% of the vertical interval for blanking purposes.

By tracking out sync pulses, the electronic complexity would be increased, and the special effects (freeze frame, slow motion, etc.) would be hard to obtain. These two drawbacks indicate that the method has more academic importance than a practical one.

The best chances for considerable playing-time extension is the use of CLV-type discs. Here the speed of rotation is inversely proportional to the readout diameter. As a result, more information can be stored on the disc, enabling a maximum play time of about 1 hour per side (if 30 minutes is the play time for the CAV disc). On the other hand, these discs can be played only in a continuous (i.e., normal forward) mode. CAV and CLV discs can be played on the same player. Compatibility does not exist between discs for the PAL and NTSC television systems. It is intended, however, that the same type of discs can be used for PAL and SECAM. An optical recorder that could be used to record CAV and/or CLV masters is shown in Figure 7.6(a).

The principle of the CLV disc recording is shown in Figure 7.6(b). Suppose that for some radius R, the smallest pit (a minimum length and width within the storage capacity of the recording disc and the capability of the player system to detect the data) occupies the central angle AOE and the recorded wavelength thus occupies the double central angle α. For the double radius, $2R$, on the same sector, two pits (two wavelengths) of the same size can be recorded. And on the radius $4R$, for the same central angle, four pits (wavelengths) can be recorded. A set of radial lines A through E shows the relationship of recording positions from one track to another. For the sake of comparison, it should be noted that in this example, for the CAV disc, the pits occupy the same central angle, AOE, for the same temporal frequency.

Because linear (tangential) velocity, $v = R\omega$, is constant,

$$R\omega = R_o\omega_o = R_i\omega_i \tag{7.5}$$

or

$$\frac{R_o}{R_i} = \frac{n_o}{n_i} \tag{7.6}$$

where n is number of rotations, and the subscripts o and i indicate "outside" and "inside."

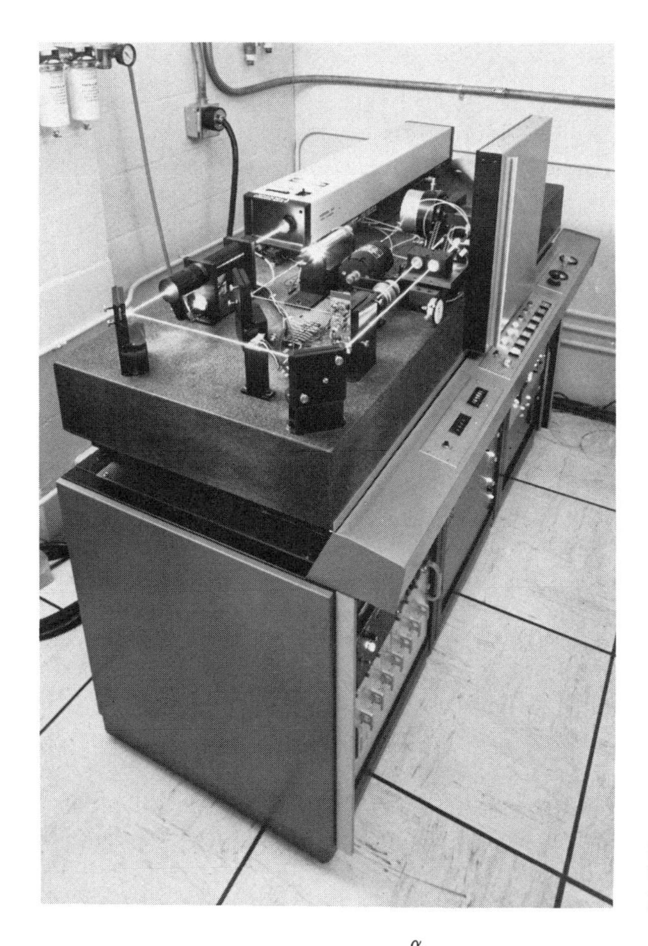

Figure 7.6(a) Optical recorder for CAV and CLV discs. (Courtesy of Discovision)

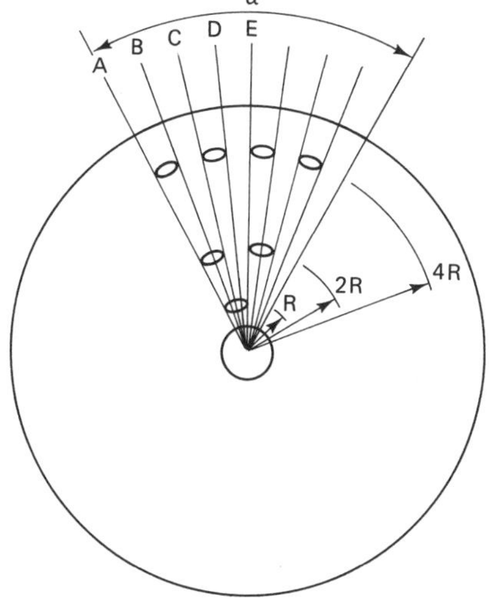

Figure 7.6(b) Arrangement of pits on the CLV disc.

In polar coordinates, the spiral of the recording track is given by

$$\rho = k\phi \tag{7.7}$$

If physical limitations are such that there are b tracks/mm, then $[k = 1/(2\pi b)]$ $\rho = \phi/2\pi b$. The total length of the spiral tracks, starting at radius R_i and finishing at radius R_o, is

$$L = \int_{\phi_i}^{\phi_o} \rho \, d\phi \tag{7.8}$$

where $\phi_i = 2\pi bR_i$ and $\phi_o = 2\pi bR_o$. Finally,

$$L = \pi b(R_o^2 - R_i^2) \tag{7.9}$$

The total capacity, C, is

$$C = \frac{L}{\gamma_{\min}} = \frac{L}{1/a} = \pi ab(R_o^2 - R_i^2) \tag{7.10}$$

where the minimal recorded wavelength is $\gamma_{\min}$ (i.e., a pits/mm).

The last result can be obtained as

$$C = \frac{A_r}{A_{i1}} = \pi ab(R_o^2 - R_i^2) \tag{7.11}$$

where $A_r = \pi(R_o^2 - R_i^2)$ is the recorded area on the disc surface, and $A_{i1} = (1/a)(1/b)$ is the smallest information area.

The maximal capacity, $C_{\max}$, would be (given R_o) for $R_i = 0$:

$$C_{\max} = \pi abR_o^2 \tag{7.12}$$

Thus it can be seen that, compared with the CAV disc, capacity is doubled with the CLV disc. But the condition $R_i \to 0$ is impractical, and it is not possible to obtain this capacity. For $R_o = 2R_i$, the capacity of the CAV disc is maximal $[C_{\max} = (\pi/2)\, abR_o^2]$. For the same conditions ($R_o = 2R_i$) the capacity of the CLV disc is

$$C' = \tfrac{3}{4}\,\pi abR_o^2 = \tfrac{3}{2}\, C_{\max}(\text{CAV}) \tag{7.13}$$

which is a 50% increase in the disc capacity.

In practice, the capacity of the CLV disc is

$$\tfrac{3}{4}\,\pi abR_o^2 < C(\text{CLV}) < \pi abR_o^2 \tag{7.14}$$

In the CLV disc system, data pits are formed along a track with substantially uniform size so that there are more pits on the outer tracks than on the inner tracks. The track revolutions hold a variable number of frames, depending on the pit capacity of the track. With this technique the playing time is considerably extended, but the CLV technique has drawbacks: first, the change of speed of rotation which must be specially controlled (the ratio of the maximal and minimal rpm's is equal to R_o/R_i), and second, these discs can be played only in the normal forward mode (i.e., no special effects).

In Ref. 6 it was suggested that a disc with this type of recording can be adapted to be rotated at a uniform angular velocity, and could be obtained using a separate memory for freeze frame display. Frames are divided into data units that correspond to a few horizontal lines, and frames on the same track are interlaced so that the data units of each frame are distributed evenly around the track revolution. Obviously, some parts of the disc capacity would be lost to data formatting. But the main problem is related to the MTF of the pickup stylus. Because of this, the signal-to-noise ratio would vary significantly with the radius.

7.4 SIGNAL PROCESSING

7.4.1 Classification of Signal Processing Techniques and Fidelity Criteria

When the physical limitations are fixed and the efficiency of use of the disc surface is determined, the remaining approach for extended play is signal processing. Reducing, for example, the luminance and chrominance bandwidth, the FM carrier can be lowered so that a 1-hour per side of the videodisc can be obtained using 900 rpm (15 rps) instead of 1800 rpm for the CAV disc. Special attention should be paid to the spectrum ordering in the channel: luminance, chrominance, and audio signals are placed on the frequency-sharing principle.

Signal processing techniques can be classified in a number of ways, as indicated in Table 7.1. Mainly, the videodisc system is used for pictorial image (natural scenes of objects or two-dimensional displays that are normally viewed by human observers) recording, but they can also be used for nonpictorial, two-dimensional arrays of data, such as radar range versus velocity data, not normally viewed by an observer [7]. Color pictures are normally recorded on the videodisc, although monochrome and multispectral information can also be recorded. Continuous-tone images are normally considered for the videodisc recording, and binary-valued imagery such as computer graphics data or black/white facsimile data can also be recorded.

Based on the type of signal processing, the techniques can be classified as analog or digital. In an analog system the analog TV signal (voltage) is processed such that the processed signal's analog bandwidth is reduced compared to a

TABLE 7.1 CLASSIFICATION OF SIGNAL PROCESSING TECHNIQUES

Type of imagery	Pictorial, nonpictorial
Photometric content	Monochrome, color, multispectral
Amplitude scale	Binary, continuous tone
Type of signal processing	Analog, digital
Processing dimensionality	Line-scan processing
	Intraframe processing
	Interframe processing

Source: Ref. 7, Chap. 21.

normally scanned video signal (Figure 7.7a). In a digital signal processing system, the analog image signal is first digitized and then processed such that the processed signal's digital rate is reduced compared to the input digitized signal. Therefore, after D/A conversion the analog bandwidth is reduced compared to a normally scanned signal (Figure 7.7b). Upon playback, decompression is performed, also in a digital manner: a similar configuration is used. Here it is assumed that the videodisc channel is an analog channel (although on–off recording is performed), but if this channel were a digital one, the signal would be recorded in digital form and there would be no need for D/A conversion before recording. In some ways a Multiplex Analog Component (MAC) method is in this group.

Figure 7.7c shows the block diagram for image bandwidth compression based on the vision model. The basic idea in subjective image bandwidth compression is to transform the image with a vision model [8], to apply a coding algorithm, and then, after passing it through the videodisc system, to decode and invert the image from the perceptual domain back into the intensity domain. The straightforward approach is simply to quantize the vision model's output. But the resulting quantization noise will produce pseudoedges in the image which will limit the degree of compression. One possible solution is to dither the signal to break up these edges. More commonly, a coding algorithm is chosen that will spatially decorrelate the quantization noise. Coding in the Fourier domain is a common means of accomplishing this. Finally, there are processing systems that process individual scan lines, systems that process areas of an image, and systems that process sequences of frames of continuous motion imagery.

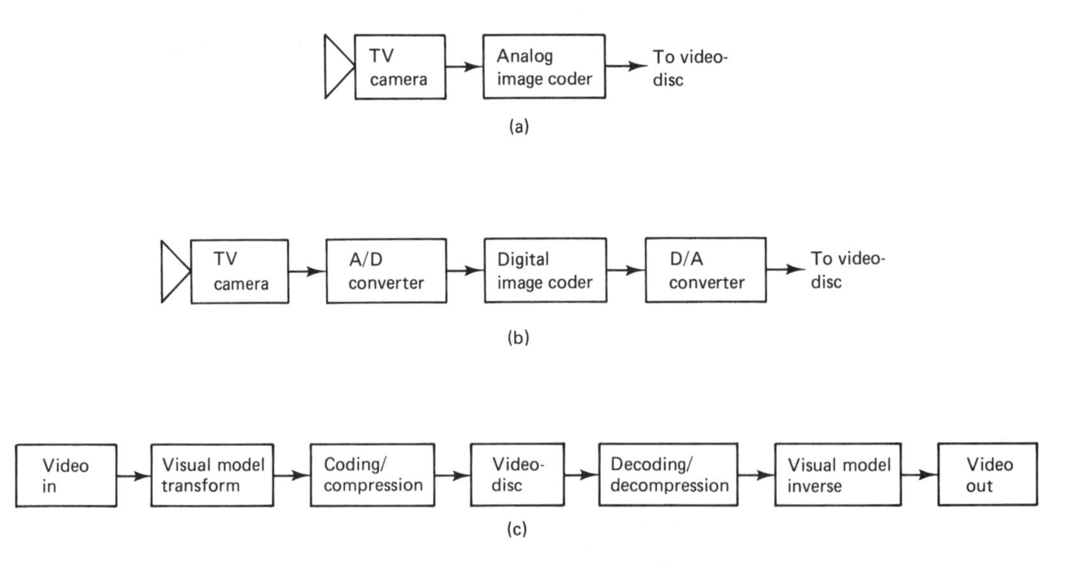

Figure 7.7 Compression processing techniques: (a) analog processing; (b) digital processing; (c) image bandwidth compression with vision model.

In some (image) signal processing systems errors in the reconstructed image can be tolerated. In this case a fidelity criterion can be used as a measure of system quality. Examples of objective fidelity criteria are the root-mean-square (rms) error between the input image and output image, and the rms signal-to-noise ratio of the output image. Suppose that the input image (or image that would be obtained before signal processing) is defined by $f(x, y)$ over area A (for a color image, three functions would be over area A). For any value of x and y inside area A, the error between an input pixel, $f(x, y)$, and the corresponding output pixel, $g(x, y)$, is

$$e(x, y) = g(x, y) - f(x, y) \tag{7.15}$$

The squared error averaged over the image array is

$$\overline{e^2} = (1/A) \iint e^2 (x, y)\, dx\, dy = (1/A) \iint [g(x, y) - f(x, y)]^2\, dx\, dy \tag{7.16}$$

and the rms error is defined as

$$e_{\text{rms}} = \sqrt{\overline{e^2}} \tag{7.17}$$

We also consider the difference between the output and input images to be "noise," so that each output signal (pixel) consists of an input signal (the corresponding input pixel) plus noise (the error), that is,

$$g(x, y) = f(x, y) + e(x, y) \tag{7.18}$$

The mean-square signal-to-noise ratio (SNR) of the output image is defined as the average of $g^2 (x, y)$ divided by the average of $e^2 (x, y)$ over the image area. In other words,

$$(\text{SNR})_{\text{ms}} = \frac{\iint_A g^2 (x, y)\, dx\, dy}{\iint_A [g(x, y) - f(x, y)]^2\, dx\, dy} \tag{7.19}$$

The rms value of SNR is then given by

$$(\text{SNR})_{\text{rms}} = \frac{\left(\iint_A g^2 (x, y)\, dx\, dy\right)^{1/2}}{\left(\iint_A [g(x, y) - f(x, y)]^2\, dx\, dy\right)^{1/2}} \tag{7.20}$$

where the variable term in the denominator is the noise expressed in terms of the input and output images.

Although this measure has a good physical and theoretical basis, it often correlates poorly with the subjectively judged distortion of the image. Much of the reason for this is due to the fact that the human visual system does not process the image in a point-by-point fashion but extracts certain spatial, temporal, and chromatic features for neural coding. Image quality assessment can be viewed as the search for a metric that will reflect these subjective properties of the image and provide engineers with objective criteria they can use in the design of image-processing systems [9].

Because the output images are to be viewed by people, it is more appropriate to use a subjective fidelity criterion corresponding to how the images appear to human observers. The human visual system has peculiar characteristics, so that two pictures having the same amount of rms error may appear to have drastically different visual qualilties. An important characteristic of the human visual system is its logarithmic sensitivity to light intensity, so that errors in dark areas of an image are much more noticeable than errors in light areas. The human visual system is also sensitive to abrupt spatial changes in gray level so that errors on or near the edges are more bothersome than errors in background texture. The subjective quality of an image can be evaluated by showing the image to a number of observers and averaging their evaluations. One possibility is to use an absolute scale such as [10, Chap. 6] the following:

1. *Excellent.* The image is of extremely high quality, as good as you could desire.
2. *Fine.* The image is of high quality, providing enjoyable viewing; interference is not objectionable.
3. *Passable.* The image is of acceptable quality; interference is not objectionable.
4. *Marginal.* The image is of poor quality and you wish you could improve it; interference is somewhat ojectionable.
5. *Inferior.* The image is very poor, but you could watch it; objectionable interference is definitely present.
6. *Unusable.* The image is so bad that you could not watch it.

Another possibility is to use the pair-comparison method, where observers are shown images two at a time and asked to express a preference. Both methods have advantages and disadvantages.

If the assumption is made that the neural image produced by the retina is transmitted on the optic nerve with uniform emphasis in its perceptual dimensions, human vision models can be used to assess image quality in a straightforward fashion by applying error measures to the image after transformation to the perceptual domain. An error metric applied in this manner could, at least theoretically, properly weight such image characteristics as edge fidelity, image contrast, color fidelity, and so on. The success of the resulting image quality measure would then depend on the validity of the vision model and which metric was used in the perceptual domain. If the image were transformed to a truly uniform perceptual domain so that a unit change is equivalent on any range and in any perceptual dimension, an absolute deviation measure would be appropriate. That some experimental results indicate a squared deviation measure is better [11] implies that the perceptual domain used placed greater emphasis on larger deviations. It is generally true, however, that the choice of the error metric is less important than the choice of the vision model. It was shown experimentally [12] that images do not change in their relative rankings for three different metrics: the maximum absolute deviation, the mean absolute deviation, and the root-mean-squared deviation.

7.4.2 Modulation Bandwidth Reduction Techniques

As shown in Chapter 5, because of the scanning process involved in TV signal formation, normally scanned television is not continuous, but rather discrete: it contains a train of discrete components separated by regions of little signal energy. Based on this, an experimental monochrome television bandwidth reduction technique called *frequency interlace* was developed [13]. In this system, the television spectrum is halved by folding over the upper half of the spectrum onto the lower half and interleaving the spectral components. Figure 7.8 illustrates the spectrum interface technique. Figure 7.8c shows the spectrum in the videodisc channel, one possible solution. The FM carrier f_o is between $f_b = f_m$ and f_w,

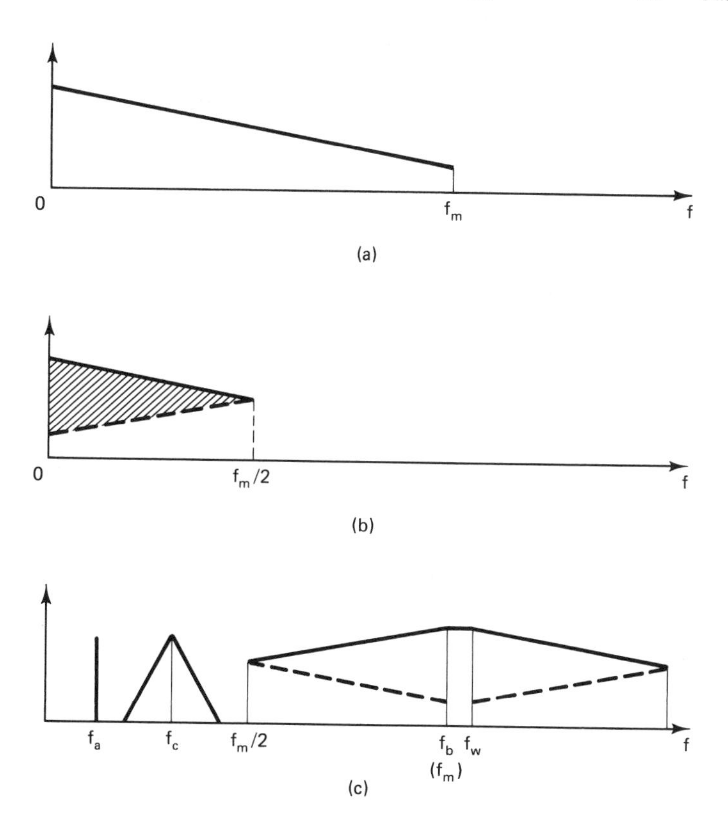

Figure 7.8 Spectrum interlace technique: (a) original video spectrum; (b) folded video spectrum; (c) spectrum in the videodisc channel.

where f_m is the maximum frequency in the original television spectrum, f_b the FM carrier corresponding to the black level, and f_w the FM carrier corresponding to the white level in the picture. If a color information carrying signal is interlaced with a bandwidth reduced luminance signal, it must be placed on a separate bandwidth, the FM carrier is marked as f_c and QAM (quadrature amplitude modulation) can be used as for a normal TV signal (with either symmetrical or nonsymmetrical quadratures channels). The audio carrier(s), f_a, is also marked in Figure 7.8c.

The spectral folding can be performed by conventional modulation and filtering methods. Reconstruction requires comb filter technique to extract the interleaved components. If the reconstruction process is not performed accurately, a low-frequency line crawling effect will result in the displayed image. The frequency interlace system is conceptually simple to implement. The reduction in analog transmission bandwidth is limited to a factor of 2:1.

7.4.2.1 Color Signal Bandwidth Reduction. The spectra of the chrominance signal regularly consists of line spectra whose spacing is line frequency f_H. By interleaving between the upper and lower sidebands of the chrominance signal, the spacing can be lowered to $f_H/2$. Now, interleaving chrominance and folded luminance spectra results in a spacing of $f_H/4$. This is illustrated in Figure 7.9a. It should be noticed that spacing f_H is between originally adjacent components in the upper or lower half of the luminance or in the upper or lower half of the chrominance signal; spacing $f_H/2$ is between adjacent luminance or adjacent chrominance components; spacing $f_H/4$ is between two adjacent components, shown in Figure 7.9a.

One way to interleave between the upper and lower sidebands (Figure 7.9b) of the chroma signal is to heterodyne down the color information. If a chrominance signal is down-converted with a carrier $s_r(t)$ whose frequency

$$f_r = f_{sc} + \left(\frac{n}{2} + \frac{1}{4}\right)f_H = f_{sc} = \frac{2n + 1}{4} f_H \tag{7.21}$$

is taken to be $f_r = f_{sc} - \frac{1}{4}f_H$ $(n = 0)$, the low sideband of the heterodyne converter will have a spectrum as illustrated in Figure 7.9c. If a chrominance signal $s_{sc}(t)$ is

$$s_{sc}(t) = C \cos [\omega_{sc} t + \theta(t)] \tag{7.22}$$

and a reference carrier is

$$s_r(t) = 2 \cos (\omega_{sc} - \tfrac{1}{4}\, \omega_H)t \tag{7.23}$$

then

$$s_{sc}(t) \times s_r(t) = C \cos [2\omega_{sc} t + \theta(t)] + C \cos [\tfrac{1}{4}\, \omega_H t + \theta(t)] \tag{7.24}$$

The low-pass filter output is then

$$s(t) = C \cos [\tfrac{1}{4}\, \omega_H t + \theta(t)] \tag{7.25}$$

In Figure 7.9c, the first negative spectral component of the signal $s(t)$ is marked with an asterisk. This component gives a $3f_H/4$ component.

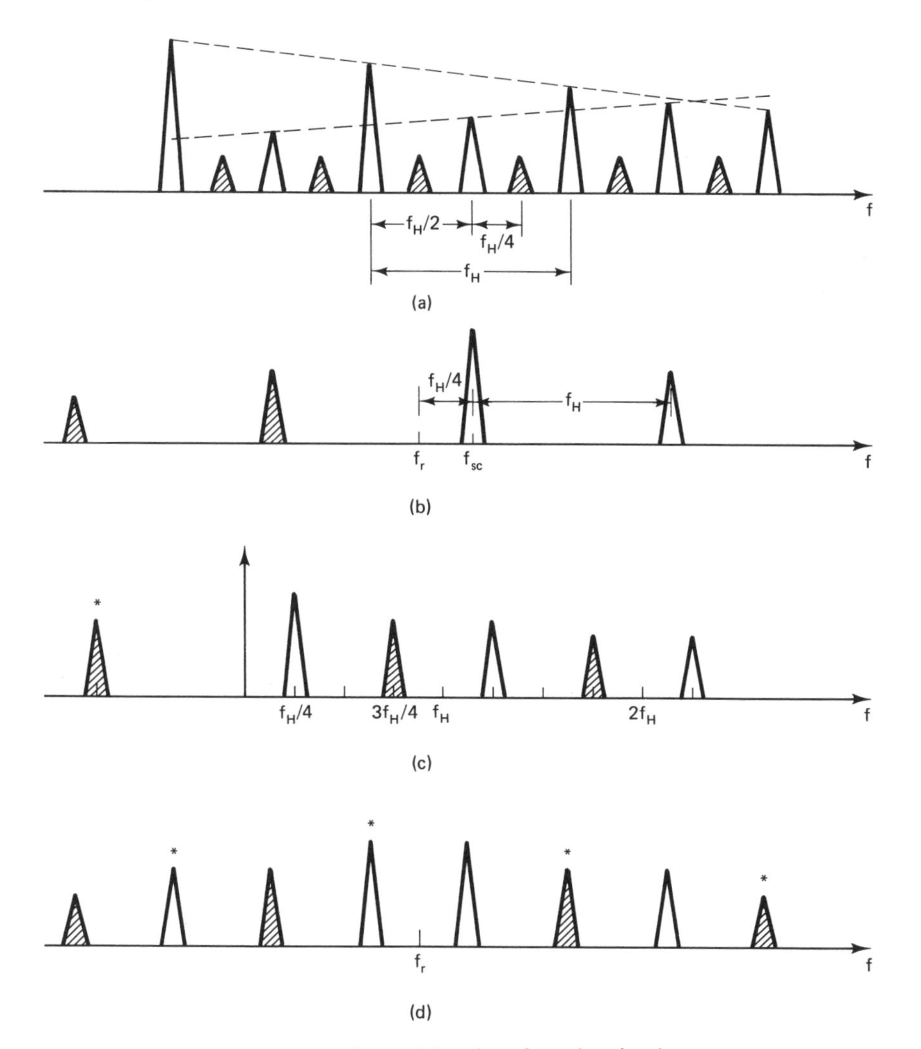

Figure 7.9 Spectral interlace for color signals.

The signal $s_{sc}(t)$ can be regenerated from the signal $s(t)$, for example, if $s(t)$ is up-converted by a balanced modulator. If the reference signal is now

$$s'_r(t) = 2\cos\left[\left(\omega_{sc} - \tfrac{1}{4}\,\omega_H\right)t + \phi_0\right] \tag{7.26}$$

where ϕ_0 is the initial phase, the up-converted signal is

$$\begin{aligned} s(t) \times s'_r(t) &= C\cos\left[\omega_{sc}t + \theta(t) + \phi_0\right] \\ &\quad + C\cos\left[\left(\omega_{sc} - \tfrac{1}{2}\,\omega_H\right)t - \theta(t) + \phi_0\right] \end{aligned} \tag{7.27}$$

Thus the up-converted signal consists of the original chrominance signal $s_{sc}(t)$ and a spurious signal whose spectra are interleaved with the original chrominance spectra (Figure 7.9d). The original chrominance signal $s_{sc}(t)$ can be obtained by a comb filter that eliminates the spurious spectra. In Figure 7.9d, spurious components are marked with an asterisk.

The spectral folding can be performed on luminance and chrominance signal separately. The spectrum in the videodisc channel would be similar to that in Figure 7.8c, except that the chrominance signal would occupy half of the bandwidth compared to that in Figure 7.8c.

7.4.3 Spatial Resolution and Frame Rate Reduction Techniques

One of the simplest methods of bandwidth reduction is to reduce the inherent bandwidth requirements of an image sensor (camera) by limiting its spatial resolution and frame repetition rate. The possibility of a bandwidth reduction is indicated by observations. First, there is a large amount of statistical redundancy or correlation in normal images. For example, two points that are spatially close together tend to have nearly the same brightness level and color. Second, there is a large amount of psychovisual redundancy in most images. That is, a certain amount of information is irrelevant and may be eliminated without causing a loss in subjective image quality, and a large amount may be removed without causing a complete loss of detail. The extent to which spatial resolution and frame rate can be reduced is often a difficult decision. It involves factors of visual image quality and photometric imaging accuracy.

7.4.3.1 Spatial Resolution Reduction. In some applications fine resolution may not be required over the entire area of an image. For example, a facsimile storage of photographs can be evaluated based on the concept of variable spatial resolution. Variable spatial resolution can also be used as a means of bandwidth reduction for facsimile and television transmission/storage systems when an object of interest is constrained to lie within the center region of an image [7, Chap. 21]. Also, the peripheral vision resolution is degraded from foveal vision for a human observer.

Figure 7.10 contains an example of variable spatial (horizontal) resolution technique in which the image is divided into three zones. Resolution in the central zone (area A_1) is set at the finest level and the corresponding maximal temporal

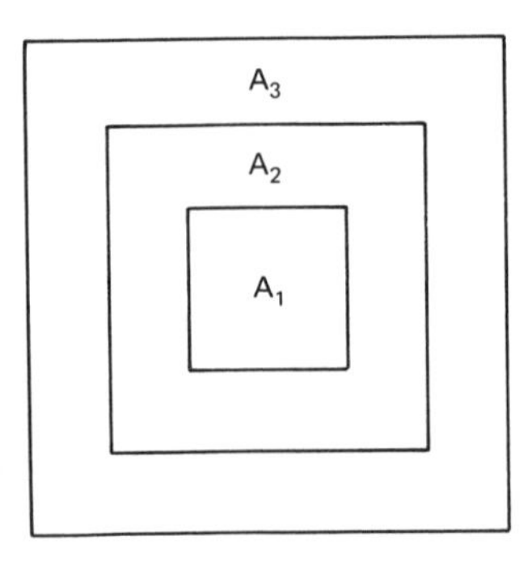

Figure 7.10 Example of variable spatial resolution.

frequency is f_m; that is, the corresponding bandwidth B_1 is $B_1 = f_{m1}$. Resolution in the first ring is low, say one-half the finest resolution; the corresponding bandwidth is $B_2 = f_{m2}$; and so on.

If the lowest FM carrier is taken as

$$f_0 = f_b \simeq 2f_m \tag{7.28}$$

the maximal temporal instant frequency is

$$f_M = 2f_m + \Delta f \tag{7.29}$$

where Δf is the maximal frequency deviation.

Suppose now that the maximal recording spatial frequency is defined and is kept the same for recording signals f_{m1} and f_{m2}. To get a proper frequency from the disc, the rotational speed, N, should be related as

$$N_2 = \frac{f_{m1} + \Delta f}{f_{m2} + \Delta f} N_1 \tag{7.30}$$

where the maximal frequency deviation is taken to be constant. A similar relationship can be written for N_3 (area A_3). In this example, the average rpm, N_{av}, is

$$N_{av} = \frac{A_1}{A} N_1 + \frac{A_2}{A} N_2 + \frac{A_3}{A} N_3 \tag{7.31}$$

where $A = A_1 + A_2 + A_3$.

As an example, consider the case when $f_{m2} = f_{m1}/2$, $f_{m3} = f_m/4$, $\Delta f \ll f_{mi}$, $i = 1, 2, 3$, and $A_1 = 25\%$ of A ($A_1/A = 0.25$), $A_2/A = 0.25$, $A_3/A = 0.5$; then

$$N_{av} = \tfrac{1}{2}N_1 = 0.5N_1 \tag{7.32}$$

That is, the playing time would be doubled. In a CAV disc system, the disc would rotate with an N_{av} rpm and, for example, a two-line memory would be needed for the signal processing. Color and audio information should be processed properly, also. The color subcarrier and audio carriers can be chosen to be between 0 and f_{m3}, for example.

The simplified version of the variable spatial resolution technique would be to reduce maximal resolution by half, for example. For the previous example it would be to take $A_1 = A_3 = 0$ and $A_2 = A$. Then if the smallest spatial wavelength on the disc is kept the same in the CAV disc system, the rotational speed would be half the normal speed; thus the playing time would be doubled.

Besides horizontal resolution, the vertical resolution can be reduced in the TV signal processing also. This can be achieved by tracking out some number of lines, for example, every second. This would also yield effective bandwidth reduction; that is, it would yield to the extension of playing time. The total spatial resolution reduction is obtained by combining horizontal and vertical resolution reduction.

It should be noted that the vertical resolution reduction achieved by tracking out every second line is equivalent to the field rate reduction with the ratio 2:1. During playback, the disc can be rotated (CAV) at normal speed or at one-half speed (compared to the system with one frame per revolution). In the first case each track is read twice, and lines from two pictures (frames) are recorded alternately: 1212121. . . (numbers 1 and 2 correspond to the lines of two pictures on the same track). A two-line memory in the player must be used. Now, freeze-frame can easily be performed where, if the half-speed method is used, an entire frame must be stored to achieve freeze-frame.

The spatial resolution acuity of a viewer is degraded substantially for moving objects: a viewer cannot easily focus to the fine detail of a moving object unless the movement is regular. Rapidly moving objects are often blurred as a result of poor temporal response of the image sensor and display. In a TV system, it is possible to reduce spatial resolution (in frame processing), as compared with stationary objects, without any apparent loss of image quality [14]. Unfortunately, the principle is much more easily stated than implemented. A fundamental problem is that of detecting when an object is moving and with what rate it is moving. Presumably, the spatial resolution for moving objects would be inversely proportional to the object's velocity [7, Chap. 21].

Experiments with real-time television systems indicates that a human viewer requires a relatively long time (700 to 800 ms) to adopt to the content of a new scene [15]. During this time, the resolution of transmitted images can be reduced substantially without noticeable degradation.

7.4.3.2 Frame Rate Reduction. The reproduction of natural motion scenes by television is possible mainly because of the persistence of the retinal receptors. The illusion of motion in a picture can be maintained at rates considerably lower than TV standards of 25 or 30 frames per second, which is required to depict rapid motion smoothly. Since the brain retains the image of the illumination pattern for about 0.1 s after the source of illumination has been removed from the eye, a minimum of about 10 frames per second can simulate natural motion. However, with normal television, low frame rates lead to picture flicker. The effect of flicker can be made unobjectionable for low-frame-rate television by a multiple display of frames at a higher field rate [16]. This requires some type of frame storage system at the receiver. In many cases involving rapid motion, a large fraction of the picture elements correspond to stationary background. It has been shown experimentally that only about 10% of the picture elements change more than 8% of brightness from frame to frame [15,17,18]. Thus both psychovisual properties and statistical correlations may be explored to obtain extended play.

In experiments performed to determine whether a lower frame rate would be acceptable from the standpoint of image breakup, a low-frame-rate intermediate (a 24-frame-per-second motion picture was the basis) was made by discarding some of the frames of the original movie. The remaining frames were repeated to produce a 24-frame-per-second film [19]. In some scenes, such as a conversation closeup, 12 frames per second prove satisfactory, but in another scene with a train moving across the screen, some jerkiness was evident at the same frame rate. At lower frame rates, all scenes showed more motion breakup. The simplest reconstruction was to print each intermediate frame two or more times as needed. The more complicated ways of frame combing showed only slight improvement over the simplest method of frame repetition.

The skip-frame extended-play videodisc system has been successfully designed and operated [20]. The method used was to skip every other video frame while recording the full audio information on two channels. Therefore, the recorded disc had two audio channels and one picture channel. Upon playback, each picture frame is played twice while each audio portion is played once.

The mastering of the skip-frame extended play is shown in Figure 7.11. Its basic operation is as follows. The audio signal is applied to a 33.4-ms audio delay line (a delay of approximately one frame) before being frequency modulated by the 4.2-MHz VCO. The input is also applied to the 4.8-MHz VCO and the outputs of the VCOs are linearly summed.

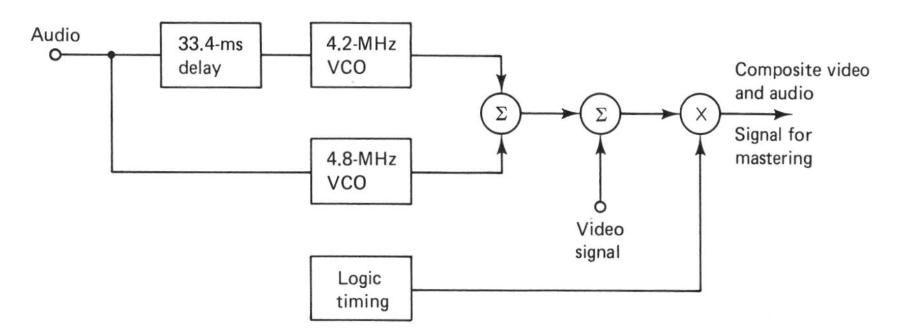

Figure 7.11 Skip-frame extended-play block diagram: mastering.

The audio signals are then summed with the video signal to provide the full composite video and audio signals. In the mastering process, every other frame of the composite signal is cut on the disc. During the frame time slot when there is no cutting, a movable mirror is used to position the write beam such that a continuous spiral can be written. It is obvious that the radial velocity of the head must be one-half normal velocity. This is accomplished by a gear change in the lead-screw drive.

A block diagram of the audio section of the playback system is shown in Figure 7.12. The composite video and sound input is applied to two bandpass filters, one at 4.2 MHz and the other at 4.8 MHz. Both are then demodulated into audio signals. Here the selected channel is dictated by whether the frame is being viewed for the first time or has been viewed once. This is done through logic circuits. Also, a stop-motion logic signal is generated after the first viewing of the picture in order to repeat the frame. One audio channel is selected while the other is blocked and a provision for squelching the selected audio signal is also provided. The selected audio signal (if not squelched) is applied to the VCO, where it frequency modulates the 4.5-MHz audio subcarrier. A lead-screw bias

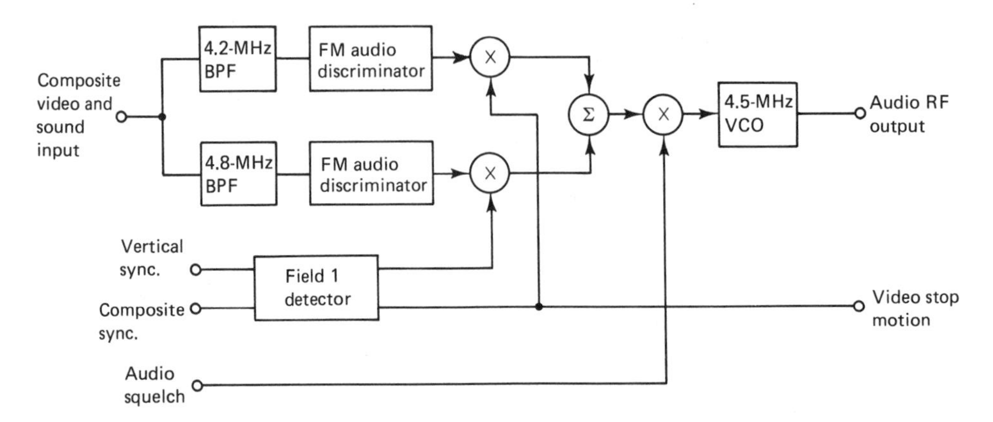

Figure 7.12 Skip-frame extended-play playback: audio section.

signal has been provided to slow the lead-screw motor, which produces one-half of the normal radial drive to the player arm.

The reason for demodulating the audio subcarrier and remodulating again is due to synchronizing problems that occurred during development. If channel switching occurred on the RF signals at 30 Hz (one frame rate), a 30-Hz noise component would be audible at the TV set. Whenever the RF channels are not exactly in phase, a change in frequency would be seen by the demodulator in the TV set, resulting in noise at 30 Hz. The composite signal formatting can also be performed in some other way. In ways similar to frame rate reduction, the field rate reduction techniques can also be used to extend playing time.

7.4.4 Picture Interlace Techniques

A three-dimensional representation can be used for conventional TV picture formation (Figure 7.13). The model is also valid for film (movie) projection systems [21,22]. The conventional way for three-dimensional representation is to consider the hybrid coordinate system; two-dimensional distance coordinates (x-y coordinates) and one time coordinate (t). The time scale is linear, and frames are placed in discrete points on distances corresponding to the frame repetition time (T_F). This representation originates from the film projection system representation.

Another method is to change the third coordinate to a statistically equivalent distance (d) instead of a time coordinate. An autocorrelation coefficient, ρ, for two points (pixels) at distance d is

$$\rho = e^{-\alpha d} \tag{7.33}$$

where α is a constant.

Experimental results [23] show that frame-to-frame autocorrelation is the same type of function as in frame statistics. It follows that the frame-to-frame distance (d_F) can be expressed compared to the line-to-line distance (d_L). It is

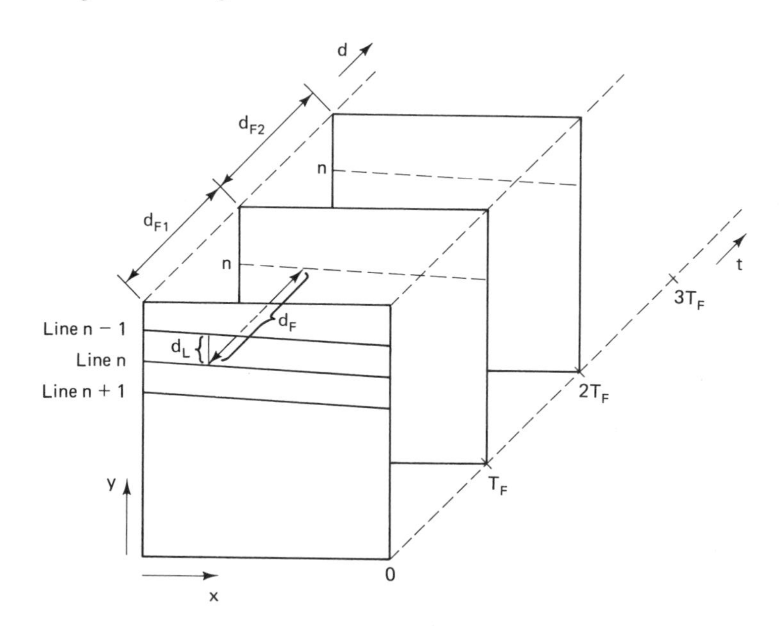

Figure 7.13 Three-dimensional video signal volume.

obvious that, in general, the frame spacing will vary because the statistical frame distance (d_{F1}, d_{F2}, . . .) varies with inter- and intraframe statistics variation.

Finally, dimensions in the three-dimensional model can be modified based on the visual phenomena. For simplification in presentation and, in some instances, lack of knowledge, the phenomena influence is considered disjoint. For example, the 2:1 line interlacing ratio used in conventional television was adopted because it provided a simple means of obtaining a 2:1 bandwidth reduction without objectionable flicker or image breakup. Basically, in this part of the chapter we are dealing with statistical and psychovisual redundancy reduction techniques. This can be presented as three-dimensional volume reordering. For example, the frame or line rate reduction corresponds to the data tracking from the three-dimensional volume or knoblike volume processing. Basic motivation for the use of this kind of signal processing is that videodisc systems utilize (frame) storage (at the player). The only possible attribute of the knoblike (interlacing) processing is its ability to provide a mechanism of bandwidth reduction, or extended play, without appreciable image breakup.

7.4.4.1 Line Interlace. In general, the field rate of the interlace system is set at the frame rate of the noninterlaced system. The bandwidth is reduced n:1 by scanning every nth line of a picture during one field, and scanning the remaining lines in the next $n - 1$ fields. Two problems are inherent in line interlacing systems: the effects of image breakup and line-crawling disturbances attributable to the sequential display of fields.

The static and dynamic resolution of a scene in the line interlacing system ranges from the resolution reduced by the interlace factor to theoretically the same resolution as that for noninterlaced television. If the frame rate is so low that the persistence of the eye is negligible beyond a single field presentation, the resolution of the scene will be reduced by the interlace factor because only a fraction of the total lines of a picture will be displayed at one time. It is assumed that there is no physical storage of fields at the receiver. However, the image jump and blur of the displayed image will not be affected by the interlacing.

For the videodisc system in which past fields are repeatedly displayed, the resolution of the picture will theoretically be the same as that for interlaced television. A reduction in resolution may be caused by "pairing" of adjacent lines because it is difficult to accurately space the interlaced lines of fields.

Figure 7.14 shows that for objects moving perpendicular to the line, the image will be smeared over several lines as a result of storage (repetition) [7, Chap. 21]. The object moving down the picture in the noninterlaced system moves in discrete jumps between fields. The same object reproduced by the

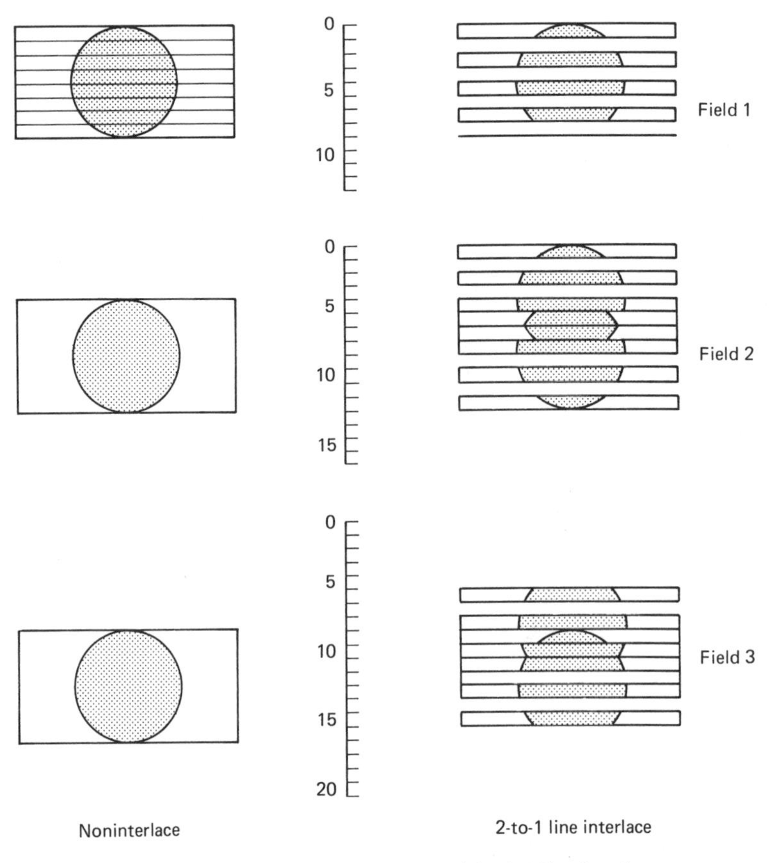

Figure 7.14 Image smearing caused by 2:1 line interlace.

interlaced system appears to be spread out spatially since some of the lines behind the movement of the objects are erased one field late for 2:1 line interlace. This phenomena exists for higher-order line interlacing: the image will be broken up by line interlacing, but the image jump of noninterlaced television will be reduced by the interlace ratio.

Figure 7.15 can be used to illustrate line interlace crawling effects for 2:1, 3:1, and 4:1 line interlacing. Lines 1, 2, 3, and 4 represent individual scanning lines of the first, second, third, and fourth fields, respectively. For the 4:1 interlace, for example, after field 4 is traced out, field 1 appears. The number 1 line appears visually with different spacing to the number 4 line: one line space below and three line spaces above. The resulting optical illusion is that line 4 appears to shift slightly down to create line 1 first, then line 2, and so on. The direction of shift (drift) is independent of the actual direction of the vertical scan, and the drift appears whenever the line distances are not symmetrical. Because lines 1 and 2 are equally separated for standard 2:1 interlacing, the drift illusion is absent. A pseudorandom line scanning process would reduce the drift illusion.

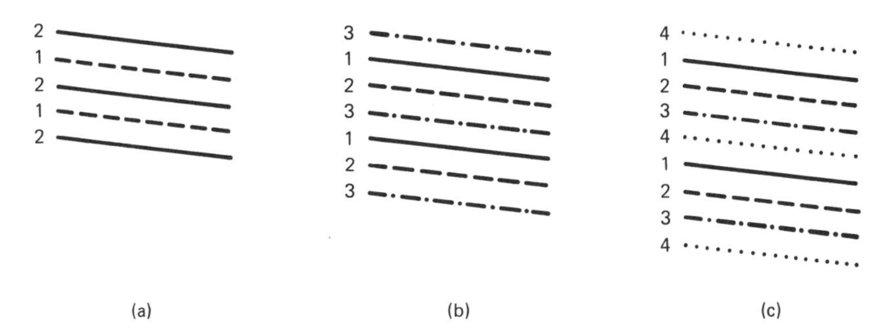

Figure 7.15 Individual scanning lines for sequential display of fields: (a) 2:1; (b) 3:1; (c) 4:1.

7.4.4.2 Dot Interlace. The dot interlace technique provides a relatively simple means of bandwidth reduction (extended play) for applications in which image motion is not great and some amount of artificial smearing of moving objects can be tolerated [24,25]. This technique is an extension of the conventional 2:1 line interlace. A picture frame is divided into several fields of picture elements (pixels) or dots that are sequentially recorded/transmitted. Fields can be chosen in an overlapping or nonoverlapping manner. Bandwidth reduction is achieved by transmitting/recording a reduced number of fields at the normal television frame rate. Figure 7.16 illustrates several dot pattern layouts for 4:1 dot interlace over a 4 $\times$ 4 block of pixels. The patterns are taken with the restriction that the first element be at the origin and that each row and column be occupied by only one dot. The module patterns are discussed in Ref. 5, Chap. 21.

Figure 7.17 illustrates the dot pattern of a single field with 4:1 dot interlace, four fields per frame, and a frame rate equal to that of noninterlace systems.

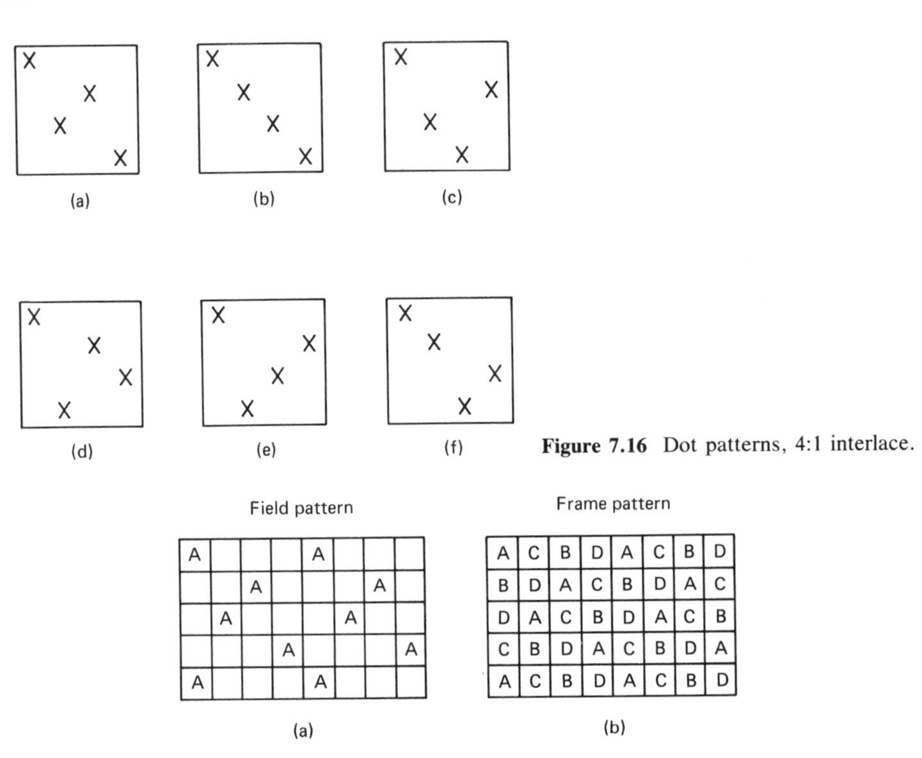

Figure 7.16 Dot patterns, 4:1 interlace.

Figure 7.17 Dot patterns interlace: (a) field pattern; (b) frame pattern.

The addition of line interlace to a dot interlace system is not generally worthwhile because of image breakup problems and line crawling. The selection of dot pattern and field presentation sequence, if there is a choice, determines the degree of optical disturbance in a reproduced picture. The example shown in Figure 7.16 is knight-like moves (as in chess) that one could use to cover the frame.

The most serious drawback of the dot interlace system of bandwidth reduction is patterning and dot-crawl optical disturbances in the picture displayed. Patterning is the optical illusion of a false texture in a picture, appearing when some of the dots are not displayed in a field period; it depends on the arrangement of dots within a frame. Because of stroboscopic effects in the presentation of successive fields and optical illusions in which lines or dots of a picture appear to crawl across the display, the crawling effect results. A repetitive structure on the dot patterns and field presentation sequences should compromise minimization of these disturbing effects without implementation complexity.

With a dot interlace system, the resolution of still scenes is reduced by the interlace ratio unless the storage on the disc is not organized so that each field that is not being replaced by "fresh" data may be read from the disc and displayed on the screen every display frame. But this will cause smearing of moving objects. The number of displayed "nonfresh" fields and the system resolution should be compromised. Also, the number of dots that are not renewed from the disc during each display frame should compromise the increase of patterning effects and decrease the dot crawl disturbance. Other signals in the channel, for example audio and pilot, should be properly arranged.

7.4.5 Variable-Velocity Scanning

The variable-velocity scanning method of bandwidth compression [26–29] is based on the facts that most of the picture information of interest to a human observer lies in the edges of the picture and that the average probability of occurrence for an edge is far less than that of a steady-state region. The total bandwidth can be reduced if the redundant information is reduced in time. The edge information can be spread out over time if the scanning velocity is modified to spend more time transmitting an edge and less time transmitting a steady-state region. The modified video signal may be recovered from the disc and appropriately processed in the player to obtain the original television signal. In principle, the velocity information may be recovered from the modified video signal so that an additional channel for a velocity signal is not needed.

To prevent line-length errors caused by camera noise, the video signal-to-noise ratio must be 60 dB. This is because of differentiation operation in the variable-velocity scanning system. This can be a strong limitation at the inner radius of the disc. Also, high-frequency, low-amplitude detail will be degraded when the system is designed for higher bandwidth reduction because the system is designed to pass the highest frequency that occurs when a full-amplitude chessboard is scanned. A higher-frequency low-amplitude signal is filtered out of the recorder (transmitter) unit. Thus the usefulness of this method for the videodisc system is limited by this degradation coupled with the high video signal-to-noise ratio requirements.

7.4.6 TV Signal Bandwidth Reduction

This technique is used in the capacitive videodisc system. Because of the mechanical limitations associated with the capacitive systems, it would be difficult, if not impossible, to record and play back standard TV signals. The signal bandwidth reduction technique itself is called a "Buried Subcarrier Encoding System" [30].

The luminance bandwidth can be chosen to be less than specified by TV standards. This will result in (horizontal) resolution reduction. If the line and frame rate are kept the same, the discrete line structure of the spectra will remain the same as for the standard TV signal, so the interlace technique can still be used.

The first question to answer is which bandwidth should be chosen for the luminance signal. It seems that between 2 and 3 MHz would be satisfactory for some applications. If the luminance bandwidth of the system is chosen to be f_{max} (see Figure 7.18a), the lowest FM video carrier frequency would have to be $2f_{max}$ in order to avoid baseband interference. When the video information

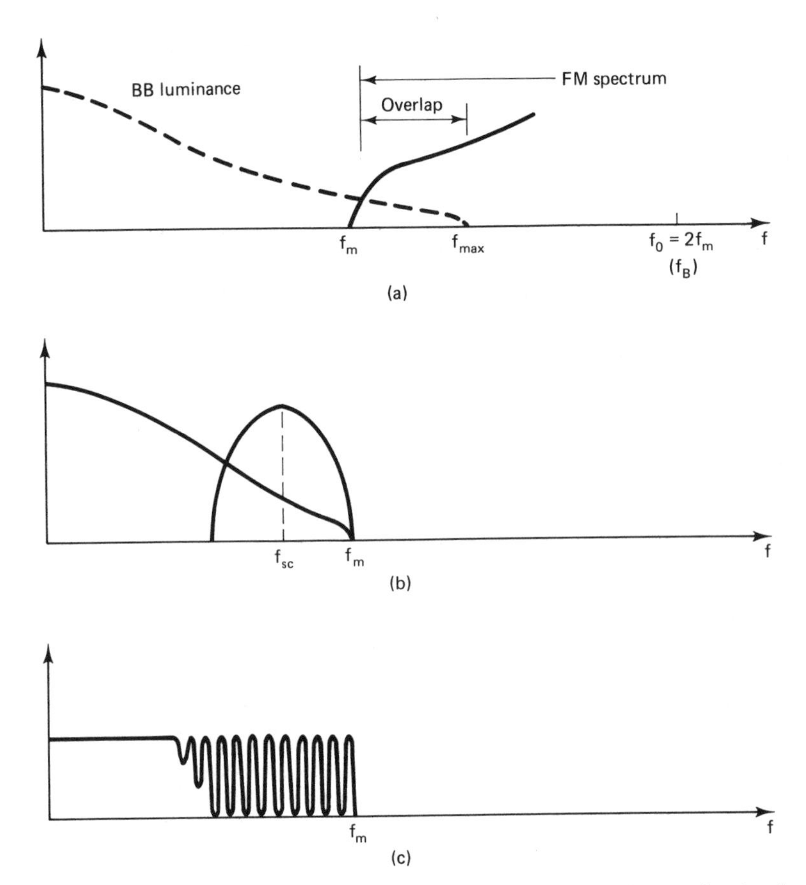

Figure 7.18 TV signal bandwidth reduction: (a) baseband luminance signal and low-sideband FM signal; (b) spectrum of video signal; (c) luminance comb filter.

is entirely luminance information, the carrier f_0 can be lowered. The baseband interference components, caused, for example, by asymmetry in the system, are proportional to the amplitudes of the original baseband components, and typically a TV signal has low-amplitude, higher-frequency components. Also, the high-frequency components carry the edge information in the picture, and the interference will appear only on the edge. But because of the masking effect, the human eye is less sensitive to the distortion around abrupt changes (edges). Thus some overlapping of the FM spectra and possible baseband distortion in the signal can be accepted. If $f_0 = 2f_m$ (corresponds to black level in the image) is less than $2f_{\max}$, the baseband components from 0 to f_m will not overlap with the FM video spectra (Figure 7.18a); the assumption is that the modulation index is so small that the first sidebands barely exist. However, if the chroma carrier or audio carrier(s) are placed in the region between f_m and $f_{\max}$, they will cause unacceptable visible continuous beats in flat areas of the picture [30,31].

Thus frequencies f_0 (or f_m) and $f_{\max}$ (if different than f_m) should be chosen to compromise perceptual distortion (interference components), resolution in the reproduced picture, and desired playing time. But the playing time cannot be calculated directly once the smallest pit size on the disc and f_0 are known. The rotational speed can be obtained when the highest recorded spatial frequency is correlated to the highest instantaneous temporal frequency. The maximal instantaneous frequency deviation, Δf, and thus the upper carrier frequency, is defined by the signal-to-noise ratio in the channel. The SNR ratio can be improved by preemphasis and deemphasis of the higher frequencies. The preemphasis causes the video FM signal to swing higher than the white level and lower than the black level. The optimization involves the parameters of desired SNR, white-level carrier frequency (or Δf), and preemphasis; it is assumed that the carrier-to-noise ratio remains above the FM threshold.

The color information can be processed relative to the luminance information on either a time- or frequency-sharing basis. But since they must be matrixed together (if placed in separate channels), the amplitudes, frequency responses, and noise characteristics of the two channels should be matched. An interlace technique allows the color signal to be placed in the luminance band and recovered in the player. The system parameters that should be defined, related to the color information, are bandwidth of the color signals, B_c, and color subcarrier, f_{sc}.

Experimental results shows that bandwidth $B_c = 0.3$ to 0.5 MHz for the I and Q or U and V components (if QAM is used) can be satisfactory for low-resolution systems. To avoid interfering baseband components, f_{sc} can be taken as (Figure 7.18b)

$$f_{sc} = f_m - B_c \tag{7.34}$$

The chrominance signal can be combed and placed on a carrier which is an odd multiple of $f_H/2$; for example, $f_{sc} = 195\, f_H/2 \simeq 1.53$ MHz is being used in the capacitive grooved systems [30].

When FM is used, the chrominance carrier frequency can be lowered. Lowering the subcarrier frequency improves the SNR for a given chrominance carrier level: the noise power density in an FM system is proportional to the square of the difference in frequency from the FM carrier. But for lower frequencies, the noise from the disc tends to be higher. Thus the optimization involves compromising between FM noise and noise from the disc.

The frequencies above $f_{sc} - B_c$ are combed, and the band below $f_{sc} - B_c$ is uncombed (Figure 7.18c). Typically, the comb filter is formed by delaying the video by one horizontal line in a delay line (e.g., a glass or CCD). Subtracting the delayed signal from the undelayed signal produces chroma. The luminance information, which tends to be the same on successive lines, cancels. Adding delayed to undelayed video produces the luminance component. The chroma cancels because of the 180° phase shift between successive lines of chroma signals. Because of the addition of successive lines, horizontal edges in the picture will be poorly defined since one line will have a luminance value that is the average of the preceding and succeeding lines. To correct for this loss of resolution in the vertical direction, the circuit from Figure 7.19 can be used.

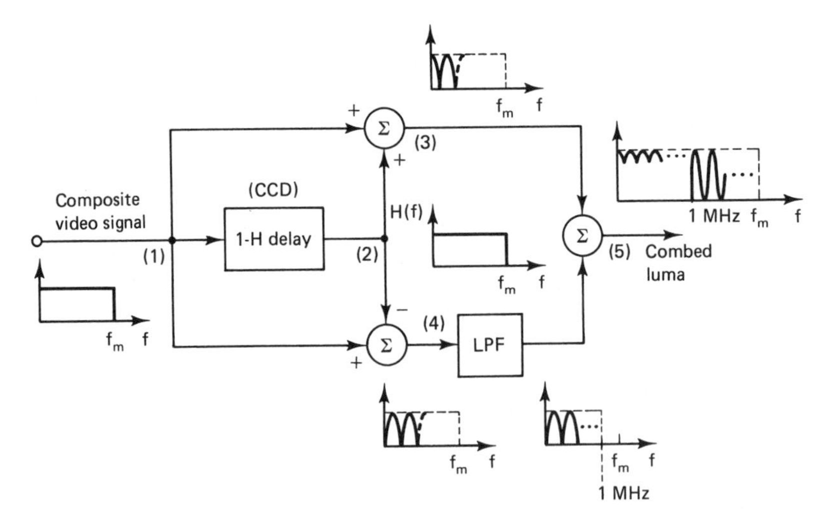

Figure 7.19 Vertical resolution improvement.

By subtracting the 1H delayed video signal from the undelayed signal, a so-called vertical detail signal is extracted. This signal contains the line-to-line difference information, which is removed by combing. The vertical detail signal is low-pass filtered to remove the chroma information and added back to the combed luminance signal. This makes the vertical picture resolution the same as it was before combing. The only net loss in resolution results on diagonals [30]. Obviously, the LPF path in Figure 7.19 is added to compensate for not desired combing at low frequencies in point 3. And the combing at low frequencies in luma channel-point 3, is obtained because the CCD delay line delays low frequency components also. This would not happen if instead of the CCD, glass delay line is used: the glass delay line is basically a bandpass filter while CCD can be considered a low-pass filter. But, the glass delay line could not be used at the low frequencies required in this case: 1-2 MHz. Thus, the practical options are: use CCD delay line and compensate in luma for combing below 1 MHz, or use some heterodyne techniques with the glass delay line.

The next parameters to be defined are audio carriers f_{a1} and f_{a2} and corresponding maximum deviations Δf_{a1} and Δf_{a2}. These also require careful consideration. The optimization for the audio carriers involve:

1. The audio carriers should not directly interfere with the video; thus they should be placed below f_m or higher than $f_w + f_m$. Higher bandwidth location is not desirable from the point of view of total bandwidth compression.
2. Baseband video may cause interference in the audio channels. It is, therefore, desirable to keep the audio carriers below $f_{sc} - B_c$ so that the relatively high level chroma carrier at f_{sc} in the baseband signal can be avoided.
3. The very low frequencies are also undesirable because of large luminance

components at the low frequencies of the baseband (large interference component in the presence of the asymmetry, for example). Also, there is more disc noise at low frequencies. For these reasons, frequencies above 0.5 MHz are desirable for the audio subcarriers.

4. To lower the visibility in the picture caused by the audio interference from the components $f_0 \pm f_{a1}$, these components should be interleaved with the video spectral components (they cannot remain interleaved as the carrier f_0 is frequency modulated). Thus the audio carrier should be chosen to be an odd multiple of the line frequency, $nf_H/2$. In addition, f_{a1} and f_{a2} can be moved from those positions for $f_H/4$ in opposite directions to further reduce interference; at least interference between the two audio channels is reduced.

The audio encoding is the same as for standard TV signal recording in the videodisc system (duty-cycle modulation, etc.). The two audio signals can have a bandwidth of 20 Hz to 20 kHz, or somewhat reduced. Each signal should be preemphasized with a playback deemphasis time constant τ_a (e.g., 75 ms). The maximum deviation of ± 50 to ± 100 kHz at frequencies f_{a1} and f_{a2} would be satisfactory.

The encoding process is illustrated in Figure 7.20. Using the comb filter technique, the luminance and chrominance signals are separated from the standard

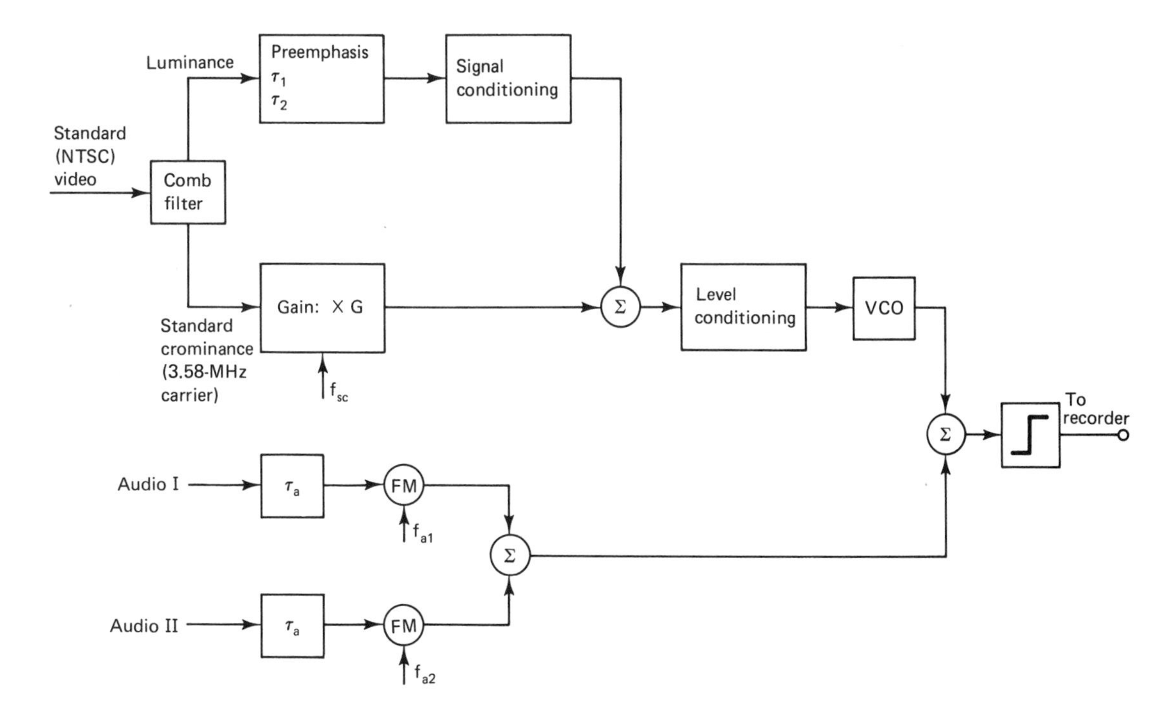

Figure 7.20 Block diagram of the buried subcarrier encoding system.

video signal, a NTSC signal in this example. The luminance signal from the comb filter is preemphasized with a two-breakpoint network with time constants τ_1 and τ_2. After level conditioning, this signal is then added to the chroma signal. The chroma signal portion of the comb filter is bandpass filtered at $\pm B_c$ about 3.58 MHz (for NTSC) with rejection teeth at multiples of f_H (odd multiples of $f_H/2$). The output of the chroma comb filter is translated to f_{sc} and is amplified G times that of a standard chroma signal (equivalent to preemphasis) before it is added to the luminance signal. This increase in chroma increases the chroma SNR during playback.

After signal conditioning the resultant sum is applied to the VCO. The sync tip f_s and carrier frequencies f_b and f_w correspond to black and white levels in the picture, respectively.

The two audio signals are first preemphasized for a time constant τ_a and then applied to the FM modulators, with audio-carrier frequencies f_{a1} and f_{a2}. The two audio carriers have a maximum deviation of $\pm\Delta f_a$ and are first summed together, then summed with the video FM signal. The levels at which the audio signals are added to the video carrier signal depend on the recorder and are typically 26 dB below the video carrier (about 20 dB before limiting—limiting lowers the level 6 dB). After the signal passes the hard limiter, duty-cycle modulation is obtained.

The parameters for the grooved capacitive system [30] are:

$\tau_1 = 0.64\ \mu s$, $\tau_2 = 0.16\ \mu s$ (the maximum preemphasis is approximately 12 dB at frequencies above 1 MHz)

$$\begin{aligned} f_{sc} &= 195 f_H/2 \simeq 1.53 \text{ MHz} \\ G &= 3 \\ f_m &= 2 \text{ MHz} \\ f_{max} &\text{ is between 2 and 3 MHz} \\ B_c &= 0.5 \text{ MHz} \\ f_s &= 4.3 \text{ MHz—sync tip} \\ \left.\begin{aligned} f_b &= 5 \text{ MHz} \\ f_w &= 6.3 \text{ MHz} \end{aligned}\right\} &\text{ carriers} \\ \Delta f &= 1.3 \text{ MHz} \end{aligned} \tag{7.35}$$

With preemphasis corne frequencies of 3.9 MHz and 6.9 MHz (black level is still 5 MHz)

$$\begin{aligned} f_{a1} &= 91 f_H/2 \simeq 716 \text{ kHz} \\ f_{a2} &= 115 f_H/2 \simeq 905 \text{ kHz} \\ \Delta f_a &= \pm 50 \text{ kHz} \\ \tau_a &= 75\ \mu s \end{aligned} \tag{7.36}$$

The decoding process is illustrated in Figure 7.21, and it is similar to standard TV signal decoding (Figure 6.7). After hard limiting, the input signal from the disc is passed through filters. The audio signals are separated by a low-

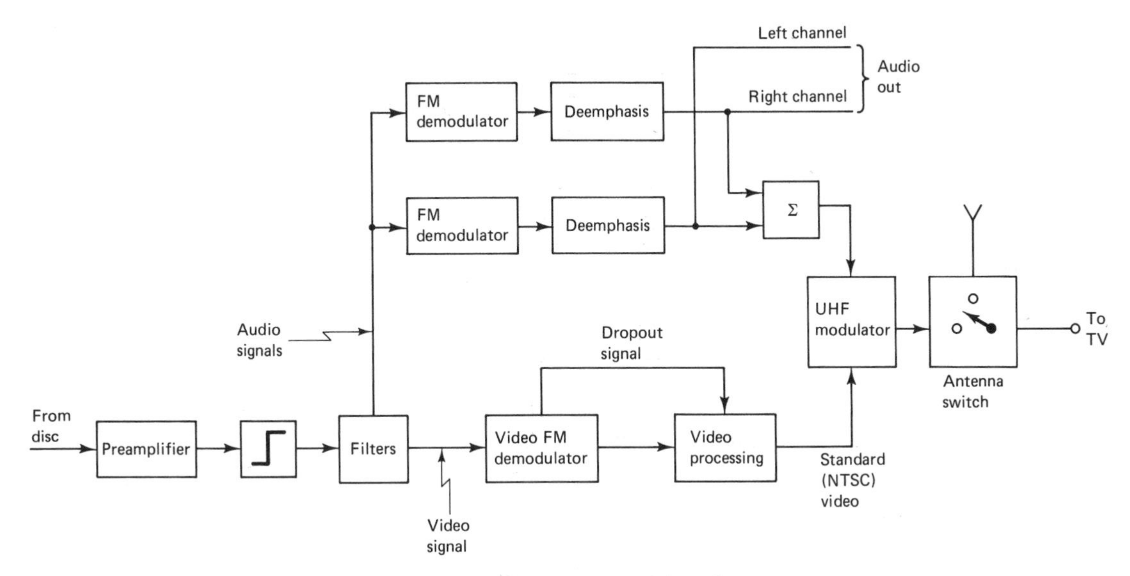

Figure 7.21 Block diagram of decoding system.

pass filter; the bandwidth includes f_{a1} and f_{a2}. After being demodulated, audio signals are deemphasized and can be used as left and right channel signals or can be used or applied to the UHF modulator.

The video FM is separated by the bandpass filter and demodulated by pulse-counting techniques. The signal from the filter is first limited and frequency doubled. The constant duty-cycle modulation (using a one-shot, for example) is performed so that the video signal can be separated by using an LPF (bandwidth $0 - f_{\max}$ or $0 - f_m$). This block also includes a dropout detector.

The video processing unit:

1. Converts chroma of f_{sc} to a standard format (e.g., NTSC with chroma at 3.58 MHz). A heterodyne technique based on AM can be used.
2. Combs filters to separate luminance and chrominance information.
3. Improves vertical resolution (combing effect at low frequencies in the luminance signal is counseled by vertical detail circuit). Really, compensates for losses in the luminance signal at frequencies below 1 MHz.
4. Combines converted chroma and luma into a standard (NTSC) TV signal.

The obtained TV signal is then combined with audio signal to form a standard TV composite signal so that it can be played on an ordinary TV set. An antenna switch allows the same set to be used for videodisc play or ordinary broadcast video. Dropout and time-base-error correction is the same for the standard TV signal recording system.

It should be mentioned that instead of standard preemphasis/deemphasis techniques, SNR can be improved by a level-adaptive technique: low-level signals are boosted more than higher-level signals. If this technique (compandor/expander) is used, the limitation of frequency deviation can be avoided and will result in lower signal distortion. Namely, with normal preemphasis circuits the higher-level, higher-frequency components would cause a large instant deviation in frequency. Because the highest spatial recording frequency is limited, the maximal instant deviation frequency should be limited also.

7.4.7 Predictive and Transform Compression

The goal of picture compression for the videodisc system, for a given gray scale and spatial resolution capabilities of a receiver or user, is to represent the picture by as low a bandwidth as possible. The degree of signal compression that can be achieved when the given picture is to be received by a system, the spatial and gray scale resolution of which exceeds or at least matches those of the picture, is rather limited and we will not discuss it here. If the spatial and gray scale resolution of the user (or the receiver) are inferior to those of the original picture, a certain amount of distortion is permitted between the original picture and the version that can be reproduced from the recorded signal on the disc. For a given amount and type of distortion, the question is: What techniques would allow us to record a picture with reduced bandwidth (and possibly, gray scale)? Also, for a given amount of distortion, the question is: Is there a minimum bandwidth and gray scale for the signal recording (to represent the picture)?

The first step in picture compression is to represent a picture by a decorrelated signal(s). This step is widely used and analyzed in the area of digital image processing [7,10,32]. In this section we briefly introduce predictive and transform compression techniques. Both techniques have advantages and disadvantages. In general:

1. The major advantage of predictive compression techniques is the ease and the economy with which they can be implemented in hardware.
2. Transform compression techniques are less vulnerable than predictive compression techniques to channel noise: the error effect on the reconstructed picture is distributed all over the picture, making it less objectionable from a visual standpoint. In a predictive compression technique, because of the feedback in the system, the error propagates to neighboring picture elements. This creates an unpleasant effect in the reconstructed picture.
3. Transform compression techniques are less sensitive to variations from one picture to another. (In digital signal processing, they have superior coding performance at lower bit rates.)

The predictive and transform compression techniques can be combined to boost the attractive features of each. In digital image processing, this technique would have some advantages [33]. The predictive and transform compression techniques can be used after the video signal is converted to digital form (Figure 7.7). After digital processing is performed, an analog signal is generated and then recorded on the disc. In the player, the signal from the disc is again converted to digital form. After decompression, the original video signal is reconstructed. This method can be effective in the future, when special-purpose integrated circuits being developed to support this technique are available.

7.4.7.1 Predictive Compression. Because the picture signals $s(x, y)$ are correlated, it is possible to derive an estimate or prediction $\hat{s}(x, y)$ for a given picture element (point) in terms of the other picture elements. The difference $s(x, y) - \hat{s}(x, y)$ is the prediction error $e(x, y)$ for the picture element and is usually called the *differential signal.*

The best estimate $\hat{s}(x, y)$, best in the sense that it minimizes the mean-square estimation error, is, in general, a nonlinear function of the picture elements that are used to form the estimate [34]. Usually, for reasons of mathematical tractability, the additional constraint of linearity is imposed on the form of the estimate. In such cases the prediction obtained is the best "linear prediction."

A block diagram of a third-order predictive system is shown in Figure 7.22. The prediction [22] is

$$\hat{s}(t) = a_1 s(t - \tau) + a_2 s(t - T_L) + a_3 s(t - T_F) \tag{7.37}$$

where a_1, a_2, a_3 = coefficients
τ = delay that corresponds to the elements resolving distance
T_L = line period
T_F = frame (field) period

Coefficients a_1, a_2, and a_3 are such that the mean-square prediction error

$$E\{[s(t) - \hat{s}(t)]^2\} = E[e^2(t)] \tag{7.38}$$

is minimized:

$$\frac{\partial E[e^2(t)]}{\partial a_k} = \frac{\partial E\{\{s(t) - [a_1\, s(t - \tau) + a_2 \cdots + a_3 s(t - T_F)]\}^2\}}{\partial a_k} = 0 \tag{7.39}$$

On differentiation one obtains

$$E[s(t)] = E[\hat{s}(t)] = a_1 E[s(t - \tau)] + a_2 E[s(t - T_L)] + a_3 E[s(t - T_F)] \tag{7.40}$$

$$E[s(t)s_k(t)] = a_1[s(t - \tau)s_k(t)] + a_2 E[s(t - T_L)s_k(t)] + a_3 E[s(t - T_F)s_k(t)] \tag{7.41}$$

$$s_k(t) = s(t - \tau),\ s(t - T_L),\ s(t - T_F) \tag{7.42}$$

The optimal weighting values a_1, a_2, and a_3 are then found from the simultaneous solution of the three linear equations in terms of the means $E[s_k]$ and the cross moment $E[s_j s_k]$ between all of the pixels involved in the predictions. Using the optimal values of a_1, a_2, and a_3, the minimum mean-square prediction difference is found to be

$$\sigma_e^2 = E[s^2(t)] - \sum_{k=1}^{3} a_k E[s(t)s_k(t)] \tag{7.43}$$

For the signal reconstruction, the same predictor is used as in the system that generates the differential signal $e(t)$. If there is no channel noise, the original

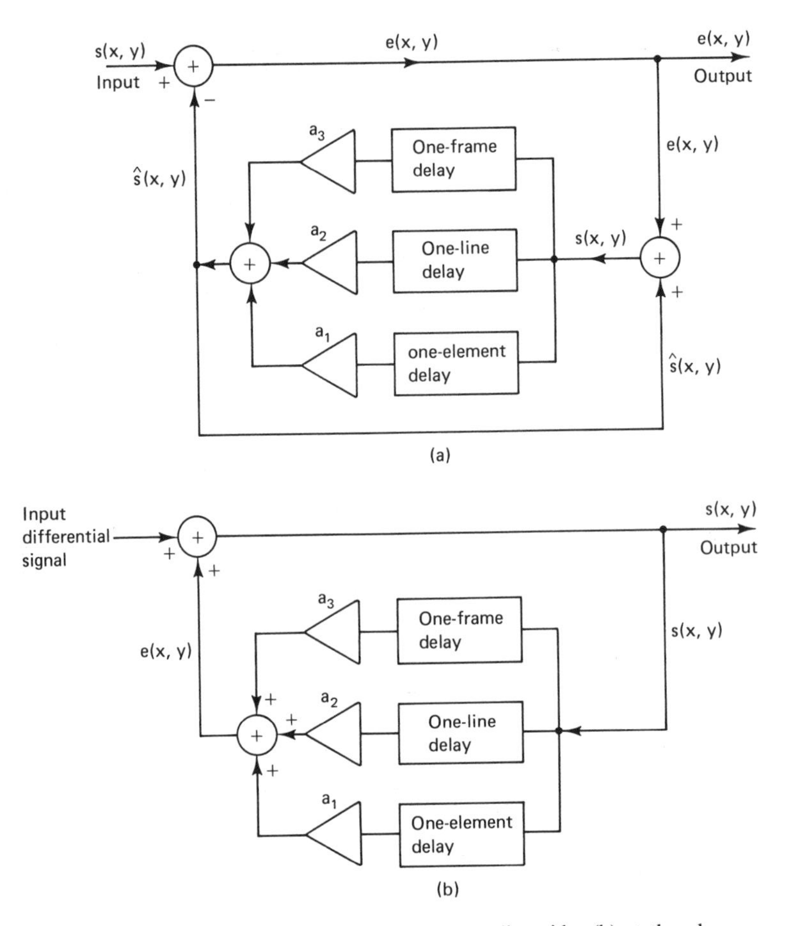

Figure 7.22 Predictive system: (a) at the recording side; (b) at the player.

signal will be reconstructed. The higher-order predictor can be used, but analysis shows [22] that at least in the general case, it is not justified; the gain does not justify the implementation complexity.

7.4.7.2 Transform Compression. Some reversible linear transformations are able to represent a picture by uncorrelated symbols (data). In general, the uncorrelated signals must be ranked according to the degree of significance of their contribution to both the information content and the subjective quality of the picture. The less important signals can be replaced (as in low-pass filtering; the spectral components can be obtained by Fourier transform). This makes possible a major degree of picture compression.

Consider the case where we know a priori that all the pictures in a collection for recording are sinusoidal. Any picture can be reproduced if we know the spatial frequency, orientation, amplitude, and phase. These are unique properties of the picture and thus uncorrelated. For a more general class of pictures, the amount of uncorrelated data may be quite large. Then the elements of the uncorrelated data should be ranked in order of their importance. The higher-order terms may, for example, represent higher spatial resolution effects in the picture. These high-resolution terms may be deleted from the data if the user (receiver) of the picture has limited spatial resolution capability. Such deletions from the uncorrelated data contribute significantly to picture compression [32, Chap. 5].

For a restricted class of pictures, it is easy to find a set of uncorrelated parameters that represent a picture and from which it can be reproduced. For a more general class of pictures, the problem becomes more difficult. One technique for representing pictures by uncorrelated data is to expand a picture in terms of a family of orthonormal functions. The coefficients of expansion are the desired data. This follows from the following statement (Karhunen–Loeve, KL, transformation for continuous pictures).

Let $-A/2 \leq x \leq A/2$, $-B/2 \leq y \leq B/2$ define a region ϕ of the xy-plane. Let $\phi_{mn}(x, y)$ be a complete family of orthonormal functions defined over the region ϕ. A random field $s(x, y)$ may then be expanded in region ϕ as follows:

$$s(x, y) = \sum_{m=0}^{\infty} \sum_{n=0}^{\infty} A_{mn}\phi_{mn}(x, y) \tag{7.44}$$

where the summation on the right-hand side converges to $s(x, y)$ in some sense and where

$$a_{mn} = \int_{-B/2}^{B/2} \int_{-A/2}^{A/2} s(x, y)\phi_{mn}(x, y)\, dx\, dy \tag{7.45}$$

For zero-mean random fields the functions $\phi_{mn}(x, y)$ that result in uncorrelated data a_{mn} must satisfy the following integral equation:

$$\int_{-B/2}^{B/2} \int_{-A/2}^{A/2} R(x, y, x', y')\phi_{mn}(x', y')\, dx'\, dy' = \gamma_{mn}\phi_{mn}(x, y) \tag{7.46}$$

where

$$\gamma_{mn} = E[a_{mn}^2] \tag{7.47}$$

and $R(\cdot)$ is the autocorrelation function for given class of pictures. The proof is beyond the scope of this discussion and can be found in Ref. 32, Chap. 5.

Nonoptimal transformations can also give satisfactory results and have advantages: it is easier to implement a suboptimal transform than, for example, the KL transform. Figure 7.23 illustrates an example of the two-dimensional Fourier transform, performed optically [35]. It is obvious that in the transform domain, the area that contains important details is reduced.

In the area of digital image processing many transforms are examined for data compression purposes [7,10,36,37]. Among them are the Fourier, Haar, Walsh–Hadeamard, slant, and sine and cosine transforms.

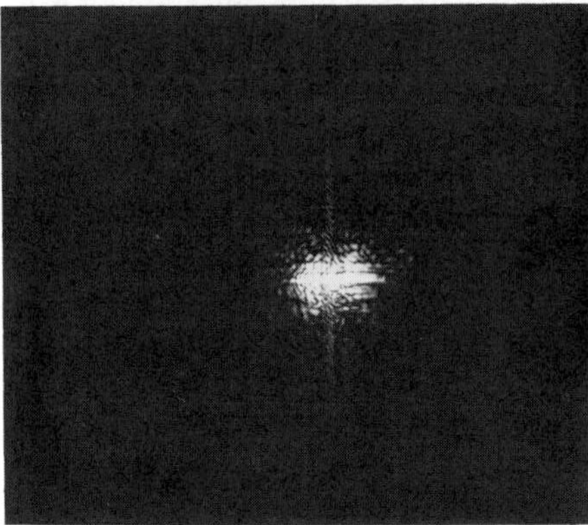

Figure 7.23 Fourier two-dimensional transform: (a) original picture of Sonja; (b) two-dimensional Fourier transform.

7.4.8 General Remarks

A number of extended play methods have just been described. A comprehensive rating of these systems is quite difficult because of the many factors involved in assessing their performance and implementation complexity. The choice should be made under particular circumstances. Sufficient information is now available on the theoretical and subjective coding performance to permit a rational choice between comparative extended-play methods as a function of implementation complexity. For now, the most widely used extended play methods are CLV and TV signal bandwidth reduction ("Buried Subcarrier Encoding System").

There are a few more things that should be pointed out. One of them concerns the preemphasis/deemphasis principles. The preemphasis is used in the FM scheme to diminish the effect of triangular noise in the demodulated signal. But this will cause higher-frequency deviations for higher frequencies present in the modulating (video) signal. The swings in the video amplitude caused by the preemphasis can be amplitude limited, for example, at -60 and $+145$ IRE units for the luma in order to reduce maximal instant frequency deviation. The limiting levels should compromise the desired gain in the SNR, caused by limiting distortion in the signal reproduced. If the luminance signal is limited in this way, then when the chrominance signal is added, the resultant sum is again limited. The limiting levels should be moved slightly apart; for example, −65 and +150 IRE units. This is to avoid relimiting the luminance signal in the presence of the chroma signal, which may cause some undesired cross products before application to the video FM modulator.

The SNR improvement in an FM system with preemphasis/deemphasis circuits depends on the corner frequencies and the relative positions in the bandwidth. If the input noise to the FM discriminator is a white noise with the two-sided power spectral density N_0, the two-sided spectral density of the output noise power $S_{no}(f)$ is (see Chap. 4)

$$S_{no}(\omega) = \left(\frac{k_d}{A_c}\right)^2 \omega^2 N_0 \tag{7.48}$$

where k_d is a constant of the discriminator and A_c is the carrier amplitude.

The mean-square value of the total output noise power $(N)_o$ in the bandwidth 0 to f_m is

$$(N)_o = \int_{-f_m}^{f_m} df = \left(\frac{k_d}{A_c}\right)^2 (2\pi)^2 \frac{2N_o}{3} f_m \tag{7.49}$$

If a deemphasis circuit with the transfer function $H(j\omega)$

$$H(j\omega) = \frac{1 + j(\omega/\omega_1)}{1 + j(\omega/\omega_2)} \tag{7.50}$$

is now used, the equivalent output noise power $(N)_{oe}$ is

$$(N)_{oe} = \int_{-f_m}^{f_m} |H|^2 S_{no}\, df \tag{7.51}$$

The ratio of noise compression ρ is

$$\rho = \frac{(N)_{oe}}{(N)_o} = \left(\frac{\omega_2}{\omega_1}\right)^2 \left\{1 + 3\left[\left(\frac{\omega_1}{\omega_m}\right)^2 - \left(\frac{\omega_2}{\omega_m}\right)^2\right]\left(1 - \frac{\omega_2}{\omega_m} \arctan \frac{\omega_m}{\omega_2}\right)\right\} \tag{7.52}$$

The SNR improvement is $1/\rho$, and is given in Appendix 7.A.

The SNR can be even more improved using a compression/expansion technique: when recording, large signals are compressed (Figure 7.24) and during playback, the opposite operation is performed. This technique can be used in both the video and audio channels. Besides SNR improvement, the harmonic distortion will be lowered in this way by better control of the instantaneous frequency deviations.

Assuming that the input noise of the FM discriminator is white noise, as discussed previously, the output SNR for the luminance signal is

$$\left(\frac{S}{N}\right)_o = 3\left(\frac{\Delta f}{B_L}\right)^2 \left(\frac{C}{N}\right)_i^{30} \left(\frac{30\text{ kHz}}{B_L}\right) \cdot \overline{g_{nr}^2} \tag{7.53}$$

where B_L is the bandwidth of the luminance signal, Δf is its peak frequency deviation, $(C/N)_i^{30}$ is the input carrier-to-noise ratio (CNR) measured over a

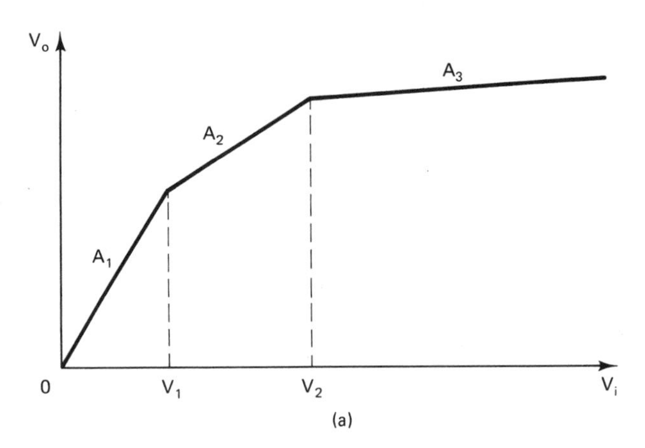

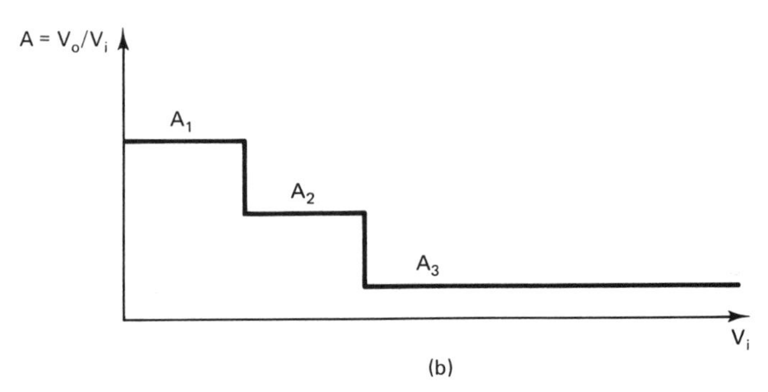

Figure 7.24 Compression of the large signals: (a) voltage transfer characteristic; (b) gain versus input amplitude characteristic.

bandwidth of 30 kHz, and g_{nr} is normalized video signal with maximum value equal to 1.

Maximum output SNR for a white picture (g_{nr} = const. = 1) is $\overline{g_{nr}^2} = 1$, and minimal for the 100% contrast black-and-white picture:

$$g(t) = \tfrac{1}{2} + \tfrac{1}{2}\cos \omega t \tag{7.54}$$

Then $\overline{g_{nr}^2} = \frac{3}{8}$.

If the chrominance bandwidth is B_c and the subcarrier is f_{sc}, the output SNR, $(S/N)_{\text{och}}$ is

$$\left(\frac{S}{N}\right)_{\text{och}} = 3\left(\frac{\Delta f_{\text{ch}}}{B_c}\right)^2 \frac{30\text{ kHz}}{2B_c[3(f_{\text{sc}}/B_c)^2 + 1]}\left(\frac{C}{N}\right)_i^{30} \tag{7.55}$$

When the modulating chroma signal is boosted, say a_{ch} times, then:

$$\Delta f_{\text{ch}} = a_{\text{ch}}\,\Delta f \tag{7.56}$$

and thus the output SNR is higher a_{ch} times.

The output SNR in the audio channel, $(S/N)_{oa}$ is

$$\left(\frac{S}{N}\right)_{oa} = 3\beta^3\left(\frac{C}{N}\right)_{ia} \quad (7.57)$$

where modulation $\beta = \Delta f/f_m$, the input CNR, $(C/N)_{ia}$, is

$$\left(\frac{C}{N}\right)_{ia} = \frac{A_{ca}^2/2}{2\Delta f_a N_0} \quad (7.58)$$

and Δf_a is the peak frequency deviation.

In general, the color subcarrier f_{sc} is placed halfway between line frequencies f_H:

$$f_{sc} = (n + 1)(f_H/2) \quad (7.59)$$

In signal-bandwidth-reduced systems, the color subcarrier, f_{sc}^*, is lower than normal. To lower the color subcarrier, usually the reference frequency f_r obtained from the f_{sc} is used and is

$$f_r = \frac{N_1}{N_2} f_{sc} \quad (7.60)$$

where N_1 and N_2 are integers. From the hardware point of view, N_1 and N_2 should be chosen so that corresponding dividers/multipliers can easily be realized.

From

$$f_{sc}^* = f_r - f_{sc} \quad (7.61)$$

for the NTSC system ($f_{sc} = 455f_H/2$)

$$455\frac{P_1}{N_2} - 228 = n; \qquad 2P_1 = N_1 \quad (7.62)$$

In Appendix 7.B, the table is given for the n, P_1, N_1, N_2, f_r, and f_{sc}^*.

It is sometimes suitable to place audio carriers, f_{a1} and f_{a2}, $f_H/4$ shifted. For example,

$$f_a = nf_H + \tfrac{1}{4}f_H \quad (7.63)$$

If

$$f_a = \frac{M_1}{M_2} f_{sc} \quad (7.64)$$

then (Appendix 7.C):

$$\frac{M_1}{M_2} = \frac{4n + 1}{910} \quad (7.65)$$

or if

$$f_a = nf_H + \tfrac{3}{4}f_H \quad (7.66)$$

then (Appendix 7.D):

$$\frac{M_1}{M_2} = \frac{4n + 3}{910} \quad (7.67)$$

In this case, the difference (bit frequency) between audio carriers, $f_{a2} - f_{a1}$, is $(K + \frac{1}{2}f_H$, while it would be Kf_H if both audio carriers are placed in the middle of the line frequency components. The choice of M_1 and M_2 also depends on the hardware realization. If the PLL is used, a component of the reference frequency (f_{sc}/M_2) will appear in the modulated spectrum, and therefore this frequency should be above the audio bandwidth. To keep the range of the phase detector large, the divider M_1 should be as large as possible and still maintain the foregoing restriction.

REFERENCES

1. Y. Naruse and A. Olkoshi, "A New Optical Videodisc System with One Hour Playing Time," *IEEE Trans. Consum. Electron.*, Vol. CE-24, No. 3, 1978, pp. 453–457.
2. K. Bulthuis, M. G. Carasso, J. P. J. Heemskerk, P. J. Kivits, W. T. Kleuters, and P. Zalm, "Ten Billion Bits on a Disk," *IEEE Spectrum,* Aug. 1979.
3. C. Bricot and T. C. Lehureen, "Optical Disc Reader Having Diffration Minima at Non-interrogated Tracks," U.S. Patent 4,171,879, Oct. 23, 1979 (France, July 23, 1976).
4. C. Eberly, *DVA Tech. Note 727,* Oct. 20, 1980.
5. S. Braat. "Optically read disks with increased information density." Applied Optics, Vol. 22, No. 14, 15 July 1983, pp. 2196–2201.
6. J. H. Bailey and G. H. Ottaway, "Video Recording Disc with Interlacing of Data for Frames on the Same Track," U.S. Patent 4,161,753, July 17, 1979.
7. W. K. Pratt, *Digital Image Processing,* Wiley, New York, 1978.
8. D. J. Granrath, "The Role of Human Visual Models in Image Processing," *Proc. IEEE,* Vol. 69, No. 5, 1981, pp. 552–561.
9. W. A. Pearlman, "A Visual System Model and a New Distortion Measure in the Context of Image Processing," *J. Opt. Soc. Am.,* Vol. 68, No. 3, 1978, pp. 374–386.
10. R. C. Gonzales and P. Wintz, *Digital Image Processing,* Addison-Wesley, Reading, Mass., 1977.
11. J. O. Limb, "Distortion Criteria of the Human Viewer," *IEEE Trans. Syst. Man Cybern.,* Vol. SMC-9, Dec. 1979, pp. 778–793.
12. O. D. Fougeras, "Digital Color Image Processing within the Framework of a Human Visual Model," *IEEE Trans. Acoust. Speech Signal Process.,* Vol. ASSP-27, Aug. 1979, pp. 380–393.
13. E. A. Howson and D. A. Bell, "Reduction of Television Bandwidth by Frequency Interlace," *J. Br. IRE,* Vol. 20, No. 2, 1960, pp. 127–136.
14. A. J. Seyler, "The Coding of Visual Signals to Reduce Channel Capacity Requirements," *Proc. Inst. Electr. Eng.,* Vol. 109, Part C, Sept. 1962, pp. 676–684.
15. A. J. Seyler and Z. L. Budrikis, "Detail Perception after Scene Changes in Television Image Presentations," *IEEE Trans. Inf. Theory,* Vol. IT-11, No. 1, 1965, pp. 31–43.
16. R. C. Brainard, F. W. Mounts, and B. Prasada, "Low Resolution TV: Subjective

Effects of Frame Repetition and Picture Replace," *Bell Syst. Tech. J,* Vol. 46, No. 1, 1967, pp. 261–271.

17. A. J. Seyler, "Statistics of Television Frame Differences," *Proc. IEEE,* Vol. 53, No. 42, 1965, pp. 2127–2128.
18. C. Cherry, "The Bandwidth Compression of Television and Facsimile," *The Television Society Journal,* Vol. 10, 1962, pp. 40–49.
19. M. W. Baldwin, "Demonstration of Some Visual Effects of Using Frame Storage in Television Transmission," *IRE Conv. Rec.,* 1958, p. 107.
20. Yoshida, "Skip Frame Extended Play," *MCA/Philips Program Note 303.*
21. J. Isailović, "Economical Transmission of the Video Signal," Ph.D. thesis, Belgrade, 1976 (in Serbian).
22. J. Isailović, "Noise Consideration in Bit Rate Compression for TV Signals Using Frame-to-Frame Correlation," *NTZ,* Vol. 28, No. 8, 1975, pp. 277–278.
23. E. Z. Soroka, "Izmerenniy mozhkadrovoy korrelapsiy televizionnogo izobrazhenniya" izdatelstvo "Elektrosuyaz" No. 9, pp. 77–79. Moscow, 1965.
24. S. Deutsch, "Pseudo-Random Dot Scan Television Systems," *IEEE Trans. Broadcast.,* Vol. BC-11, No. 1, 1965, pp. 11–21.
25. G. G. Gouriet, "Dot Interlaced Television," *Electron. Eng.,* Vol. 24, No. 290, 1952, pp. 166–171.
26. E. E. Wright, "Velocity Modulation in Television," *Proc. Phys. Soc. (London),* Vol. 46, July 1934, pp. 512–514.
27. L. H. Bedford and O. S. Puckle, "A Velocity Modulation Television System," *J. IEEE,* Vol. 75, 1934, p. 63.
28. M. P. Beddoes, "Experiments with Slope Feedback Coder for TV Compression," *IRE Trans. Broadcast.,* Vol. BC-7, No. 2, 1961, pp. 12–28.
29. M. P. Beddoes, "Two Channel Method for Compressing Bandwidth of Television Signals," *Proc. IEEE,* Vol. 110, No. 2, 1963, pp. 369–374.
30. J. K. Clemens, "Capacitive Pickup and the Buried Subcarrier Encoding System for the RCA Videodisc," *RCA Rev.,* Vol. 39, No. 1, 1978, pp. 33–59.
31. R. N. Rhodes, "The Videodisc Player," *RCA Rev.,* Vol. 39, No. 1, 1978, pp. 198–221.
32. A. Rosenfeld and A. C. Kak, *Digital Picture Processing,* Academic Press, New York, 1976.
33. A. Habibi, "Hybrid Coding of Pictorial Data," *IEEE Trans. Commun.,* Vol. COM-22, 1974, pp. 614–621.
34. A. Popoulis, *Probability, Random Variables and Stochastic Processes,* McGraw-Hill, New York, 1971.
35. J. Isailović, "Application of Linear Transforms to Picture Coding," *Elektrotehnika,* Vol. 27, No. 1, 1978, pp. 105–113 (in Serbian).
36. E. L. Hall, *Computer Image Processing and Recognition,* Academic Press, New York, 1979.
37. H. C. Andrews, *Computer Techniques in Image Processing,* Academic Press, New York, 1970.

APPENDIX 7.A THE SNR IMPROVEMENT ($1/\rho$) IN THE FM SYSTEM WITH PREEMPHASIS/DEEMPHASIS CIRCUITS

The ratio of noise compression ρ, Eq. 7.52:

$$\rho = \frac{(N)_{oe}}{(N)_o} = \left(\frac{\omega_2}{\omega_1}\right)^2 \left\{1 + 3\left[\left(\frac{\omega_1}{\omega_m}\right)^2 - \left(\frac{\omega_2}{\omega_m}\right)^2\right]\left[1 - \frac{\omega_2}{\omega_m} \text{arcty} \left(\frac{\omega_m}{\omega_2}\right)\right]\right\}$$

can be rewritten as:

$$\sigma = \left(\frac{y}{x}\right)^2 \left\{1 + 3(x^2 - y^2)\left[1 - y \arctan\left(\frac{1}{y}\right)\right]\right\}$$

where: $x = \dfrac{\omega_1}{\omega_m}$

$y = \dfrac{\omega_2}{\omega_m}$

The SNR improvement is:

$$\frac{\left(\dfrac{S}{N}\right)_{oe}}{\left(\dfrac{S}{N}\right)_o} = \frac{(N)_o}{(N)_{oe}} = \frac{1}{\rho}$$

Numerical examples of the SNR improvements are given in Table 7.A1; selected curves are shown in Figure 7.A1.

APPENDIX 7A.1 FM IMPROVEMENT (dB) FROM PREEMPHASIS

$\frac{\omega_2}{\omega_m} \Big/ \frac{\omega_1}{\omega_m}$	∞	*1.0*	*.9*	*.8*	*.7*	*.6*	*.5*	*.4*	*.3*	*.2*	*.1*
.025	27.44	26.15	25.90	25.56	25.12	24.52	23.67	22.44	20.59	17.59	11.93
.050	21.59	20.26	20.00	19.66	19.21	18.59	17.73	16.49	14.62	11.61	5.93
.075	18.24	16.88	16.61	16.26	15.80	15.18	14.30	13.04	11.16	8.14	2.45
.100	15.92	14.52	14.25	13.89	13.42	12.79	11.90	10.63	8.73	5.70	0.00
.125	14.16	12.72	12.45	12.08	11.61	10.96	10.07	8.78	6.88	3.83	
.150	12.75	11.28	11.00	10.64	10.15	9.50	8.59	7.30	5.38	2.32	
.175	11.59	10.09	9.81	9.43	8.94	8.28	7.37	6.06	4.13	1.07	
.200	10.60	9.08	8.79	8.41	7.92	7.25	6.32	5.01	3.07	0.00	
.225	9.76	8.21	7.91	7.53	7.03	6.35	5.42	4.10	2.16		
.250	9.02	7.45	7.15	6.76	6.25	5.57	4.63	3.31	1.35		
.275	8.37	6.77	6.47	6.08	5.57	4.88	3.94	2.60	.64		
.300	7.79	6.17	5.87	5.47	4.96	4.26	3.31	1.97	0.00		
.325	7.27	5.63	5.33	4.93	4.41	3.71	2.75	1.40			
.350	6.80	5.15	4.84	4.43	3.91	3.21	2.24	.89			
.375	6.38	4.71	4.39	3.99	3.46	2.75	1.78	.43			
.400	6.00	4.30	3.99	3.58	3.05	2.33	1.36	0.00			
.425	5.64	3.94	3.62	3.21	2.67	1.95	.98				
.450	5.32	3.60	3.28	2.86	2.32	1.61	.63				
.475	5.02	3.29	2.96	2.55	2.01	1.28	.30				
.500	4.75	3.00	2.68	2.26	1.71	.99	0.00				

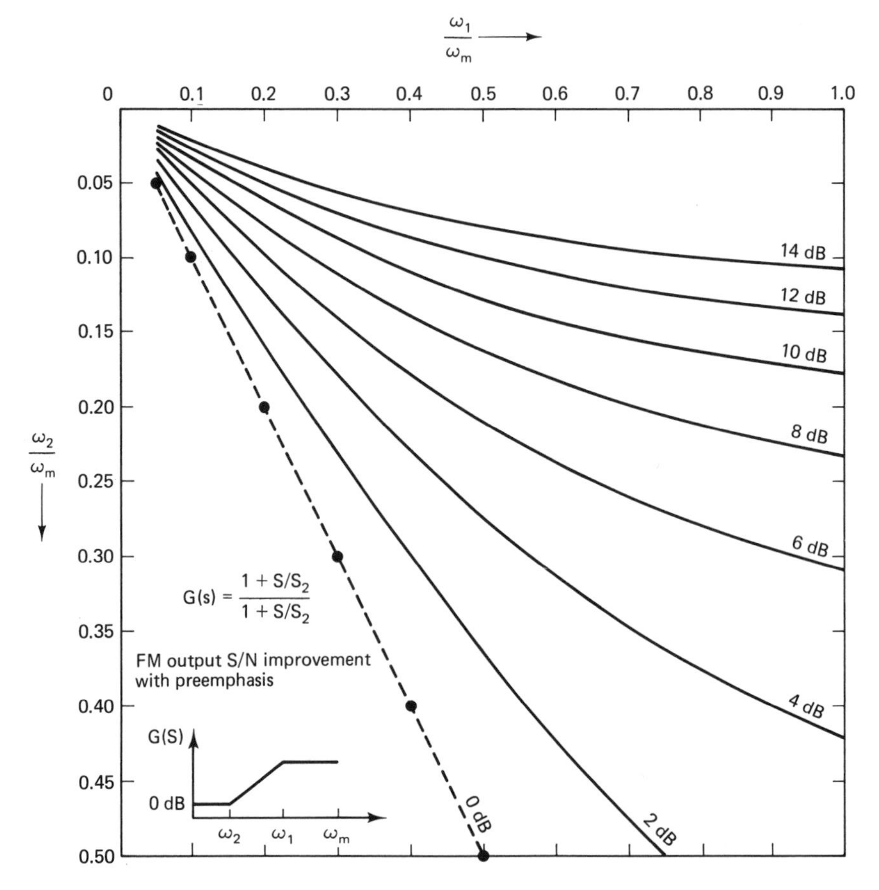

Figure 7A.1 SNR improvements.

APPENDIX 7.B COLOR SUBCARRIER FOR THE NTSC-LIKE SYSTEMS* WITH 1/2 LINE OFFSET

N	P_1	N_1	N_2	f_r (MHz)	f^*_{sc} (MHz)
63	291	582	455	4.578671	.999126
64	292	584	455	4.594405	1.014860
65	293	586	455	4.610139	1.030594
66	42	84	65	4.625874	1.046329
67	59	118	91	4.641608	1.062063
68	296	592	455	4.657342	1.077797
69	297	594	455	4.673076	1.093531
70	298	596	455	4.688811	1.109266
71	23	46	35	4.704545	1.125000
72	60	120	91	4.720279	1.140734
73	43	86	65	4.736013	1.156468

N	P_1	N_1	N_2	f_r (MHz)	f^*_{sc} (MHz)
74	302	604	455	4.751748	1.172203
75	303	606	455	4.767482	1.187937
76	304	608	455	4.783216	1.203671
77	61	122	91	4.798950	1.219405
78	306	612	455	4.814685	1.235140
79	307	614	455	4.830419	1.250874
80	44	88	65	4.846153	1.266608
81	309	618	455	4.861887	1.282342
82	62	124	91	4.877622	1.298077
83	311	622	455	4.893356	1.313811
84	24	48	35	4.909090	1.329545
85	313	626	455	4.924825	1.345280
86	314	628	455	4.940559	1.361014
87	9	18	13	4.956293	1.376748
88	316	632	455	4.972027	1.392482
89	317	634	455	4.987762	1.408217
90	318	636	455	5.003496	1.423951
91	319	638	455	5.019230	1.439685
92	64	128	91	5.034964	1.455419
93	321	642	455	5.050699	1.471154
94	46	92	65	5.066433	1.486888
95	323	646	455	5.082167	1.502622
96	324	648	455	5.097901	1.518356
97	5	10	7	5.113636	1.534091
98	326	652	455	5.129370	1.549825
99	327	654	455	5.145104	1.565559
100	328	656	455	5.160839	1.581294
101	47	94	65	5.176573	1.597028
102	66	132	91	5.192307	1.612762
103	331	662	455	5.208041	1.628496
104	332	664	455	5.223776	1.644231
105	333	666	455	5.239510	1.659965
106	334	668	455	5.255244	1.675699
107	67	134	91	5.270978	1.691433
108	48	96	65	5.286713	1.707168
109	337	674	455	5.302447	1.722902
110	26	52	35	5.318181	1.738636
111	339	678	455	5.333915	1.754370
112	68	136	91	5.349650	1.770105
113	341	682	455	5.365384	1.785839
114	342	684	455	5.381118	1.801573
115	49	98	65	5.396852	1.817307
116	344	688	455	5.412587	1.833042
117	69	138	91	5.428321	1.848776
118	346	692	455	5.444055	1.864510
119	347	694	455	5.459790	1.880245
120	348	696	455	5.475524	1.895979
121	349	698	455	5.491258	1.911713
122	10	20	13	5.506992	1.927447

N	P_1	N_1	N_2	f_r (MHz)	f^*_{sc} (MHz)
123	27	54	35	5.522727	1.943182
124	352	704	455	5.538461	1.958916
125	353	706	455	5.554195	1.974650
126	354	708	455	5.569929	1.990384
127	71	142	91	5.585664	2.006119
128	356	712	455	5.601398	2.021853
129	51	102	65	5.617132	2.037587
130	358	716	455	5.632866	2.053321
131	359	718	455	5.648601	2.069056
132	72	144	91	5.664335	2.084790
133	361	722	455	5.680069	2.100524
134	362	724	455	5.695803	2.116258
135	363	726	455	5.711538	2.131993
136	4	8	5	5.727272	2.147727
137	73	146	91	5.743006	2.163461
138	366	732	455	5.758741	2.179196
139	367	734	455	5.774475	2.194930
140	368	736	455	5.790209	2.210664
141	369	738	455	5.805943	2.226398
142	74	148	91	5.821678	2.242133
143	53	106	65	5.837412	2.257867
144	372	744	455	5.853146	2.273601
145	373	746	455	5.868880	2.289335
146	374	748	455	5.884615	2.305070
147	75	150	91	5.900349	2.320804
148	376	752	455	5.916083	2.336538
149	29	58	3 5	5.931817	2.352272
150	54	108	65	5.947552	2.368007

* The CED system ("Buried Subcarrier Encoding Systems") is obtained for n = 97 (f^*_{sc} = 1.534091 MHz).

APPENDIX 7.C AUDIO CARRIERS WITH $f_H/4$ SHIFT (EQ. 7.63)

For the phase-lock loop design common notation for loop coefficients is P and Q (Eq. 7.65: P = M_1, Q = M_2)

N	P	Q	f_{a1} (MHz)	f_{aR1} (kHz)
18	73	910	.28715031	3.9336
19	11	130	.30288458	27.5350
20	81	910	.31861884	3.9336
21	17	182	.33435310	19.6678
22	89	910	.35008737	3.9336
23	93	910	.36582163	3.9336
24	97	910	.38155590	3.9336
25	101	910	.39729016	3.9336
26	3	26	.41302442	137.6748

N	P	Q	f_{a1} (MHz)	f_{aR1} (kHz)
27	109	910	.42875869	3.9336
28	113	910	.44449295	3.9336
29	9	70	.46022721	51.1364
30	121	910	.47596148	3.9336
31	25	182	.49169574	19.6678
32	129	910	.50743001	3.9336
33	19	130	.52316427	27.5350
34	137	910	.53889853	3.9336
35	141	910	.55463280	3.9336
36	29	182	.57036706	19.6678
37	149	910	.58610132	3.9336
38	153	910	.60183559	3.9336
39	157	910	.61756985	3.9336
40	23	130	.63330412	27.5350
41	33	182	.64903838	19.6678
42	13	70	.66477264	51.1364
43	173	910	.68050691	3.9336
44	177	910	.69624117	3.9336
45	181	910	.71197543	3.9336
46	37	182	.72770970	19.6678
47	27	130	.74344396	27.5350
48	193	910	.75917823	3.9336
49	197	910	.77491249	3.9336
50	201	910	.79064675	3.9336
51	41	182	.80638102	19.6678
52	209	910	.82211528	3.9336
53	213	910	.83784954	3.9336
54	31	130	.85358381	27.5350
55	17	70	.86931807	51.1364
56	45	182	.88505234	19.6678
57	229	910	.90078660	3.9336
58	233	910	.91652086	3.9336
59	237	910	.93225513	3.9336
60	241	910	.94798939	3.9336
61	7	26	.96372365	137.6748
62	249	910	.97945792	3.9336
63	253	910	.99519218	3.9336
64	257	910	1.01092645	3.9336
65	261	910	1.02666071	3.9336
66	53	182	1.04239497	19.6678
67	269	910	1.05812924	3.9336
68	3	10	1.07386350	357.9545
69	277	910	1.08959776	3.9336
70	281	910	1.10533203	3.9336
71	57	182	1.12106629	19.6678
72	289	910	1.13680055	3.9336
73	293	910	1.15253482	3.9336
74	297	910	1.16826908	3.9336
75	43	130	1.18400335	27.5350

N	P	Q	f_{a1} (MHz)	f_{aR1} (kHz)
76	61	182	1.19973761	19.6678
77	309	910	1.21547187	3.9336
78	313	910	1.23120614	3.9336
79	317	910	1.24694040	3.9336
80	321	910	1.26267466	3.9336
81	5	14	1.27840893	255.6818
82	47	130	1.29414319	27.5350
83	333	910	1.30987746	3.9336
84	337	910	1.32561172	3.9336
85	341	910	1.34134598	3.9336
86	69	182	1.35708025	19.6678
87	349	910	1.37281451	3.9336
88	353	910	1.38854877	3.9336
89	51	130	1.40428304	27.5350
90	361	910	1.42001730	3.9336
91	73	182	1.43575157	19.6678
92	369	910	1.45148583	3.9336
93	373	910	1.46722009	3.9336
94	29	70	1.48295436	51.1364
95	381	910	1.49868862	3.9336
96	11	26	1.51442288	137.6748
97	389	910	1.53015715	3.9336
98	393	910	1.54589141	3.9336
99	397	910	1.56162568	3.9336
100	401	910	1.57735994	3.9336
101	81	182	1.59309420	19.6678
102	409	910	1.60882847	3.9336
103	59	130	1.62456273	27.5350
104	417	910	1.64029699	3.9336
105	421	910	1.65603126	3.9336
106	85	182	1.67176552	19.6678
107	33	70	1.68749979	51.1364
108	433	910	1.70323405	3.9336
109	437	910	1.71896831	3.9336
110	63	130	1.73470258	27.5350

APPENDIX 7.D AUDIO CARRIERS WITH $\frac{3}{4}f_H$ SHIFT (EQ. 7.66)

(P and Q as in App. 7.3)

N	P	Q	f_{a2} (MHz)	f_{aR2} (kHz)
18	15	182	.29501745	19.6678
19	79	910	.31075171	3.9336
20	83	910	.32648597	3.9336
21	87	910	.34222024	3.9336
22	1	10	.35795450	357.9545
23	19	182	.37368876	19.6678

N	P	Q	f_{a2} (MHz)	f_{aR2} (kHz)
24	99	910	.38942303	3.9336
25	103	910	.40515729	3.9336
26	107	910	.42089155	3.9336
27	111	910	.43662582	3.9336
28	23	182	.45236008	19.6678
29	17	130	.46809435	27.5350
30	123	910	.48382861	3.9336
31	127	910	.49956287	3.9336
32	131	910	.51529714	3.9336
33	27	182	.53103140	19.6678
34	139	910	.54676566	3.9336
35	11	70	.56249993	51.1364
36	21	130	.57823419	27.5350
37	151	910	.59396846	3.9336
38	31	182	.60970272	19.6678
39	159	910	.62543698	3.9336
40	163	910	.64117125	3.9336
41	167	910	.65690551	3.9336
42	171	910	.67263977	3.9336
43	5	26	.68837404	137.6748
44	179	910	.70410830	3.9336
45	183	910	.71984257	3.9336
46	187	910	.73557683	3.9336
47	191	910	.75131109	3.9336
48	3	14	.76704536	255.6818
49	199	910	.78277962	3.9336
50	29	130	.79851388	27.5350
51	207	910	.81424815	3.9336
52	211	910	.82998241	3.9336
53	43	182	.84571668	19.6678
54	219	910	.86145094	3.9336
55	223	910	.87718520	3.9336
56	227	910	.89291947	3.9336
57	33	130	.90865373	27.5350
58	47	182	.92438799	19.6678
59	239	910	.94012226	3.9336
60	243	910	.95585652	3.9336
61	19	70	.97159079	51.1364
62	251	910	.98732505	3.9336
63	51	182	1.00305931	19.6678
64	37	130	1.01879358	27.5350
65	263	910	1.03452784	3.9336
66	267	910	1.05026210	3.9336
67	271	910	1.06599637	3.9336
68	55	182	1.08173063	19.6678
69	279	910	1.09746490	3.9336
70	283	910	1.11319916	3.9336
71	41	130	1.12893342	27.5350
72	291	910	1.14466769	3.9336

N	P	Q	f_{a2} (MHz)	f_{aR2} (kHz)
73	59	182	1.16040195	19.6678
74	23	70	1.17613621	51.1364
75	303	910	1.19187048	3.9336
76	307	910	1.20760474	3.9336
77	311	910	1.22333901	3.9336
78	9	26	1.23907327	137.6748
79	319	910	1.25480753	3.9336
80	323	910	1.27054180	3.9336
81	327	910	1.28627606	3.9336
82	331	910	1.30201032	3.9336
83	67	182	1.31774459	19.6678
84	339	910	1.33347885	3.9336
85	49	130	1.34921312	27.5350
86	347	910	1.36494738	3.9336
87	27	70	1.38068164	51.1364
88	71	182	1.39641591	19.6678
89	359	910	1.41215017	3.9336
90	363	910	1.42788443	3.9336
91	367	910	1.44361870	3.9336
92	53	130	1.45935296	27.5350
93	75	182	1.47508723	19.6678
94	379	910	1.49082149	3.9336
95	383	910	1.50655575	3.9336
96	387	910	1.52229002	3.9336
97	391	910	1.53802428	3.9336
98	79	182	1.55375854	19.6678
99	57	130	1.56949281	27.5350
100	31	70	1.58522707	51.1364
101	407	910	1.60096134	3.9336
102	411	910	1.61669560	3.9336
103	83	182	1.63242986	19.6678
104	419	910	1.64816413	3.9336
105	423	910	1.66389839	3.9336
106	61	130	1.67963265	27.5350
107	431	910	1.69536692	3.9336
108	87	182	1.71110118	19.6678
109	439	910	1.72683545	3.9336
110	443	910	1.74256971	3.9336

eight

Audio Signal Recording

8.1 INTRODUCTION

In the early stage of videodisc technology development, attention was concentrated on video signal recording. The audio was something "too simple" to think about. But in the later stage of technology development, audio is getting more attention. In analogy with video signals, two questions have been repeated frequently: first, how to get better quality audio reproduction from the disc, and second, how to gain flexibility for the audio.

In general, quality improvement can be obtained using conventional analog signal processing methods for audio records or audio tapes. Basically, noise reduction systems are used. There are two groups of these systems:

1. With signal preprocessing before recording
2. Without signal preprocessing

Preprocessing before signal recording assumes an inverse process during playback. In fact, before recording an audio signal is passed through a compressor and through an equivalent expander in the playback chain. Frequently, the terms *coding* and *decoding* are used, although analog methods are also used. For the already recorded audio on the videodisc, if existing noise does not allow satisfactory playback by manipulation of the reproduction system bandwidth, noise can be reduced. In this process, signal components below noise level cannot be separated. This means that when those signal components, say 40 or 50 dB below the program level, are a vital part of the sound, the quality of the reproduced sound will be degraded.

In general, the systems with signal preprocessing are not compatible with systems without preprocessing or between themselves.

Digital signal processing is another approach for quality improvement. The era of digital audio started several years ago in the field of high-fidelity sound. Up to the tape mastering stage, digital technology was generally thought to provide only marginal improvements in the quality of the sound. In fact, it is felt that in the recording, mixing, and mastering stages, digital technology has no discernable effect on sound quality.

The digital influence, however, becomes clearly evident during such further processing steps as copying, multitrack remixing, and editing—all of which take place long after the original recording session. At these steps, digital technology provides noticeable improvements in sound quality through complete elimination of noise, distortion, echo, wow, and flutter. Digital technology also prevents progressive losses in succeeding sound generations when copying and remixing. When digital audio is employed in recording and processing, the sound quality attained in the resulting analog records then depends solely on the degree of sophistication in the manufacturing of the disc.

It can be expected that in the future the audio chain will be digital from analog-to-digital (A/D) converters just after the microphones until digital-to-analog (D/A) converters just before speakers (in the consumer's home). Later, power D/A, digital speakers, and digital microphones might be developed.

Digitization of the voice signals can also be used to gain flexibility. For example, some length of the audio (i.e., 20 s), can be digitized with possible involvement of the compressed audio techniques, and incorporated with the video information; thus performing a sound over still-frame.

Also, by combining the latest in sound techniques with microcomputer technology, the system allows programmers to tailor audio/visual messages in unique ways. First, program producers can combine stereophonic sound and picture. Second, producers can create audio/visual programs to interact in much the same way as a computer interacts with a data file.

The dual-track system provides two separate and controllable audio channels. This allows a single disc to play two languages or address two audiences: an explanation for the doctor can be recorded on one channel while guidance for the nurse is recorded on the other; a customer presentation on one channel and selling tips for the sales representative on the other; background for a teacher together with more detailed material for the student.

8.2 SPEECH PERCEPTION

The ultimate recipients of information in an audio communication link are usually human beings. Their perceptual abilities dictate the precision with which audio data must be processed and transmitted. These abilities essentially prescribe fidelity criteria for reception and, in effect, determine the channel capacity necessary for the transmission of voice messages. Consequently, it is pertinent to inquire into the fundamental mechanism of hearing and to attempt to establish the capabilities and limitations of human perception [1]. However, for any transmission system to benefit from prior knowledge of the information source, knowledge must be put into a tractable analog form that can be used in the design of signal processing operations.

Speech information, originating from a speaker, traversing a transmission medium, and arriving at a listener, might be considered in a number of coding stages. On the transmitter side, the stage might include the acoustic wave, the muscular forces manipulating the vocal mechanism, or the physical stage and excitation of the tract. On the receiver side, the information might be considered in terms of the acoustic-mechanical motions of the hearing transducer, or in terms of the electrical pulses transmitted to the brain over the auditory nerve. Characteristics of one or more of these codings might have applications in practical transmission systems.

The complete mechanism of auditory perception is far from being adequately understood [1, p. 108]. Even so, current knowledge of ear physiology, nerve electrophysiology, and subjective behavior makes it possible to relate certain auditory functions among these disparate realms. Such correlations are facilitated if behavior can be quantified and analytically specified.

Pitch is that subjective attribute which admits a rank ordering on a scale ranging from low to high. As such, it correlates strongly with objective measures of frequency. One important facet of auditory perception is the ability to ascribe a pitch to sounds that exhibit periodic characteristics.

Another aspect of perception is binaural centralization. This is the subjective ability to locate a sound image at a particular point inside the head when listening with earphones. If identical clicks (impulses of sound pressure) are produced simultaneously at both ears, normal listeners believe that the sound image is located exactly in the center of the head. If the clicks at one ear are produced a little earlier or with slightly greater intensity than at the other ear, the sound image shifts toward the ear. The shift continues with increasing intervals of time or intensity difference until the image moves completely to one side and eventually breaks apart. One then begins to hear individual clicks at both ears.

The subjectively determined minimum audible pressure for pure (sine) tones is shown in Figure 8.1. This is an important curve for noise reduction considerations. Binaural thresholds of audibility for a variety of periodic pulse trains differ from this curve and depend on the pulse rate and durations [1, p. 138].

Literature on hearing comprises a large amount of data on subjective responses to speech and speechlike stimuli. There are, for example, determinations on the ear's ability to discriminate features such as vowel pitch, loudness, format frequency, spectral irregularity, and the like. Such data are particularly important in establishing criteria for the design of speech transmission systems and in estimating the channel capacity necessary to transmit speech data.

Frequency-domain representation of speech information appears advantageous from two standpoints. First, acoustic analysis of the vocal mechanism shows

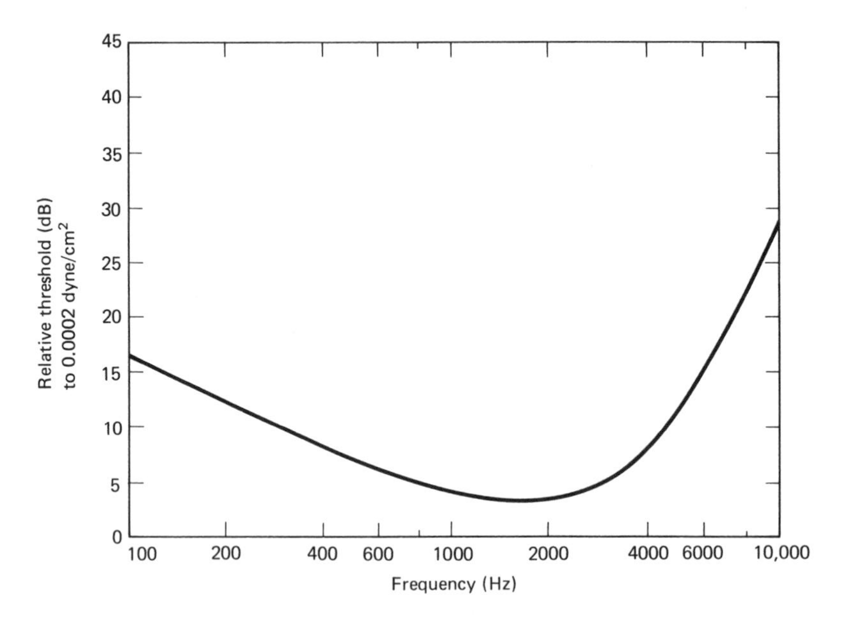

Figure 8.1 Threshold sensitivity.

that the normal mode or natural frequency concept permits concise descriptions of speech sounds. Second, clear evidence exists to prove that the ear makes a crude frequency analysis at an early stage in its processing. Presumably, then, features salient in frequency analysis are important in production and perception, and consequently hold promise for efficient coding and noise reduction. Experience supports this notion.

Further, the vocal mechanism is a quasi-stationary source of sound. Its excitation and normal modes change with time. Any spectral measure applicable to the speech signal should therefore reflect temporal features of perceptual significance as well as spectral features. Something other than a conventional frequency transform is indicated.

In speech applications, it is usually desirable for short-time analysis to discriminate vocal properties such as voiced and unvoiced excitation, fundamental frequency, and format structure. The choice of the analyzing time window $h(t)$ determines the compromise made between temporal and frequency resolution. A time window short in duration corresponds to a broad bandpass filter. It may yield a spectral analysis in which the temporal structure of individual vocal periods is resolved. A window with a duration of several pitch periods, on the other hand, corresponds to a broad bandpass filter. It may yield a spectral analysis in which the temporal structure of individual vocal periods is resolved. A window with a duration of several pitch periods, on the other hand, corresponds to a narrower bandpass filter. It may produce an analysis in which individual harmonic spectral components are resolved in frequency.

Distributions of the absolute root-mean-square speech pressure in these bands measured 30 cm from a speaker's mouth producing continuous conversational speech are shown in Figure 8.2. Bandpass filters with bandwidths one-half octave wide below 500 Hz and one octave wide above 500 Hz were used [2]. The integration time was 1/8 s.

Auditory masking of the ear is commonly exploited in noise reduction techniques. Whenever one sound is being heard, it reduces the ability of the listener to hear another sound, but this is not as easily quantified. In the case of steady-state white noise masking the audibility of pure tones, useful and repeatable data are available. White noise raises the threshold of hearing for pure tones by a level that is dependent on the frequency of the tone. A curve showing this increase in the audible threshold is shown in Figure 8.3. The curve shows a general trend. At higher frequencies the tone has to be increased in amplitude compared to a 1-kHz tone in order to be heard. More noise spectra are contributing to masking as the frequency goes up. Although not equally apparent from the curve, only noise spectra in a frequency band centered on the tone contribute to the masking of that tone [3].

These results are well supported by physical evidence. The hearing mechanism

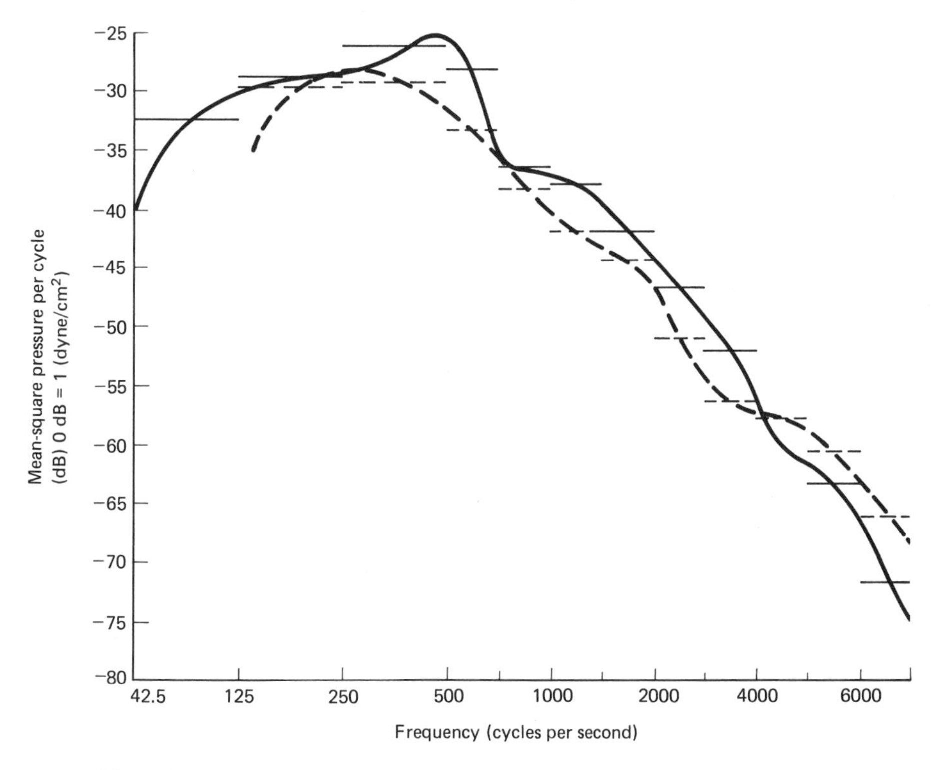

Figure 8.2 Long-time power density spectrum for continuous speech measured 30 cm from the mouth. Solid line: composite of six men; dashed line: composite of five women.

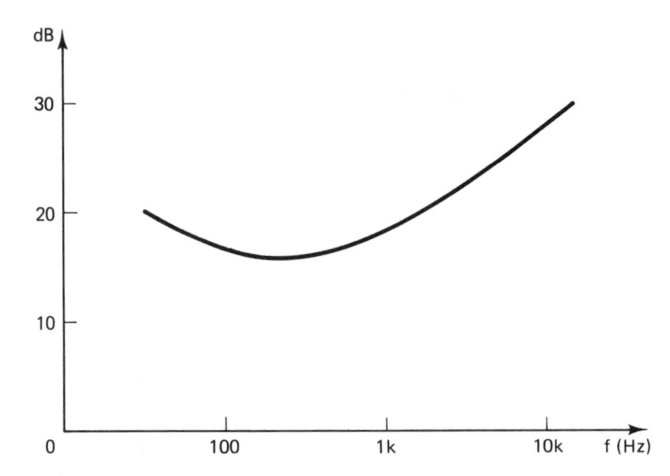

Figure 8.3 Difference between tone intensity and noise spectrum level at the tone frequency for masking.

in the ear involves the basilar membrane, which is approximately 30 mm long by 0.5 mm wide. The nerve endings giving the sensation of hearing are spaced along the membrane so that the ability to hear at one frequency is not masked at another frequency when the two frequencies are well separated. White noise can excite the entire basilar membrane since it has spectral components at all frequencies. For any single frequency, therefore, there will be a band of noise spectra capable of simultaneously exciting the nerve endings that are responding to the single frequency—and masking occurs.

8.3 AUDIO PLUS VIDEO

For the standard videodisc application, each audio channel occupies between 100 and 250 kHz of videodisc channel bandwidth. The amplitude and location of the audio carriers are chosen so that the influence of the audio to video channel and vice versa is minimized. The amplitude of the audio carrier should be low enough compared to the video carrier so that no visible Moiré patterns can be found in the picture reproduced. To satisfy this, a difference of between 20 dB–26 dB the two modulated carriers is usually enough. The carrier frequencies f_0 and f_a are chosen so that any audio interference from the components $f_0 \pm f_a$ into the video will be interleaved with the video spectral components. This lowers the visibility in the picture considerably, even though they cannot remain interleaved as the carrier is frequency modulated. The offset can be either a half or a quarter of the line frequency: $f_a = (2k + 1)f_H/2$ or $f_a = (2k \pm \frac{1}{2})f_H/2$. Similar consideration is valid for both audio channels. Also, because of the existing nonlinearities in the videodisc systems, the influence of the beat component $f_0 \pm (f_{a2} - f_{a1})$ should be considered.

Dynamic range of the record is limited by disc surface noise. An average signal-to-noise ratio in the audio channels is approximately 60 dB, including preemphasis/deemphasis improvement.

Experimentally, it is found that the influence of the dropouts is satisfactorily compensated if the audio signal is kept constant during the dropout period. This can be performed, for example, with the sample-and-hold circuit.

In some applications, high-quality audio is especially desired. This includes, for example, musical shows and classical music programs "decorated" with visual events. In the first case, both video and sound performance are boosted, while in the second case video is "relaxing" and audio is of primary importance. To improve sound quality, an analog or digital approach can be used when the analog audio channel noise reduction technique is performed.

8.4 NOISE REDUCTION TECHNIQUES

Characteristics of the noise introduced by the videodisc, among other things, determine the efficiency of the noise reduction techniques. Typically, the noise spectrum in the audio channel of the videodisc rises at high frequencies.

Audio channels in the Laservision (LV) systems can be at first approximation, considered as the narrowband FM audio channels. In the FM domain, FM carriers 2.3 MHz and 2.8 MHz for example, noise can be closely approximated by the white noise. After FM demodulation a triangular noise is the dominant noise. The bandpass filters are used to separate FM audio channels, and because the narrow bandwidths even otherwise small multiplicative noise or particular hardware design [4] can cause the noise in the audio channel to rise at both low and higher frequencies, similar to the ear's sensitivity. If the loudness contour is subtracted from the noise spectrum, the resultant perceived spectrum is almost flat with frequency. Thus a noise reduction system must operate over an entire spectrum uniformly. Suboptimal results can be obtained when the system operates over part of the used audio spectrum.

The dynamic range of the videodisc record audio is approximately 60 dB. With present techniques, wide-dynamic-range material will be reduced in range by adding hiss. The recorded dynamic range of music, using contemporary analog techniques, is approximately 80 dB, and using digital techniques is 85 to 90 dB. *Compansion* is a technique that squeezes a wide dynamic range to a range that can be accommodated in a medium and after transmission/playback restores the original range. This is a noise reduction system.

For already recorded discs, a noise reduction method is needed that can get rid of the noise that is already present in the record.

8.4.1 Dynamic Noise Reduction Systems

A dynamic noise reduction system is a noncomplementary noise reduction system which depends on two principles for its operation [5,6]. The first is that the level of noise we hear from our audio system depends directly on the system bandwidth and how wide a range of frequencies it is capable of reproducing. For the white noise, that is, noise whose amplitude is uniform over the frequency bandwidth, the total noise, and hence the signal-to-noise ratio (voltage), is directly proportional to the square root of the system bandwidth. For example, if the system bandwidth is reduced from 20 kHz to 800 Hz, the aggregate S/N ratio changes by

$$20 \log \sqrt{20/0.8} = 10 \log (20/0.8) = 13.98 \text{ dB} \tag{8.1}$$

The second principle is based on a psychoacoustic phenomenon known as *auditory masking*. Restricting the frequency response to eliminate the noise is not enough to produce a satisfactory noise reduction system. If the bandwidth is kept limited, the music would sound dull and lifeless. Our ability to hear noise is strongly dependent on the program material that is simultaneously present. With no other sound stimuli present, we can hear very faint noises—the proverbial pin dropping, for example. But as the program sound level starts to increase, it "masks" or hides any low-level noises. The threshold for hearing the noise is raised and now noises have to be much louder in order to be heard. The system bandwidth can be increased to allow all the program harmonics to be faithfully reproduced, but because of this masking effect, the increase in noise that accompanies an increase in bandwidth cannot be heard. The influence of high-frequency noise in the presence of the dominant low-frequency tone can be reduced by preemphasis during recording and complementary deemphasis during playback.

Experimental measurements of the ability of pure tones to mask noise show that extremely high sound pressure levels (SPLs) are required to raise the noise threshold level and provide masking. The most effective tone frequencies are between 700 Hz and 1 kHz near the natural resonance of the ear, and even then, SPLs higher than 75 dB are needed for masking noise at a 16-dB SPL. However, these data apply to pure tones—as soon as the tone acquires distortion, or frequency modulation, or transient qualities, the masking ability increases dramatically. Typical music and speech can be regarded as excellent masking sources. The broadband spectral components and high concentration of energy around 1 kHz for most musical instruments (Figure 8.4) improve the noise-masking ability by more than 30 dB compared to a pure 1 kHz tone. By comparing the frequency spectrum of musical instruments with the ear sensitivity curve (Figure 8.1), we can see that this high energy content is precisely where it needs to be to provide effective masking. Most of the ear's subjective response to noise is within the region from 1 to 6 kHz. Above 6 kHz the response of the ear decreases rapidly.

The general arrangement of the dynamic reduction system is shown in Figure 8.5. IC chips are made on the same principle [3], for example, LM1111. Two lowpass filters (for stereo) are placed in the audio path with −3-dB bandwidths controlled by the amplitude and frequency of the incoming signal. Each filter response is felt below the corner frequency (−3 dB) with a smooth single-pole roll-off above the corner frequency (−6 dB/octave slope) for any control setting. Cascading the two filters will give a −12-dB/octave slope. At the same time the minimum bandwidth curve frequency for each filter should be increased by a

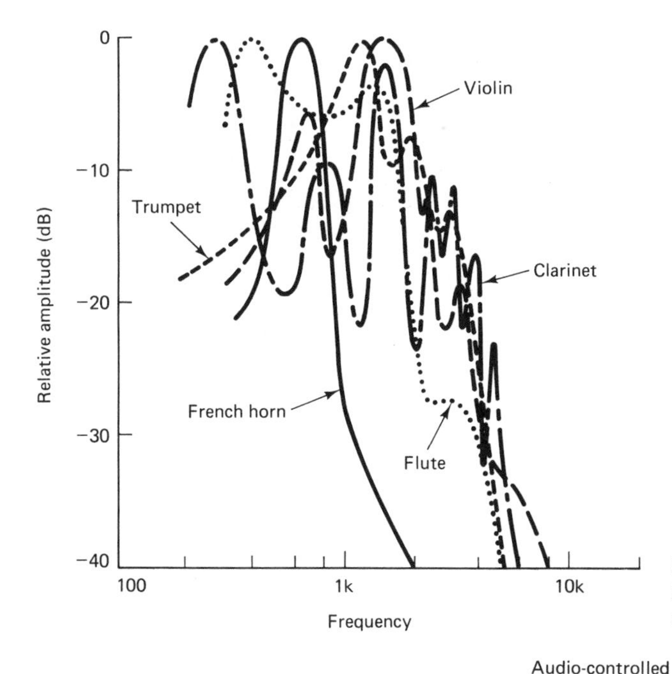

Figure 8.4 Typical spectra of musical instruments: (a) french horn; (b) flute; (c) trumpet; (d) violin; (e) clarinet.

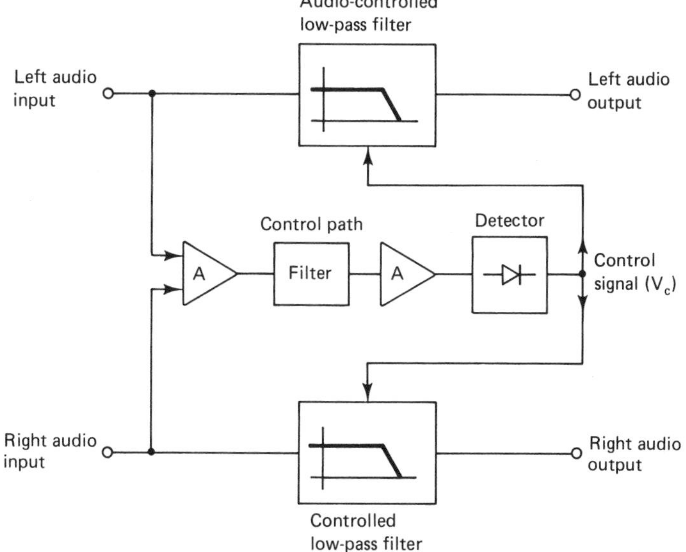

Figure 8.5 Stereo noise reduction system.

factor of 1.56, so that the resulting two-pole low-pass filter still has a minimum −3 dB bandwidth at approximately 800 Hz. Now the noise reduction can be as much as 18 dB versus 14 dB noise reduction with a single-pole filter [3].

One possible solution for the low-pass filter is shown in Figure 8.6a, and it consists of a variable transconductance (g_m) block driving an op-amp integrator.

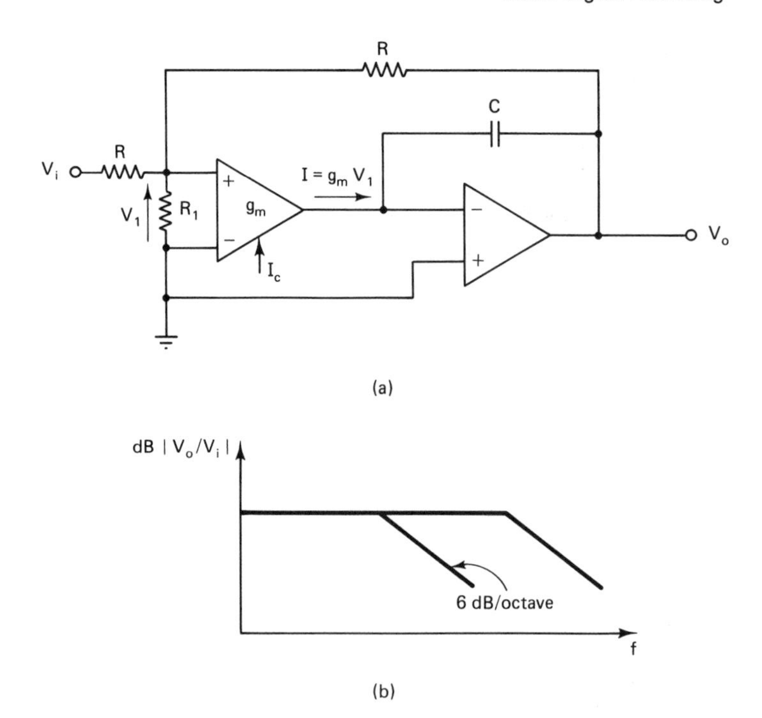

Figure 8.6 (a) Variable single-pole low-pass filter; (b) frequency response.

The transfer function is

$$V_o/V_i = A(s) = -1/(1 + sCk/g_m) \tag{8.2}$$

where $k = 2 + R/R_1$

$g_m = g_m(I_c) = cI_c,\ c = \text{const.}$

This expression follows from the obvious relations

$$V_o = -g_m V_1/sC \tag{8.3}$$

and

$$V_1/R_1 = (V_i - V_1)/R + (V_o - V_1)/R \tag{8.4}$$

The asymptotes of the transfer function are shown in Figure 8.6b. Since the filter's transconductance g_m is controlled by I_c, we have a variable single-pole low-pass configuration with unity gain below the corner frequency.

The filter in the control path is a high-pass filter with a -3 dB corner frequency at approximately 6 kHz and a -12 dB/octave roll-off slope (Figure 8.7). An optional notch at 19 kHz is included when the source material contains the FM pilot tone, which would otherwise increase the minimum bandwidth above 800 Hz when the detector threshold is set at the noise floor. Basically, the presence of high-frequency energy in the music (audio) source is used to control the audio bandwidth. When the presence of high frequencies are detected in the control path, this means that simultaneously large levels of energy are

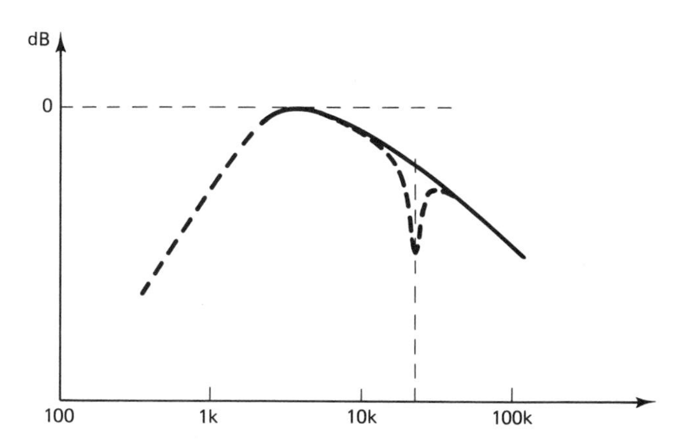

Figure 8.7 Frequency/amplitude response of the control path.

present in the critical masking frequency range. Therefore, the audio bandwidth can be safely increased as necessary to prevent audible impairment of the music while the noise remains completely masked.

At approximately 1.6 kHz, an additional slope of −20 dB/decade is employed to prevent low-pass filters from being activated by extremely large amplitudes at low frequencies, drums, for example. A control path should be compensated for the relatively fast decrease in spectral energy with an increase in frequency. If the detector does not follow the rising edge of the signal (music) change, the distortion will be caused by loss of the high-frequency components coupled by a too-slow increase of the low-pass filter bandwidth.

The rise time of music signals depends on the type of instrument used to generate the sound. For example, the English horn reaches 60% of its peak amplitude in 5 ms. For some other instruments, this time varies between 50 and 200 ms, and for the sound made by clapping the hands, the rise time is 0.5 ms. To minimize the HF rise-time losses, the typical rise time used is 0.5 ms.

The rise time represents only half of the problem relating to fast changes of the signal. It is also important to know the response of the detector to the drop in energy in the HF portion of the signal spectrum. If the bandwidth changes are slow, a cracking noise will be heard. On the other hand, if the bandwidth changes are too fast, the higher harmonics which last longer will be attenuated.

It was demonstrated by an experiment that the exponential change in the bandwidth is the most desirable one for the preservation of the harmonic amplitudes. The human ear does not usually register a weak noise that follows the strong signals within the first 150-ms period. The typical value is the decay to 10% of a stationary signal value in about 50 ms. Often, a capacitor used for the filtering purpose in the detector circuitry is sufficient.

8.4.2 Noise Reduction Systems with Signal Preprocessing

The dynamic range of the audio recorded on the videodisc is limited by disc surface noise (Figure 8.8). For reference, the dynamic range of a digital master

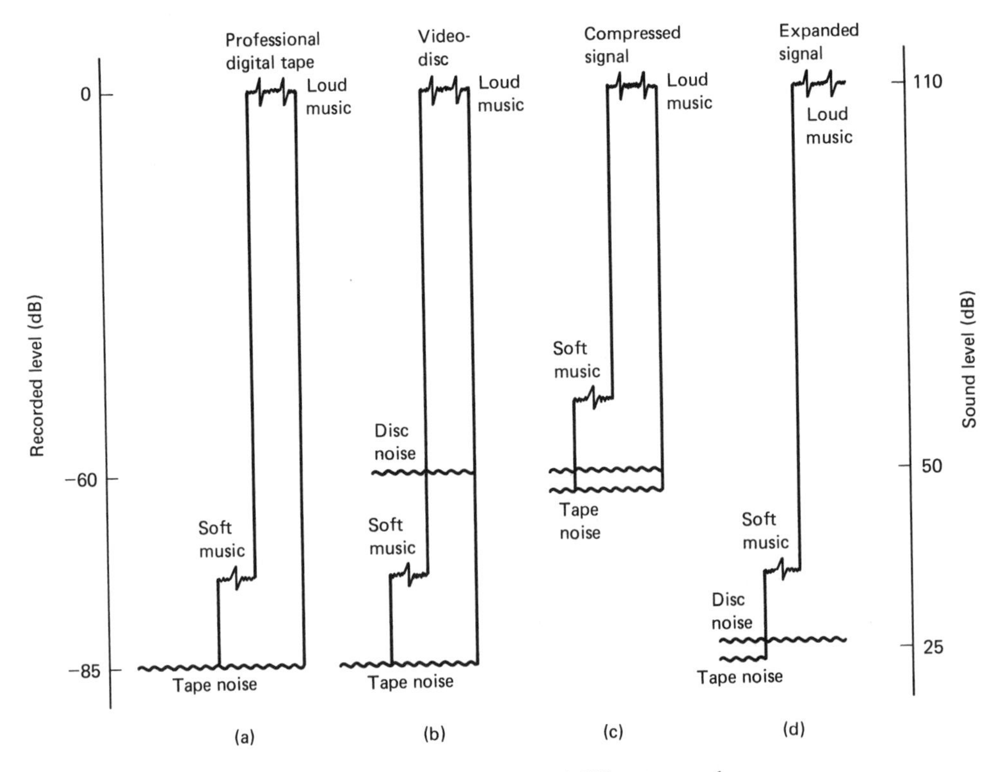

Figure 8.8 Dynamic range of different records.

tape is shown. The average level of the loud music, the average level of the soft music, and the tape noise floor are marked. The range from loud music to tape noise is 85 dB. If the tape is well mastered, the soft music passages are recorded sufficiently above the tape noise. Thus when soft passages are replayed, the music is heard but not the noise. If the loudest average music is adjusted to be reproduced at a sound level of 110 dB in the home, the tape noise floor is only 25 dB. In nearly all instances, the tape noise is below the average background noise of the listening room and becomes inaudible. The digital recorder thus allows playback of music at levels achieved in live performances while eliminating tape noise from the listening process [7].

When the same music is transferred to a videodisc, with a smaller dynamic range, the music is no longer heard in a noise-free manner. In Figure 8.8b, the dynamic range of the videodisc is indicated as 60 dB from the loudest average music to the disc surface noise. If the music on the digital tape is transferred directly to the disc, the soft music passages lie at or below the level of the record surface noise. Upon playback this music is contaminated with unwanted sounds of the vinyl surface. On a sound-level basis, when the loud music is at 110 dB, the disc noise is now 50 dB and can generally be heard in a listening-room environment. To avoid this noise, the level of the soft program material relative to the loud music may be increased. This would be an arbitrary form of compression which reduces the dynamics of the music on the original tape. While the result may still be musically acceptable, the original dynamics cannot be accurately recaptured.

The peak signal-to-noise ratio of the digital tapes is approximately 95 dB per channel. For a two-channel record obtained by the mixing of 24 recorded signals is

$$95 - 10 \log 24 = 81 \text{ dB} \tag{8.5}$$

This is a little over 20 dB higher than the signal-to-noise ratio for the videodisc audio. The videodisc signal-to-noise ratio can be improved, as shown in Figure 8.8c. The softest music passages are increased, so it is placed on the disc comfortably above the record surface noise. Although the dynamics are reduced from the original tape, the process is directly reversible. Such a disc could not claim compatibility, and in fact would be objectionably noisy to most listeners. Using a reversible process in a playback decoder the full dynamic of the original digital source tape can be recovered (Figure 8.8d). The loud and soft music and tape noise can be placed at precisely the same levels as on the original master tape.

In this way, two processes are combined:

1. Compression during recording
2. Expansion during playback

The terms *compansion* and *companding* are commonly used to define both processes.

The basic requirements and goals for the compensation systems are:

1. Increase S/N for about 10 to 25 dB
2. Good transient response
3. Minimal pumping of noise or program
4. Independence of signal level
5. No alteration of frequency response

When an encoded disc is played back without a decoder, it can be compatible or not. Compatibility here is not identical; that is, the encoded disc has modified the program dynamics but in a manner that is not obvious with most audio material. Thus compatible (coded) discs can be played on both players, with and without expanders/decoders. If played on a regular player, it should give approximately the same quality as regular discs. When played on expander/decoder, coded discs should have a corresponding increase in the S/N and thus high-quality reproduction. For all compensation systems, processing should be based on psychoacoustic considerations.

The principal block diagram of the compander noise reduction system is shown in Figure 8.9 [8]. A variable-gain circuit, in general, has a nonlinear

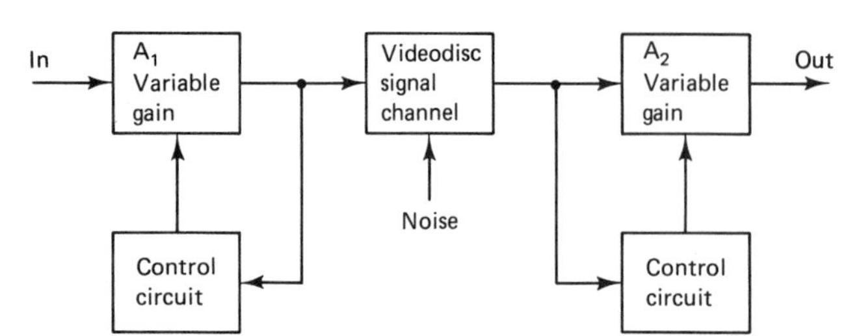

Figure 8.9 Compander noise reduction system.

input/output transfer function. Two variable-gain circuits in Figure 8.9 should give in overall linear transfer function:

$$A_1 \times A_2 = \text{const.} \tag{8.6}$$

where $A_i = A(L, f)$, $i = 1, 2$, and L is the average signal level.

There are a number of the compander noise reduction systems, for example, Dolby [9], DBX [10], and CX [11] (for compatible expansion). Dolby encoded programs when played nondecoded are overly bright. DBX without decoding is overly bright and contains much dynamic error, causing pumping and low-frequency modulation of HF tones. CX is almost a compatible system.

The principal block diagrams of the compressor and expander are shown in Figure 8.10, while compression and expansion curves for the CX system [11] for a single-channel 1-kHz sine wave are shown in Figure 8.11. The compressor and expander are the inverse of one another. This is assured by placing the same control circuit in the feedback path for the compressor and in the feed-forward path for the expander.

The 2:1 rate for high-level signals is rapidly converted to 1:1 for low-level signals. Because of the "knee" (−40 dB) in the compression and expansion curves, they should ideally be matched, but a mismatch of up to 6 dB is almost unnoticeable. The reduction of noise with a companding system is a dynamic function; when the music is soft, maximum compression and reexpansion occur to provide maximum noise reduction. With loud music, very little compression and expansion are required since the noise is naturally masked by the music. As this implies, the noise floor is continuously moving depending on the level of the music program material. A good companding system must handle this motion smoothly so that noise modulation is not audible.

Filtering in the control path is very critical. A very long time constant in the control path will allow only very slow transitions of the noise as a function of time, which would not be noticeable to listeners. The problem with such a system is that it will also respond slowly to rapid change in musical program dynamics, rendering it unable to follow the typical attacks and decays of the music. A fast attack and relatively fast decay will result in a quick response to music dynamics. But this system will move the noise floor about in a rapid manner and produce noticeable noise modulation, which is heard as a "swishing" sound. Thus a circuit is needed that can alternate between fast and slow operations in response to a change in music dynamics and also ignore small changes that

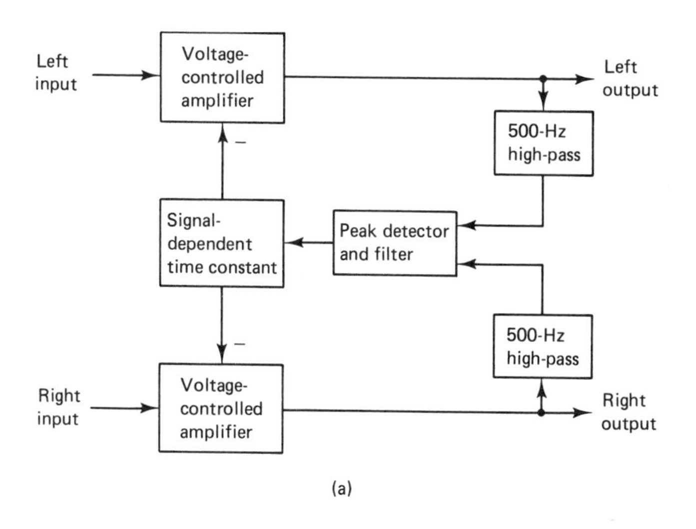

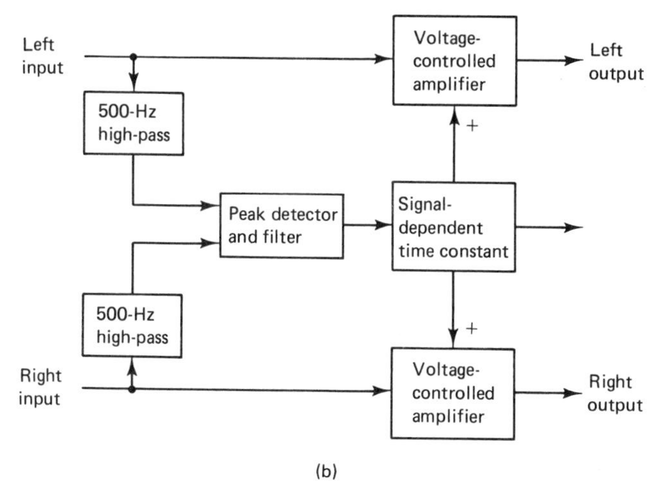

Figure 8.10 Compander block diagrams: (a) CX compressor; (b) CX expander.

could induce modulation distortion. The modulation distortion of the audio program material is produced when a continuously fast-acting circuit tracks small changes in the musical program rather than reaching an acceptable average level.

As an example, consider the CX control path [7,11]. A filter circuit with fast-acting 1-ms attack and 10-ms decay is used. If used alone, performance would be unacceptable due to pumping and distortion. Because of this, four separate filters are subsequently summed:

Filter 1: high pass with dead band and positive peak rectifier, $T_1 = 30$ ms
Filter 2: low pass, $T_2 = 2$ sec
Filter 3: dead band and positive peak rectifier, $T_3 = 30$ ms
Filter 4: dead band and negative peak rectifier, $T_4 = 200$ ms

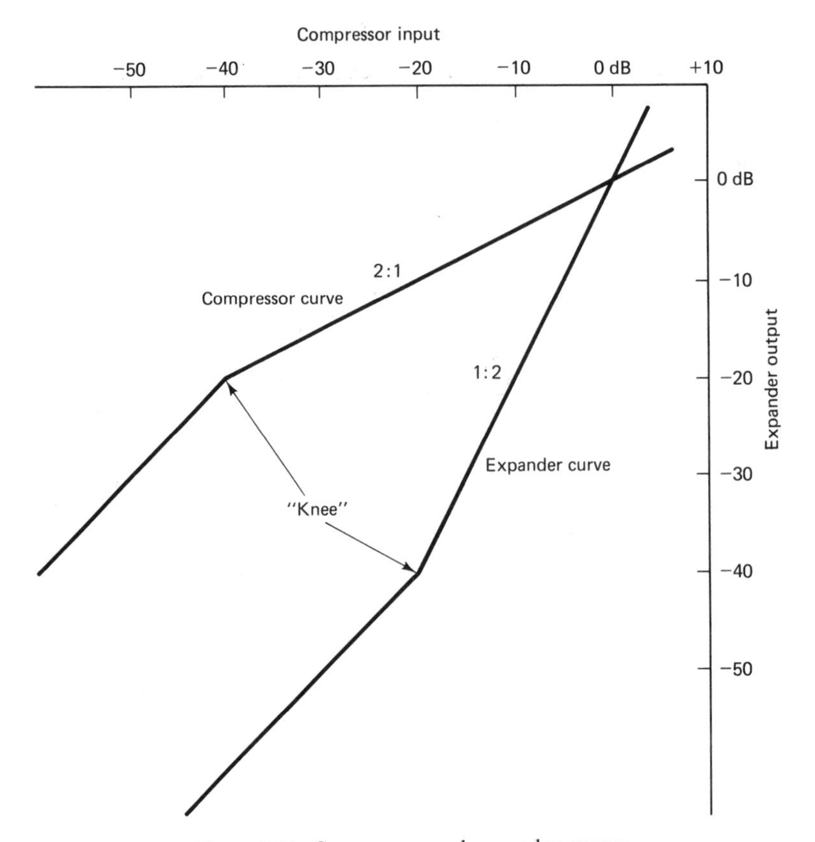

Figure 8.11 Compressor and expander curves.

The companding system reduces background videodisc noise but does not, in general, reduce burst or pulse-type noise: for example, dropout-like noise or deep scratches on the (capacitive) disc surface.

8.5 AUDIO SIGNAL DIGITALIZATION

In brief, the advantages of coding a signal digitally are: digital representation offers ruggedness; efficient signal regeneration; easy encryption; the possibility of combining transmission and switching functions; and the advantage of uniform format for different types of signals [12,13]. The price paid for these benefits is a need for increased bandwidth.

The main applications of digital audio in videodisc systems are:

1. Incorporate reliable storage of (20 s, for example) sound-over-still frame using compressed audio techniques
2. Compact disc (CD) system with high-performance audio playback

Many dramatic results in speech compression are dependent on the utilization of fundamental features such as excitation descriptions, vocal-tract resonances, or articulatory parameters [1]. Another approach is a straightforward encoding and reconstruction of the acoustic-time waveform by means of the closely related discrete-time, discrete-amplitude representations known as *pulse-code modulation* (PCM), *differential pulse-code modulation* (DPCM), *and delta modulation* (DM). A straightforward waveform coding approach is attractive, at least from the point of view of coder complexity, and is the most generally applicable technique to signal coding. Besides the foregoing quantifying techniques, adaptive versions (APCM, ADPCM, and ADM) have been developed recently.

8.5.1 Pulse-Code Modulation

In 1938, the concept of PCM was defined and its advantages recognized [14]. Waveform coding by PCM [15–17] involves the following steps:

1. *Time discretization (sampling)*. The bandlimited waveform is sampled at a rate of at least $2f_m$ hertz, the Nyquist frequency, where f_m is the highest frequency contained in the spectrum (waveform). This is also called the *sampling theorem*. The minimum sampling rate can be less than $2f_m$ if the lowest signal frequency is nonzero [16].
2. *Amplitude discretization (quantizing)*. Quantizing is the process of transforming a continuous (in the amplitude domain) slope signal into a set of discrete output states: the amplitude of each signal sample is quantized into one of 2^n levels, not necessarily equally spaced. This implies an information rate of n bits per sample and an overall information rate of $2f_m n$ bits per second for a low-pass filtered signal.
3. *Coding*. Coding is the process of assigning a digital code word to each of the output (quantized) states.

In the decoder, the binary words are mapped back into amplitude levels, and the amplitude-time pulse sequence is low-pass filtered with a filter whose cutoff frequency is also f_m.

The basic processes in the PCM are illustrated in Figure 8.12. A continuous bandlimited signal $s(t)$, in this example positive only, is sampled with sampling frequency f_0 (Figure 8.12a):

$$f_0 = 1/T_0 \geq 2f_m \tag{8.7}$$

The quantizer step size denoted by Δ:8 levels are marked in Figure 8.12a, that is, a 3-bit quantization is performed. A corresponding binary signal is shown in Figure 8.12b and quantizing error in Figure 8.12c.

If the number of quantizing levels is large and the signal does not overload the quantizer, the quantizing error has the uniform distribution

$$p(e) = 1/\Delta, \qquad -\Delta/2 < e < \Delta/2 \tag{8.8}$$

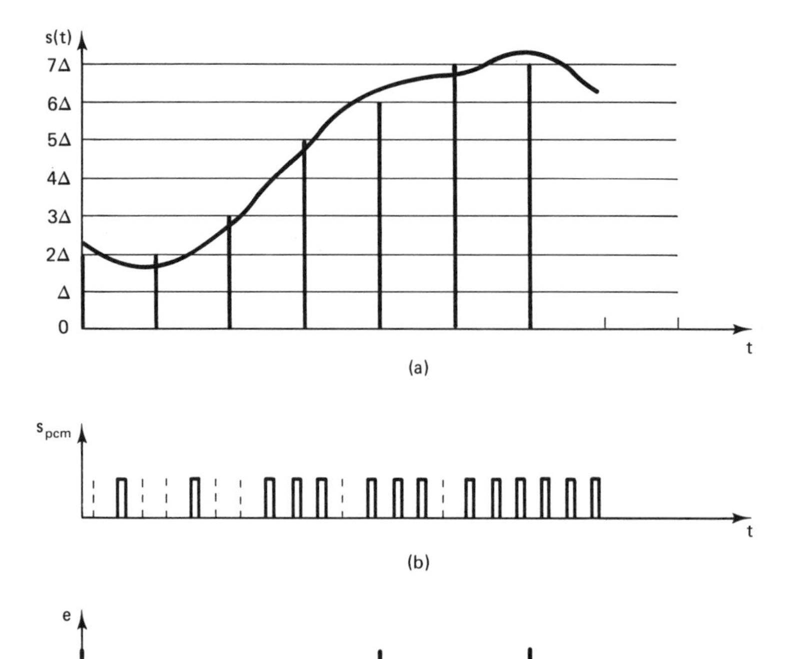

Figure 8.12 PCM: (a) sampling and quantization; (b) coded signal; (c) quantizing error.

The mean-square value of the quantization error is

$$\int_{-\Delta/2}^{\Delta/2} e^2 p(e)\, de = \Delta^2/12 \tag{8.9}$$

If the rms value of the input signal is S_{rms}, the signal-to-noise (quantizing error) ratio is

$$\mathrm{SNR} = S_{\mathrm{rms}}^2/(\Delta^2/12) \tag{8.10}$$

For the input signal $s(t)$ whose probability density function (PDF) $p(s)$ is modeled by a zero-mean Gaussian function, if the quantization is fine enough (say, $n > 6$) to prevent signal-correlated patterns in the error waveforms, the SNR is:

$$\mathrm{SNR} = 6n + \mathrm{const.} \tag{8.11}$$

The sampling theorem states that if a continuous, bandwidth-limited signal contains no frequency components higher than f_m, the original signal can be recovered without distortion if it is sampled at a rate of at least $2f_m$ samples per second. This is illustrated in Figure 8.13. Sampling is basically a convolution process (in the frequency domain); a sampled signal spectrum (Figure 8.13b) is obtained as the convolution of the continuous signal spectrum (Figure 8.13b

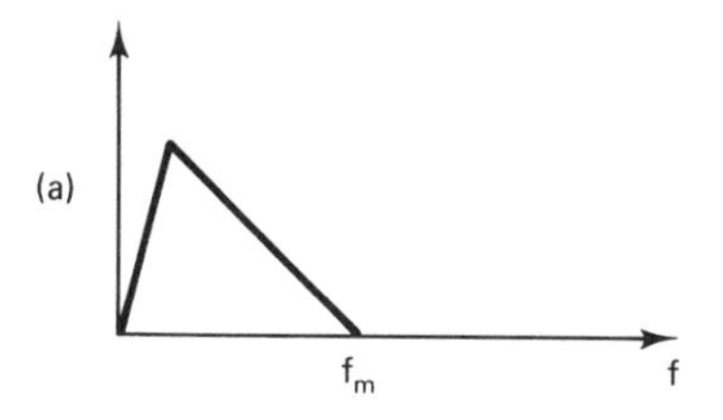

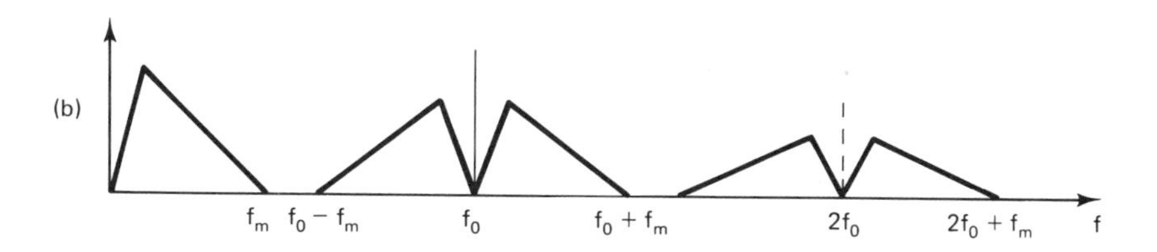

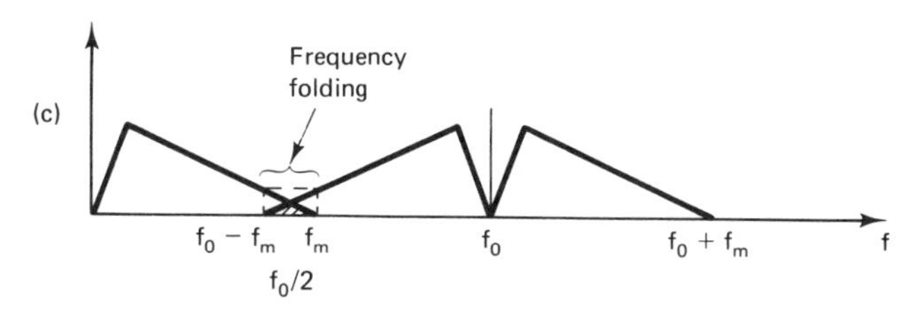

Figure 8.13 Sampling theorem illustration: (a) continuous signal spectrum; (b) sampled signal spectrum, $f_0 \geqslant 2f_m$; (c) aliasing error, $f_0 < 2f_m$.

shows the spectrum for positive frequencies) and a sampling pulse spectrum (not shown in Figure 8.13). When the sampling rate is less than $2f_m$, in the original signal bandwidth, 0 to f_m, folded frequency components are obtained and the original signal cannot be reconstructed undistorted by a lowpass filter (Figure 8.13c). This distortion is referred to as *aliasing noise,* or as in optics as *Moiré patterns*. The effect of an inadequate sampling rate on a sinusoid is illustrated in Figure 8.14; an alias frequency in the recovered signal results. The alias

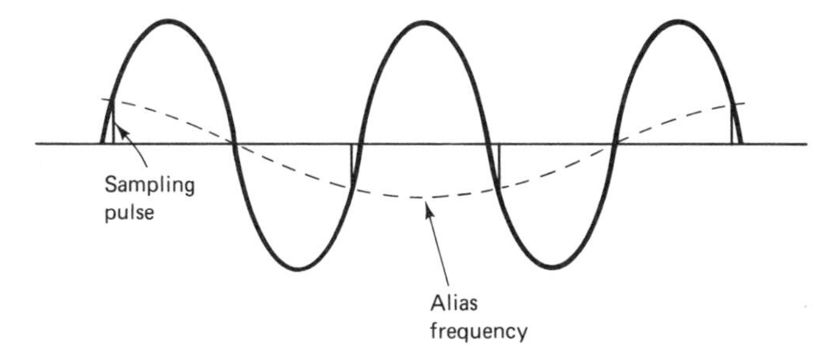

Figure 8.14 Alias frequency caused by inadequate sampling rate.

frequency can be significantly different from the original frequency. From Figure 8.14 it can be seen that if the sinusoid is sampled at least twice per cycle, as required by the sampling theorem, the original frequency is preserved.

It is found that a white-noise-like error sequence from the quantizer is less objectionable, perceptually, than a signal-dependent error waveform. When demands on speech quality are not severe, crude quantizers ($n < 6$) are of interest, but then in the waveform of a quantization error, undesirable signal-dependent patterns are present. In this situation a pseudorandom noise sequence can be added to the speech that is to be quantized; subsequent subtraction of the pseudorandom sequence from the quantizer output provides a white-noise-like error sequence and also preserves the original value of SNR. The technique is most useful in the narrow but practically significant range $4 \leq n \leq 6$ [18].

In general, the nonstationarity of the speech signal leads to quantizer-input mismatches, especially when several speakers or circuits are to be handled by a single encoder/decoder system. Mismatches between the quantizer range and signal power entail obvious deteriorations of quantizer performance, reflected by values of SNR that are lower by definition than what is predicted by equation (8.11). Two, not mutually exclusive, solutions to the mismatch problem are used [12]:

1. Nonuniform quantization [17,19,20]
2. Adaptive quantization

The advantage of nonuniform quantization increases with the ratio of peak to rms value (crest factor) of the signal. Adaptive strategies combined with differential encoding have proved to be extremely fruitful from an application viewpoint [21,22].

8.5.2 Differential Pulse-Code Modulation

Many classes of information signals, including speech, which are sampled at the Nyquist rate exhibit a very significant correlation between successive samples (S_i and S_{i-1}). One consequence of this correlation is that the variance of the first difference

$$e = S_i - S_{i-1} \tag{8.12}$$

is smaller than the variance of the signal itself; hence fewer bits will be required to describe the error samples than the input samples. This was the motivation for the DPCM encoding [23]. Basically the same method of data compression is linear prediction [24]. The idea is to form an estimate of the input sample signal and quantize the difference between the signal and its estimate, or error signal. Estimators typically include linear prediction networks, both adaptive and nonadaptive and single or multiple integrators. DPCM and DM can be considered as special cases of predictive quantizing, the latter using merely a 1-bit quantizer for the error signal.

Prediction of the signal requires knowledge of input signal statistics. In nonadaptive prediction, data are built into a fixed feedback network. In adaptive prediction, the network is changed as the input signal changes its characteristics.

A predictive quantizing system is shown in Figure 8.15, where marked signals are:

S_i: input signal samples
$\hat{S}_i$: estimate of the output sample
e_i: error signal
$\tilde{e}_i$: quantized error signal
$\tilde{S}_i = e_i + \hat{S}_i$: locally reconstructed signal
$\tilde{e}_i'$: corrupted version of the error signal, after e_i is coded into a prescribed digital format, transmitted, and decoded
s_i: reconstructed signal

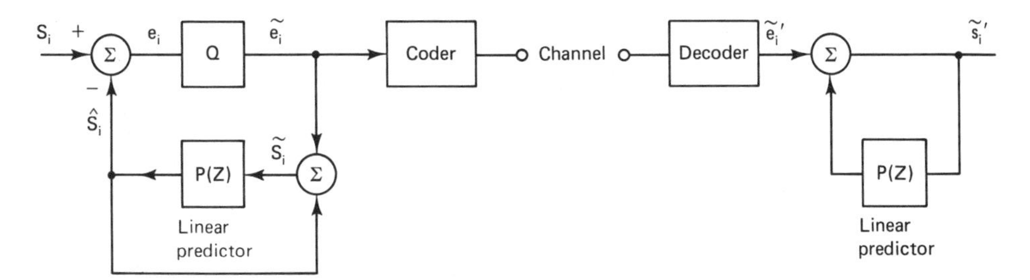

Figure 8.15 Predictive quantizing system.

It can be shown that the quantization mask in the transmitted error signal is identical to the quantization noise in the reconstructed signal. Normally, quantization noise samples are

$$\begin{aligned} q_i &= e_i - \tilde{e}_i \\ &= S_i - \hat{S}_i - \tilde{e}_i \\ &= S_i - \tilde{S}_i \end{aligned} \tag{8.13}$$

This means that quantization noise does not accumulate in the reconstructed signal. This should not be misconstrued with the channel noise, which does accumulate because of the feedback in the receiver.

For an Nth-order predictor, the signal estimate (prediction) is formed from a linear combination of past values of the reconstructed input signal:

$$\hat{S}_i = \sum_{j=1}^{N} a_j \tilde{S}_{i-j} \tag{8.14}$$

The variance of the error signal is

$$\sigma_e^2 = E[e_i^2] = E[(S_i - \hat{S}_i)^2] \tag{8.15}$$

Differentiation of e^2 with respect to a_j and setting the resulting equations to zero gives

$$\begin{aligned}
\rho_1 &= (1 + 1/R)a_1 + a_2\rho_2 + a_3\rho_3 + \cdots + a_N\rho_{N-1} \\
\rho_2 &= a_1\rho_1 + (1 + 1/R)a_2 + a_3\rho_1 + \cdots + a_N\rho_{N-2} \\
&\vdots \\
\rho_N &= a_1\rho_{N-1} + a_2\rho_{N-2} + a_3\rho_{N-3} + \cdots + (1 + 1/R)a_3
\end{aligned} \tag{8.16}$$

when $R = \sigma_i^2/E[q^2]$ is the signal-to-quantizing noise ratio, $\sigma^2 = E[S_i^2]$ is the signal variance, and $\rho_j = E[S_iS_{i-j}]/\sigma^2$ is the signal autocovariance. The minimum of σ_e^2 can be written as [25]

$$\sigma_e^2 = \sigma^2\left[1 - \sum_{j=1}^{N} a_j\,(\rho_j/(1 + 1/R)\right] \tag{8.17}$$

Adaptive predictive coding has been used to reduce signal redundancy in two stages: first by a predictor that removes the quasi-periodic nature of the signal, and second by the predictor that removes formant information from the spectral envelope [26]. Predictive quantizing can be implemented with adaptive quantization as well as with adaptive prediction.

8.5.3 Delta Modulation

The basic principle of DM was described for the first time in a French patent [27] issued in 1946. More detailed descriptions by de Jager and Libois [28,29] of many aspects of delta modulation appeared in 1952.

Although an inexpensive digital encoding system, the DM system remains experimental because [13]:

1. Many delta modulation systems suffered from idle noise caused by an asymmetry of the two current sources normally used in delta coders. If these current sources are not perfectly matched, an audible idling pattern appears in pauses between words of conversation that, when demodulated, produces an audible sound that is very disturbing.
2. The dynamic range of delta coders operating at bit rates comparable to those used in PCM systems was inadequate. Voice signals of low amplitude were quantized very coarsely.

Therefore, despite the attractive simplicity of these delta codes, the drawbacks had prevented much use. The situation began to change when more refinements were suggested. In 1963, the first method for improving the dynamic range was proposed [30]. Later, more propositions followed [31–35].

A block diagram of the DM system is shown in Figure 8.16. Note its basic similarity with the DPCM network in Figure 8.15. Important differences are the use of a two-level (1-bit) quantizer in DM and the replacement of a general predictor network with a simple integrator (which can be regarded as a first-

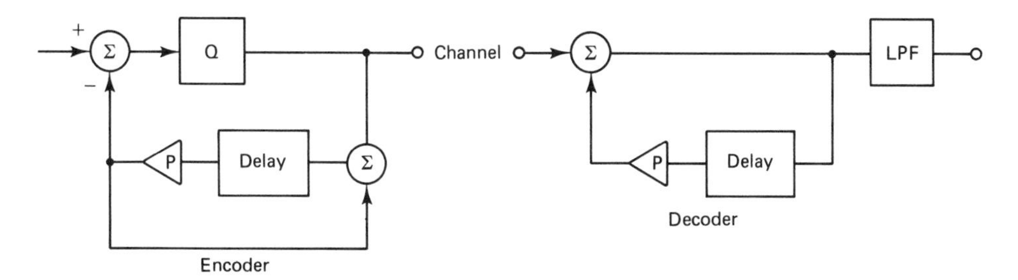

Figure 8.16 Delta modulation system.

order predictor). The difference between the incoming signal and its predicted value is quantized to 1 bit for transmission. The feedback loop in the encoder may be described by the transfer function

$$H_1(Z) = \beta Z/(1 - \beta Z) = \beta Z + \beta^2 Z + B^3 Z + \cdots \tag{8.18}$$

when $Z = \exp(-sT)$ represents the unit delay operator and T is the sampling interval. Therefore, the output of the feedback loop is the weighted sum of all past inputs by T.

The transfer function of the decoder loop is

$$H_2 = 1/(1 - \beta Z) = \beta Z + \beta^2 Z^2 + \beta^3 Z^3 + \cdots \tag{8.19}$$

Thus the decoder output is the sum of all past points. The signal can be low-pass filtered or integrated to estimate the input signal. The low-pass filter in a DM circuit rejects the out-of-band quantization noise in the high-frequency staircase output function, and this has no parallel in DPCM or PCM with the Nyquist sampling [12].

The terms *nonadaptive* and *linear* are used synonymously for the coder, which approximates an input time function by a series of linear segments of constant slope, in anticipation of delta modulators, where the slope of the approximating function is variable or "adaptive."

The principle of nonadaptive DM is illustrated in Figure 8.17. The local estimate provided by the integrator, $\hat{s}(t)$, is the staircase function. The step size, Δ, typically chosen, is small compared to the input signal magnitude. The band-limited input signal is sampled at a rate $f_0 = 1/T$, which is much higher than the Nyquist frequency. The basic DM principle can be formalized in these equations:

$$b_i = \operatorname{sgn}(s_i - s_{i-1}) \tag{8.20}$$

$$\hat{s}_i = \hat{s}_{i-1} + \Delta b_i \tag{8.21}$$

Two types of distortion can occur in the DM: granular distortion (noise) and slope overload. Granularity is determined by the step size and it refers to a situation where the staircase function $s(t)$ bends around a relatively flat segment of the input function, with a step size that is too large relative to the local slope characteristics of the input. Slope overload is caused by the ability of the encoder to follow the signal when its slope magnitude exceeds the ratio of step size to

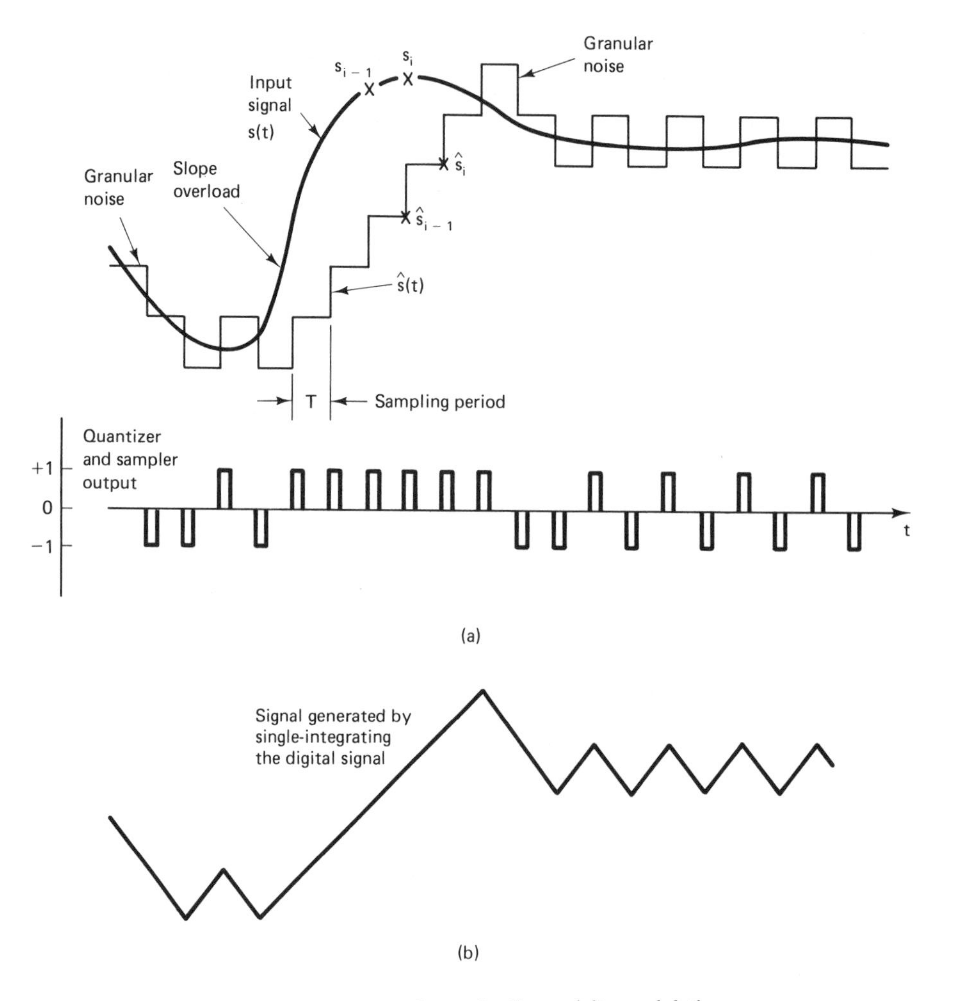

Figure 8.17 Waveforms for linear delta modulation.

sampling period:

$$|ds/dt| > \Delta/T \tag{8.22}$$

Perceptually, more overload noise power is tolerable than granular noise power. During granular distortion, the samples of the error input tend to be uncorrelated and the error signal power spectrum tends to be uniform.

For given statistics of input and input signal slope, granular distortion can be made smaller by using a small step size, and slope overload can be reduced by using a large step size by running the sampler (clock) faster. The optimum step size, Δ_{opt}, which will maximize SNR is given by [31]

$$\Delta_{opt} = \langle(s_i - s_{i-1})^2\rangle^{1/2} \ln(f_0/f_m) \tag{8.23}$$

where $<\alpha>$ is the mean value of α. For useful DM, the oversampling index $F = f_0/2f_m$ is generally much greater than 1.

The signal generated by a single integrator is also shown in Figure 8.17. If positive and negative slopes are not equal (not shown in Figure 8.17), idle noise can be generated. This can be eliminated by matching in magnitude the positive and negative currents in the DM coder [36].

DM performances, for example dynamic range and SNR, can be significantly improved using adaptive methods. Both instantaneous and so-called syllable adaptive techniques for compounding or adaptation are used. This is well documented in literature [12, 32, 37–39]. A DM system for voice and data signals is also presented in references [40, 41].

8.5.4 Comparison of Techniques

Although only three basic coding techniques, PCM, DPCM, and DM, are used for audio signal coding, the number of modulations is much larger [42]. Only a few comparisons will be presented here and may be used as guidelines of performance.

The signal-to-quantization error ratio SNR is frequently used as a measure of quantizer performance. In Figure 8.18, an SNR is related as a function of bit rate for three coders [12]: logarithmic PCM, ADPCM (with adaptive quantizer and a simple first-order predictor), and ADM (with a 1-bit memory). The results are from computer simulations for two bandwidths: 200 to 3200 Hz and 200 to 2400 Hz. For PCM and ADPCM, the respective sampling frequencies f_0 were 8 kHz and 6.6 kHz. The product nf_0 determined the bit rate. For ADM, f_0 was variable and numerically equal to the bit rate.

It can be seen that ADPCM always has an SNR gain over PCM: between 8 and 12 dB over PCM, and variable gain over ADM. The SNR comparison between ADM and PCM is dependent on bit rate. This is because SNR increases as the cube of the bit rate for ADM, while the increase is exponential for PCM.

SNR is an inadequate performance measure for speech or video coding. Signal-dependent (or signal-correlated) noise or distortion does not have the same annoyance value as independent additive noise of equal variance. For example, the perceptual quality of ADM is largely controlled by granular errors, although slope overload distortion constitutes the major component of total error power. If $n \leq 6$, the quantizing error in PCM waveform has noticeable signal-dependent components which are perceptually annoying and redistributing a given error variance into signal independent patterns by dithering is helpful. Signal-dependent errors in PCM become significant at higher bit rates as well, if nonuniform quantization is employed. It has been noted [21] that the quality of speech output in an ADPCM coder, relative to log-PCM quality, is better than what is predicted by the SNR curves in Figure 8.18. The perceptual effect involved is that ADPCM errors have a greater proportion of low-frequency (slope-overload) distortion, and therefore are correlated more with speech than with PCM noise: the error

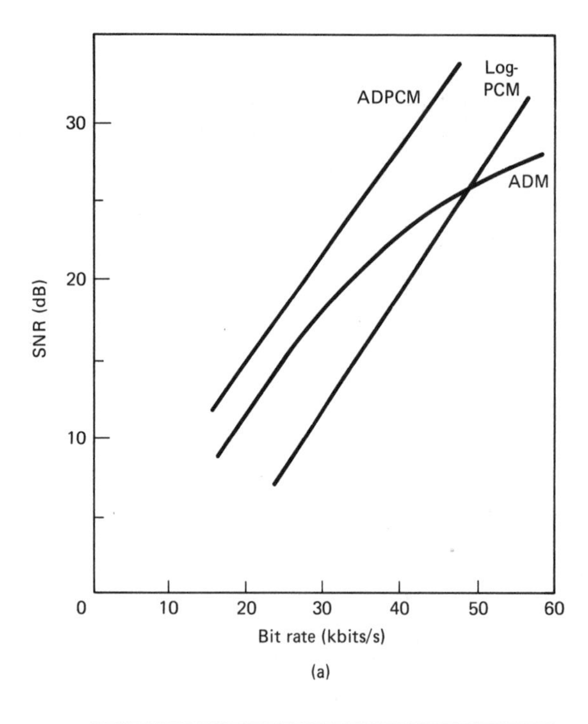

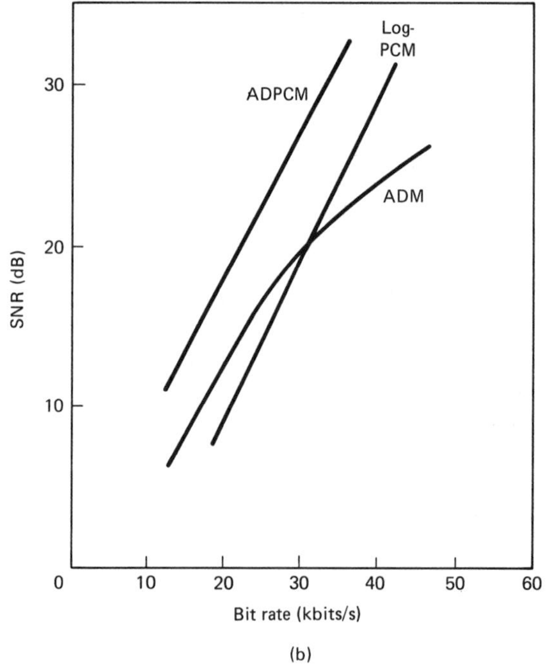

Figure 8.18 SNR as a function of bit rate: (a) bandwidth 200 to 3200 Hz; (b) bandwidth 200 to 2400 Hz.

spectrum in log-PCM is relatively white, while in ADPCM, it drops at the high-frequency end (it is matched to the long-term spectrum of speech itself). In ADPCM noise there is a certain pitch-related business. The suppression of quantizing noise during the silent intervals of speech is better in ADPCM than in log-PCM.

Different levels of perceptual distortion can be used as indications. Among those are distortion that is just perceptible, distortion that is annoying, and distortion that reduces intelligibility.

Channel error is an important parameter of the videodisc system. The feedback action in DPCM and DM tends to be a certain propagation of channel errors in the decoder output [43]. But from a perceptual viewpoint, DPCM is more tolerant than PCM to randomly occurring bit errors. This is because the greater magnitude of a typical PCM error spike makes it more annoying, in spite of the fact that it does not propagate in time. The high sampling rate of DM contributes to a perceptual advantage when compared to equal bit rate DM and PCM with a given percentage of digital errors. The error propagation problem in DM is alleviated in large measure by the use of an imperfect interpretor. Within the class of DM coders, instantaneously adapting delta modulators are more vulnerable to channel noise than are slowly compounding or linear coders. A number of syllabical compounded delta modulators have been designed for use over noisy channels [12]. The effect of dropouts on digital speech codes is currently being investigated.

Digital techniques can be used for audio noise cancellation. For example, adaptive predictive decomposition can be achieved utilizing an adaptive transversal predictor to estimate and cancel the correlated signal components [44].

8.6 DIGITAL AUDIO IN ANALOG VIDEO CHANNEL

Special features of the videodisc system, especially freeze-frame, can be combined with compressed audio to give even more "special features." For example, some reasonable length of audio, say between 20 s and a few minutes, can be stored in the video channel and then reproduced while the still frame is displayed. Instead of one frame, a few frames can be repeated. Recorded audio can be read from the disc once, at a higher frequency, placed in local memory (RAM), and then read from the RAM at the proper rate. Or it can be read in smaller pieces if the local memory capacity is limited but reproduced continuously. On the disc, (digital) audio can occupy one or more video frames.

There is more than one way to record digital audio on the videodisc—mixed with the video signal. It can be performed on a time- or frequency-shared basis. The time-sharing principle will be discussed here because frequency sharing has already been discussed in connection with standard TV signal recording.

A straightforward method would be to record direct digitally, for example, using a frequency modulation (FM) code (Figure 8.19a). Instant frequency change (idealized case) is shown in Figure 8.19b. For example, if "1" is represented

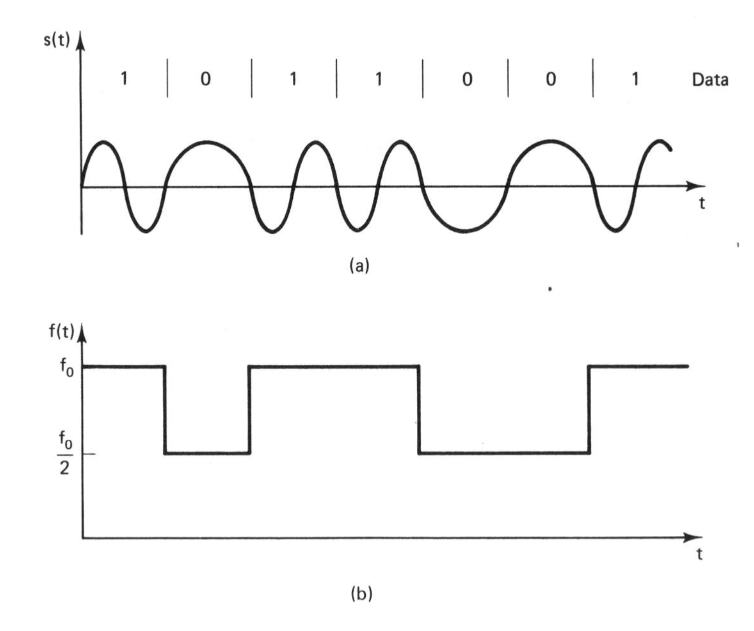

Figure 8.19 FM recording: (a) FM signal; (b) frequency change.

with one period of frequency, $f_0 = 8$ MHz. The main problem with this approach is that the player has to be modified to compensate for the dropout compensation influence (see Figure 8.20). When a dropout pulse is detected, incoming decoding data are squelched. For example, if $f_w = 9.3$ MHz, $f_b = 8.1$ MHz, and $f_s = 7.6$ MHz (Figure 8.20a), $f_d = 7.2$ MHz. Thus one of the binary states, 0 in the example in Figure 8.19a, will be considered a "dropout."

With a lower signal-to-noise ratio, binary data can be stored in the video, as shown in Figure 8.20b (this is an idealized case; the pulses are square although the bandwidth in the system is limited). Similarly, digital data can be placed in the video channel using multilevel (block) coding [45,46]. A simple idealized case with four levels is shown in Figure 8.21.

More discussion on digital data recording on the videodisc is given in Chapter 9. But in short, using multilevel coding as illustrated in Figure 8.21, a higher bit rate can be obtained; discs can have both digital and picture recorded on the same disc; mastering is on the same mastering machine, and the color burst can be used to synchronize a decoder.

Digital (audio) signal can also be recorded:

On the bottom of the sync pulses
In the undisplayed portions of each line
In undisplayed lines in each frame

The last two methods are based on TV standards, not on how the master is displayed on the screen.

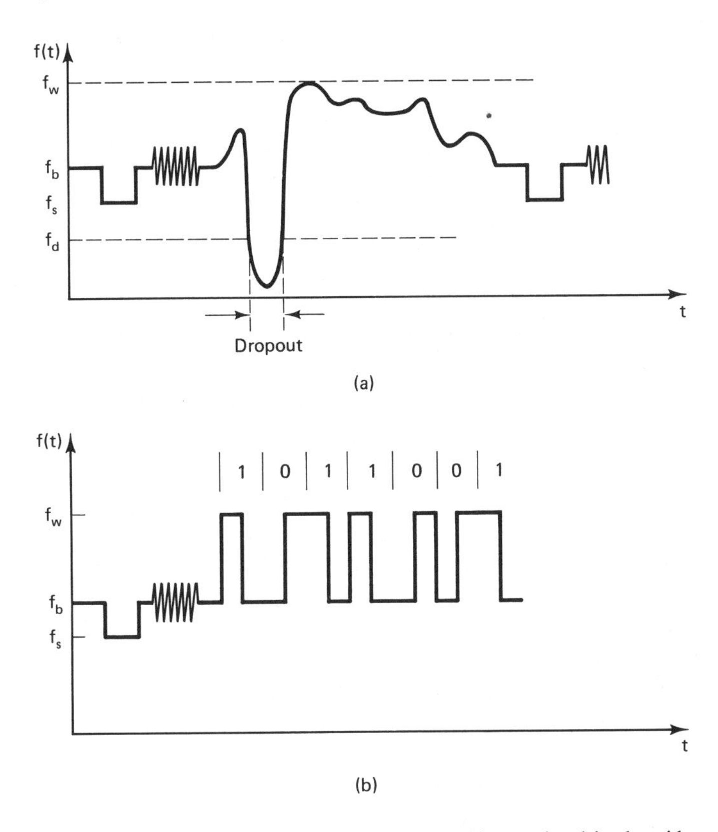

Figure 8.20 (a) Dropout in the video FM; (b) binary signal in the video.

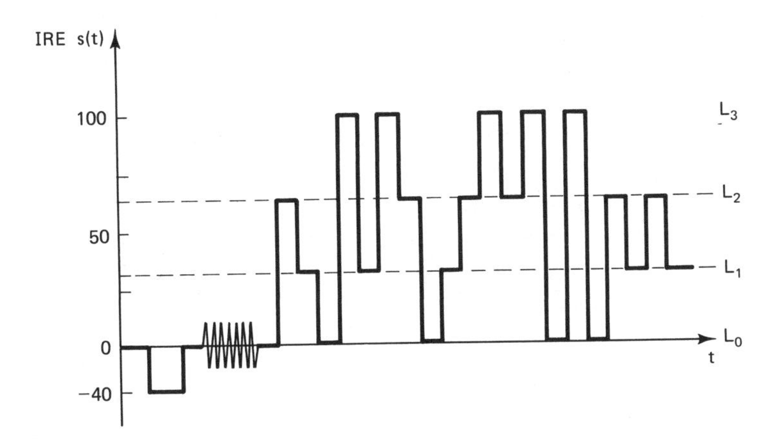

Figure 8.21 Equivalent video signal for the multilevel coding (four levels).

8.7 DIGITAL AUDIO–DIGITAL CHANNEL: COMPACT DISC

Using the same videodisc technology, up to 75 minutes per disc side of high-quality stereo music can be stored on a disc only 12 cm in diameter. A corresponding player can be substantially miniaturized compared with a conventional record player and therefore justifies its commonly used name, the *compact disc* (CD). This can be attractive in areas of limited space (e.g., automotive CD). The CD system offers all the advantages of digital techniques and the digital approach to signal processing, including significant improvement of dynamic range in music passages.

The first optical digital recording was done by companies already heavily involved in the videodisc technology. In mid-1978 DiscoVision in conjunction with Pioneer, which had no disc manufacturing facilities at that time, conducted nine months of research on discs [47, p. 27]. Direct digital format and the optical discs were used. In 1979 NV Philips demonstrated the CD. After an agreement between Philips and Sony in 1979, announced in June 1980, a common system standard was defined. This standard gradually became the world standard. In the joint research programs between Philips and Sony, Philips investigated the basic operating principles and designed the hardware. Sony's contribution centered mainly on the development of software, including the signal processing method. Interchangeability with videodisc systems was not included in the design goals of the CD system. In 1982 CD systems were introduced in Japan.

From the beginning, CD systems offered such specifications as:

Signal-to-noise ratio more than 90 dB
Dynamic range greater than 90 dB
Flat frequency response, 20 to 20,000 Hz
Stereo channel separation greater than 90 dB
Harmonic distortion less than 0.05%
No rumble
No wow or flutter (quartz crystal precision)
No intermodulation

To maintain a constant linear velocity as the pickup scans the digital track, the rotation speed of the disc continuously varies (from 500 rpm at the inside to 200 rpm at the outside edge). Tracking, decoding, and rotation speeds are precisely synchronized with a central clock generator, which itself is governed by information encoded in the track on the disc. As a result, wow and flutter are no longer measurable. The servo systems are the same as in the CLV videodisc systems.

The improvement in sound quality is in essence obtained by accurate waveform coding and decoding of the audio signals, and, in addition, the coded audio information is protected against disc errors [48]. In digital audio, signal-to-noise ratio depends on the number of bits per "word." In the CD system each bit contributes about 6 dB to the accuracy of signal reproduction. Thus a signal-to-quantizing noise ratio of over 90 dB is attained. Any noise heard while listening to a good CD record either stems from the original recording or from some other component in the stereo system.

An essential function of the digital interface for the digital audio is to allow the transfer of digital audio samples. Additional data have to be transmitted along with the digital audio samples, for a number of reasons: synchronization, flag bits, possible error protection, user-definable data, and so on. These additional data are transmitted along with the digital audio samples, usually called auxiliary audio data. Also, information relating to equipment status and control has to be exchanged. This information relating to equipment status and control is called *control data*. The data streams must be synchronized: the digital audio samples, auxiliary audio data, and control data have to refer to the same (fixed) time reference. This function is usually called a *synchronization function*. In the CD system, test and reliable recovery of the synchronization functions are necessary in case of errors.

8.7.1 Digital Audio Modem

The digital audio encoding system is shown in Figure 8.22. Two audio channels are processed similarly. The input audio signal is first filtered to minimize aliasing error. Sample-and-hold circuits and analog-to-digital (A/D) converters perform sampling, quantization, and coding. The time multiplexer (MUX), properly synchronized, converts parallel data to serial. The same process is performed twice more. First, digital audio, including error correction, is mixed with the control data. The resultant serial output is applied to the modulator. The modulator output is mixed with sync data and serial output is obtained, with MSB first and LSB last.

The digital decoding system is shown in Figure 8.23. Serial audio data, corrupted with noise and dropouts, are detected from the disc. After signal conditioning, proper timing is recovered to synchronize basically the inverse operation performed in the encoder. A shaped data stream is demodulated and then error correction is performed. In general, the modulator and demodulator can be realized with a look-up table in a ROM. Two digital audio channels are obtained from the serial data by demultiplexing. Digital-to-analog (D/A) converters generate continuous signals. Low-pass filters separate original audio spectra.

Requirements for the channel code, in which the signal is stored on the disc, are sometimes contrary. The nonreturn-to-zero (NRZ) signals from the A/D converter have no limited distance between transients (run length) and cannot be stored directly on the disc. To regenerate the clock from the signal after readout, the signal must have a sufficient number of transients and the maximum run length must be as small as possible. On the other hand, to minimize intersymbol interference, the minimum run length should be as large as possible.

The modulation code must be dc-free, because the low frequencies of the spectrum give rise to the noise in the servo system. The error propagation of the modulation system must be as small as possible.

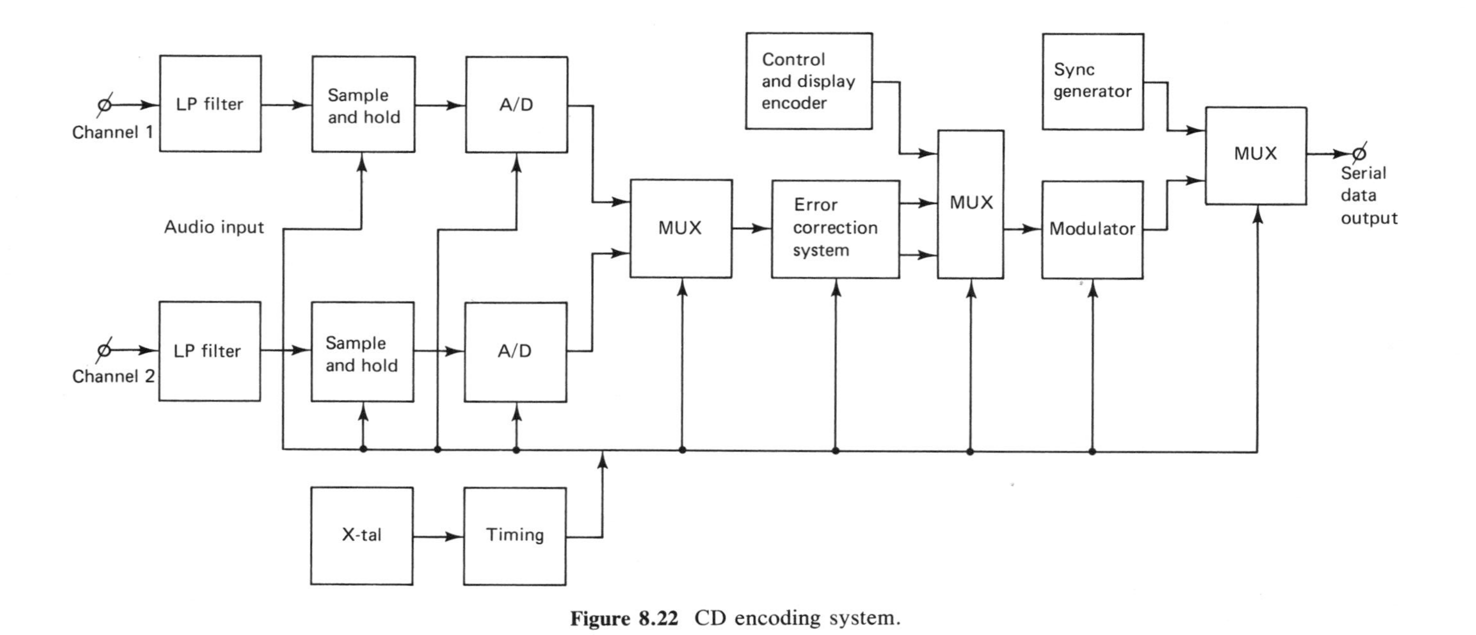

Figure 8.22 CD encoding system.

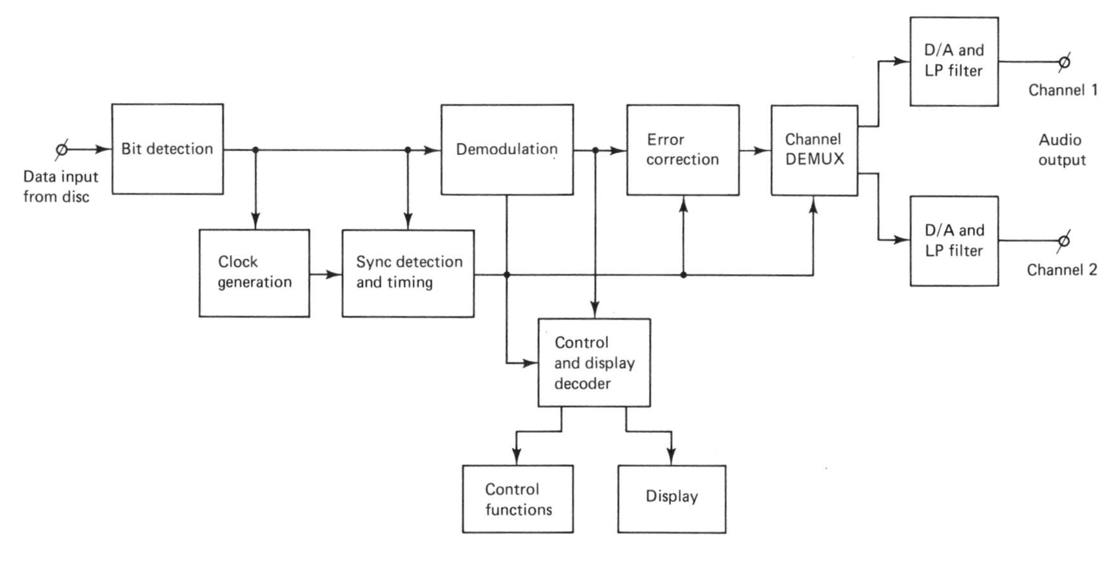

Figure 8.23 CD decoding system.

8.7.2 Standards, Disc Structure, and Specifications

To allow digital audio systems to interface and interchange with one another, it is necessary to standardize the systems. The main specification and structure of the optical digital audio disc are shown in Table 8.1 and Figure 8.24 respectively. The disc label is shown in Figure 8.25.

The thickness of the digital audio disc is the same as a normal videodisc: for example the transparent layer 1.2 mm, single sided disc, and 10 to 30 μm plastic protection coating. Disc diameter is approximately 120 mm. The extended play is obtained by CLV format, and the linear (scanning) velocity is 1.2–1.4 m/s. The recorded element parameters, pit sizes, track pitch, etc., are the same as in the $\lambda/4$ reflective videodisc systems. This means that there are over 20,000 tracks in a signal area 33-mm wide. Signal format is chosen to satisfy the communications requirements of the CD digital audio system. This is discussed in the next paragraph.

8.7.3 Signal Processing

Block diagrams showing the various signal operations at the transmitting and receiving ends of the CD digital audio systems are given in Figure 8.22 and Figure 8.23. It can be seen that, as is usual in a communications system, some of the signal operations at two ends are inverse.

Sampling is the first step in the analog-to-digital (A/D) conversion. There are many considerations involved in the choice of the sampling rate [49]. The sampling rate must be at least twice the highest audio frequency in the band. Room must be left to insert a sharp-cutoff filter required to eliminate alias

TABLE 8.1 CD SPECIFICATIONS

Disc	
Playing time: single side, 2 channels	Approx. 60 min.
Sense of rotation seen from reading side	Anti clockwise
Scanning velocity (2 channels)	1.2–1.4 m/s
Track pitch:	1.6 μm
Diameter:	120 mm
Thickness:	1.2 mm
Center hole diameter:	15 mm
Recording area:	46 mm–117 mm
Signal area:	50 mm–116 mm
Material:	Transparent material with 1.5 refractive index, such as acryl-plastic
Minimum pit length:	0.833 μm (1.2 m/sec.) to 0.972 μm (1.4 m/sec.)
Maximum pit length:	3.05 μm (1.2 m/sec.) to 3.56 μm (1.4 m/sec.)
Pit depth:	Approx. 0.11 μm
Pit width:	Approx. 0.5 μm
Optical system	
Standard wavelength:	λ = 780 nm (AlGaAs laser)
Focal depth:	approx. ±2 μm
Numerical aperture NA	0.45
Signal format	
Number of channels:	2 channels (4-channel recording is also possible at twice the present rotational speed.)
Quantization:	16-bit linear quantization
Quantizing timing	Concurrent for all channels
Sampling frequency:	44.1 kHz
Channel bit rate:	4.3218 Mb/sec.
Data bit rate:	2.0338 Mb/sec.
Data-to-channel bit ratio:	8:17
Encoding	2's complement
Error correction code:	CIRC* (with 25% redundancy)
Modulation system:	EFM**

* *C*ross *I*nterleave *R*eed-Solomon *C*ode

** *E*ight to *F*ourteen *M*odulation

components in the pass band. Finally it is desirable that the sampling frequency is compatible with television (NTSC, PAL, SECAM, and sampling frequency of 13.5 MHz for the digital TV), film, and other currently used digital audio transmission systems.

The sampling frequency is quartz-crystal-controlled and is equal to 44.1 kHz, which allows a recorded audio bandwidth of 20 kHz. In the PAL, the

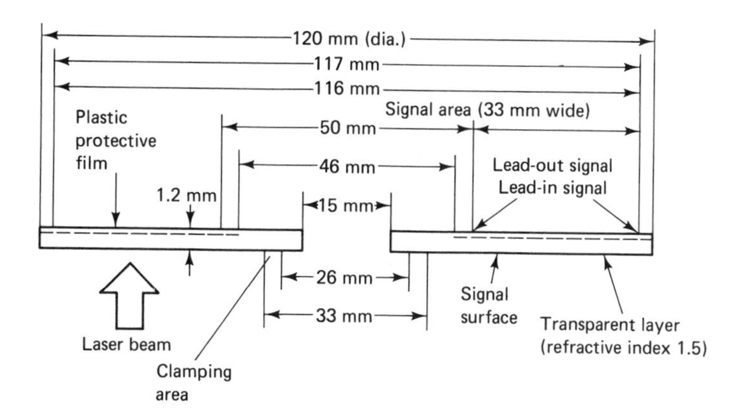

Figure 8.24 Structure of the compact disc.

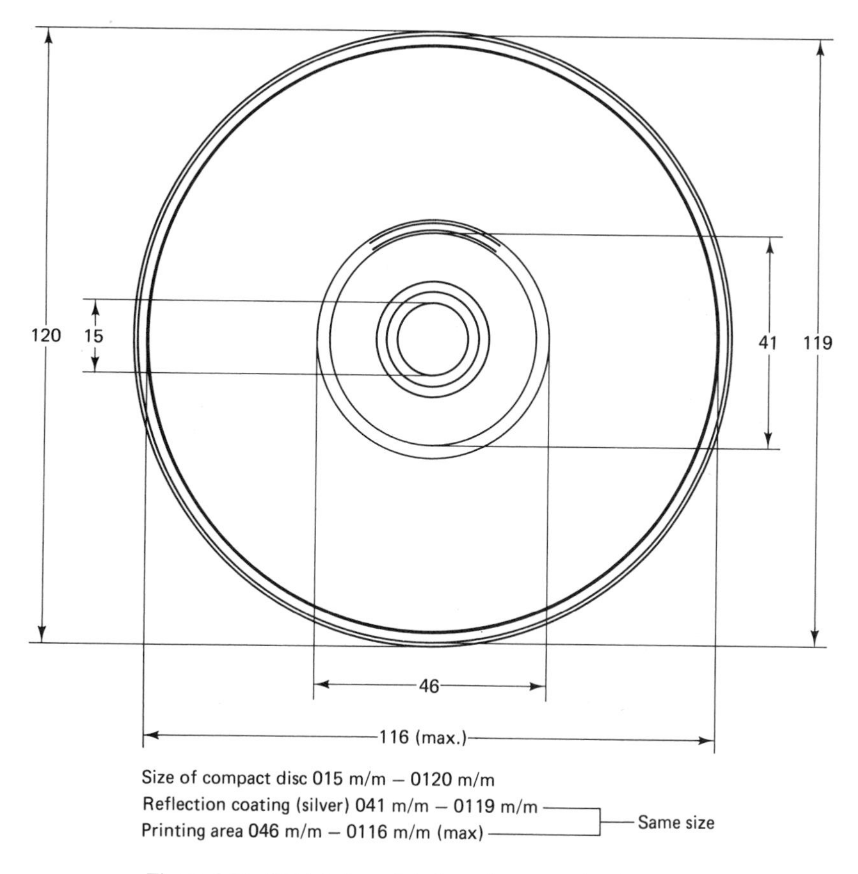

Figure 8.25 CD label—printed on the protective layer side.

television standard is

$$\frac{625 - 37}{625} \times 3 \times 15625 = 44.1 \text{ kHz} \tag{8.24}$$

where 625 is the number of lines in a PAL picture, 37 is the number of unused lines, 3 is the number of audio samples recorded per line, and 15625 Hz the line frequency [50].

The samples of both channels are uniformly quantized to 16 bits [51]. The number of bits per sample is a compromise between the quantizing noise and the play time. The A/D output signal is in the non-return-to-zero (NRZ) form, and 2's complement code.

8.7.3.1 Eight to Fourteen Modulation (EFM) Code. For the optimal use of any communication (recording) channel, input signal must be properly tailored. In Section 9.5 more discussion is included on the channel coding for the optical recording. Some of the most critical properties required for the channel code in the CD systems are [52]

Selfclocking: The CD is a one-dimensional channel, and the bit clock must be regenerated from the signal read-out from the disc. Thus, the maximum distance between transients in the signal, also called run length, must be as small as possible.
Maximal capacity (playtime): In all optical replica disc systems, recording density and system resolution are limited by the player which is imposed to the economical requirements. This requires the minimum run length as large as possible.
Servo: The channel code must be dc free, because the low frequencies of the spectrum can disturb the tracking control servo systems.
Error propagation: Must be as small as possible.

The NRZ signals from the A/D converter may have a high dc content and the run length is not limited and cannot be used on the optical disc. Therefore, the NRZ signals have to be converted into another channel code which should meet previous requirements. In the CD system the EFM channel code is used. This is a fixed length group code, achieved by bit mapping [52–54].

In the (m,n) group code, in general, a group of m input bits is converted, mapped, in a group of n bits, $n > m$. For a unique mapping, only 2^m out of a possible 2^n digital words should be selected according to some criteria. Then, for example, the NRZM can be applied.

The EFM code is a (8, 14) group code: a group of 8 bits (also called a byte or symbol) is mapped into 14 channel bits. Thus, the input list has $2^8 = 256$ different digital words and the output list has $2^{14} = 16384$ different digital words. To satisfy run length requirements, the constraint is imposed on the output code patterns: the digital words with more than 2, but less than 10 consecutive zeros, are selected. After elimination, 267 words out of 16384 meet this condition. But, this is more than the 256 words needed for the original 8-bit input: 11 code words out of 267 are further eliminated, and thus ($2^8 \times 14$) size ROM is needed in conversion.

8.8.3.2 Frame Format. The problem with the EFM code is that the 14-bit sequences cannot be run after the other without violating the constraints of at least 3 and at most 11 consecutive ones and zeros. By inserting 3 properly chosen merging bits between 14-bit blocks, the run-length requirements can again be satisfied while at the same time suppressing the lower signal frequencies, Figure 8.26.

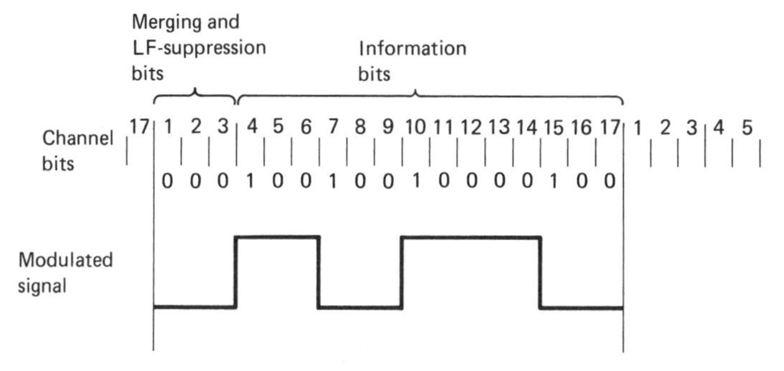

Figure 8.26 Sequence of the EFM: 3 merging bits are added to satisfy the run–length requirements.

As already discussed, in the CD system, both audio channels (left and right) are sampled with a frequency of 44.1 kHz. Each sample is represented in 16 bits using uniform quantization. Then, the audio samples are gathered in frames of 12 audio samples each, 6 samples from the left audio channel (L_n) and 6 samples from the right channel (R_n), as shown in Figure 8.27a. Now each sample of 16 bits consists of 2 bytes or symbols, so that each frame can also be viewed as consisting of 24 audio bytes. In the CD encoder, the bytes of a number of consecutive frames are scrambled and parity bytes are added such that disc errors can be corrected (or detected if correction fails).

In the channel modulator, the data stream is split up into frames. The data and error correction words are each split up into two 8-bit blocks, which are fed into the modulator circuit, where each block is converted into 3 + 14 channel bits. The 14 bits are EFM code words while 3 bits are connecting bits for interfacing EFM code words and suppressing dc. Each frame contents:

a synchronization pattern of 24 bits
12 data words of 16 bits each
4 error correction parity words of 16 bits each
a control and display symbol of 8 bits.

The total number of channel bits per frame is shown in Table 8.2, and Figure 8.27b).

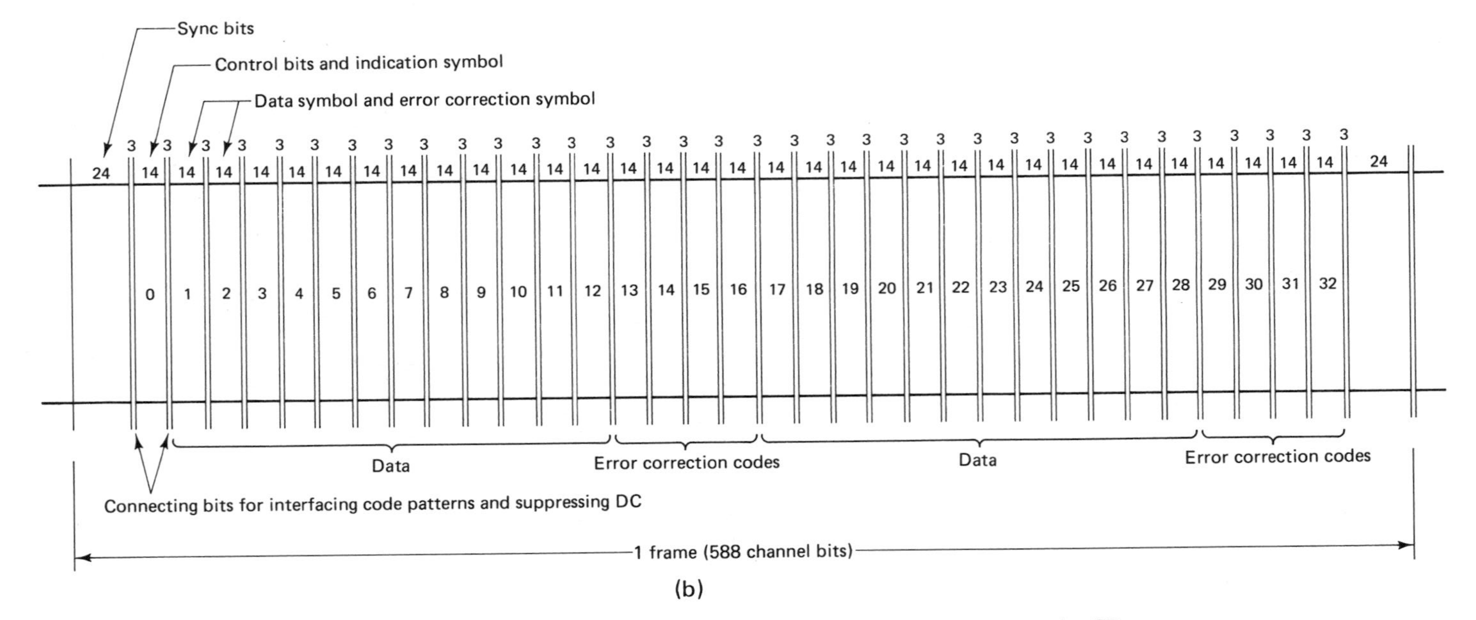

Figure 8.27 CD frame. (a) audio (data) symbols; (b) final frame structure on the CD.

TABLE 8.2 FRAME FORMAT

	Data bits	*Channel bits*
Synchronization		24
Control & display	1 × 8 = 8	1 × 14 = 14
24 data symbols	12 × 2 × 8 = 192	12 × 2 × 14 = 336
8 error correction symbols	4 × 2 × 8 = 64	4 × 2 × 14 = 112
Merging and low frequency suppression		34 × 3 = 102
Total frame	264	588

8.7.4 Error Correction

Because of noise and dropouts in the videodisc system, digital data on the disc should be correctable. The replicated videodisc system is a one-way channel in which error probabilities can be reduced with error-correcting codes but not by error detection and retransmission. For example, it is too late to ask for retransmission after the metalized plastic disc has been made and errors are detected when the disc is read. But error detection and rewriting (retransmission) can reduce the error probability introduced in the recording process. Error detection is a much simpler task than error correction and requires much simpler decoding equipment. Also, error detection with rewriting is adaptive; recording of redundant information increases when errors occur.

Two fundamentally different types of codes are block codes and tree codes [55]. In block coding a continuous sequence of information digits are grouped into k-symbol blocks, and each possible information block is associated with an n-tuple of channel symbols, where $n > k$. In the decoder, these tuples and code words are decoded independently of all other code words. It is said that n is the code or block length. The encoder for a tree code processes the information continuously and the encoding rules are most conveniently described by means of a tree graph. A convolutional code is a subset of the tree codes. Block codes and convolutional codes have similar error-correcting capabilities and the same fundamental limitations. Almost all known block and tree codes are linear codes. The two important properties of linear codes are: the sum (mod 2) of any two code words is a code word, and the all-zero word is a code word.

Given the system performance requirements, the first (error correcting) coding problem is how to select and design a code that gives appropriate reliability and efficiency. The second problem would be the design of the encoder, which will determine the code word from the message, and the decoder, which will determine the code word from the received word and then determine the message from the code word.

In a decoder, for each received word a *syndrome* is calculated; a received word is a code word if and only if its syndrome is zero. The syndrome of the error pattern is an $(n - k)$-tuple of parity-check failures. The syndrome is the sum of the locally generated checks and received checks. It depends only on the error pattern, not on the transmitted code word. There are 2^{n-k} syndromes and the same number of the correctable error patterns. An analogy for the syndrome can be found in the next example. Suppose that P_1, P_2, ..., P_m are numbers of pages in each of m chapters in a book, and that T is the total number of pages. The calculation of numbers of pages can be checked by using "syndrome S":

$$S = P_1 + P_2 + \cdots + P_m - T = 0 \tag{8.25}$$

Usually, a feedback shift register is used as a syndrome register. This can produce an effect known as (infinite or catastrophic) *error propagation*. An uncorrectable channel error may cause the syndrome register to enter a state in which, in the absence of additional channel errors, the decoder will continue to decode incorrectly forever. This is possible because the output of the nonlinear feedback shift register with nonzero initial conditions may be periodical if an all-zero sequence is applied at its input.

A cyclic code is a linear code with the additional property that every cyclic shift of a code word, by any number of places—in either direction—is a code word. Every code word in an (n, k) code can be associated with a polynomial of degree $n - 1$ or less, and no polynomial of degree $\geq n$ corresponds to a code word. The generator polynomial of a cyclic (n, k) code divides $x^n + 1$ and has a degree of $n - k$.

The Bose–Chaudhuri–Hocquenghem (BCH) codes are a class of cyclic codes. They are random-error-correcting codes and are the best known large class of codes. Codes exist for all integers m and t such that:

$$\begin{aligned} n &= 2^m - 1 \\ d &= 2t + 1 \ (d = \text{minimum distance}) \\ m - k &\leq mt \ (\text{usually, an equality}) \end{aligned} \tag{8.26}$$

A very important subclass of BCH codes are the Reed–Solomon (RS) codes. These are cyclic codes whose symbols are binary m-tuples rather than bits. The length is equal to $2^m - 1$ symbols, or

$$n = m(2^m - 1) \text{ bits} \tag{8.27}$$

A t-error-correcting RS code needs $2t$ check symbols. In a t-error-correcting RS code, the following error patterns are correctable:

1 burst of total $(t - 1)m + 1$ bit
2 bursts of total length $(t - 3)m + 3$ bits
⋮
i bursts of total length $(t - 2i + 1)m + 2i - 1$ bits

Decoding of RS codes is only slightly more complex than for BCH codes. In addition to the locations of the errors, their values (m-tuples) must also be determined. Decoder cost is relatively insensitive to m and increases very erratically with t. Because steps of decoding are independent, pipelining (to depth 4 or 5) can be used effectively to increase speed.

For the videodisc channel, a burst-error-correcting code is of particular interest. A burst of errors of length b is a sequence of error bits, the first and last of which are 1s. The burst-correcting capability of a code is defined as one less than the length of the shortest uncorrectible burst. The burst-detecting capability of a code is defined as one less than the length of the shortest undetectable burst. Four ways to construct burst-correcting codes are: fire codes, interleaving, Reed–Solomon codes, and computer search (nonconstructive).

Symbol interleaving or interlacing is the technique of forming an (n_i, k_i) code with burst-correcting capability from the cyclic (n, k) code. The parameter i is referred to as the *interleaving degree*. Block interleaving rather than symbol interleaving, such that symbols of a block are adjacent on the channel, can also be performed.

Reed–Solomon codes and interleaved Reed–Solomon codes are the most powerful of the known class of multiple-burst-correcting codes. Because of this, the RS codes are used in CD systems. The basic idea behind all burst-correcting codes used in compact disc systems is that the bits involved in the decoding of a particular digit are spread on the disc so that only one, or at most a few, are affected by a single burst of errors. The simplest way of accomplishing this is by symbol or block interleaving. Figure 8.28 shows an example of block interleaved code with a total of 96 bits in 4-bit groups.

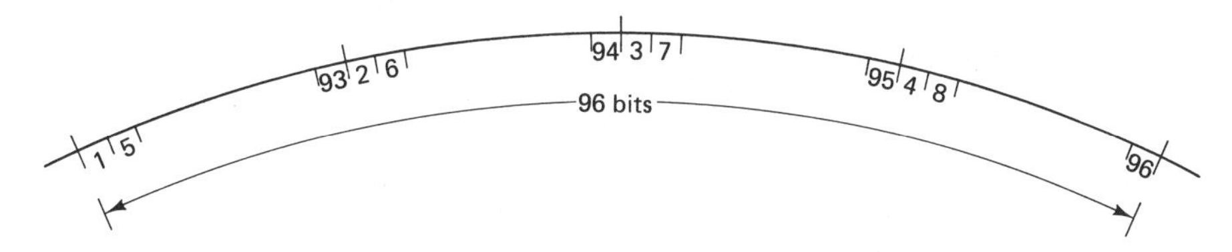

Figure 8.28 Interleaved code on the disc surface for $n = 96$.

8.7.4.1 CIRC Error Correction. In the CD systems the Cross Interleaved Reed–Solomon Code (CIRC), which copes with both random and burst errors, is used [56–59]. This code corrects the bulk of errors encountered and helps detect errors it cannot correct. The effect of these is minimized by filtering. The CIRC can fully correct a burst of up to 8232 consecutive recorded bits, corresponding to a 2.3-mm band of tracks on the disc [60]. It can also detect and partially correct errors in up to 28224 bits, or a 7.8-mm band of tracks. This is obtained by interpolation: the new samples are inserted instead of the erroneous ones.

CIRC has an efficiency of 3/4, this means that 3 data bits will result in 4 bits after encoding. The decoder complexity of the CIRC code has been reduced considerably by splitting up the decoder into two main parts: An integrated (LSI) circuit and associated random-access memory. The signal format has been designed in such a way that 4 channels are possible in the future, without changes in the decoder chip.

The CIRC consists of a C_1 and C_2 Reed–Solomon Code (Figure 8.29): C_1 is the Reed–Solomon Code with (n = 32, k = 28) and C_2 is the same code with (n = 28, k = 24).

The horizontal blocks between C1 and C2, represent 8-bit-wide delay lines of unequal length (interleaving). Before the C2-encoder a delay of one symbol is inserted in the even words to facilitate concealments in simplified decoder versions.

After the C1-encoder a delay of one symbol (8 bits) is inserted in the even symbols (scrambling). In short, the decoder operates as follows [52]:

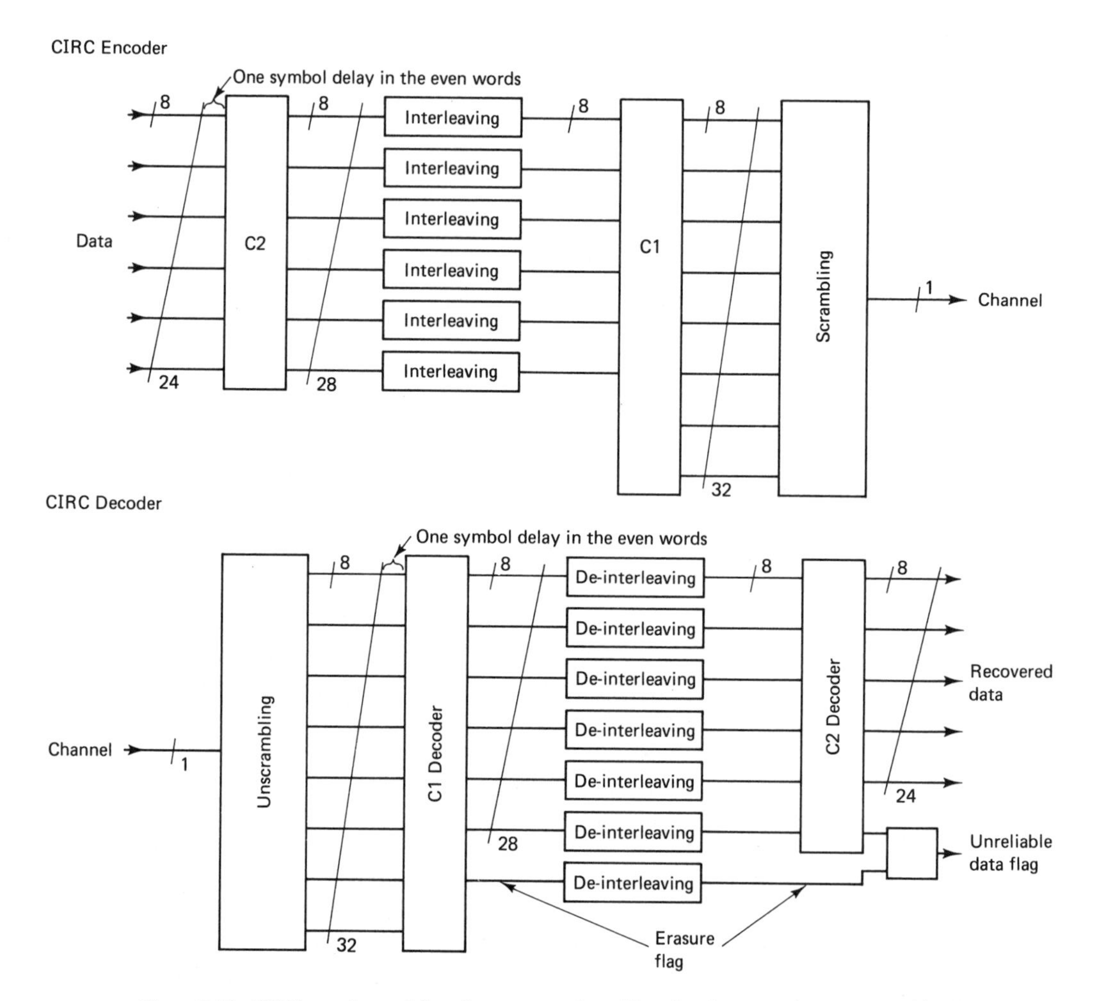

Figure 8.29 CIRC encoder and decoder: cross code and interleaving procedures are combined.

The C1-decoder accepts 32 symbols of 8 bits each, from which 4 parity symbols are used for C1-decoding. The parity is generated according to the rules of Reed–Solomon coding and because of that the C1-decoder is able to correct a symbol error in every word of 32 symbols. If there is more than one erroneous symbol, then regardless of the number of errors, the C1-decoder detects that it received an uncorrectable word. If this is the case, it will let all 28 symbols pass through uncorrected, but an erasure flag is set for each symbol to mark that all symbols from C1 are unreliable at that moment.

Because the delay lines between the C1 and C2-decoders are of unequal length, the symbols marked with an erasure flag at one instant, arrive at different moments at the C2-decoder input. Thus the C2-decoder has for every symbol an indication whether it is in error or not: if a symbol does not carry an erasure flag it is error free. If no more than 4 symbols carry an erasure flag, then the C2-decoder can correct a maximum of 16 frames. In case even the C2-decoder cannot correct, it will let the 24 data symbols pass through uncorrected, but marked with the erasure flags originally given out by the C1-decoders.

8.7.5 Word of Caution in the CD Specs Interpretation

Sometimes, in the digital vs. analog audio comparison, numbers can be taken almost out of context. Thus, it may result in misleading conclusions [61]. The psychoacoustic phenomena make the situation even worse because of masking involved. Things should be taken for what they are defined, nothing more or less. For example, the peak-to-peak values should not be confused with the rms values.

For example, the dynamic range of an n-bit digital audio system is 20 log $(2_n - 1)$ and it reflects the difference between a largest and smallest possible value. The noise is not (directly) incorporated in this. Also, in the signal-to-noise ratio comparison, the quantizing noise is the noise considered for the digital audio. In the analog audio there is discernible information buried in the noise floor, while in the digital systems there is no signal below the noise.

8.8 THE CD DIGITAL AUDIO MODULATION IN THE LASERVISION VIDEODISC CODING FORMATS

The basic idea of the addition of digital audio in the Laservision (LV) coding format is to use the (noise) low-end frequency range up to 1.75 MHz [62]. Thus, the frequency sharing principle is applied. The basic block-diagrams of the combined LV signals and digital audio encoder and decoder are shown in Figures 8.30 and 8.31, respectively.

There are two basic interference problems: digital audio to video, and LV signals to digital audio crosstalks. The digital audio signal amplitude and its modulation index on the main carrier have an important role in the disturbance

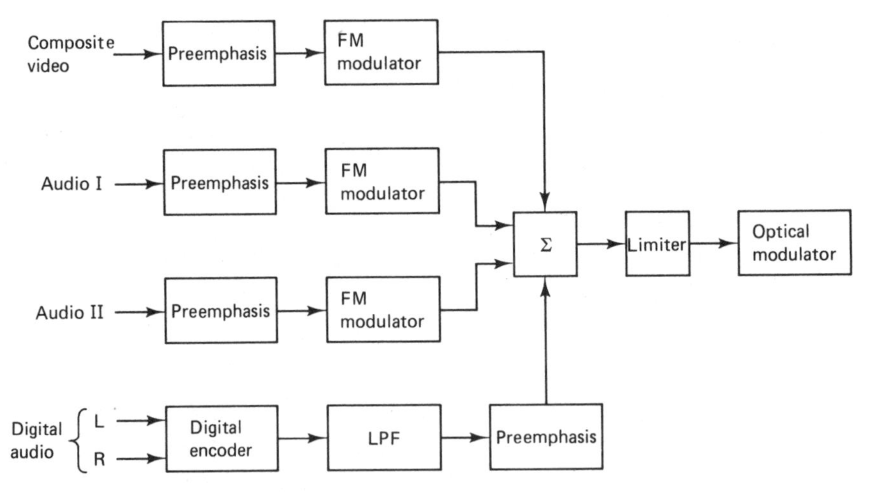

Figure 8.30 Combined LV signals and digital audio: encoder.

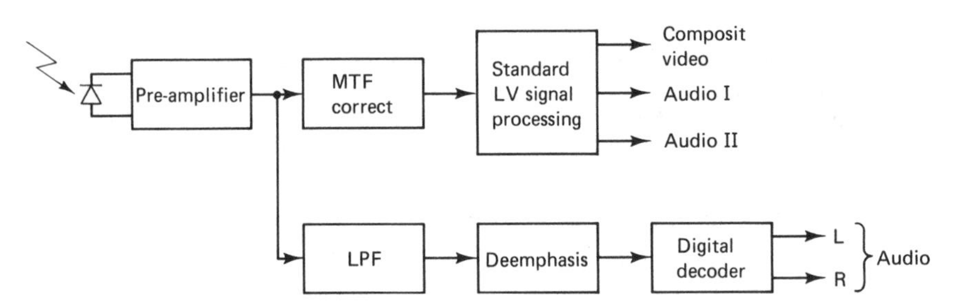

Figure 8.31 Combined LV signals and digital audio: decoder.

of the digital audio signal in the video picture. The low modulation index of the digital audio signal, required for the low interference visibility (in the picture), results in a poor signal-to-noise ratio. Also, in the optical recording, the S/N ratio of the low frequencies (approximately below 500 kHz), deteriorates as a result of the laser noise. To improve S/N ratio preemphasis/deemphasis is used for the digital audio signal. The amplitude of the digital audio signal must be 26 dB below the main carrier.

To illustrate the interference of the LV signals to the digital audio signal, the frequency spectra of the NTSC and PAL LV formats are shown in Figure 8.32. Obviously, the interference caused by the second order sideband J_2 has to be removed. One possible solution is to generate a second harmonic of the color subcarrier, $2f_{sc}$, and use it to reduce second-order sideband of the subcarrier, f_{sc} [62]. A corresponding block-diagram is shown in Figure 8.33 and spectral components are illustrated in Figure 8.34. The chroma band is filtered out from the composite video signal. The second harmonic is generated by the multiplier

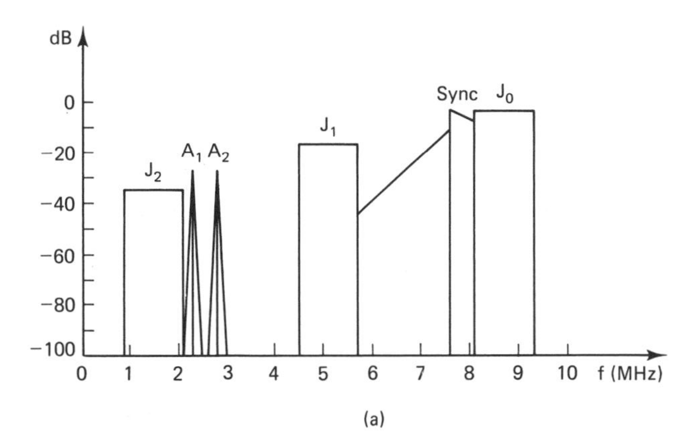

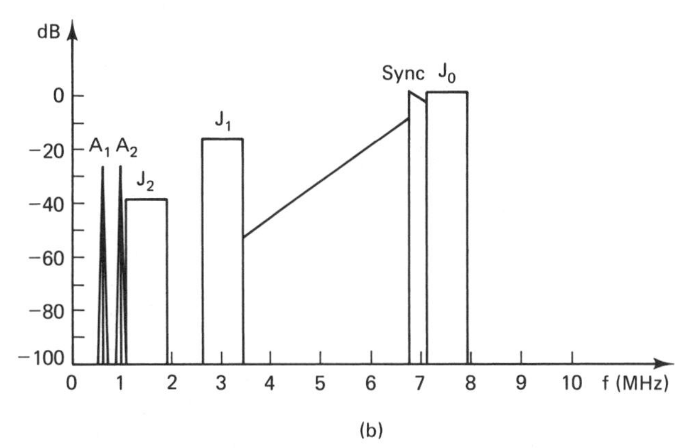

Figure 8.32 Frequency spectra of the LV formats. (a) NTSC; (b) PAL.

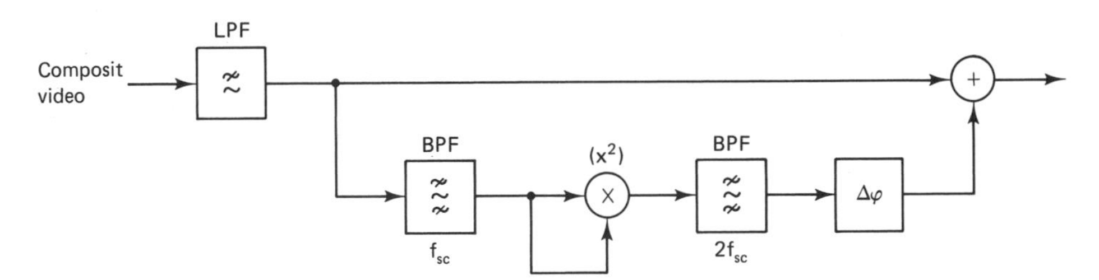

Figure 8.33 Reduction of LV signals interference to the digital audio: block diagram.

(square circuit) and separated by the bandpass filter. This frequency-doubled chroma signal is added with the correct phase and amplitude to the original composite video signal. After FM the chroma J_2 component will be reduced, possibly counseled.

In Figure 8.35 the channel spectra of the combined LV signals and digital audio for the NTSC and PAL are shown. It can be seen that in the PAL video format the analog audio carriers have to be removed.

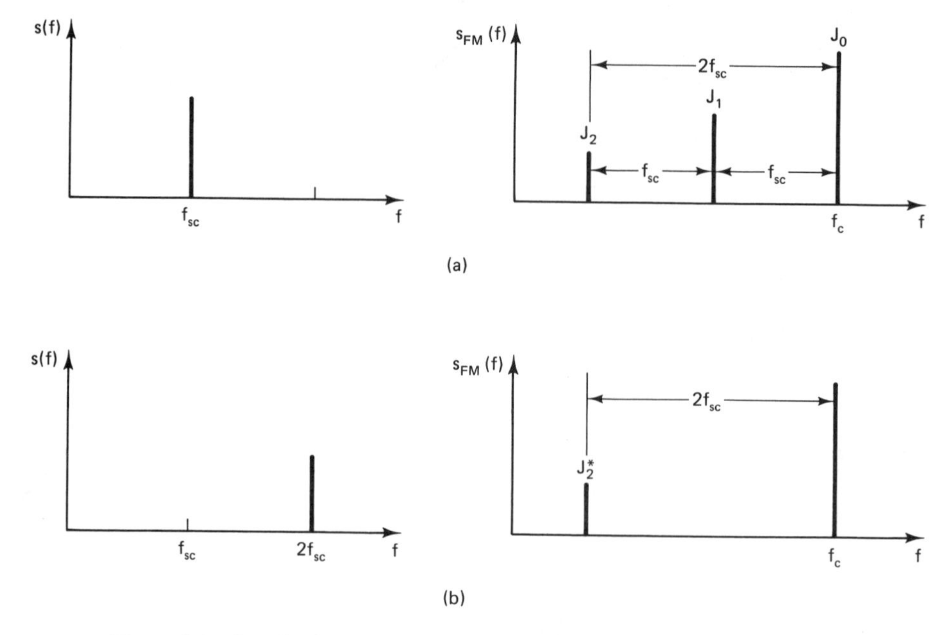

Figure 8.34 J_2 reduction. (a) f_{sc} (input) and J_1/J_2 (channel spectra); (b) $2f_{sc}$ and first sideband of it (channel spectra).

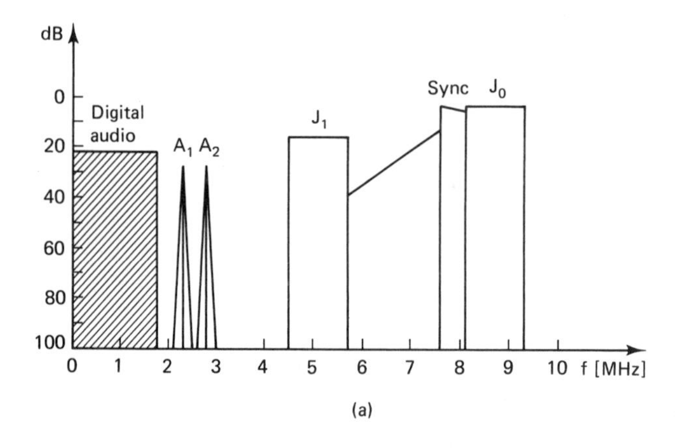

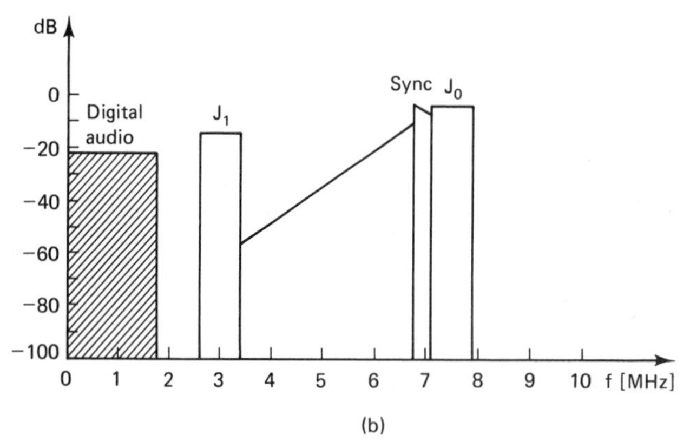

Figure 8.35 Channel spectra of the combined LV signals and digital audio for the (a) NTSC format; (b) PAL format.

REFERENCES

1. J. L. Flanagan, *Speech Analysis Synthesis and Perception,* 2nd expanded ed., Springer-Verlag, Berlin, 1972.
2. H. K. Dunn and S. D. White, "Statistical Measurements on Conversational Speech," *J. Acoust. Soc. Am.,* Vol. 11, 1940, pp. 278–288.
3. M. Giles, *Audio Noise Reduction and Masking,* National Semiconductor Corporation, 1981.
4. H. M. Horowitz, personal communication, Oct. 1985.
5. M. Giles, "On-Chip Stereo Filter Cuts Noise Without Preprocessing Signals," *Electronics,* August 11, 1981, pp. 104–108.
6. M. Giles, J. Wright, "A Noncomplementary Audio Noise Reduction System," *IEEE Trans. on Cons. Electr.,* Vol. CE-27, No. 4, Nov. 1981, pp. 626–630.
7. L. A. Abbagnaro, "The CX Noise Reduction System for Records," *Audio,* Feb. 1982.
8. J. Isailović, *Dynamic Noise Reduction in Analog Audio Systems,* R. O. Produkcija Gramofonskih Ploca RTV, Belgrade, December 1981 (in Serbian).
9. R. Dolby, "A 20 dB Audio Noise Reduction System for Consumer Applications," *J. Audio Eng. Soc.,* Vol. 31, No. 3, March 1983, pp. 98–113.
10. K. Shinohara, M. Fuse, and Y. Komatsu, "The Development of Bipolar IC's for dBX Noise Reduction System," *IEE Trans. on Cons. Electr.,* Vol. CE-28, No. 4, Nov. 1982, pp. 553–561.
11. *CX Low Cost Expander-Model E-1016,* CBS Technology Center, Apr. 1981.
12. N. S. Jayant, "Digital Coding of Speech Waveforms: PCM, DPCM and DM Quantizers," *Proc. IEEE,* Vol. 62, No. 5, 1974, pp. 611–632.
13. H. R. Schindler, "Delta Modulation," *IEEE Spectrum,* Vol. 7, Oct. 1970, pp. 69–78.
14. A. H. Reeves, French Patent 852,183, Oct. 3, 1938.
15. B. M. Olivie, J. R. Pierce, and C. E. Shannon, "The Philosophy of PCM," *Proc. IRE,* Vol. 36, Oct. 1948, pp. 1324–1331.
16. H. S. Block, *Modulation Theory,* D. Van Nostrand, Princeton, N.J., 1953.
17. P. F. Panter and W. Dite, "Quantization Distortion in Pulse Count Modulation with Non-uniform Spacing of Levels," *Proc. IRE,* Vol. 39, Jan. 1951, pp. 44–48.
18. L. R. Rabinar and J. A. Johnson, "Perceptual Evaluations of the Effects of Dither on Low Bit Rate PCM Systems," *Bell Syst. Tech. J.,* Sept. 1972, pp. 1487–1497.
19. B. Smith, "Instantaneous Companding of Quantized Signals," *Bell Syst. Tech. J.,* 1957, pp. 653–709.
20. J. Max, "Quantizing for Minimum Distortion," *IRE Trans. Inf. Theory,* Vol. IT-6, Mar. 1960, pp. 7–12.
21. P. Cummiskey, N. S. Jayant, and J. L. Flanagan, "Adaptive Quantization in Differential PCM Coding of Speech," *Bell Syst. Tech. J.,* Sept. 1973, pp. 1105–1118.
22. C. S. Xydeas, M. N. Faruqui, and R. Steele, "Envelope Dynamic Ratio Quantizer," *IEEE Trans. Commun.,* Vol. COM-28, No. 5, 1980, pp. 720–728.
23. C. C. Cutler, "Differential Quantization of Communications," U.S. Patent 2,605,361, July 29, 1952.

24. P. Elias, "Predictive Coding," *IRE Trans. Inf. Theory,* Vol. IT-1, Mar. 1955, pp. 16–33.
25. R. A. McDonald, "Signal-to-Noise and Idle Channel Performance of Differential Pulse Code Modulation Systems—Particular Applications to Voice Signals," *Bell Syst. Tech. J.,* Vol. 45, 1966, pp. 1123–1151.
26. B. S. Atol and M. R. Schroeder, "Adaptive Predictive Coding of Speech Signals," *Bell Syst. Tech. J.,* 1970, pp. 1973–1986.
27. E. M. Deloraine, S. Va Miero, and B. Djordjevitch, "Méthode et système de transmission par Impulsions," French Patent 932,142, Aug. 1946.
28. L. J. Libous, "Un Nouveau procédé de modulation codé à la modulation en delta," *L'Onde Electron.,* Vol. 32, Jan. 1952, pp. 26–31.
29. F. de Jager, "Delta Modulation, a Method of PCM Transmission Using the 1-Unit Code," *Philips Res. Rep. 7,* 1952, pp. 442–466.
30. M. R. Winckler, "High Information Delta Modulation," *IEEE Int. Conv. Rec.,* Part 8, pp. 260–265.
31. J. E. Abate, "Linear and Adaptive Delta Modulation," *Proc. IEEE,* Vol. 55, Mar. 1967, pp. 298–308.
32. A. Tomozawa and H. Kaneko, "Companded Delta Modulations for Telephone Transmission," *IEEE Trans. Commun. Technol.,* Vol. COM-16, Feb. 1968, pp. 149–157.
33. S. J. Brolin and J. H. Brown, "Companded Delta Modulation for Telephony," *IEEE Trans. Commun. Technol.,* Vol. COM-16, Feb. 1968, pp. 157–162.
34. J. A. Greefkes and F. de Jager, "Continuous Delta Modulation," *Philips Res. Rep. 23,* 1968, pp. 233–246.
35. N. S. Jayant, "Adaptive Delta Modulation with One Bit Memory," *Bell Syst. Tech. J.,* Vol. 49, Mar. 1970, pp. 321–342.
36. P. P. Wang, "Idle Channel Noise of Delta Modulation," *IEEE Trans. Commun. Technol.,* Vol. COM-16, Oct. 1968, pp. 737–742.
37. R. Steele, *Delta Modulation Systems,* Pentech, London, 1975.
38. N. S. Jayant, *Waveform Quantizing and Coding,* IEEE Press, New York, 1976.
39. C. K. Un, H. S. Lee, and J. S. Song, "Hybrid Companding Delta Modulation," *IEEE Trans. Commun. Technol.,* Vol. COM-29, No. 9, 1981, pp. 1337–1343.
40. J. O'Neal, "Waveform Encoding of Voiceband Data Signal," *Proc. IEEE,* Vol. 68, Feb. 1980, pp. 232–247.
41. T. H. Lee and C. K. Un, "A Performance Study of Delta Modulation Systems for Voiceband Data Signals," *IEEE Trans. Commun.,* Vol. COM-29, No. 9, 1981, pp. 1324–1329.
42. J. L. Flanagan et al., "Speech Coding," *IEEE Trans. Commun.,* Vol. COM-27, No. 4, 1979, pp. 710–737.
43. J. B. O'Neal, "Predictive Quantizing Systems for the Transmission of Television Signals," *Bell Syst. Tech. J.,* May–June 1966, pp. 689–719.
44. J. E. Paul, "Adaptive Digital Techniques for Audio Noise Cancellation," *IEEE Circuits Syst. Mag.,* Vol. 2, No. 4, 1979, pp. 2–7.
45. J. Isailović, "Binary Data Transmission Through TV Channels by Equivalent Luminance Signal," *Electron. Lett.,* Vol. 17, No. 6, 1981, pp. 233–234.

46. J. Isailović, "Binary Data Transmission Through TV Channels by Equivalent Chrominance Signal," *SPIE Conf.*, Los Angeles, Feb. 18–22, 1982.
47. J. Isailović, "Videodisc and Optical Memory Technologies," Prentice-Hall, 1985.
48. J. B. H. Pean, "Communications Aspects of the Compact Disc Digital Audio System," *IEEE Commun. Mag.*, Vol. 23, No. 2, Feb. 1985, pp. 7–15.
49. R. J. Youngquist, "Editing Digital Audio Signals in a Digital Audio/Video System," *SMPTE J.*, Dec. 1982, pp. 1158–1160.
50. T. Doi, Y. Tsuchiga, and A. Iga, "On Several Standards for Converting PCM Signals into Video Signals," *J. Audio Eng. Soc.*, Vol. 26, No. 9, Sept. 1978, pp. 641–649.
51. B. A. Blesser, "Digitization of Audio," *J. Audio Eng. Soc.*, Vol. 26, No. 10, Oct. 1978, pp. 739–771.
52. "Draft Compact Disc Digital Audio System," Now under discussion in the International Electrotechnical Commission as Document 60A (Central Office) 82.
53. K. A. Immink, "Modulation Systems for Digital Audio Discs with Optical Readout," *IEEE Int. Conf. on Acoust., Speech, and Sig. Process.*, Atlanta, GA, March 30–April 1, 1981, pp. 587–589.
54. J. P. J. Heemskerk and K. A. Schouhamer Immink, "Compact Disc: System Aspects and Modulation," *Philips Tech. Rev. (Special Issue)*, Vol. 40, No. 6, 1982, pp. 157–164.
55. W. W. Peterson and E. J. Weldon, *Error-Correction Codes,* 2nd ed., MIT Press, Cambridge, Mass., 1972.
56. L. B. Vries and K. Odaka, "CIRC—The Error Correcting Code for the Compact Disc," *The AES Premier Conference—The New World of Digital Audio,* New York, June 3–6, 1982.
57. L. M. H. E. Driessen and L. B. Vries, "Performance Calculations of the Compact Disc Error Correcting Code on a Memoryless Channel," *Int. Conf. Video and Data Recording,* University of Southampton, April 20–23, 1982.
58. H. Hoeve, J. Timmermans, and L. B. Vries, "Error Correction and Concealment in The Compact Disc System," *Philips Tech. Rev. (Special Issue)*, Vol. 40, No. 6, 1982, pp. 166–172.
59. T. Doi, "Error Correction for Digital Audio Recordings," *The AES Premier Conference—The New World of Digital Audio,* New York, June 3–6, 1982.
60. S. Miyaoka, "Digital Audio Is Compact and Rugged," *IEEE Spectrum,* March 1984, pp. 35–39.
61. P. B. Fellgett, "The Digital Dilemma," *Studio Sound,* Nov. 1981, pp. 88–90.
62. K. A. S. Immink, A. H. Hoogendijk, and J. A. Kahlman, "Digital Audio Modulation in the PAL and NTSC Laservision Video Disc Coding Format," *IEEE Trans. on Cons. Electr.,* Vol. CE-29, No. 4, Nov. 1983, pp. 543–550.

nine

Other Applications of Videodisc Systems

9.1 INTRODUCTION

In the early stage of videodisc technology development, the aim was to store a video signal on the disc, together with a (stereo) audio signal, and reconstruct them from the disc, all this for the reasonable length of the recorded program. Once the technology was established, many other applications were developed.

A single prerecorded videodisc program can be resequenced and redirected to serve a variety of audiences. This is based on videodisc programming. Thus the military, industry, and aerospace technology have a new training tool. Catalogs in stores and museums can be used in a more convenient and sophisticated way. For libraries, discs might store both text and picture. Desired output can be obtained from a large encyclopedia stored on discs, combined with hard copy or CRT display.

The videodisc technology offers enormous storage capabilities, random access, cheap reproduction, long life, and flexibility. Those characteristics are obviously useful in disc applications as a computer peripheral. Also, pictures, voice, alphanumerical data, and logic can be stored on discs with equal ease and in any desired proportion. As a computer peripheral, videodisc can be used as either an analog or a digital channel.

9.2 VIDEODISC PROGRAMMING

The videodisc player is a stand-alone system, which when connected to a television set or television monitor, comprises a self-contained video retrieval and display system. But the videodisc player is programmable and the exact sequence and display of information presented to the viewer can be predetermined by the program designer.* For example, it is possible to program audiovisual materials in multiple-choice, question-and-answer format, to select different program segments to meet the needs of a specific audience, to review some segments while skipping others, to provide two narrations, and so on.

In general, the videodisc is the storage medium for the video and audio information processed by the videodisc player. Only special-mode discs will be considered here. The microprocessor control usually offers:

Programmable memory (e.g., 1024 bytes)
Memory retention (e.g., 72 hours)
Frame number display

The integration of a microcomputer into the player design makes possible the many play and search-out display functions of which the player is capable. The internal microcomputer controls all phases of player operation, processing both external command functions and internally generated command, control, and status signals.

Any program is based on the following player properties:

To play until a specified advance frame
To stop (freeze frame)
To search a particular frame
To reject and turn to the beginning position
To choose one, both, or no audio channel

A user program for the player consists of a series of commands that when stored in the random access memory (RAM) and executed by the microcomputer, causes the player to operate in a predefined way. Commands are the functional instructions that the programmer may use to develop an application program. Many of the commands are direct counterparts of buttons on the remote control unit (RCU): for example, AUDIO 1, DSP FRM (frame display), and so on. The numerical parameters required by commands are specified by their associated arguments. An argument may specify an integer data value, a frame number, a memory location, or a register number. These arguments may denote time delays or register content data, such as the acceptable range of numeric values entered by the viewer in response to input commands.

A user program, when stored in memory, resembles a continuous string of characters. In general, succeeding commands are not processed until the function specified by the "current" command has been completed. There can be exceptions. For example, the play command instructs the videodisc player to begin playing audio/video material and to continue until terminated by the program. Program execution continues while the play is in progress.

* To avoid possible confusion, the so-called "player levels" are not discussed here.

9.2.1 Program Loading and Operating Modes

User programs may be loaded into RAM in one of three ways:

1. From program coded on the videodisc (program dumps)
2. Manually (from a remote control unit)
3. Under the control of an attached host computer

A dump is a computer program that is written on one audio track during the disc mastering process, in addition to audio and video information. Program dumps may be loaded into RAM from a videodisc. There can be multiple dumps on the same disc under user program control; each may overwrite a specified RAM capacity.

Computer programs may be written into RAM through use of the remote control unit. The RCU usually duplicates most of the front-panel controls of the player, but may contain additional keys for entering program instructions and data. RCU thus permits a programmer to modify existing programs and tailor them to their unique requirements or to design and debug player control programs prior to their becoming finalized as program dumps on the videodisc.

The RCU for the optical videodisc player manufactured by Discovision Associates is shown in Figure 9.1. Aspects of player operation that can be accomplished through the use of the RCU, including the loading of user programs into minicomputer RAM, may also be performed via the external computer interface, a special connector that permits interconnection of the player and a host computer. The player may be operated in one of three modes:

1. Manual
2. Automatic
3. Programming

Manual mode is initiated by loading a disc, for example, by mounting a disc and closing the player's protective cover, and initiating a play command. The disc may be played from beginning to end without any further action by the viewer; suppose that the loading of a program dump is inhibited. The viewer may, however, at any time vary the presentation of the material through use of the front-panel controls or the buttons on the RCU. Manual mode terminates if videodisc reject, automatic mode, program dump, or programming mode is selected.

In automatic mode, operation of the player is controlled by the application program stored in RAM. This mode is automatically initiated by the successful loading of a program dump or manually selected by RCU. Viewer input during automatic mode can be limited, for example, to the following [1]: audio control, frame display, viewer response (if the application program has been written to allow viewer interaction), mode termination, and so on. The viewer may terminate automatic mode and return the player to manual mode.

Programming mode provides the user with direct access to both the "addressable" RAM and user registers of the player's microcomputer. In this mode,

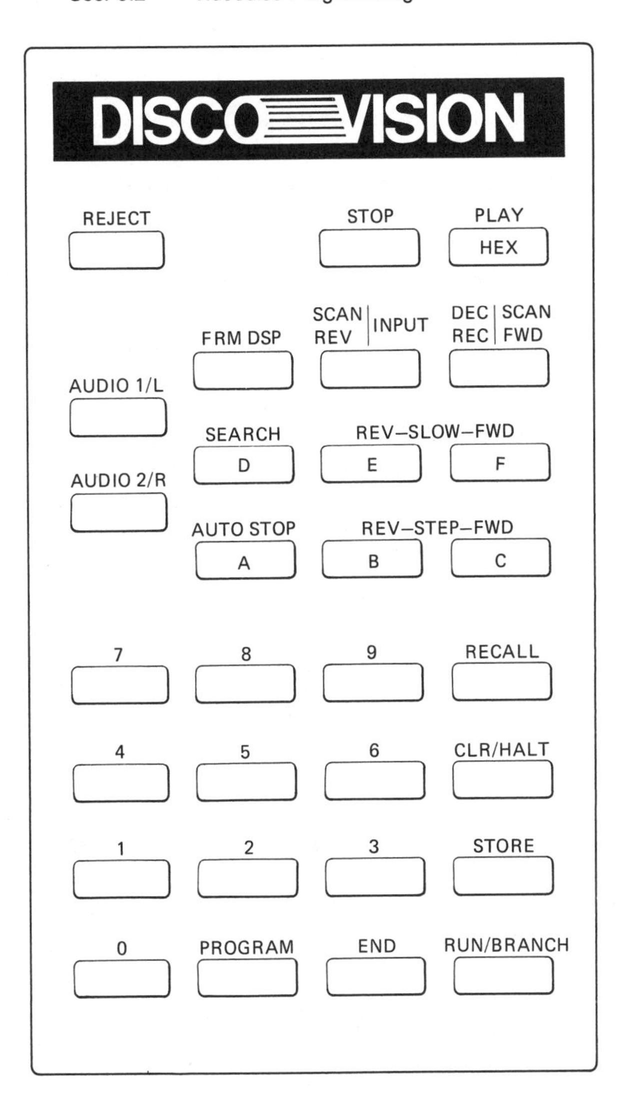

Figure 9.1 RCU program-entry push-buttons.

the user may enter commands, arguments, and register data to construct a program that will subsequently control the operation of the player in automatic mode. During programming mode, the television set can display the numerical values and command mnemonics entered from the RCU. Programming mode is normally succeeded (ENDed) by manual mode.

9.2.2 Sample of a Simplified Program Block Diagram

Programming capabilities depend on the particular videodisc system, or, more precisely, on the capabilities of the particular player type. The following examples are illustration for the general case.

Let's examine a typical programming situation for a sample videodisc on the subject of personnel training. The block diagram is shown in Figure 9.2. Students are to be tested on material recorded on the disc in frames N_1 to N_3, and five scenes after a frame N_2 need to be watched for, say 6 s each. After a student answers the question successfully (in this example only one question is considered), a scene recorded on frames N_8 to N_{10} should be shown. In that part, only one frame, N_9, requires a few seconds for student analysis. The disc also contains recorded supporting material for better understanding of the material for which students should be tested.

After the program starts, it searches to the beginning of the desired video program segment, at frame N_1. Then the program plays frames N_1 to N_3 in normal playing mode, except five frames after N_2 which are shown in a freeze-frame mode for 6 s each. A frame after N_3, $N_3 + 1$, program freezes to give students time to think, and possibly for questions that can be displayed or asked through the audio channel. In the latter case, the stop audio mode can be used, or the freeze frame will be $N_3 + X$, where X is number of frames displayed for the time when question is asked. Of course, the question can be displayed and/or pronounced.

The user (student) has a choice of four responses:

1. Directs the student through a corrective sequence recorded starting at frame N_4, containing eight pictures with 20 s of audio (possibly, an oral explanation), and back to review part of the main presentation.
2. The correct response causes the video program to continue.
3. This response indicates a lack of understanding of the material; it directs the user through a corrective sequence recorded on frames N_6 through N_7 and back to the program continuation (as when the answer is response 1).
4. Directs the user back to review the main presentation after passing through the corrective sequence, which contains 500 frames played in the slow-motion mode.

A fifth response can be used when a question is asked through the audio channel and there is a possibility of missing a part of the question.

The main program continues displaying frames N_8 to N_{10} with a 10-s stop at frame N_9. The program ends at frame N_{10} when play stops.

For some applications more players can be used—a player's network. One example is shown in Figure 9.3: a combination of five players. Four monitors and a screen are used as displays. Four monitors are combined to produce a large display: one example of the three letters JRI is shown in Figure 9.3a. Monitors can also display a separate item at the same time, sequentially or in combination. In parallel, on the screen a corresponding title, for display on the monitors, is shown; the title changes every 10 s. Instead of a title, an item or picture can be displayed.

A simplified program block diagram is shown in Figure 9.3b. When all programs are started, all five players will be locked on the same frame, N_1. With

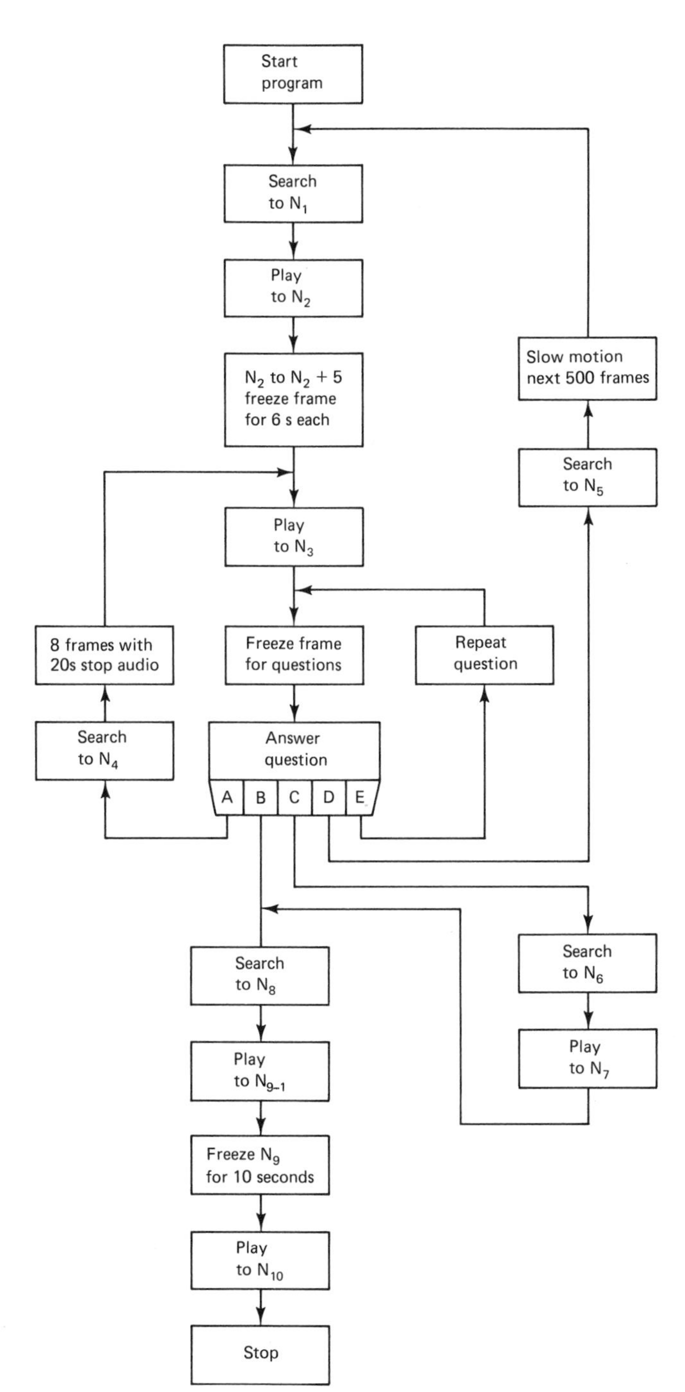

Figure 9.2 Sample program block diagram.

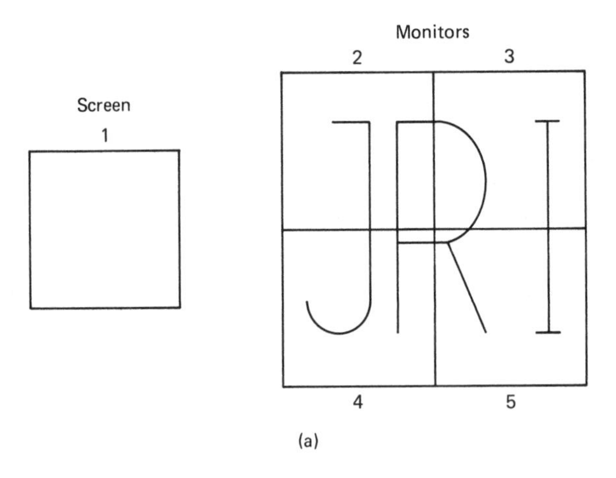

(a)

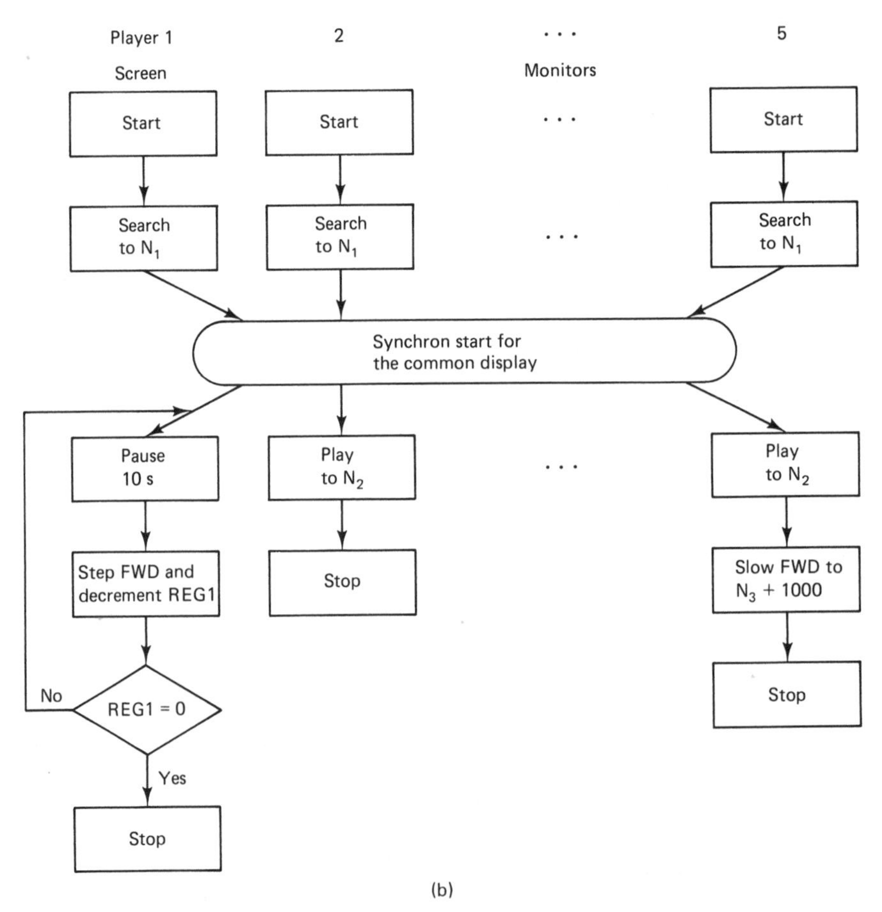

(b)

Figure 9.3 Multiplayer programming: (a) four monitors and one display screen; (b) corresponding simplified program block diagram for the five players.

a synchron start command, all five players start synchron execution of the corresponding programs. In the example shown, the first player displays consecutive frames on the screen for 10 s each. The second player is always in the normal playing mode. The fifth player is in the normal playing mode, except for the last 1000 frames.

Obviously, for this type of multiplayer application, the proper preproduction procedure should be performed. Adequate formatting of the material for the videodisc is one of the needs.

9.3 INFORMATION STORAGE AND RETRIEVAL

The videodisc is probably the most powerful information-storage medium yet devised. Basically, the plastic videodisc is a read-only memory (ROM), economically used in mass production. The same technology can be used for programmable ROM (PROM) systems. In the future, an erasable data storage is expected to follow from this technology.

Besides a consumer market (application) there are many other applications of the videodisc for information storage and retrieval. Disc itself can be used in different ways: direct play, program stored on the disc, or interactive (programming) discs. The following are among the main applications of videodisc systems as a source of video material:

Education
Training; flight simulators
Electronic publishing
Libraries, encyclopedias, books, and magazines
Catalogs: stores, museums
Hotels, airlines, ships at sea
Medicine; hospitals
Office: word processing
Government, military, law enforcement
Interactive videodisc technology
Games
Home computers
Point of sale
Digital image processing: analog input
High-definition TV
Three-dimensional TV
Transferring a video picture onto a motion-picture film
Specific applications: teletext, program for deaf and dumb

There are many ways that the videodisc can be used as an educational tool. The straightforward approach is to use the videodisc as a source of recorded material that can be displayed in desired order and as long as it is needed. The next step would be to use the videodisc's flexibility more. Information related to a single subject can be presented at different levels of complexity, depending on the user's needs and interests (the "intelligent disc"): for example, the same disc for technicians, students, or engineers, or for patients, nurses, or doctors. Information for the technicians could focus on, say, how to realize devices. The segment for engineers would examine how to design devices, and why. Students could pick and choose from information in both parts. A more sophisticated approach would be to combine the system's flexibility with programming or interactive programming.

The audiovisual material can be programmed in a question-and-answer format. After reviewing some material, the question(s) is asked and a choice of answers is offered. Further (dis)play depends on the answer chosen. Also, correct answers can be recorded so that they are available to instructors only. For example, answers can be recorded on special selected frames which are displayed only on special request, password-like. Stop audio can be included to improve system efficiency.

The videodisc used for training is usually associated with the industrial market. Many industries can be affected by disc technology, among them, automotive, aerospace, insurance, utilities, electronics, merchandising, extraction and processing, and medical. But not only industry can utilize a videodisc as a training tool. Home use or, in general, "how to" instructions are a training tool. Instead of reading instructions, a housewife or technician can watch as the instructor shows them, both visually and orally, how to cook or to assemble or install an appliance or part.

Generally speaking, any activity that requires frequent information searches of a large data file and/or large distribution of reports is a candidate for videodisc use. Large numbers of books or catalogs can be stored on records where many copies of each are required for wide distribution. Libraries can offer their patrons a mass computer retrieval service of printed information. Access to a library's centralized information and retrieval facility containing published material stored electronically in the information bank's computer is possible. Patrons can interrogate the library's computer, which will enable them to punch up the desired information for display on a cathode-ray tube (CRT). Conversion from CRT presentation to hard copy could be accomplished via a copy machine attached to the CRT screen. Also, articles from magazines, or whole magazine(s), can be placed on disc. Discs containing encyclopedia material can be programmed for simplified access to the desired subject.

Catalogs of merchandise or artwork can be stored on disc where many copies of each are required for wide distribution. For merchandising, electronic shopping at home as well as at store counters is possible. The videodisc catalog offers live action, demonstrations, fashion shows, including sequential display of similar designs from different houses for direct comparison, and so on. Usually, disc programming is used for videodisc catalogs. As with the printed version, customers can browse through the disc catalog or use a table of contents. As an example, suppose that the content displayed is:

1. —
2. —
.
.
.
5. Clothing
.
.
.

After pressing button 5 or a corresponding button, the program will search for the chapter "clothes" and freeze at that frame:

1. —
2. —
.
.
.
3. Children's clothing
.
.

After pressing button 3 the program will search for the subchapter "children's clothing" and freeze at that frame. This can continue, branching in mode subsections. Finally, say under "trousers," items can be displayed in slide mode, in normal playing mode, or in a combined mode. A similar approach can be used for catalogs of artwork (as well as for encyclopedias).

Instead of transmitting films to rooms through expensive closed-circuit systems, hotels and motels are able to provide programming for their guests via videodiscs for use on an in-room video player, or can provide, through existing master antenna systems, a complete entertainment package through the use of control video players operating on unused TV channels. Videodisc players and discs can also be used for in-flight movies on planes and ships at sea. Besides movies, other material can be offered: educational programs, games, musical shows, and so on.

Videodisc application in the field of medicine varies from potential archival storage and retrieval of x-rays to simulation of open heart surgery. Storing x-rays on videodiscs would reduce significantly the amount of space needed at present for this purpose. Physicians would be able to store and access the desired x-rays immediately and view them on a CRT monitor. Of course, other medical material can also be recorded on disc. A videodisc/microcomputer unit can teach complex skills. A medical student can simulate surgery using a light pen, for example. The visual information is from the disc, and programming—answer evaluation and branching, among others—is from the disc or from the outside (micro) computer.

The videodisc system provides an alternative way to publish the countless reports needed in business and industry and may well constitute the first step toward electronic documentation—the deescalation of office paperwork—by turning out computer data, for example, on low-cost discs instead of paper. For word processing, the disc can contain word-processing source information on a fraction of its surface. The rest of the surface, or the entire surface, can hold huge vocabularies, grammar rules, or biographical facts that would be drawn upon automatically to check a writer's input [2]. As a part of an electronic office network, a multidisc-player, multiuser system can be developed. The arrangement would combine the fast, complex calculation capabilities of computers with interactive disc features. Extensive disc libraries, like today's magnetic-tape libraries, in a setting where a business already has extensive control of computer time-sharing operations, could be housed in control locations to be called up from remote terminals, physically accessed, and searched for required information. Fewer discs would be required than tapes, and access time would be shorter.

In governmental areas, videodisc can be used for archiving and as an educational and training tool, particularly in the armed forces. Increasingly, complex weapons require better training media. In a test program [2] the Army found that soldiers using discs learned their lessons up to 35% faster, took tests 23% faster, and completed jobs up to 30% faster. In an archival use, the IRS, for example, could store tax returns on disc. Law enforcement is another area in which the videodisc can meet specific requirements; the disc is an inexpensive and efficient means of maintaining, updating, and transmitting important information and records such as fingerprints, mug shots, MOs, aliases, and the like.

Interaction is one of the very powerful properties of videodisc systems. This means that users can pick and choose what they want from the material on the disc by giving simple (microprocessor) commands. One example is possible application of the interactive videodisc technology for simulators such as those used in pilot training. Instead of simulating, particular aircraft areas can be photographed from different angles and in different seasons and recorded on discs. More than one disc can be played at the same time for the same display. The computer controls the images on the disc to create an approximation of flight experiences based on the user's computer-replicated responses. The same approach can be used for driving testing and training. For this use, streets are photographed.

A home computer can be used to control disc interaction, and there are many other possible applications of the disc/home computer combination, for example: the telephone book or real estate listings on disc. Searching through information stored on a disc can be faster and easier than searching through a book. Also, video games can be improved. Instead of little stick figures, fully animated knights or aliens or trolls can be shown [2]. New variations can be added to the rules. Another videodisc computer interface application is in the area of computer software storage. The disc can be used to store entire software packages, including several different programs designed to control the operation of a computer and to perform diagnostic routines.

The videodisc can be used as a memory band for any information that is capable of being displayed as an image on a cathode-ray tube or in digital form. Thus, for digital image processing facilities, videodisc can be used as an analog data base.

A high-definition television (HDTV) system is able to project a high-resolution picture onto a large screen [3,4]. The HDTV can be considered as a third step in the four-step TV evolution: monochrome TV, color TV, high-definition wide-screen TV, and stereoscopic TV. The last three steps are still under development, and TV is currently at step two. A highly defined and pleasing large-screen representation may be made with the standard systems (NTSC, PAL, or SECAM), with comb filtering, digital noise reduction, and extended luminance bandwidth. But this is still far from ideal in sharpness, realism, and visual impact. More psychophysical evaluation has to be conducted in HDTV research and development. The results of a subjective evaluation test relating picture size and aspect ratio show that as the picture becomes large, the 5:3 or even 2:1 ratio becomes more preferable over the 4:3 ratio of current television systems. Also, a large picture viewed from a great distance seems more impressive than a small picture viewed from a small distance, even if the viewing angles are the same. When picture size is large, sharpness is an important factor of picture quality. Resolution, brightness, and contrast all contribute to the sharpness of a television picture. The resolution necessary for a TV system is closely related to visual acuity and viewing distance. Vertical resolution depends on the number of scanning lines. If the viewing distance is three times the picture height, more than 1000 scanning lines are desirable in HDTV so that adjacent scanning lines become indistinguishable. But more experiments have to be conducted to find out when additional lines in a television picture do not make a difference to the eye. The HDTV videodisc system should ultimately include a picture pickup camera with high resolution and a large-screen display device. The TV signal bandwidth is between 15 and 30 MHz, and proper recording on the disc should be performed. For example, the inside radii should be larger than one-half of the outside radii (required for maximal capacity) in order to perform high-frequency output. Another approach would be to use more than one revolution per picture (in the CAV mode). The CLV mode can also be used. For some applications only the luminance signal can be recorded: printed page display, for example.

The final aim of high-fidelity television is stereoscopic television. Binocular stereoscopic TV using the parallax effect is comparatively easily attained, but there remain some problems of picture quality and visual fatigue, and the result could not be described as high-fidelity TV. An ideal high-fidelity stereoscopic TV which does not cause eye fatigue and in which concealed portions of the picture can be viewed by changing the viewpoint is difficult to attain because of the necessary bandwidth and line camera [4]. Further, if a small image were to be viewed in stereoscopic TV, it would be like viewing miniatures, and the feeling produced would be unnatural.

There are a number of mechanisms that the brain uses to perceive depth. There is shading, perspective, and occlusion—the fact that close items can block the view of farther ones, for example. To prevent each eye from knowing what the other is seeing, glasses can be used. The brain can determine depth by analyzing the angle that both eyes form in focusing on an object; this mechanism is called *binocular vision*. Figure 9.4 shows one example of a three-dimensional TV signal recording on videodisc. Left- and right-eye pictures, produced with a three-dimensional (twin-lens) camera, are transferred to videotape. A three-dimensional video processor electronically colors the left-eye image red and the right-eye image cyan (blue-green). The images are then superimposed. The output is a standard (NTSC, PAL, SECAM) signal that is recorded in the usual manner on a disc. During playback on a conventional color TV set with three-dimensional TV glasses, the scene can be viewed three-dimensionally in full color. Instead of red and bright-green filters over the eyes, a color three-dimensional image can be combined through polarizing filters rotated 90° apart for the two eyes' views. The audience wears polarized glasses with appropriate polarization, the 90° rotations preventing the wrong eye from seeing what it shouldn't. There are also a number of methods of presenting three dimensions without glasses. Some involve vertical shots and are called *parallax stereograms,* and some use large lenses and are called *autostereograms*. For some of them, varied fixed viewing positions are required.

It has been shown [5] that the videodisc as a source for video-to-film transfer offers great flexibility in its application to European as well as U.S. television standards, offering at the same time the advantage of great compactness of the equipment necessary. Some of the technical and mechanical complications inherent in tape-to-film transfer are completely eliminated when using the videodisc as a source. While it is easy to take a still photograph of a single static video picture from a TV screen, it is much more difficult to film a sequence of such pictures without sacrificing their content or the precision of their movements. The design of a camera having a fast shutter and pull-down time, equal to the field flyback time of the video image, roughly 1.5 ms, poses difficult mechanical problems. But even if a camera with less than 2 ms of closed-shutter time is used, the problem is not yet solved because the sampling rate of the camera (24 frames per second) is different from that of the TV system (30 frames per second). It is thus not possible in continuous operation to synchronize the camera shutter with the flyback cycle of the video image.

On the disc surface, the line sequence that forms the framework of the television picture is recorded on the spiral, and with the same chronology: odd lines first, then even lines, but end to end, not interlaced as on a TV screen. The discs available for special-mode play can be used for video-to-film transfer. Usually, a multiple videodisc scanning is used, so that transfer time increases from 4 to 6 times over real time with European video standards, and from 3 to 4.6 times for U.S. video standards [5]. The same videodisc transfer equipment can be used with equal effectiveness with European or with U.S. standards and as easily with 35-mm as with 16-mm film. For the European 25-frame-per-second standard there is almost a strict correspondence between motion-picture and

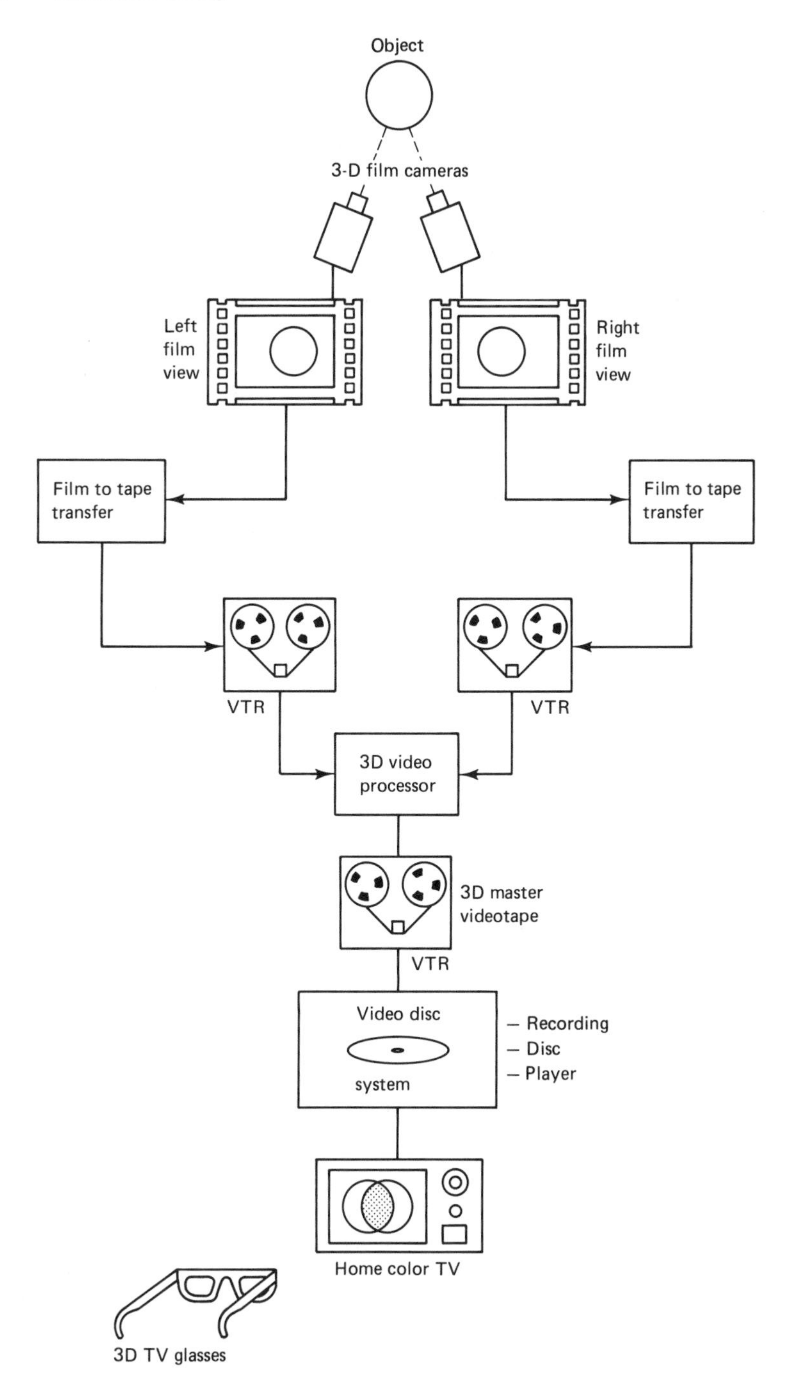

Figure 9.4 Three-dimensional TV signal processing.

video sampling rates. The player can be programmed so that each frame is read four times before proceeding to the next frame. The first scanning can be used to apply the red component of the image to a very high performance black-and-white TV monitor, the second scan is used for the blue components, and the third scan for the green one. A camera positioned in front of the monitor photographs and superimposes the consecutive color extractions through a synchronized rotating color sector filter disc, recombining them as a color image on a single film frame. The fourth sector of the four-sector filter is an opaque sector, and it coincides with the fourth scan and with the closed-shutter sector of the camera. The only purpose is to provide sufficient time for film pull-down in the camera. Thus transfer time is four times real time. The use of primary color filters to reconstitute the three color components of the video picture on color film lead to some restrictions in the choice of the monochrome CRT because its phosphor must emit the three components with adequate spectral intensity. The final transfer onto color film is then made in a processing laboratory by means of the process customarily used for the reissue of color film stored on black-and-white separation stock. It is easy to derive from this system a variation applicable to NTSC (30 frames per second) specifications [5].

Special programs for the deaf and dumb can be recorded on disc. Programming can also be performed for such programs. If needed, a teletex can be recorded similar to its normal use in TV channels.

9.4 DIGITAL DATA STORAGE

In channel coding, the data to be recorded or transmitted are modified to obtain the highest information density permitted by the limiting characteristics of the channel. In videodisc systems, the disc can be used as either an analog (video) or a digital channel for data recording. There are economic advantages in exploiting existing videodisc production facilities for the publication of machine-readable information by encoding digital information on the standard video signal [6]. Also, encoding information into video-formatted frames provides the ability to intermix digital and video information on the same disc. Frames of video, when accessed, are directed to a video monitor, and frames of digital information will be directed first to a microcomputer and then to the desired output device. The videodisc is very suitable for direct digital recording [7] as the recorded signal is always in on–off form. In general, for a direct digital recorded disc, the player has to be modified to reconstruct the data properly.

Error identification and correction is one of the main problems for successful use of videodisc technology for the storage and retrieval of digital information. For example, the dropouts that presently occur in the production of videodiscs for entertainment purposes, although often visible, are seldom offensive to viewers. However, the sensitivity of digital information to errors is much greater, since the loss or change of a single bit could cause a computer program to malfunction or an index to return incorrect information [8].

9.4.1 Binary Data Transmission Through TV Channels

The existing videodisc systems were designed in such a way that the synchronization of the readout system is based, among other things, on the horizontal and vertical synchro pulses in a composite video signal. This means that through direct use of the existing system, about one-third of the total capacity of the videodisc would be lost, because that percentage is necessary for memorizing all synchro pulses.

Direct digital data recording in the video channel rather than in the FM channel has many advantages:

1. Unlike schemes that use the FM channel, the recording frequencies can be kept in the video range so that special modifications are not necessary within the player to eliminate false sync signal detection.
2. The color burst (subcarrier) can be used to synchronize a decoder clock using conventional integrated circuits, thereby avoiding the need to have a phase-locked loop track the data.
3. Digital data and pictures can be recorded on the same discs and mastered on the same mastering machine.

The two main problem areas to be overcome are lack of frequency response and a high incidence of disc defects. The first problem is best tackled by multilevel block coding and the second by error-correction techniques. A multilevel code is a block code that takes n bits of data and records them as a word of 2^n possible parameters (amplitude, phase, frequency) of the carrier. Since the word (symbol) rate is reduced by a factor of $1/n$, the frequency response of the channel is enhanced.

The video channel can be used fully or in part for digital data. For example, either luminance and/or chrominance signals can carry digital data. Because of the lack of three different methods (standards) for chrominance signal generation, luminance and chrominance signal generation based on binary data will be discussed separately.

9.4.1.1 Luminance Signal Generation [9]. The problem here consists of the following. It is necessary, on the basis of binary data with bit rate R_b, to create a luminance signal $Y(t)$ according to existing standards. This signal is to be processed as usual and memorized on the videodisc as a part of the composite signal, to be read out by existing video readout systems. The signal-to-noise ratio in the videodisc channel is about 40 dB; the bit rate R_b should be as large as possible.

Figure 9.5 displays the simplified block diagram of the luminance signal channel [10,11]. Basically, this is a multilevel-coding, M-ary ASK. The input data are inserted into the buffer serially, the bit rate being R_b. The buffer capacity is n bits. The readout of the buffer is parallel and performed every T_s seconds, where $T_s = n/R_b = nT_b$, where T_b is a bit time.

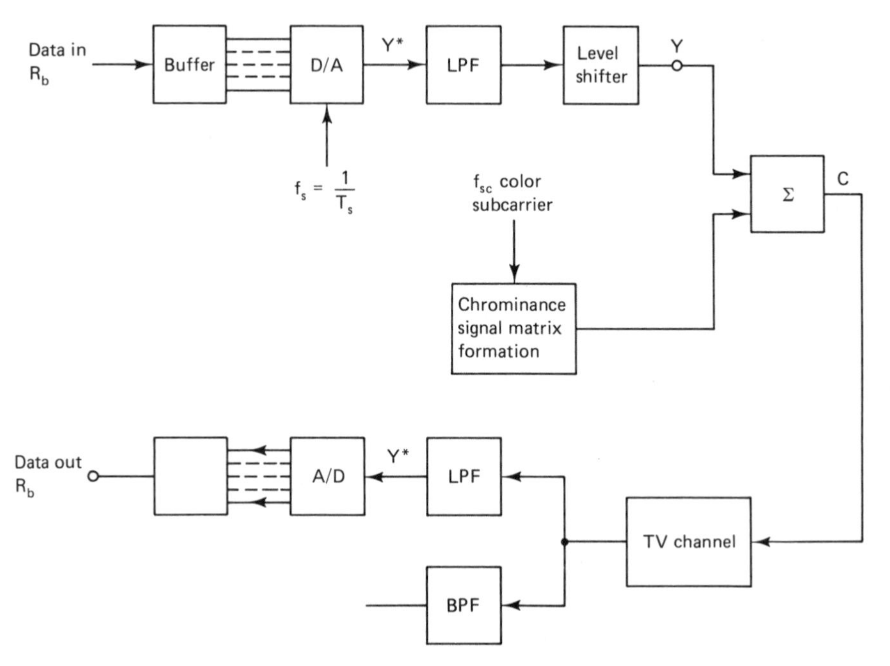

Figure 9.5 Simplified block diagram of the luminance (M-ary) signal generation.

At the output of a D/A converter, a discrete (in the amplitude domain), continuous signal $Y^*(t)$ is obtained with a total of 2^n levels L_k, $k = 0, 1, \ldots, 2^{n-1}$. The low-pass filter bandwidth B should ensure that (1) the signal spectrum at the filter output corresponds to that of the standard luminance signal (i.e., not have spectral components higher than the maximal frequency of the luminance signal f_m), and (2) the signal energy within the bandwidth B is sufficient to make a reliable level detection (A/D conversion).

On this basis it follows that $B \leq f_m$. The second condition is fulfilled if [12] $B \geq f_s$. It was found experimentally that the signal reproduction quality for frequencies exceeding f_{sc} is often unsatisfactory, especially when the readout is performed for the inside radius of the disc. In the same manner, the reference frequency of the TV channel is the subcarrier frequency f_{sc} (where $f_{sc} \simeq 3.6$ MHz in 525-line systems and $f_{sc} \approx 4.3$ MHz for 625 lines). Therefore, we select $\frac{3}{2}f_s \leq f_{sc} = B$, where, due to synchronization, it is desirable that f_s and f_{sc} be multiples of the common frequency.

In the level-shifter circuit, the absolute value of the maximal and minimal levels is adjusted according to existing TV standards. The obtained luminance signal $Y(t)$ is then combined with the (possible) chrominance signal to form a composite video signal.

After processing through the standard TV channel, the luminance and chrominance signals are separated in the conventional manner using a bandpass filter (BPF) for the chrominance and a low-pass filter (LPF) for the luminance.

The bandpass of the LPF is also B; the possible difference from the LPF in the coder is that here the subcarrier could be suppressed. The luminance signal $Y^*(t)$ obtained at the LPF output is then converted back to digital form.

As already mentioned, the signal-to-noise ratio in the videodisc channel is approximately 40 dB. Experimentally, it was found that four levels ($n = 2$) can be produced fairly well, although 8 and 16 levels can also be distinguished.

Since $R_b \leq (2n/3)B$, for $n = 2$ we have $R_b = \frac{4}{3}B$. If the audio signal is transmitted in the video channel, it is convenient to take $B = f_{sc}$. But if the entire video channel is vacant, B can be 6 or 7 MHz, depending on the standards. For example, for a system with 525 lines, $f_{sc} = 3.6$ MHz, in the first case it would be $R_{b\max} = \frac{4}{3}f_{sc} = 4.8$ Mbits/s, and in the latter case for, say, $B = 5.4$ MHz, it is $R_{b\max} = 7.2$ Mbits/s. If $B = (3/2)f_{sc}$, then $R_{b\max} = 2f_{sc}$ Mbits/s.

To investigate the possibility of level detection at $B = B_{min}$, a computer simulation was carried out for the case when the level change was made after T_s seconds. In addition, the unfavorable level orders were selected: L_0, L_2, L_1, L_3, L_1, L_2, L_0, L_1, L_2, L_3, L_1, L_2, L_2, L_0, L_2, L_0, ..., as in Figure 9.6. It was assumed that the mean value was equal to zero when the diagram in Figure 9.6 was plotted. On the basis of the results obtained, it is clear that for $B = B_{\min} = \frac{3}{2}f_s$, a reliable level discrimination is possible, but it is necessary to carefully select the decision levels because of the overshoots and oscillations around particular levels. For example, the decision level between L_2 and L_3 should reside between the signal level corresponding to L_2 in the sequence L_0, L_2 and the signal level corresponding to L_3 in the sequence L_2 and L_3 in the input data. The methods presented can be used for any TV system, including the broadcasting system.

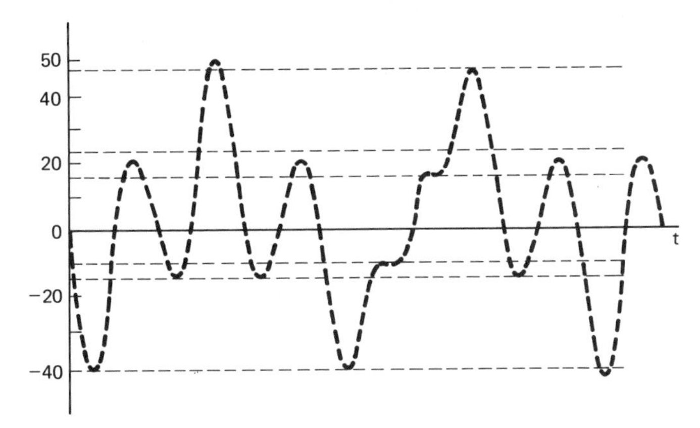

Figure 9.6 Output signal for $B = B_{\min}$.

9.4.1.2 Chrominance Signal Generation [13]. The problem considered herewith consists of the following. It is necessary on the basis of binary data with rate R_b to create a chrominance signal $C_h(t)$ according to the existing standards. This signal is to be processed as usual and recorded on a videodisc as part of a composite signal to be read out by existing video readout systems. The signal-to-noise ratio in the videodisc channel is about 40 dB (or 60 dB for the carrier-to-noise ratio for a bandwidth of 30 kHz). The bit rate R_b should be as large as possible. In general, the chrominance signal carrying binary (digital) information can be generated according to one of the standards NTSC, PAL, or SECAM. The NTSC video signal coding processing will be considered first. Some of the basic assumptions for the NTSC composite video signal relevant for us are:

1. The luminance signal is filtered to a bandwidth of about 4.2 MHz (−2 dB).
2. The I and Q signals are bandlimited to about 1.3 MHz and 0.5 MHz, respectively.
3. The color subcarrier is at $f_{sc} = 3.58$ MHz.
4. The I and Q signals amplitude modulate subcarrier signals that are 90° out of phase with respect to each other [quadrature amplitude modulation (QAM)].

The chrominance signal is summed with the luminance signal, $Y(t)$, to form a composite video signal. The luminance signal can carry monochrome picture information or digital (binary) information.

Figure 9.7 displays a simplified block diagram of the solution proposed. Basically, this is a multilevel coding, combined phase and amplitude shift keying. Or, more precisely, the technique is quadrature amplitude shift keying (QASK), because there are two channels: in-phase (I) and in-quadrature (Q) channels. The input data are inserted into the buffer serially, the bit rate being R_b. The buffer capacity is n bits. The readout of the buffer is parallel and performed every T_s seconds, where $T_s = n/R_b = nT_b$, T_b being a bit time. The parallel output is split into two parts: n_i bits for the I channel and $n_q = n - n_i$ bits for the Q channel.

At the output of D/A converters, discrete (in the amplitude domain), continuous signals I^* and Q^* are obtained with a total of 2^{n_i} levels L_k, $k = 0, 1, 2, \ldots, 2^{n_i} - 1$, for the I^* signal and 2^{n_q} levels L_l, $l = 0, 1, \ldots, 2^{n_q - 1}$. The low-pass filter bandwidths B_I and B_Q should ensure that (1) the signal spectrum at the filter outputs corresponds to that of the standard I and Q signals (i.e., not have spectral components higher than the maximal frequency of the color signals $f_{\max I}$ and $f_{\max Q}$, and (2) the signal energies within the bandwidths B_I and B_Q are to be sufficient to make a reliable level detection (A/D conversion).

On this basis it follows that $B_I \leq f_{\max I}$ and $B_Q \leq f_{\max Q}$. The second condition is fulfilled if [12] $B \geq \frac{3}{2} f_s$. The outputs of the product modulators are:

I channel: $s_I(t) = I^* \cos(\omega_{sc} t + \alpha_0)$
Q channel: $s_Q(t) = Q^* \sin(\omega_{sc} t + \alpha_0)$

After filtering and summing the standard chrominance signal, $C_h(t)$ is obtained:

$$C_h(t) = I^* \cos(\omega_{sc} t + \alpha_0) - Q^* \sin(\omega_{sc} t + \alpha_0) \tag{9.1}$$

or

$$C_h(t) = A(t) \cos[\omega_{sc} t + \alpha_0 + \phi(t)] \tag{9.2}$$

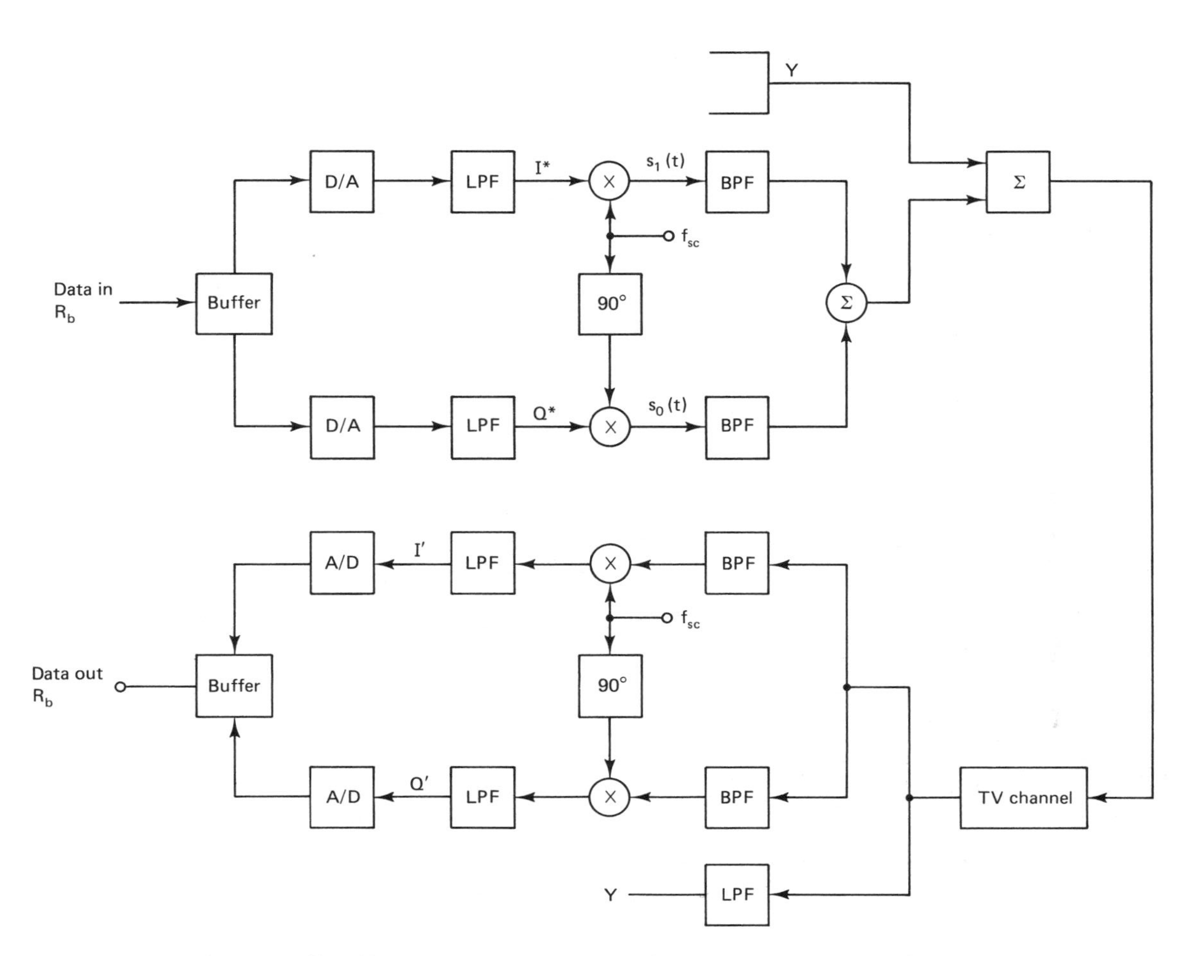

Figure 9.7 Simplified block diagram of the chrominance (M-ary) signal generation.

where

$$A(t) = I^{*2} + Q^{*2}$$

and

$$\tan \phi(t) = Q^*/I^*$$

The chrominance signal $C_h(t)$ obtained is then combined with the luminance signal to form a composite video signal.

After processing through the standard TV channel, the luminance and chrominance signals are separated in the conventional manner using a bandpass filter (BPF) for the chrominance and a low-pass filter (LPF) for the luminance. The baseband signals are obtained by product detectors and selected by low-pass filters, signals I' and Q'. The bandpass of the LPFs is the same as in the coder. The signals I' and Q' obtained at the LPF outputs are then converted back to digital form. These signals are sampled by the A/D converters at an appropriate rate and phase. The sampled value is compared to preset threshold values ("slicing levels") and a decision is reached as to which symbol was transmitted based on the sampled value of the signal received. At the output the signal is obtained back in a digital form.

The total number of signals M is typically the square of an even number $K(K = 2N)$, and the quadrature amplitudes $I = a_i$ and $Q = b_j$ take on equally likely values of id and jd, with $i, j = \pm 1, \pm 3, \ldots, \pm(K - 1)$. In the phase amplitude plane, the QASK signals form concentric squares of signal points (or phasors), with each pair of adjacent points separated by $2d$. In Figure 9.8 an example of this signal point structure for QASK-16 ($M = 16$) or 4-bit QASK is shown. Corresponding 4-bit words in Gray code are assigned.

A signal presentation similar to that in Figure 9.8 is typical for QAM systems, where I and Q signals are equally bandlimited. In the NTSC system these signals are bandlimited: to about 0.5 MHz in the case of the color-difference signal containing the green-purple color-axis information, the Q signal; and to

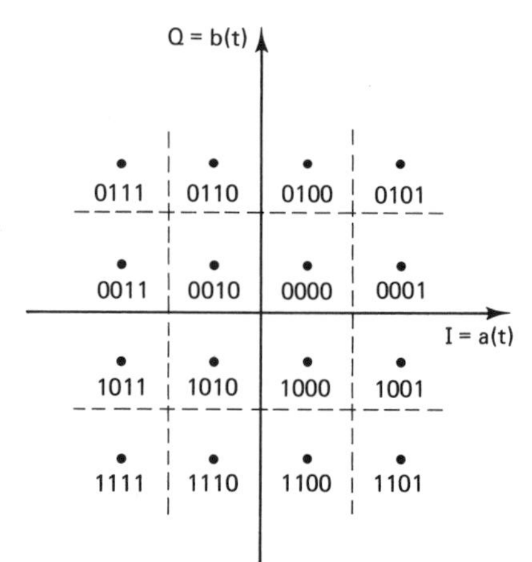

Figure 9.8 Four-bit QASK represented as 16 discrete subcarrier amplitudes and phases.

about 1.3 MHz in the case of the color-difference signal representing the orange-cyan color-axis information, the I signal. The instantaneous phase of the chrominance signal represents the hue of the scene at the moment, and its amplitude—relative to the brightness signal amplitude—is a measure of saturation.

Thus the I signal can carry more information than the Q signal. In first approximation only the I signal can be taken to carry information, while the Q signal can be retained constant.

As already mentioned, the signal-to-noise ratio in the videodisc channel is approximately 40 dB. Experimentally it was found that four levels ($n = 2$) can be fairly well produced in each signal, although 8 and 16 levels can also be distinguished.

Four signal points can be represented as in Figure 9.9, if the Q signal is constant (zero in this example). Corresponding 2-bit words in Gray code are assigned. The subcarrier can have two discrete amplitudes, d and $3d$, and two discrete phases, 180° in difference. Since

$$R_b \leq \tfrac{4}{3} n B_I \qquad \text{for } n = 2 \tag{9.3}$$

we have $R_b = \frac{8}{3} B_I$. For $B_I \simeq 1.3$ MHz, the maximal bit rate would be $R_{b\max} \simeq$ 3.5 Mb.

Intersymbol interference (ISI), noise, and poor synchronization cause errors and the transmitting and receiving filters should be designed to minimize the errors. The ISI can be eliminated by using pulses with raised cosine frequency characteristics.

PAL system. The most significant difference in signal processing in the PAL system compared to the NTSC system is that the U and V signals are equally bandlimited to about 1 MHz. This means that signals U and V should be used equally; a state diagram similar to that in Figure 9.8 would be appropriate. All other is analog to NTSC signal processing (Figure 9.7), discussed earlier.

SECAM system. Practically all the time chroma signals would be generated in the same way.

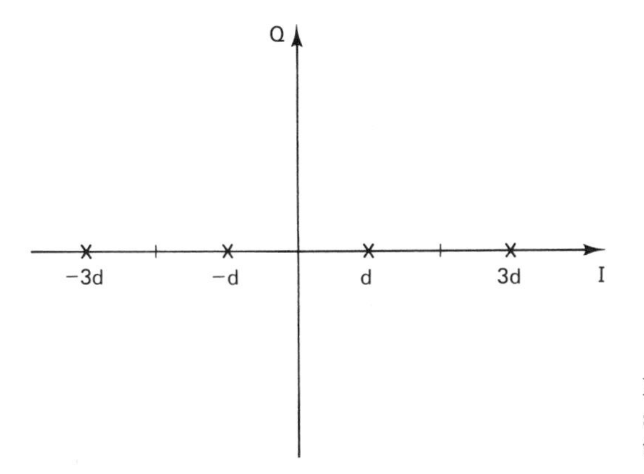

Figure 9.9 Two-bit signal points represented as two discrete subcarrier amplitudes and two discrete phases.

The choice of (M-ary) signaling method must be determined for the particular videodisc system.

Coherent phase-shift keying (CPSK) and differential phase-shift keying (DPSK) are two techniques used often in digital communications. The literature abounds with analyses of their performance under a variety of conditions [14–24]. Techniques for determining the degradation in M-ary CPSK operating in the presence of ISI, additive Gaussian noise, and imperfect carrier phase, as well as the performance of M-ary DPSK subject to ISI and additive Gaussian noise, are also presented in the literature [25]. The chief reasons for the widespread use of these techniques are simplicity of implementation, superior performance over the additive Gaussian noise channel, minimal bandwidth occupancy, and minimal envelope variation. The relative performance of CPSK and DPSK systems is well understood only in the presence of additive Gaussian noise. In this case, the detection efficiency of DPSK is known to be about 1 dB (in S/N) below that of CPSK for binary modulation, and this degradation approaches 3 dB for multilevel systems. In CPSK, however, the generation and extraction of a local carrier-phase reference at the receiver is required. A coherent phase estimate is usually obtained using phase-locked-loop (PLL) techniques, and because of the frequency instabilities and phase flitter inherent in transmitter and receiver systems, carrier recovery loop bandwidths cannot be made arbitrarily small. Consequently, in practice, a noisy phase estimate is obtained and only partial coherent reception can be claimed. The reason for using DPSK is its immunity from slow carrier-phase fluctuations; therefore, the phase recovery problem inherent in CPSK is avoided. However, the detection efficiency of DPS may approach that of CPSK under noise phase estimation conditions and intersymbol interference (ISI).

9.4.2 Digital Data–Digital Channel

Videodiscs are being developed solely as entertainment mass storage devices. But they are very suited for digital signaling: the information can be encoded in a form most suitable for digital recording and with no video compatibility. The term video will be used in this application to indicate the (same) technology. Practically, slight modifications of the recording and playback electronics are required, and a very high density, low-cost, archival mass storage medium is obtained. Total capacity per storage unit could be as high as 15 to 30 billion bits.

Some of the advantages of the digital videodisc technology are: (1) cheap (mass) media, (2) reduction in physical floor space, and (3) direct rapid access to the full content. Huge capacity is important for I/O peripherals. This technology, combined with microprocessors or home computers, can be used in many applications. In digital images processing, data and algorithms for processing (programs) can be stored on-site, for example.

Digital information is recorded the way similar to that for analog (FM);

information is contained in the pit pattern. Clock and data are derived from the read waveform on the disc, and the transitions in the signal must occur frequently enough to provide synchronization pulses for the free-running clock. On the other hand, consecutive transitions must be far enough apart to limit the interference to an acceptable level for reliable detection. Because of this, binary data are encoded into coded binary sequences that correspond to waveforms in which the maximum and minimum distances between consecutive transitions are constrained by prescribed coding rules. The maximal disc capacity is limited by the smallest distance between adjacent edges that can be detected accurately. For the particular system, the smallest pit size is specified in advance. Then the maximal number of pits on the disc is defined. The information capacity is then dependent on the coding rules, that is, on the number of bits per pit (i.e., the smallest spatial period). Some of the codes used most frequently are discussed in Chapter 6, and more discussion follows.

The maximum and minimum distances between consecutive transitions correspond to the minimum and maximum pulse rates in the waveform, respectively. For any given data density, it is desirable to constrain this range of pulse rate to be as narrow and as low as possible, whenever a (highly) nonlinear read-write process causes irregular read signal amplitudes and results in phase-shift errors [26]. Usually, ac-coupling elements are included in the channel. A dc component in the waveform results in a nonzero average value of the amplitude and causes charge accumulation of any of these elements. The errors in signal detection can be reduced by reducing the signal distortion caused by the ac coupling networks. Although, based on the MTF, it can be said that the optical part of the videodisc system is dc coupled, ac coupling is usually performed in the electronic part of the channel. The accumulated charge, or any digit in a binary-coded sequence, is the difference between the numbers of positive and negative pulses in the corresponding waveform up to that digit, because the accumulated charge increases by one unit for a positive pulse and decreases by one unit for a negative pulse from digit to digit in the waveform.

9.5 CODES FOR OPTICAL RECORDING

First, we will define some terms (Ref. 26). A digital signal is a sequence of discrete, discontinuous voltage pulses. Each pulse is a signal element. Binary data is transmitted by encoding each data bit into signal elements. If the signal elements all have the same algebraic sign, then the signal is unipolar. In polar signaling, one logic state is represented by a positive voltage level, and the other by a negative voltage level. The data signaling rate (or data rate) of a signal is the rate, in bits per second (b/s), at which that data is transmitted. The duration or length of a bit is the amount of time it takes for the transmitter to emit the bit; for a data rate R, the bit duration is 1/R. The modulation rate, in contrast, is the rate at which signal level is changed. This depends on the nature of the digital encoding. The modulation rate is expressed in bauds, which is signal elements per second. Finally, the terms mark and space, for historical reasons, refer to the binary digits 1 and 0, respectively.

9.5.1 Code Parameters and Evaluation Criteria

Coding sequences may be denoted by the designed parameters (d, k, c) corresponding to the following three constraints [26,27]:

1. d (digits) is the number of 0's between any two consecutive 1's (in the coded sequence) in the shortest run length (sequence).
2. k (digits) is the number of 0's between any two 1's (in the coded sequence) in the longest run length (sequence).
3. The accumulated charge at any digit position in the sequence is bound by $\pm c$ units.

If group of the m input bits is converted in a group of n bits, the ratio m/n is called the *rate of the code*. If the bit interval is T(seconds), the time intervals for the minimal and maximal distance of the change are $T_{min} = (m/n)(d + 1)$ T(seconds), and $T_{max} = (m/n)(k + 1)T$(seconds). The density ratio (DR) is a measure of recording efficiency and is given by

$$\mathrm{DR} = \frac{T_{min}}{T} = \frac{\text{data density}}{\text{highest recorded density}} = \frac{m}{n}(d + 1) \tag{9.4}$$

Also, an important code parameter is the quantity W:

$$W = \frac{m}{n} T \tag{9.5}$$

The reciprocal quantity is a clock rate (CLR). The parameters for some codes are given in Table 9.1.

There are two important tasks involved in interpreting digital signals at the receiver. First, the receiver must know the timing of each bit. That is, the receiver must know with some accuracy when a bit begins and ends. Second, the receiver must determine whether the signal level for each bit position is high (1) or low (0).

TABLE 9.1 CODE PARAMETERS

Code	*m*	*n*	*m/n*	*d*	*k*	T_{min}	T_{max}	*DR*	*W*	*CLR*
NRZI	1	1	1	0		T		1	T	—
FM	1	2	0.5	0	1	$T/2$	T	0.5	$0.5T$	$2/T$
MFM	1	2	0.5	1	3	T	$2T$	1	$0.5T$	$2/T$
GCR	4	5	0.8	0	2	$0.8T$	$2.4T$	0.8	$0.8T$	$1.25T$
3PM	3	6	0.5	2	11	$\frac{3}{2}T$	$6T$	1.5	$0.5T$	$2T$

Source: Ref. 28.

A number of factors determine how successful the receiver will be in interpreting the incoming signal: the signal-to-noise ratio (S/N), the data rate, and the bandwidth of the signal. With other factors held constant the following statements are true [26]:

An increase in data rate increases bit error rate (the probability that a bit is received in error).
An increase in S/N decreases bit error rate.
Increased bandwidth allows increased data rate.

"Error rate" is defined here as the number of errors in a given group of bits, divided by the number of bits, as the ratio of errored bits to transmitted/recorded bits in some time interval.

Each of the many digital modulation codes has certain desirable and undesirable characteristics. The best choice for minimizing the probability of error is the code whose characteristics best match the channel. Some of the key parameters of concern in evaluating a potential modulation code are:

Signal spectrum—Several aspects of the spectrum are important. A lack of high-frequency components means that less bandwidth is required for transmission. On the other hand, with direct current (dc) in the signal, there must be direct physical attachment of transmission components; with no dc component, alternating current (ac) coupling is possible.
Spectral zeros—These are the frequencies at which the code has no energy. The theoretical minimum bandwidth at which a baseband signal of rate 1/T can be transmitted without error is the Nyquist bandwidth 1/2T (where T is the bit period). For a practical high-density recording channel, the typical bandwidth runs 1.3–1.4 times 1 divided by twice the minimum transition interval. It may also be possible to put a pilot signal (a tone) at a spectral zero that can be used for synchronization without interfering with the data.
Signal synchronization capability—The need to determine the beginning and end of each bit position may require a separate clock channel (lead) to synchronize the transmitter and receiver. Some coding schemes avoid this problem.
Maximum dc content—This is expressed as the ratio of the dc content of a worst case pattern to the peak signal value. It is 1 for NRZ.

The digital sum variation (dsv) is the running integral of the area beneath the coded sequence: in computing the dsv, the binary levels are assumed to be ± 1. If the dsv of the code can grow indefinitely, the code has dc content; if the dsv is bounded, the code is dc free. Alternatively an accumulated charge, c, can be used. If accumulated charge at any digit position in the sequence is bounded by ± 1 c units, the code is dc free.

Signal interference and noise immunity—Certain codes exhibit superior performance in the presence of noise. This is usually expressed by the bit error rate.

Error propagation span—The span of source data bits over which more than one decoding error will be made for a single error in decision on the channel amplitude.

Eye opening—The width of the opening of the eye pattern should be expressed in terms of the source data bit period and the height in terms signal amplitude for random data. It can be expressed both as an RMS value for random data and as a worst case value. For the purposes of this criteria, it should be assumed that the equalizing, unless otherwise stated, is flat to half the maximum transition rate and essentially zero at the second harmonic.

Error-detection capability—Many signaling schemes have an inherent error-detection capability, which is: the probability of detecting signal errors.

Pattern sensitivity—Are there any patterns in the data which would significantly increase the raw bit error rate?

Cost and complexity—Although digital logic continues to drop in price, this factor should not be ignored. Sometimes, the tradeoff in S/N improvement vs. $ spent for development must be considered.

The minimum transition interval (minimum run length), maximum run length and the code rate are already discussed.

9.5.1.1 Spectral Density for Some Codes. The normalized spectral density, $S_n(f)$, for some codes are [29]:

1. For NRZ code:

$$S_n(f) = \frac{\sin^2 t}{t^2} \tag{9.6}$$

2. For FM:

$$S_n(f) = \frac{\sin^4(t/2)}{(t/2)^2} \tag{9.7}$$

3. For MFM:

$$\begin{aligned} S_n(f) = 23 &- 2\cos t - 22\cos 2f - 12\cos 3t \\ &+ 5\cos 4t + 12\cos 5t + 2\cos 6t - 8\cos 7t \\ &+ 2\cos 8t/[2t^2(17 + 8\cos 8t)] \end{aligned} \tag{9.8}$$

where $t = \pi f T$.

The analysis is performed for the Markov source of the first order, and for the same probabilities of 1's ($p = 50\%$) and 0's.

Spectra for two MFM codes, the Miller and Jordan codes, are presented in Figures 9.10 and 9.11 respectively [30]. Some logical conclusions are obvious:

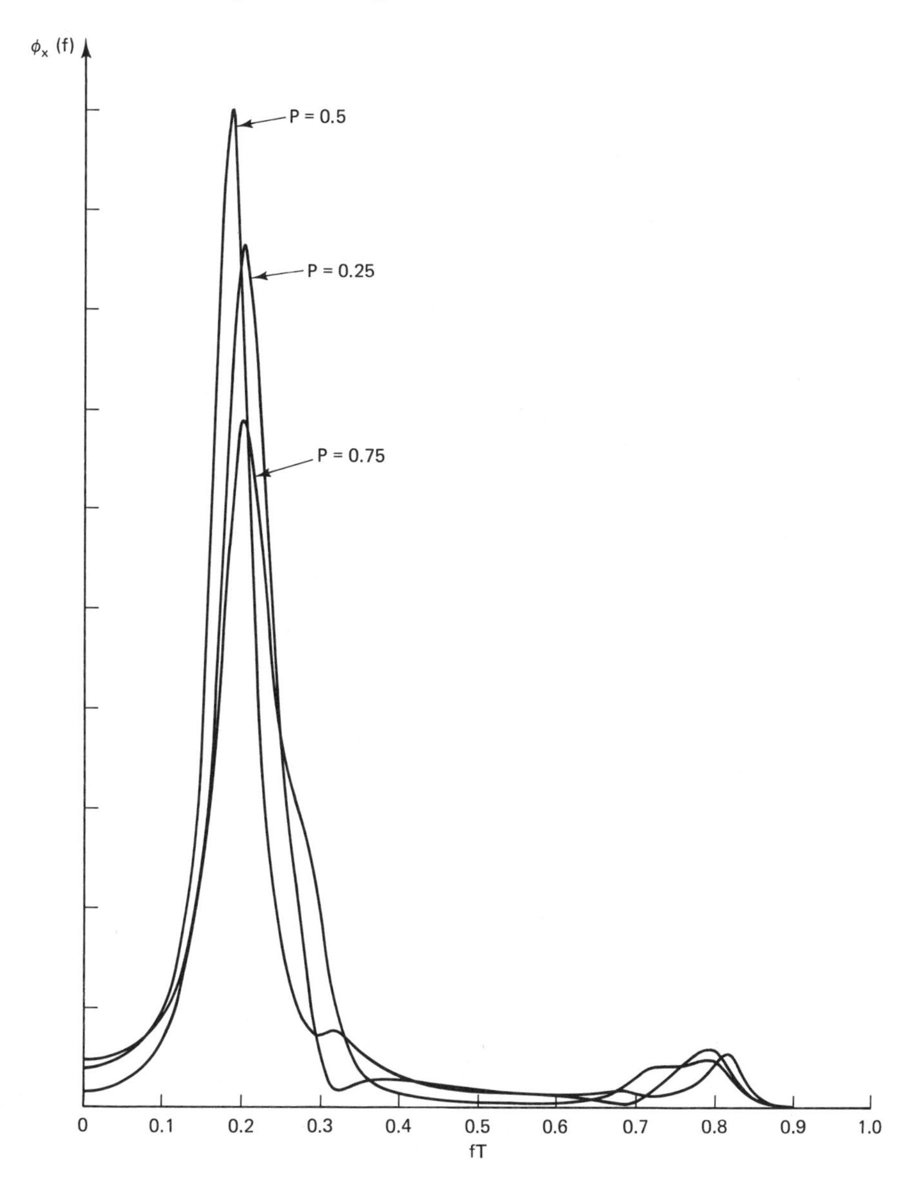

Figure 9.10 Spectra of the Miller code for the various probabilities of 1's.

1. The lowest-frequency occupancy for the Miller code is for an equal probability of 1's and 0's.
2. The lowest-frequency occupancy for the Jordan code is when 0's are dominant.
3. For the Miller code, spectra are similar for dominant 1's and dominant 0's in the coming data stream.

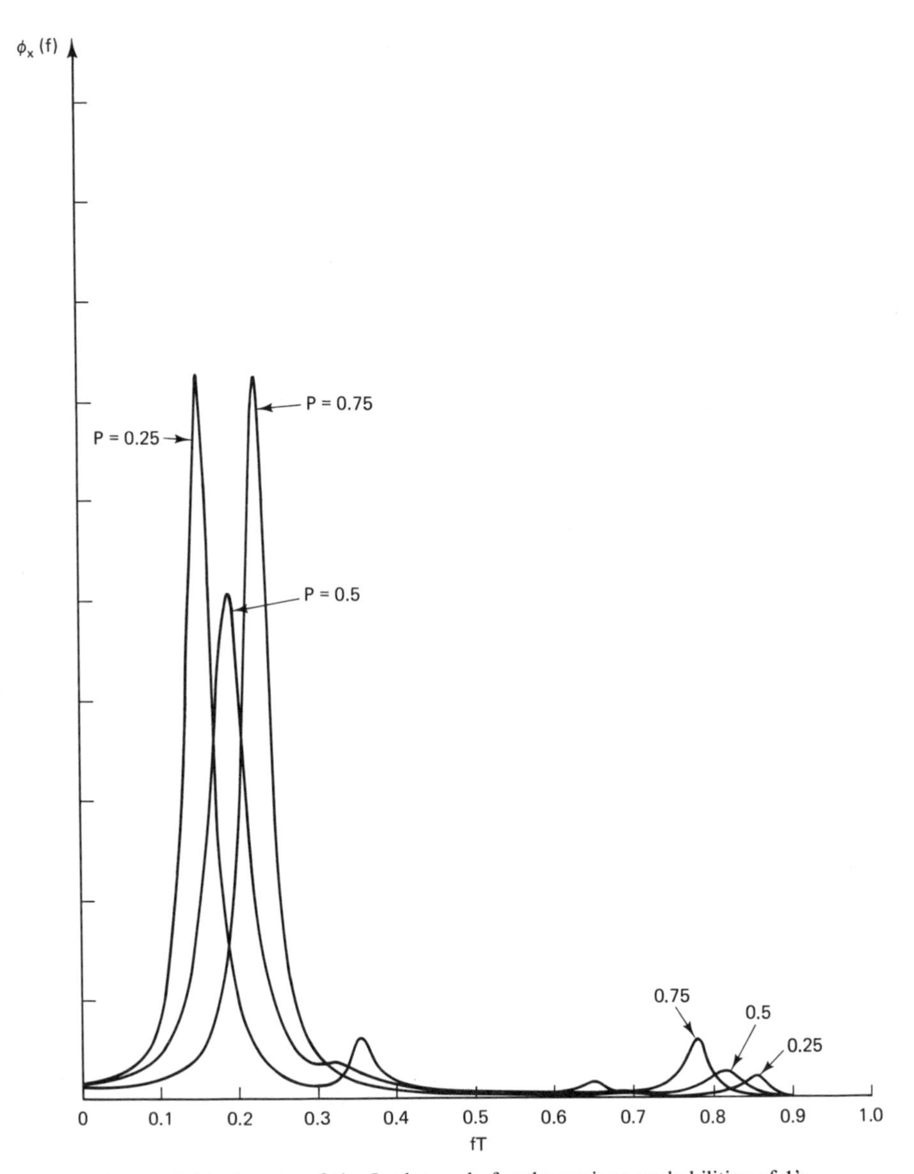

Figure 9.11 Spectra of the Jordan code for the various probabilities of 1's.

4. For the Jordan code, the spectrum contains higher-frequency components for fewer 0's in the input data.
5. The lowest dominant frequency for the Jordan code ($p = 0.25$) is lower than the corresponding frequency for the Miller code ($p = 0.5$). This is because $p = 0.5$ includes streams of 0's and/or 1's, and not only interchanging 0's and 1's (1010101...), when the dominant frequency for the Miller code would be the lowest.

Two codes can be combined [31] to get a low spectrum in the channel:

Dominant 0's. Use the Jordan code.
Close to equal probability of 0's and 1's. Use the Miller code [32].
Dominant ones. Use the inverted Jordan code (rules for coding 1 and 0 are interchanged).

Figure 9.12 shows a three dimensional-like presentation for the spectra of Jordan code.

The coder for MFM is relatively simple, while the decoder requires a slightly more sophisticated system. The decoder realization for the Miller code is given, for example, in Refs. 32 and 33, and for the Jordan code in Refs. 34 and 35. Figure 9.13 shows the realization of a decoder for two MFM codes, the Miller

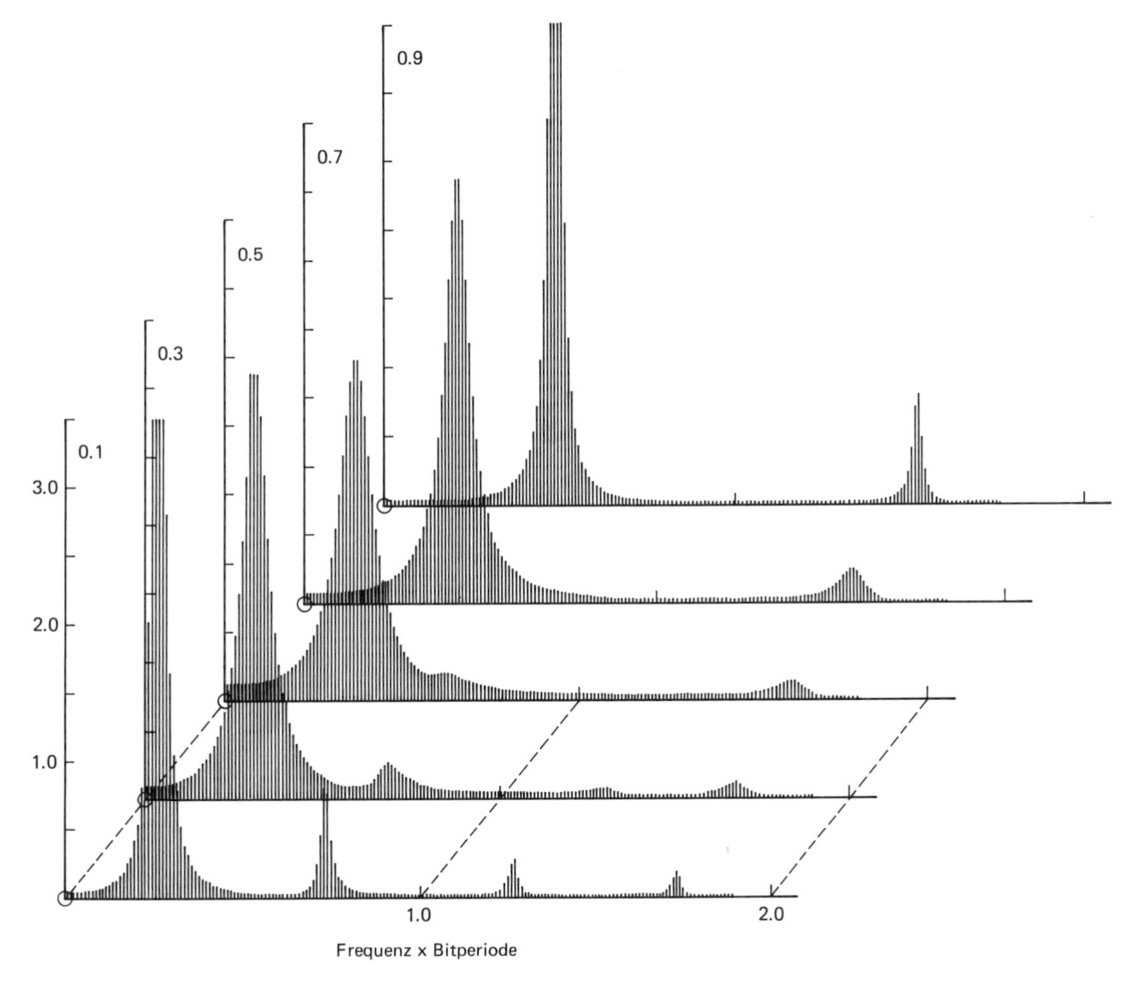

Figure 9.12 Three dimensional-like presentation for the spectra of Jordan code.

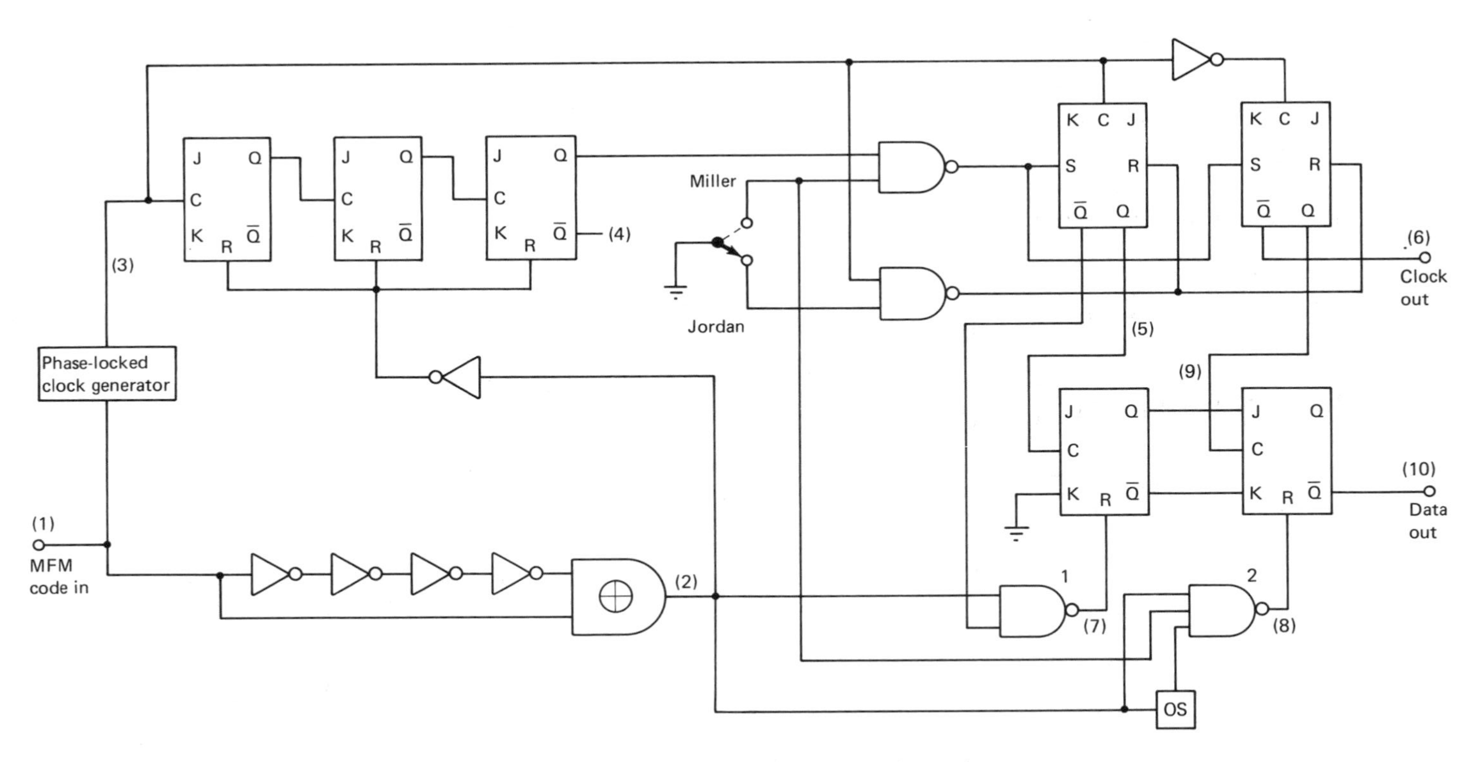

Figure 9.13 Decoder for the Miller and Jordan codes.

and Jordan codes, enabling the decoding of both codes. Figure 9.14 shows waveforms at the characteristic points when the code switch is in the Jordan code position. The input signal (1) is in the Jordan code. The transition detector gives output pulses (2) whenever the input signal contains a transition. An inspection of the encoded data stream immediately gives the observer a certain amount of information, for example: (a) when the distance between two transitions equals the length of one data cell, the observer knows that the second transition represents a binary 1 in the original NRZ code; (b) when the distance between two transitions is two bit cells in length, the observer knows that (i) a series of 0's is present in the NRZ code and (ii) both transitions occur between data cells.

A "zero detector" gives an output pulse (4) when the distance between two adjacent transitions is two bit-cell times. This can happen only when there is a stream of 0's in the NRZ signal. This pulse is used to synchronize the clock generator. The pulse can occur only during the last quarter of a bit-cell time. The zero detector can be used for data dropout indication (location).

NAND gate 1 selects all transitions in the middle of a bit-cell time (7). A one-shot (OS) and NAND gate 2 select all pulses at a distance of one bit-cell time from the previous transition (8). Together, signals 7 and 8 represent all 1's in the original NRZ signal. This is used (7 and 8) to regenerate the original NRZ (10) through the output stage flip-flop. Similarly, we can draw the waveforms at the characteristic points when the code switch is in the position "Miller code."

The code switch may be realized in the form of a logical network so that

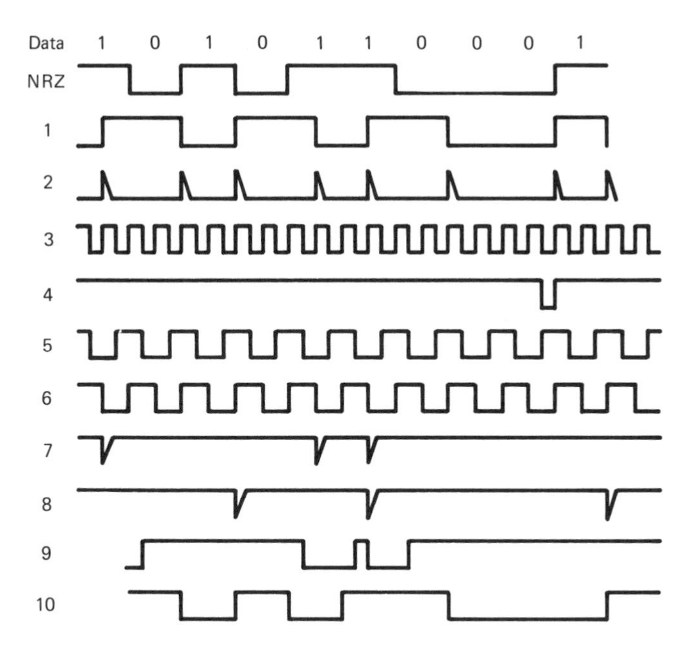

Figure 9.14 Waveforms at characteristic points.

its position during the decoding time of any sequence may be determined by a digital word at the beginning of a sequence. At coding, the sequence is tested for high transmission density of the contents (e.g., 10101 content suitable for the Miller code) and for low transition density of the contents (suitable for the Jordan code). The selection of the code is made depending on the information content of the sequence.

9.5.2 The Codes

The basic codes can be classified as belonging to one of four basic family groups: Non-Return-to-Zero (NRZ), Return-to-Zero (RZ), Biphase and Delay Modulation (DM) codes. They are defined in Table 6.1 and depicted in Figure 6.14.

Extra transitions can be added to NRZ in order to make it run-length limited. This assures clock extraction for any data input. Enhanced NRZ is then obtained. Two variations of ENRZ have been described in the literature (not shown in Figure 6.14). In both versions an extra interval is inserted after eight (or seven) code bits. In ENRZ-parity, the added bit creates an odd parity with the previous code bits, in ENRZ-complement the added bit is the complement of the last code bit. Enhancement can be applied to any NRZ code.

Randomized NRZ is another modification of the NRZ. Straight NRZ data are processed through a randomizer before being recorded to increase the number of level transitions and thus reduce the dc component. The code is not run-length-limited. There is always the bit pattern which could exactly nullify the randomizing process to produce unacceptable long strings of "1s" or "0s". A typical bit randomizer is shown schematically in Figure 9.15.

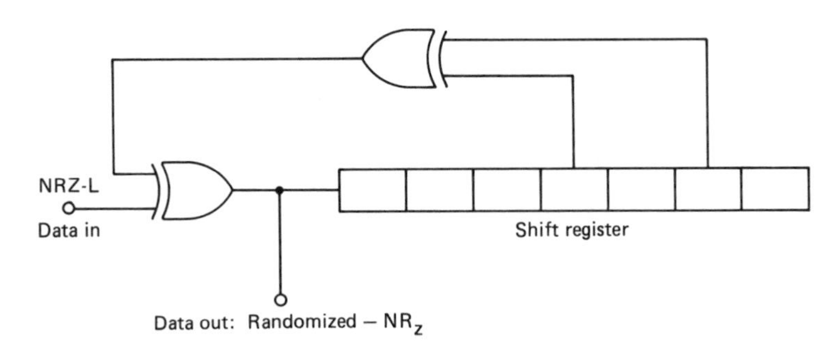

Figure 9.15 Randomizer for NRZ code.

9.5.2.1 More of Double Density Codes. Besides Miller and Jordan codes, two more double density (three frequency) codes must exist. This is obvious from the following discussion. Sync patterns are:

101—for Miller Code
010—for Miller Complement Code.

and

000—for Jordan Code
111—for Jordan Complement Code.

This can be marked as 101/010 for Miller and 000/111 for Jordan code. Thus, two more sets are left:

001/110

and

100/011

9.5.2.2 Group Codes. Codes such as NRZ, NRZI, and Biphase operate on the principle of one symbol for one bit. However, codes could be constructed whereby groups of bits are coded with unique patterns. Such codes are called group codes, and 4/5 code could be viewed as such.

A similar class of codes is adaptive codes, whereby groups are viewed, as before, but the waveform symbol depends not only on the group presently being coded, but also on previous and subsequent groups. The distinction between the two classes is vague, however. Miller and Jordan codes can be considered as adaptive codes.

Franaszek's paper (Ref. 36) is an attempt to provide a means of producing optimum group codes, given the constraints of minimum and maximum interval-reversal times.

There are several modifications of the Miller code. Zero modulation—ZM (Ref. 27), and Miller Square—M^2 (Ref. 37) are examples: the basic Miller code (Ref. 38) is modified so that dc is reduced and yet retains unambiguous code patterns which allow non-error propagating decoding. The ZM and M^2 codes are defined in Table 9.2 and depicted in Figure 9.16. The advantages of M^2 over ZM are that only 3 bits of memory are needed and that a density ratio, DR, of unity is maintained. The bandwidth required for this dc free code is, in consequence, little greater than for NRZ. Negligible base-line wander or eye-pattern closure occurs with M^2. The effective worst-pattern signal-to-noise ratio expected is 3–5 dB better than in Miller code; in high density digital recording this difference becomes significant.

9.5.2.2.1 Three Phase Modulation (3 PM) Code. The 3-phase modulation 3PM (Ref. 28) attempts to get 3 bits per cycle ($1\frac{1}{2}$ bits per minimum zero crossing). It achieves this by allowing 3 phases of the basic $\frac{1}{2}$ cycle waveform. The data stream is split into groups of three bits, which are then encoded into a six-bit word for recording. The code is adaptive, in that the bit pattern for the word currently being encoded depends on the previous and the following word.

The algorithm of the 3PM code is similar to fixed length block encoding and decoding with one additional rule. The algorithm converts three-bit data

TABLE 9.2 CODING RULES OF THE MODIFIED MILLER CODES

Zero modulation code
The bit stream to be encoded is broken into sequences of three types: (a) Any number of consecutive ones. (b) Two zeros separated by either no ones, or an odd number of ones. (c) Two zeros separated by an even number of ones. Sequences type (a) and (b) are coded as in normal Miller code. In sequences type (c), ZM encodes the zeros in the Miller manner, but the ones are encoded as though they were zeros but with alternate transitions deleted.
Miller^2 code
The bit stream to be encoded is broken into sequences of three types: (a) Any number of consecutive ones. (b) Two zeros separated by either no ones, or an odd number of ones. (c) One zero followed by an even number of ones (terminated by a zero not counted as part of the sequence). Sequences type (a) and (b) are coded as in normal Miller code. Sequences type (c) have the transition corresponding to the final '1' inhibited.

groups into six-bit code groups. The encoding is explained by looking at Figure 9.17 and Table 9.3. The space allowed for a word of three data bits is divided into six equidistant positions: P 1 to P 6.

Minimum two zeros are maintained between adjacent ones. The boundary position (P 6) is occupied by zeros in all code words. In a sequence of words, where a one occurs at P 5 of the present word and also at P 1 of the following word the d = 2 condition would be violated. The special rule of the 3PM code

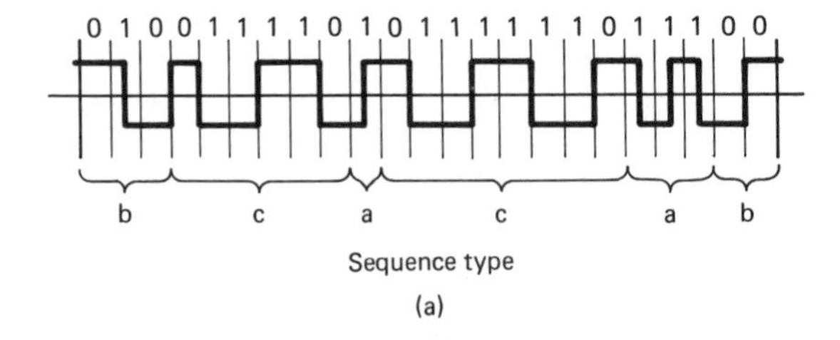

(a)

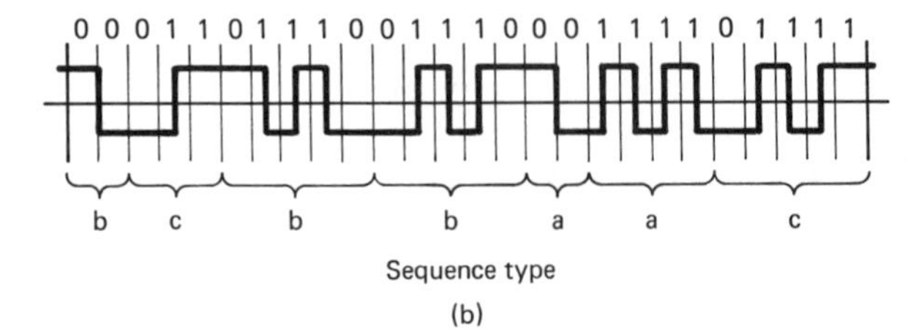

(b)

Figure 9.16 Code waveforms. (a) ZM; (b) M^2.

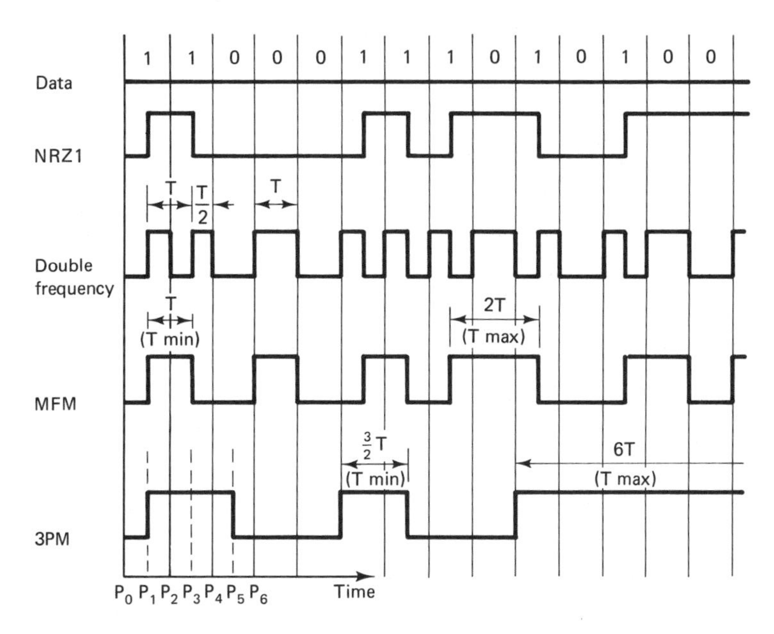

Figure 9.17 Data waveforms.

TABLE 9.3 INITIAL ENCODING TABLE FOR 3M CODE

Binary data word			*Transition positions*					
			P1	*P2*	*P3*	*P4*	*P5*	*P6*
0	0	0	0	0	0	0	1	0
0	0	1	0	0	0	1	0	0
0	1	0	0	1	0	0	0	0
0	1	1	0	1	0	0	1	0
1	0	0	0	0	1	0	0	0
1	0	1	1	0	0	0	0	0
1	1	0	1	0	0	0	1	0
1	1	1	1	0	0	1	0	0

provides that in this case the P 5 transition of the present word and the P 1 transition of the following word will not be written in their original locations but will be replaced by a single transition at P 6. The two original transitions, P 5 and P 1, will be merged into one transition at P 6. The P 6 position in the initial encoding table is reserved for this merging operation.

The results of the final encoding, after the merging rule has been carried out, is shown in Table 9.4. Here all combinations of binary data words of three bits each are shown together with the influence of adjacent data words. If the preceding word ends in P 5, a P 1 transition of the present word will be shifted to the sixth position of the previous word denoted by P 6′. Similarly, if the

TABLE 9.4 3M CODE: FINAL ENCODING WITH MERGING

Binary data word			*Influence of adjacent words*		*Transition positions*						
			Preceding	*Following*	*P6'*	*P1*	*P2*	*P3*	*P4*	*P5*	*P6*
0	0	0	X	0	0	0	0	0	0	1	0
0	0	0	X	1	0	0	0	0	0	0	1
0	0	1	X	X	0	0	0	0	1	0	0
0	1	0	X	X	0	0	1	0	0	0	0
0	1	1	X	0	0	0	1	0	0	1	0
0	1	1	X	1	0	0	1	0	0	0	1
1	0	0	X	X	0	0	0	1	0	0	0
1	0	1	0	X	0	1	0	0	0	0	0
1	0	1	1	X	1	0	0	0	0	0	0
1	1	0	0	0	0	1	0	0	0	1	0
1	1	0	1	0	1	0	0	0	0	1	0
1	1	0	0	1	0	1	0	0	0	0	1
1	1	0	1	1	1	0	0	0	0	0	1
1	1	1	0	X	0	1	0	0	1	0	0
1	1	1	1	X	1	0	0	0	1	0	0

1: Yes
0: No
X: Don't care

following word starts with P 1, a P 5 transition of the present word will be shifted to P 6. The result of this merging rule is that any number of octal data words can be catenated, while simultaneously maintaining the d = 2 condition everywhere in the sequence. The encoding logic that implements this rule is simple. It has to look back to the P 5 position of the previous word and look ahead to the P 1 position of the following word, thus dealing with 9 positions simultaneously. This way, the words are chained to one another in a natural way, preserving the fixed block length property of the code.

9.5.2.2.2 Group Coded Recording (GCR). In the Group Coding Recording (GCR) code, also known as "4/5 code" and "run-length-coded NRZ" four data bits are represented by a five-bit pattern (Table 9.5). From the 32 possible combinations of 5 bits, those that begin or end with more than one 0 and those that have more than two 0's internally can be eliminated, leaving 17, from which one can be discarded to produce the 16 unique patterns required. This one can then be used as a special pattern, for checking or error detection, as it obeys the constraints and is therefore detectable.

9.5.2.2.3 Other Fixed Length Codes. In a 4:6 code [39], digital data is encoded into words having two valued signals in which six of such signals represents four input data bits. Three of the six signals are positive and three are negative, resulting in an alphabet of words which are of zero dc average level and equal power. These words are transmitted over a channel, as by being optically recorded and then played back therefrom. Table 9.6 shows mapping.

TABLE 9.5 FOUR-FIVE GCR

Decimal value	*Hexadecimal value*	*Data words*	*Code words*
0	0	0000	11001
1	1	0001	11011
2	2	0010	10010
3	3	0011	10011
4	4	0100	11101
5	5	0101	10101
6	6	0110	10110
7	7	0111	10111
8	8	1000	11010
9	9	1001	01001
10	A	1010	01010
11	B	1011	01011
12	C	1100	11110
13	D	1101	01101
14	E	1110	01110
15	F	1111	01111

TABLE 9.6 ENCODING/DECODING TABLE FOR 4:6 CODE

	Binary				*H—Code (4:6)*						
Decimal	*A*	*B*	*C*	*D*	*U*	*V*	*W*	*X*	*Y*	*Z*	*Signal*
0	0	0	0	0	1	1	1	0	0	0	S1
1	0	0	0	1	0	0	0	1	1	1	$\overline{S1}$
2	0	0	1	0	1	1	0	1	0	0	S2
3	0	0	1	1	0	0	1	0	1	1	$\overline{S2}$
4	0	1	0	0	1	1	0	0	1	0	S3
5	0	1	0	1	0	0	1	1	0	1	$\overline{S3}$
6	0	1	1	0	1	1	0	0	0	1	S4
7	0	1	1	1	0	0	1	1	1	0	$\overline{S4}$
8	1	0	0	0	1	0	1	1	0	0	S5
9	1	0	0	1	0	1	0	0	1	1	$\overline{S5}$
10	1	0	1	0	1	0	1	0	0	1	S6
11	1	0	1	1	0	1	0	1	1	0	$\overline{S6}$
12	1	1	0	0	1	0	0	1	1	0	S7
13	1	1	0	1	0	1	1	0	0	1	$\overline{S7}$
14	1	1	1	0	1	0	0	1	0	1	S8
15	1	1	1	1	0	1	1	0	1	0	$\overline{S8}$
FSW*	—	—	—	—	1	0	0	0	1	1	S9

* FSW—frame sync word

In some systems it is more convenient to reserve two signals for frame synchronization in order to avoid sensitivity to data pattern. The symbols S1 and S1 of Table 9.6 will allow the breaking of false signal lock (repetitive signals which have adjacent bits which can be taken as good zero average words). A signal is reserved for the frame sync word and another signal is available for special word that is not shown. Seventeen of the eighteen available signals or codes in the alphabet are thus utilized. These 17 signals are indicated by the numbers 0 to 15 and FSW in the column headed "decimal" in the encoding/decoding table shown in Table 9.6. The 4:6 code is shown in the table headed H-code. The binary words ABCD and the H-code words UVWXYZ as well as the signals to which they correspond are all shown in the same row of the encoding/decoding table.

In the (5,6) code [40] the disparity of codeword is the excess of logical 1's over logical 0's. The (5,6) code, suitable for high density recording, achieves zero response at dc by use of codewords with either zero disparity or ± 2 disparity in alternation. In Table 9.7 the codewords are given in NRZ notation: 0 represents a pulse level of -1 and 1 a pulse level of $+1$. The first 20 codewords have zero disparity, and state 1 = state 2. The remainder are codewords with disparity of $+2$ (state 1) and their complements with disparity -2 (state 2). Zero disparity words do not cause the encoder to change states. Codewords with non-zero disparity make the encoder alternate between complementary states of opposite disparity.

TABLE 9.7 CODEWORD DICTIONARY FOR (5,6) CODE

k	*Input word*	*Codeword state 1*	*k*	*Input word*	*Codeword state 1*
1	00011	000111	17	11000	110001
2	00101	001011	18	11001	110010
3	00110	001101	19	11010	110100
4	00111	001110	20	11100	111000
5	01001	010011	21	00000	011011
6	01010	010101	22	00001	110110
7	01011	010110	23	00010	111001
8	01100	011001	24	00100	110011
9	01101	011010	25	01000	100111
10	01110	011100	26	10000	101101
11	10001	100011	27	01111	101011
12	10010	100101	28	10111	110101
13	10011	100110	29	11011	111010
14	10100	101001	30	11101	011101
15	10101	101010	31	11110	101110
16	10110	101100	32	11111	010111

In Figure 9.18 the power spectral density (PSD) is shown for 5:6 and 3PM codes. In the octal coded binary (OCB) code, data are coded in groups of three [41]. A binary 2/3 rate code, another group coding algorithm, is presented in Ref. 42. (Table 9.8).

9.5.2.3 Synchronous Variable Length Group Codes. The code conversion tables of encoding and decoding operations in dk-limited, variable-length coding systems wherein the (d,k) constraints are (1,7) (1,8) and (2,7) are shown in Table

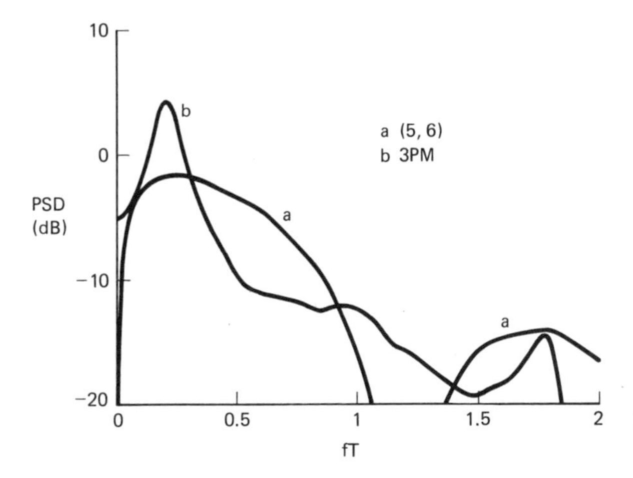

Figure 9.18 Power spectral density for 5:6 and 3PM codes.

TABLE 9.8 CODEWORD DICTIONARY FOR (2/3) CODE

Decimal	*Data word*		*Code word*		
0	0	0	1	0	1
1	0	1	1	0	0
2	1	0	0	0	1
3	1	1	0	1	0

9.9. For the (1,7) and (1,8) codes there are three encoded bits for two original bits, this 3/2 ratio being constant. Similarly, in the case of the (2,7) code, the ratio between the numbers of encoded bits and original bits is 2/1.

In the case of the (1,8) code, the entire code dictionary includes 16 code words, whose lengths vary from 3 to 9 bits, in multiples of 3. In the case of the (1,7) and (2,7) codes, the code dictionary includes 7 code words with lengths varying from 2 bits to 3 bits, in multiples of 2. If information were to be encoded with an equivalent bit-per-symbol value in a run-length-limited coding system having fixed word lengths, the size of the code dictionary would increase in orders of magnitude due to the relative inflexibility of coding in a fixed-length, run-length limited system. This would greatly increase the complexity of the apparatus needed for table lookup operations or equivalent encoding and decoding functions. A (4,9) code has a code dictionary of 512 words in a fixed-length format but only six words in a variable-length format.

The group code (4,9) is given by Franaszek [36] as an example of a run-length limited code. Coding commences by the viewing of the first two data bits. If the pair is 00, the third bit is viewed also, and the three bits together are then coded. The parameter values are:

$$N = 3, d = 4, k = 9, M = 3$$

In the (1,11) Rice code [41], the parameters are:

$$N = 1\tfrac{1}{2}, d = 1, k = 11, M = 6$$

TABLE 9.9 SOME OF THE VARIABLE-LENGTH (d,k) CODES

A) Code conversion table for (1,7) code

NRZ	*1,7*
0 1	X 0 0
1 0	0 1 0
1 1	X 0 1
0 0 0 1	X 0 0 0 0 1
0 0 1 0	X 0 0 0 0 0
0 0 1 1	0 1 0 0 0 1
0 0 0 0	0 1 0 0 0 0

X = Complement of last symbol

B) Code conversion table for (1,8) code

	Encoded words	*Original or decoded words*	*L*	
	0 1 0 X X X X X X	0 0 X X X X	0 1 0	
	1 0 0 X X X X X X	1 1 X X X X	0 1 0	
	1 0 1 0 0 0 X X X	1 0 0 0 X X	1 0 0	
	1 0 1 0 1 0 X X X	1 0 0 1 X X	1 0 0	
X = Don't	0 0 1 0 1 0 X X X	1 0 1 0 X X	1 0 0	
care	0 0 1 0 0 0 X X X	1 0 1 1 X X	1 0 0	
	0 0 0 1 0 0 X X X	0 1 0 0 X X	1 0 0	
	0 0 0 0 1 0 X X X	0 1 1 0 X X	1 0 0	
	0 0 0 0 0 1 0 1 0	0 1 1 1 0 0	1 1 0	
W = 9	0 0 0 0 0 1 0 0 0	0 1 1 1 0 1	1 1 0	
N = 3	0 0 0 1 0 1 0 1 0	0 1 1 1 1 0	1 1 0	
α = 2	0 0 0 1 0 1 0 0 0	0 1 1 1 1 1	1 1 0	
	0 0 1 0 0 1 0 1 0	0 1 0 1 0 0	1 1 0	
	0 0 1 0 0 1 0 0 0	0 1 0 1 0 1	1 1 0	
	1 0 1 0 0 1 0 1 0	0 1 0 1 1 0	1 1 0	
	1 0 1 0 0 1 0 0 0	0 1 0 1 1 1	1 1 0	
		0 0 0 0 0 0		Dummy word (read out if no match found during decoding)
	←9 Bits→	←6 Bits→	←3 Bits→	

C) Code conversion table for (2,7) code

Data word:	*Code word:*
11	1000
011	001000
0011	00001000
10	0100
010	100100
0010	00100100
000	000100

9.5.3 Discussion

The code classification presented here is not rigorous from an academic point of view: it is more oriented towards future use of the background study of various codes. To select the most promising candidates for the best encoding schemes for the optical recording, a second step in this study is required: the channel parameters must be incorporated. Among others, the following parameters/questions should be compared for different codes [43–45]

1. Code synchronization
2. Code overhead
3. Theoretical maximum bit packing density
4. Maximum user data packing density for given bit error rate (BER)
5. Error propagation
6. Complexity and costs

The third possible step would be an experimental verification of the selected codes.

From the previous discussions it follows that different codes can be selected as a "best" codes for optical recording, depending on the selected criteria. In Ref. 45 it was suggested that the following codes can be taken as reference codes for optical recording: enhanced NRZ, double density, and 3PM (or 2,7) codes. Any other promising codes should be compared with these codes. The detection process has great influence on the system performance and combination code-detection technique should be considered.

The optical readout method, for example a $\lambda/4$ or $\lambda/8$ (differential method), Figure 9.19, can drastically change the total frequency response of the system. In the $\lambda/8$ readout system, the low (dc) components are significantly reduced. In the $\lambda/4$ system the ratio of the highest and low significant components in the band is extremely high.

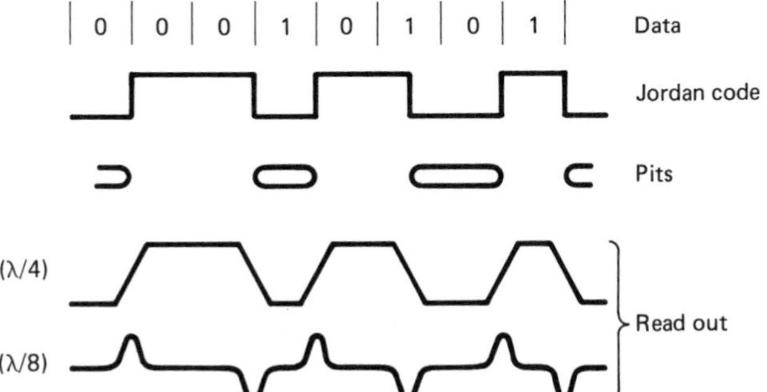

Figure 9.19 Output signal for the central ($\lambda/4$) and differential ($\lambda/8$) detection.

9.6 COMPACT DISC–READ ONLY MEMORY

The compact disc read only memory (CD-ROM) can be considered as another application of the compact disc. It is an economical, reliable optical mass storage from several megabytes to several hundred megabytes and has the possibility to combine text, data, voice, image end graphics. A world-wide standard is an obvious plus.

The CD-ROM recording and production is essentially that of replicated videodisc production. Digital data is first placed on the CD-ROM master tape, usually $\frac{1}{2}$ inch magnetic tape, density 1600 or 6250 bpi. In the premastering an overhead is added:

Sync
Header: block address and mode
Error detection code (EDC) and Error correction code (ECC)

Disc mastering and replication is discussed in Chap. 2.

The data structure of the 540 Mbytes CD-ROM is [46]:

```
ONE BLOCK : TOTAL 2352 BYTES
     SYNC : 12 BYTES
   HEADER : 4 BYTES : MINUTE/SECOND/BLOCK/MODE
                      —BLOCK ADDRESS—
USER DATA : 2K BYTES (2048 BYTES)
      EDC : 4 BYTES :
    SPACE : 8 BYTES
      ECC
 P-PARITY : 172 BYTES : REED SOLOMON CODE
 Q-PARITY : 104 BYTES : REED SOLOMON CODE
USER DATA
   1 BLOCK = 2K BYTES (2048 BYTES)
   1 DISC = 270K BLOCKS = 540M BYTES
USER DATA TRANSFER RATE
   150K BYTES/SEC
```

This is illustrated in Figure 9.20 while track structure is shown in Figure 9.21.

After error correction the bit error-rate is less than 10^{-12}. As a readout unit either adapted consumer (CD) player or special drivers are used. In either case a semiconductor laser (λ = 780–830 nm) is used. Transfer rate is 150–600 kbytes/s, and the access time, first to last block—is in the order of 1.5 s.

Major CD-ROM applications are external database reference (Library, professional databases, etc.), software publishing including works with expert systems software, electronic publishing, vertical markets (legal, medical, banking, educations, government, technical, scientific, etc.), etc. Examples of CD-ROM applications are:

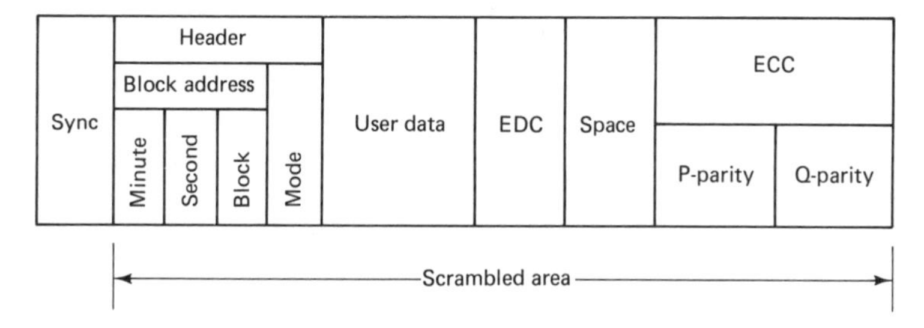

Figure 9.20 Data structure of the CD-ROM disc.

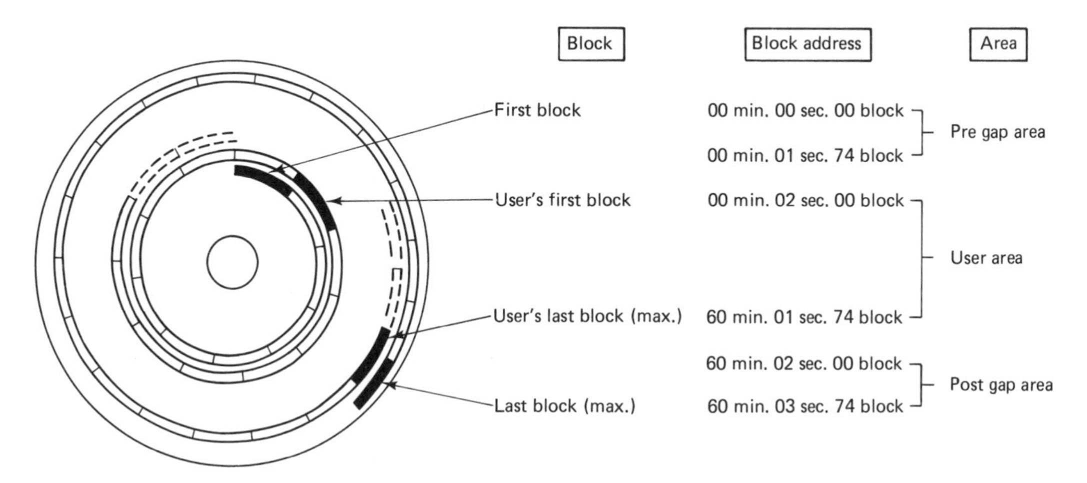

Figure 9.21 Track structure of the CD-ROM disc.

Software publishing
Electronic publishing
Computer based training, educational programs
Bibliographical references
Product catalogues
Patents
Government agency data (IRS, SEC, Defense)
Medical images
Medical database
Phone directories
Legal database (indexed jurisprudence)
Marketing statistics
Econometric databases
Engineering drawings
Navigational systems (automotive, cartographic)
Credit card verification

Payroll data
Contracts
Purchase orders, invoice history

Initial applications include both: optical disc drive as a peripheral of the PC and the drive incorporated in the large system. In either case, the retrieval software needs to be able to identify the optical disc in the drive [47].

9.7 NON-DISC FORMATS

The basic videodisc technology can be applied also in the non-disc formats as are cards, tapes, and cylinders even. Optical card is minimum cost format. Both, read-only and writable cards are possible choices [48]. For example the medical application uses a write-once card, whereas most publishing applications are read-only.

Medical uses are expected to be a major optical card market. The medical record optical card can contain:

Photograph of card owner
Medical history
Health insurance data
Signature facsimile
Electrocardiogram copy
Digitized medical imagery

Thus, text, image and data would be all on the card. Some other applications include: publishing—that is data distribution, programming for automatic test equipment, corporate memos distribution, etc.

The optical tape has the highest volumetric density, among disc or non-disc formats. A serial recording and readout are used. Thus, potential applications must be tolerant on the lower access time compared with the disc formats, for example.

REFERENCES

1. *Model PR-7820-3 Optical Videodisc Player: Programming Reference Guide,* Discovision Associates, 1981.
2. M. Edelrart, "Optical Discs—The Omnibus Medium," *Technology,* No. 1, Nov.–Dec. 1981, pp. 42–57.
3. G. D. Fink, "The Future of High-Definition Television: Conclusion of a Report of the SMPTE Study Group of High-Definition Television," *SMPTE J.,* Vol. 89, No. 3, 1980, pp. 153–161.

4. K. Hayashi, "Research and Development on High-Definition Television in Japan," *SMPTE J.*, Vol. 90, No. 3, 1981, pp. 178–186.
5. G. Broussaud and C. Tinet, "The Videodisk as a Means of Transferring a Video Picture onto Motion-Picture Film," *SMPTE J.*, Vol. 88, Apr. 1979, pp. 247–252.
6. G. C. Kenney, "Special Purpose Applications of the Optical Videodisc System," *IEEE Trans. Consum. Electron.*, Vol. 22, November 1976, pp. 327–338.
7. J. Isailović, "Preliminary Study of Digital Encoding and Error Correction Techniques," *MCA Labs Tech. Note 422*, Aug. 17, 1977.
8. C. M. Goldstein, "Optical Disk Technology and Information," *Science*, Vol. 215, Feb. 1982, p. 12.
9. J. Isailović, "Binary Data Transmission through TV Channels by Equivalent Luminance Signal," *Electron. Lett.*, Vol. 17, No. 6, 1981, pp. 233–234.
10. J. Isailović, "Apparatus for Generating Luminance Signal Based on the Binary Data," *Proc. Inf.*, Vol. 81 (Sarajevo).
11. W. R. Dakin and J. Isailović, "Digital Formatting System," U.S. Patent Appl., Oct. 1980.
12. K. S. Shanmugam, *Digital and Analog Communication Systems*, Wiley, New York, 1979.
13. J. Isailović, "Binary Data Transmission through TV Channels by Equivalent Chrominance Signal," *SPIE Conf.*, Los Angeles, Feb. 18–22, 1982 (329–19).
14. P. Stavroulakis, *Interference Analysis of Communication Systems*, IEEE Press, New York, 1980.
15. W. C. Lindsey, "Phase-Shift-Keyed Signal Detection with Noisy Reference Signals," *IEEE Trans. Aerosp. Electron. Syst.*, Vol. AES-2, July 1966, pp. 393–401.
16. S. A. Rhodes, "Effect of Noisy Phase Reference on Coherent Detection of Offset-QPSK Signals," *IEEE Trans. Commun.*, Vol. COM-22, Aug. 1974, pp. 1046–1054.
17. J. M. Aein, "Coherency for the Binary Symmetric Channel," *IEEE Trans. Commun.*, Vol. COM-18, Aug. 1970, pp. 344–352.
18. K. Shibata, "Error Rate of CPSK Signals in the Presence of Coherent Carrier Phase Jitter and Additive Gaussian Noise," *Trans. IECE (Japan)*, Vol. 58-A, June 1975, pp. 388–395.
19. S. Kabasawa, N. Morinaga, and T. Namekawa, "Effect of Phase Jitter and Gaussian Noise on *M*-ary CPSK Signals," *Electron. Commun. Jap.*, Vol. 61-B, Jan. 1978, pp. 68–75.
20. O. Shimbo, J. R. Fank, and M. I. Celebiler, "Performance of *M*-ary PSK Systems in Gaussian Noise and Intersymbol Interference," *IEEE Trans. Inf. Theory*, Vol. IT-9, Jan. 1973, pp. 44–58.
21. V. K. Prabhu, "Imperfect Carrier Recovery Effect on Filtered PSK Signals," *IEEE Trans. Aerosp. Electron. Syst.*, Vol. AES-14, July 1978, pp. 608–615.
22. S. Stein, "Unified Analysis of Certain Coherent and Noncoherent Binary Communication Systems," *IEEE Trans. Inf. Theory*, Vol. IT-10, Jan. 1964, pp. 43–51.
23. F. E. Glave, "An Upper Bound on the Probability of Error Due to Intersymbol Interference for Correlated Digital Signals," *IEEE Trans. Inf. Theory*, Vol. IT-8, May 1972, pp. 356–363.

24. B. R. Saltzberg, "Intersymbol Interference Error Bounds with Application to Ideal Band-Limited Signaling," *IEEE Trans. Inf. Theory,* Vol. IT-14, July 1968, pp. 563–568.
25. V. K. Probhu and T. Solz, "The Performance of Phase-Shift-Keying Systems," *Bell Syst. Tech. J.,* Vol. 60, No. 10, 1980, pp. 2307–2343.
26. W. Stollings, "Digital Signology Techniques," *IEEE Com. Magazine,* Vol. 22, No. 12, December 1984.
27. A. M. Patel, "Zero-Modulation Encoding in Magnetic Recording," *IBM J. Res. Dev.,* Vol. 19, No. 4, 1975, pp. 366–378.
28. G. V. Jacobi, "A New Look-Ahead Code for Increasing Data Density," *IEEE Proc. Magn.,* Vol. MAG-13, No. 5, 1977, pp. 1202–1204.
29. W. C. Lindsey and M. K. Simon, *Telecommunication Systems Engineering,* Prentice-Hall, Englewood Cliffs, N.J., 1973.
30. G. Petrovic, "Program for the Codes Spectra," Electrochemical Faculty, University of Belgrade.
31. J. Isailović, "Realization of a Decoder for Two MFM Codes," *Electron. Lett.,* Vol. 17, No. 3, 1981, pp. 117–119.
32. J. C. Mallinson and J. W. Miller, "Optimal Codes for Digital Magnetic Recording," *Radio Electron. Eng.,* Vol. 47, 1977, pp. 172–176.
33. W. E. Bentley and S. G. Varsos, "Squeeze More Data onto Mag Tape," *Electron. Des.,* Vol. 21, 1975, pp. 76–78.
34. J. Isailović, "Method and Means for Encoding and Decoding Digital Data," U.S. Patent 4,204,199, 1980.
35. J. Isailović, "A New Code for Digital Data Recording/Transmitting," Publication of the Electrical Engineering Faculty, University of Belgrade, Series: *Electron. Telecommun. Autom. 136–141,* 1980, pp. 41–51.
36. P. A. Franaszek, "Sequence-State Method for Run-Length-Limited Coding," *IBM J. Res. & Dev.,* 14, No. 4, July 1970, pp. 376–83.
37. J. W. Miller, U.S. Patent 4,027,335, May 31, 1977.
38. A. Miller, U.S. Patent 3,108,261, October 22, 1963.
39. P. H. Halpern, "High Density Data Processing System," U.S. Patent No. 3,921,210, November 18, 1975.
40. D. A. Lindholm, "Power Spectra of Channel Codes for Digital Magnetic Recording," *IEEE Transact. on Magnetics,* Vol. MAG-14, No. 5, September 1978, pp. 321–323.
41. N. D. Mackintosh, "The Choice of Recording Codes," *The Radio and Electronic Engineer,* Vol. 50, No. 4, April 1980, pp. 177–193.
42. M. Cohn, G. V. Jacoby, and A. Bates, III, "Data Encoding Method and System Employing Two-Thirds Code Rate with Full Work Look-ahead," U.S. Patent 4,337,458, June 29, 1982.
43. "Parallel Mode High Density Digital Recording—Technology Fundamentals," Bell and Howell, 1981.
44. J. Isailović, "Background Study of Various Codes for Optical Recording," Anaheim Hills, California, June 1984.

45. J. Isailović, "Codes for Optical Recording," *SPIE,* Vol. 529, Optical Mass Data Storage, Los Angeles, 1985, pp. 161–168.
46. "Note of CD-ROM System (Version 1.1)," Sony Corporation, Tokyo, Japan, May 6, 1985.
47. J. C. Gale, "CD-ROM Standards: Do the Users Really Benefit?" *CD Data Report,* Vol. 1, No. 5, March 1985, pp. 1–3.
48. *Optical Memory News,* June 1985, Issue #24, pp. 1–6.

Index

1/f noise, 270
1:7 and 1:8 codes, 429
1:7 code, 429
1:8 code, 429
2:7 code, 429
2:7 codes, 431
2/3 rate code, 428
3-2 pulldown, 272
3-db frequencies, 222
3/2 pull-down codes, 54
3PM, 431
3PM code, 424, 428
4:9 coding, 429
4/5 code, 423
4:6 code, 426, 427
5:6 code, 428

A/D converter, 376
Ablation, 288
Ablative materials, 39
Ablative thin film, 28, 49
AC coupling, 413
Access time, 432, 434
Acoustic-optical modulator, 42, 217
Acquisition range, 24
Active line interval, 37
Active lines, 145, 170, 210
Actuators, 52
Adaptive codes, 423
Adaptive prediction, 361
Adaptive predictive coding, 362
Adaptive quantization, 360
Adder, 189
Additive Gaussian noise, 412
Additive noise, 365
Address, 226
Address codes, 34
Addressing, 220
Address signals, 212, 226
Adjacent edges, 413
ADM, 357
ADPCM, 357, 365, 367
Air bearing, 24
Albative materials, 288
Alias frequency, 359
Aliasing effect, 143, 248
Aliasing error, 248, 371
Aliasing noise, 359
Aluminum, 46, 51, 57
Aluminum discs, 47
AM detector, 16
AM signal, 71, 95
AM spectra, 67, 107
AM systems, 78
AM with asymmetrical sidebands, 67
AM-to-FM conversion, 256
AM-to-PM conversion, 100
AM, 238–90, 321
AM:SSB, 76
Ambient illumination, 157
Amplifiers, 129, 133
Amplitude, 65
  characteristic, 180
  distortion, 144
  equalization, 95
  keying, 208
  modulated, 16, 239, 256
  modulation (AM), 65, 67, 214
  modulation, 100, 139, 224, 243
  modulator, 9
  shift keying, 66, 408
  spectral distribution, 208
  spectrum, 137, 207
Analog
  channel, 11, 300
  components, 5
  data base, 401
  delay, 267
  modulation systems, 67, 71, 74
  pulse modulations, 213
  signal discretization, 71
  signals, 72
  systems, 67, 69, 300
  TV signal, 300
  video channel, 49, 367
Analog-to-digital (A/D) conversion, 373
Angle modulation, 100
Angular frequency, 114, 118
  deviation, 123, 217
Angular modulation, 68, 243, 256
Antenna, 218, 220, 399
Antenna switch, 321
APCM, 357
Aperiodic signals, 67
Aperture, 129
Aperture effect, 145
Application program, 392
Archive storage, 49, 399
Argon laser, 31
Aspect ratio, 131, 137, 401
Astigmatic focus system, 25
Astigmatic focusing, 24
Astigmatic lens surface, 24
Asymmetry, 252, 253, 255, 316, 319
  components, 255
  sideband components, 193
Audible threshold, 345
Audio
  bandwidth, 29, 330, 350, 351
  carriers, 220, 223, 236, 251, 255, 304, 307, 316, 330, 346
  channels, 217, 220, 224, 329, 342, 346, 347, 377, 391
  communication link, 342
  decoding, 236
  discs, 43, 45
  encoding, 236
  frequency modulation, 222
  information, 223, 224, 307, 309
  interference, 319
  noise, 247, 248
  noise reduction, 85
  playback, 4, 356
  records, 39, 45
  signal demodulation, 236
  signals, 225, 294, 305, 309, 320, 321, 370, 390, 407

subcarriers, 213, 225, 309, 319
track, 392
weighting network, 248
Auditory
masking, 345, 348
nerve, 343
perception, 343
Autocorrelation, 114, 310
coefficient, 310
function, 137, 325
Autocovariance, 362
Automatic
flare correctors, 281
gain control (AGC), 149
mode, 392, 393
Autostereograms, 402
Auto stop, 29
Auxiliary audio data, 371
Auxiliary beams, 23, 26
Average
duty cycle, 224
noise power, 77
power, 97, 107
signal power, 77

Back porch, 134
Balance (product)
detector, 180
modulator, 68, 184
Balanced modulators, 164, 220, 305
Ballistic galvanometer, 276
Bandbase video, 9
Bandpass, 385
Bandpass filters, 119, 178, 189, 199, 309, 321, 344–45, 347, 385, 406, 410
Bandwidth, 247, 253, 257, 311, 321, 322, 327, 341, 347, 351, 356, 359, 365, 406, 408, 415
compression, 300, 315
expansion ratio, 113
reduction, 210, 306, 307, 311, 313, 314
savings, 91
Bandwidth-reduced systems, 329
Bar patterns, 146
Baseband
audio signals, 224
components, 213, 224, 316
data recording, 73
interference, 315
message signal, 90
signal, 250, 254, 410, 415
spectrum, 86
system, 64, 73
Basic codes, 422
Basilar membrane, 346
BASK, 66
Bauds, 72, 414
Beam, 24, 276
Beam splitter, 23, 51, 53
Beat
components, 225, 234, 236
frequencies, 107, 251
frequency, 163, 217, 224
interference, 245
rate, 72
Beat-frequency, 15
Bell filter, 199
Bender motors, 23
Bessel functions, 104, 118, 216, 250
Binary
data rate, 230
data transmission, 405
data, 233–76, 407, 413
digits, 72
FSK, 76
modulation, 66
quantized delay line, 261, 264
rate, 72
sequences, 230
signals, 72, 74
systems, 72, 79
transversal filter, 74, 82
Binocular vision, 402
Biphase, 230, 423
Biphase-L code, 233
Biphase-level (Manchester), 232
Biphase-mark, 230, 232
Biphase-space, 230
Bipolar or alternate-mark-inversion (AMI), 232
Birefringes, 53, 56
Bismuth (Bi), 40, 42
Bit
components, 226
error rate, 58, 416, 432
error rate (BER), 431
frequencies, 161
frequency, 330
interval, 230
packing density, 431
rate, 230, 362, 365, 367, 368, 405, 408
shift, 58
Bits, 2, 35
Black and white, 128
Black and white levels, 133
Black-and-white receivers, 153, 200
Black-and-white TV signal, 129
Blanking interval, 164
Blanking level, 133, 221, 226
Blanking period, 186
Blanking pulses, 130
Block codes, 379
Block interleaving, 381
Bose-Chaudhuri-Hocquenghem (BCH) codes, 380
Brain, 308, 343, 402
Brightness, 127, 308
level, 306
sensation, 163
signal, 411
variations, 134
Broadband chrominance receivers, 179
Broadcast
channel, 165
communications, 5
systems, 149
video channel, 148
Broadcasting
standards, 261
system, 407
TV, 135
Bruh blanking, 128
Buried subcarrier, 36, 46
encoding system, 315, 326
Burst errors, 381
Burst phase, 186
Burst signal, 178, 221, 269

Camera noise, 315
Cameras, 34, 306, 401, 402, 404
Capacitance, 262, 263
Capacitive
discs, 11, 12, 46, 218
electronic disc (CED), 36
pickup, 12, 13
probes, 289
systems, 21, 36
videodisc systems, 29, 315
Carriage drive, 52
Carrier, 105, 123, 250, 252
amplitude, 327
frequency, 215, 267, 269, 317
level, 46
signal, 88
Carrier-to-noise ratio, 113, 327, 408
Carson's rule, 108
Catastrophic distortions, 53
Cathode-ray tubes, 129, 157, 398, 401
CAV, 11, 29, 52, 226, 297, 307
CAV discs, 211, 240, 274, 275, 288, 298, 307
CAV mode, 29, 53, 401
CAV reflective discs, 218
CCD, 27, 317
CCD delay lines, 272, 318
CCIR, 134, 136, 248
CD, 58
audio channel, 236
digital audio systems, 373, 383

CD (*cont.*)
  laser, 31
  record, 371
  ROM, 12
  systems, 370, 376, 377, 381
CD-ROM, 58
CED (capacitive electronic disc), 12, 47
Center frequency, 249
Central detection, 18
Channel
  attenuation, 86
  bandwidth, 73
  capacity, 154, 342
  characteristics, 73, 82
  codes, 371, 376, 404
  error, 367
  impairments, 121
  modulator, 377
  noise, 114, 322, 323, 361
  property, 213
  symbols, 379
  transfer function, 115
Chapter codes, 54
Chapter numbers, 54, 226
Character generator, 34
Charge-coupled devices (CCDs), 263, *See also* CCD.
Chess pattern, 162
Chroma
  carrier, 316
  noise, 39
  phase errors, 16, 260
  saturation, 239
  sidebands, 234, 249
  signal, 267, 327, 320, 385
Chromacity diagram, 167
Chromacity errors, 180
Chrominance
  bandwidth, 328
  carrier frequency, 317
  channels, 157, 179, 247
  information, 242
  lock, 197
  signal phase error, 196
  signal subcarrier frequency, 186
  subcarrier, 161
  vector, 176
CIE color rendering index (CRI), 279
CIE system, 280
CIE tristimulus values, 279
Cinescope, 157, 158, 178
CIRC error correction, 381
Circular apertures, 257, 290
Circular beam, 145
Circular spot, 24
Clamping circuits, 134
Clock rate (CLR), 414
Closed-circuit TV, 135, 210
CLV, 11, 29, 326
  code, 226
  constant linear velocity, 11
  disc, 298
  format, 373
  mode, 52, 53, 401
  videodisc systems, 370
Coated discs, 26
Code
  dictionary, 429
  overhead, 431
  parameters, 414
  rate, 416
  word, 379, 380
Coded modulations, 65
Coder, 363, 407, 410, 419
Codes, 226, 413, 422
Coding, 341, 344, 356, 357, 370, 371
Coding algorithm, 300
Coherent carrier, 88
Coherent demodulation, 76, 99
Coherent light source, 23
Coherent phase-shift keying (CPSK), 412
Collimator lens, 23
Colometric values, 163
Color, 127, 306
  acuity, 154
  bars, 242
  burst, 164, 268, 270, 368, 405
  burst signals, 168, 269
  camera, 155
  correction, 34, 268
  difference signals, 156, 180, 186, 200
  diffraction theory, 18
  fidelity, 302
  hue (tine), 128
  image, 301
  information, 36, 154, 268, 304, 317
  Mach bands, 280
  monitor, 5
  picture, 255
  quality, 279
  receiver screen, 170
  rendering, 281
  rendering index, 279
  reproduction system, 279
  saturation (vividness), 128
  signal bandwidth reduction, 304
  signals, 242
  subcarrier frequency, 52, 53
  sync (burst), 128
  television, 210
  temperature, 163
  test signal, 205
  TV, 147, 401, 402
  TV signals, 152, 170
  under, 37
  value, 128
  video, 2
  video signals, 221
Color-axis, 164
Color-bar chart, 170
Color-bar patterns, 170
Color-bar signals, 170
Color-difference signals, 156, 162, 179, 183, 198, 411
Colorimetry, 152
Colorimetry distortion, 155
Comb filtering, 401
Comb filters, 142, 179, 304, 306, 317, 319, 320
Combined amplitude and PSK, 77
Combined modulation systems, 65
Comite Consultant International des Radiocommunications, 134
Communications system, 77, 373
Compact disc (CD), 356, 370, 432
Compact disc read only memory (CD-ROM), 432
Compander, 353
Compander noise reduction systems, 354
Companding, 353
Companding system, 354, 356
Compandor/expander, 322
Compansion, 347, 353
Comparator, 114
Complementary colors, 162, 173
Complementary hues, 176
Component recording, 4
Component television, 5
Composite
  color TV signal, 160, 166, 210
  color video signal, 165
  recording, 5
  signal, 170, 405, 408
  sync, 131, 278
  test signal, 205, 226
  video signal, 128, 152, 182, 199, 266, 272, 385, 405, 406, 410
Compression, 322
Compression molding, 45
Compression techniques, 210
Compressor, 341, 354
Computer
  control, 2
  graphic data, 299, 400
  interface, 29
  peripheral, 390
  programs, 392
Computer-generated signals, 71

Conductive surface, 13
Constant-amplitude continuous-phase signal, 109
Constant angular velocity, 11, 211
Constant duty cycle, 238
Constant linear velocity (CLV), 211, 370
Constant luminance principle, 155, 158
Consumer videodisc, 11
Continuous
  angular modulation, 68
  carrier, 69
  complex spectrum, 137
  messages, 66
  modulation systems, 65
  pilot signal, 225
  random noise, 247
Continuous-wave FM, 214, 236
Continuous-wave (CW) modulation, 95, 96, 117
Contrast, 128, 157
Control data, 371
Controller, 49
Conventional AM, 76
Conventional amplitude modulation (AM or CAM), 67, 89
Conventional frequency discriminators, 236
Conventional television, 311
Convolutional code, 379
Convolution theorem, 139, 209, 358
Copper substrate, 47
Corner frequencies, 327, 348
Correlated receiver, 85
Crest factor, 360
Criterion of fidelity, 67
Cross interleaved Reed Solomon Code (CIRC), 381
Crossband, 37
Crosstalk, 29, 155, 257, 288, 296, 383
CRT display, 390
CRTs, 155, 404
Crude quantizers, 360
Cueing information, 223
Cue track, 226
Cutoff frequency, 223, 239, 262, 290, 296
Cutterhead, 47, 48
Cyclic code, 380

D/A conversion, 300
Data, 230, 299, 432, 434
  base, 2, 4
  compression, 325, 360
  density, 413
  file, 342, 398
  pattern, 231
  rate, 73, 413, 415
  recording, 404, 405
  signaling rate, 413
  storage, 2
  stream, 417
  symbols, 85
  volume, 4
  window, 60
DBX, 354
DC component, 90, 413
  restoration, 147
DC free, 371, 376
DC free code, 423
DC motor, 26
DC offset, 91
DC value, 88, 224
DC variations, 134
Decibels, 247
Decision level, 407
Decoder, 353, 357, 363, 368, 379, 419
Decoding, 178, 198, 236, 320, 341, 370, 428, 429
Decompression, 323
Deemphasis, 255, 317, 332, 347, 384
Deemphasis circuit, 327
Deemphasis filter, 199
Deemphasis network, 116
Defection system, 147
Defects, 239, 240
Defocusing, 21, 24, 29, 257
Defocusing effect, 26
Delay codes, 233
Delay line, 189, 198, 236, 262, 264, 309, 317, 382, 383
Delay-line (PAL-D) receiver, 188
Delay modulation, 230
Delay modulation (DM) codes, 422
Delta coders, 362
Delta modulation (DM), 11, 76, 357, 362
Delta modulators, 363, 367
Demodulated signal, 64, 243, 246
Demodulated video, 260
Demodulation, 68, 76, 236, 252
Demodulator, 238, 371
Density ratio (DR), 414
Depth of focus, 18, 24, 30, 290
Detection process, 217
Detector signal, 256
Deterministic signals, 64, 66
Development, 253
Diamond pickup systems, 12
Differential
  coding, 360
  detectors, 21, 76, 120
  error signal, 189
  phase delay, 20
  phase error, 193, 200
  phase-shift keying (DPSK), 412
  pulse code modulation (DPCM), 65, 76, 357, 360
  signal, 323
Diffracted light, 21
Diffraction, 292
Diffraction pattern, 290, 293, 294
Diffraction-limited lens, 23, 40
Diffraction-limited system, 30
Digit rate, 78
Digital
  audio channels, 371
  audio codes, 54
  audio discs, 1, 4
  audio modem, 371
  audio signal, 384
  audio systems, 373
  carrier modulated systems, 85
  carrier modulation, 74
  carrier system, 76
  channels, 11, 58, 370, 390, 412
  coding, 213
  communications, 412
  data, 6, 368, 379, 412, 432
  data recording, 4, 368
  data storage, 2, 404
  decoding system, 371
  encoding system, 362, 414
  format, 370, 410
  frequency modulation, 66
  images processing, 322, 325, 401, 412
  information, 2
  interface, 371
  microphones, 342
  modulating systems, 71
  modulation codes, 415
  modulation methods, 77
  modulations, 65
  optical discs, 58
  phase modulation, 66
  rate, 71, 300
  signaling, 74
  signal processing system, 300, 342
  signals, 71, 76
  speakers, 342
  speech, 367
  storage, 2
  sum variation (DSV), 415
  systems, 67, 71, 383
  tapes, 353
  techniques, 367
  technology, 342
  time-base-error correction, 261, 267, 268

Digital (*cont.*)
TV, 374
videodisc technology, 412
write-once discs, 12
Digital-to-analog (D/A) converters, 371
Digitized video, 268
Diode lasers, 49
Dirac delta function, 143
Direct current, 415
Direct-read-after-write (DRAW), 8, 12, 33, 42
Dirt, 51
Disc
capacity, 288, 290
coating, 39
defects, 405
errors, 370
mastering process, 392
noise, 225, 319, 352
player, 1
production, 8, 33
replication, 39, 44, 48
simulator, 54
surface noise, 346, 351
system, 289
testers, 56
testing, 53
track, 217
Discrete messages, 66
Discrete source, 73
Discriminator, 114, 122, 220, 264, 327
Discs, 8, 33, 34, 39, 240, 400
Displays, 299
Distortion, 117, 155, 213, 224, 236, 242, 243, 245, 257, 301, 327, 342, 348, 365
Dividers, 162, 329
Document storage, 2
Dolby, 354
Domaines, 243
Dot interlace, 313, 314
Dot-pattern, 200
Dot-sequential systems, 154
Double density codes, 58, 230, 422
Double frequency, 230
Double frequency code, 121, 230
Double-sideband demodulation, 86
Double-sideband modulation, 86
Double-sideband-suppressed carrier modulation (DSB or DSBC), 67
Down conversion, 100
DRAW signal, 51
Drive, 434
Drivers, 432
Drop-frame, 35
Drop-in, 271
Dropout, 35, 46, 220, 270, 347, 367, 371, 379, 404
Dropout compensation, 270, 368
Dropout compensator, 220
Dropout detector, 321
Dropout pulse, 368
Dropout signal, 271, 272
Dropout-like noise, 356
DSB, 71, 76, 88, 91, 94
Dust, 22, 46, 51, 54, 271
Duty cycle, 58, 120, 209, 249, 253, 254
modulation, 224, 234, 236, 239, 320, 321
offset, 225
Dye discs, 28
Dynamic error, 354
Dynamic noise reduction system, 347, 348

Earphones, 343
Eccentricity, 26, 27, 46, 260, 269
Edge, 315
detection, 239
fidelity, 302
information, 315, 316
Editing, 34, 53, 342
EFM code, 376, 377
Eight to fourteen modulation (EFM) code, 376
Eight-field sequence, 183
Electrical motor, 24
Electro-optical (EO) modulator, 42, 253, 255
Electroforming, 44
Electromagnetic radiation, 154
Electromechanical recorder (EMR), 47
Electromechanical recording, 47
Electron beam, 4, 144
Electron distribution, 129
Electronically variable delay line, 261
Electronic discs CED, 11
Electronic documentation, 400
Electronic limiter, 255
Electronic mail, 35
Elliptic polarization, 28
Elliptic spots, 24
Embossing, 45
Encoder, 35, 371, 379, 428
Encoding, 37, 220, 424, 428, 429
Encoding process, 164, 234, 319
Encryption, 356
Energy, 43, 108, 109, 221, 288, 406, 415
content, 139
distribution, 138
spectrum, 137
threshold, 288
Enhanced NRZ, 422, 431
ENRZ-complement, 422
ENRZ-parity, 422
Envelope demodulation, 95
Envelope detector, 68, 76, 90
Equalization, 83
Equalizing circuit, 48
Equalizing filter, 73
Equalizing pulses, 128, 135, 148
Equipment, 39
Erasable
data storage, 397
discs, 1, 27
media, 49
PROD (EPROD), 8, 12
videodiscs, 5, 58
Erase process, 28
Erasure flag, 383
Error
correction code (DCC), 371, 379, 432
detection code (EDC), 379, 432
pattern, 380
power, 365
probability, 78
propagation, 367, 371, 376, 380, 416, 431
rate, 58, 60, 415
samples, 360
signals, 16, 24, 26, 53, 268, 269, 270, 361
visibility, 148
Error-free recording, 51
Evaluation, 53
Evaluation criteria, 414
Exchanging playing time, 3
Expander, 80, 81, 341, 354
Exponential modulation, 67, 100
Exponential process, 68
Extended play, 211, 287, 288, 299, 308, 326
External synchronization, 29
Eye, 152, 402
opening, 416
pattern, 60, 230
persistency, 162
sensitivity, 247

Facsimile data, 299
Fan-out, 34, 45
Faraday effect, 28
FCC (Federal Communications Commission), 154, 163
Feedback FM demodulators, 114
Feedback noise, 53
Fidelity criteria, 299, 301, 342

Field
  frequency, 134, 137, 142, 147
  period, 137
  rate, 311
  rate reduction, 310
  scan period, 135, 161
  synchronizing period, 182
  synchronizing pulses, 128, 137
Field-blanking period, 137
Film picture, 272
Film projection, 127
Film scanner, 273
Film-based videodiscs, 46
Filter, 114, 198, 220, 247
Fingerprints, 22
Finite set, 67, 71
First-order Bessel function, 145
First-order sidebands, 218, 234, 236, 242, 250
Fixed-length asymmetries, 250, 254
Fixed-length block encoding, 423, 426
Fixed-length distortions, 54
Flag bits, 371
Flat capacitive disc, 11, 13, 16
Flicker, 128, 308, 311
Flip-flop, 421
Flutter, 342, 370
Flying spot scanner, 272
FM
  band, 225
  carrier frequency, 43
  carriers, 220, 299, 303, 304, 306, 307, 317, 347
  channel, 405
  code, 121
  demodulation, 76, 249, 347
  demodulator, 110, 236, 238, 246
  detection, 225
  discriminator, 115, 327
  domain, 249
  modulator, 327
  modulated video signal, 234
  modulation, 37, 266
  noise, 39, 317
  pilot, 350
  pulse train, 125
  scheme, 326
  signal spectrum, 217
  signals, 108, 243, 246, 248, 249, 250, 251, 252, 256, 269, 272, 283, 317
  spectrum, 106, 109
  systems, 100, 252, 317, 327, 332
  threshold, 317
  tone-modulated signal, 106
  video, 316
Focal distance, 290
Focal plane, 24
Focused beam, 26
Focused spot, 23
Focus error, 26, 290
Focus error signal, 26
Focusing lens, 21, 40, 42, 49, 51
Focusing lens objective, 24
Focus servo, 24, 40
Folded frequency, 359
Folded sidebands, 217
Format frequency, 343
Format structure, 344
Four-five GCR, 427
Four-segment photodetector, 26
Fourier
  analysis, 139
  coefficients, 207
  components, 108
  series, 67, 103
  transform, 67, 258, 324, 325
  transform pairs, 137
Foveal vision, 306
Frame
  crawl, 278
  format, 377, 379
  frequency, 137, 139, 141, 182
  identification, 35
  numbers, 49, 273, 391
  period, 130, 161
  rate, 35, 148, 198, 306, 308, 309, 311, 312, 313, 315
  rate reduction, 308, 310
  repetition frequency, 142
  synchronization pulses, 131, 427
  time, 161
Freeze frame, 275, 278, 308, 367, 391
Freeze-frame mode, 394
Frequency
  bandwidth, 200, 348
  demodulator (FD), 68
  deviation, 122, 150, 200, 201, 242, 256, 317
  deviation index, 101
  domain, 74
  interlace, 303
  interleaving, 163
  modulation (FM), 100, 230, 271, 348
  offset, 87
  resolution, 344
  response, 145, 243, 244
  sharing, 36, 39, 64
  spectrum, 160, 234
  subcarrier, 242
Frequency-modulated carrier, 220
Frequency-modulated subcarrier, 199
Frequency-shift keying (FSK), 66, 230
FSK, 74
FSK systems, 76

Gamma amplifier, 158
Gamma correction factor, 157, 159
Gamma value, 196
Gas laser, 23, 49
Gaussian distribution, 31, 59
Gaussian function, 358
Gaussian noise, 111, 246
Glass delay line, 318
Glass substrate, 40
Granular distortion, 363
Granular noise power, 364
Graphics, 432
Gray code, 410
Gray levels, 155, 255
Gray scale, 2, 322
Green channel, 156
Grooved capacitive system, 16, 296, 320
Grooved capacitive videodisc, 47
Grooved disc, 13
Grooved track, 16
Group code, 376, 423
Group coded recording (GCR), 426
Group coding algorithm, 428
Group delay distortion, 245
Group delay response, 267
Guard band, 13

Half-line offset, 181
Hanover bars, 188
Hard limiting, 224, 253, 320
Harmonic analysis, 67, 207
Harmonic distortion, 53, 119, 245, 249, 250, 327, 370
Harmonics, 108, 249, 255, 344, 348, 351
Heterodyne color recovery system, 261, 266
Heterodyne technique, 69
Hexadecimal value, 229
Hexadecimal word, 229
High-definition television (HDTV), 5, 401
High-fidelity sound, 342
High-fidelity TV, 401
High-order sidebands, 247
High-pass filter, 220, 236
Hilbert transform, 91
Hilbert transform filter, 94
Home computers, 412
Horizontal
  blanking intervals, 147

Horizontal (*cont.*)
line synchronizing pulses, 128, 136
resolution, 307
sync, 148, 261, 262
sync instabilities, 260
sync pulses, 134, 147
Host computer, 392
Hue, 156, 260
Human
color vision, 154
interface, 3
perception, 287, 343
vision models, 302
visual spectral responses, 280
visual system, 279
Hybrid mode, 11

I and Q channels, 163
I and Q signals, 168, 408
I and Q synchronous demodulators, 179
I color-difference signal, 164
I component, 166
I signal, 411
Ideal limiter, 260
Identification signals, 210
Idle noise, 362
Illuminant, 279
Illumination, 128, 133, 293
Illumination pattern, 308
Image breakup, 311, 314
Image fidelity, 279
Image quality measure, 302
Image-processing systems, 302
Impairments, 243
Improvement factor, 117
Indent signal, 201
Industrial players, 6
Inertia, 138
Inertial systems, 139
Information
density, 2, 288, 404
plane, 23
signals, 360
storage, 397
track, 213
Injection molding, 45
Instant deviation, 322
Instant frequency, 68, 69
Instantaneous
amplitude, 101
angular frequency, 100, 107, 120, 216
duty cycle, 120
frequency, 100, 234
frequency deviation, 115, 221, 327
phase deviation, 123
temporal frequency, 317
Integrating time, 247, 345
Intelligibility, 279
Intensity-modulated laser beam, 42
Interactive videodisc, 400
Interactive video simulators, 225
Interference components, 317, 319
Interference, 255
Interlaced scanning, 135
Interlaced systems, 162
Interlace technique, 35, 36, 139, 315, 317
Interlacing, 134, 135, 381
Interleaved Reed-Solomon codes, 381
Interleaving degree, 381
Interleaving process, 160
Intermediate frequency (IF), 112
Intermodulation (IM) products, 46, 249, 251
International test signal (ITS), 226
Interpolation, 381
Intersymbol interference (ISI), 371, 411
Inverse Fourier transform, 91, 137

Jordan codes, 231, 417, 418, 419, 421, 422, 423
Jump, 16, 276
Jump back, 26

Kell's factor, 145
Kerr effect, 28
Keying, 16
Kick-back pulse, 27
KL transformation, 325
Kronecker symbol, 74

Laser, 28, 31, 42, 240
beam recording, 288
diodes, 23, 31
intensity, 42
light wavelength, 290
noise, 53, 384
power, 288
Laservision (LV), 347, 383
player, 3
recorder, 3
standard, 210, 213
systems, 35
videodisc, 383
Lead-in code, 229
Lead-in tracks, 212
Lead-out code, 229
Lead-out tracks, 212
Lead-screw drive, 309
Leading edge, 134
Lens MTF, 289
Lens numerical aperture, 24
Light
amplitude modulation, 28
beam, 275
diffraction, 18
intensity, 20
source, 279
Light-ray diffraction, 18
Limiter-discriminator, 76
Line
crawling, 314
frequencies, 137, 141, 160, 186, 198, 304, 329, 346, 376
interlacing, 312
interval, 201, 221
period, 137, 147
rate, 134
scanning, 128, 313
synchronizing pulse, 137
Linear
coders, 367
codes, 379, 380
continuous-wave modulation systems, 86
CW modulation systems, 96
distortion, 115, 180
modulation, 67, 68
polarized beam, 28
prediction, 360
production, 323
scanning, 132, 138
superposition, 69
Linearity, 236
Line-blanking interval, 137
Line-by-line switching, 191
Line-crawling, 311
Line-scanning rate, 163
Line-sequential, 154
Local carrier, 76
Local oscillator, 87
Logarithmic PCM, 365, 367
Look-up table, 371
Loudspeaker voice coil, 24
Low-pass filter (LPF), 83, 114, 117 121, 180, 236, 239, 257, 262, 304, 348, 349, 351, 359, 363, 371, 406, 408, 410
Low-pass filtering, 249, 324
Lubricant, 48
Luminance, 210, 280, 299, 319, 405
band, 36
bandwidth, 245, 35, 315
changes, 156
channel, 247
components, 225
information, 36, 166, 316, 317
level, 239, 255

signal band, 224, 234
signals, 155, 217, 224, 255, 266, 304, 318, 320, 327, 401, 405, 406, 408, 410
Luminosity, 127
Lumped constant delay lines, 264, 267
Lumped-element circuit, 15
LV signals, 384, 385

MACs, 5, 39
Magnetic field, 28
Magnetic mastering, 269
Magnetic storage, 2
Magnetic tapes, 269
Magnetic-tape libraries, 400
Magneto-optical discs, 28
Main carrier, 216, 254
Manchester codes, 121, 230
Manual mode, 392, 393
Margin analysis, 58, 61
Markov source of the first order, 416
MASK, 77
Mass data storage systems, 26
Mass production, 240
Mass replica discs, 1, 8, 28
Mass storage medium, 412
Master disc, 40, 42, 44, 48
Master recording process, 39, 49, 213
Master tape, 272, 274, 432
Mastering, 11, 12, 33, 34, 40, 46, 217, 272, 309, 342, 368
Mastering machine, 405
Matched filter receiver, 85
Material, 271
Mathematical expectation, 96
Matrix, 183
circuits, 198
process, 54
processing, 39, 44, 48
Maximal frequency deviation, 307
Maximal recording, 289
Maximal resolving spatial frequency, 21
Maximum angular frequency deviation, 103, 125
Maximum frequency deviation, 108, 221
Maximum phase deviation, 103
MCA code, 226, 233
Mean-square estimation error, 323
Mean-square value, 327
Meander of the PAL blanking, 186
Mechanical recorders, 4
Media microdefects, 33
Media parameters, 56
Melting point, 289
Melting process, 39
Memory, 2, 49, 391, 401
Message signal, 86, 123
Metal disc, 47
Metal film, 42, 288, 464
Metal mold, 34, 48
Metallic film, 46
Metallization, 39, 46
Microcomputer, 391, 392, 399
Microcomputer technology, 342
Microprocessors, 278, 400, 412
Microscope objective, 51
Microscopic lens, 42
Miller code, 231, 417–419, 421–423
Minimal optical depth, 20
Minimum recording wavelength, 296
Minimum shift keying (MSK), 85
Mirror components, 224
Mirrors, 23, 27
Mise en forme, 198
Mobile radio, 152
Modified frequency modulation (MFM), 230
Modulated
carrier, 89
index, 89
pulse train, 119
signal, 64, 65, 244, 283
signal spectra, 76
spectrum, 330
Modulating frequency, 215
Modulating signal spectra, 68
Modulating signals, 64, 65, 67, 71, 79, 86, 107, 115, 220, 238, 250
Modulation, 35, 64, 304
bandwidth, 303
code, 371
depth, 42
distortion, 355
frequency, 123
index, 107, 108, 215, 216, 242, 245, 249, 250, 283, 316, 383, 384
polarity, 148
rate, 413, 414
signal, 67
systems, 77
transfer function (MTF), 21, 218, 288
Moire components, 225, 248, 249, 250, 251
Moire effects, 248, 269
Moire patterns, 248, 346, 359
Mold, 45
Molding, 43, 46
Monochrome
picture, 153, 178, 217
receiver, 160
set, 170
signal, 153, 242
still picture, 140
systems, 268
television, 303
TV broadcast services, 161
TV, 147, 221, 401
video signal, 226, 236
Motion, 128, 131
Motion breakup, 309
Motion-picture film, 53, 129
Motor drive, 52
Multiburst signal, 205
Multichannel systems, 4
Multilevel (block) coding, 34, 77, 368, 405, 408
Multiplayer, 278
Multiple dumps, 392
Multiplex Analog Components, (MA), 300
Multiplexed analog component (MAC), 37, 38
Multiplication process, 86
Multiplicative noise, 347
Multipliers, 86, 114, 329
Multispectral information, 299
Music, 348

Narrowband chrominance receiver, 179
Narrowband AM, 123
Narrowband FM (NBFM), 109, 347
National Television System Committee (NTSC), 161
Negative frequencies, 207, 217, 344
Negative modulation, 148, 221
Negative photoresist, 43
Network, 82, 247
Neural coding and image, 302
Nickel, 44
Nickel metal molds, 44
Nickel replicas, 47
Nickel stamper, 48
Noise, 15, 31, 53, 78, 225, 245, 247, 268, 269, 270, 288, 301, 341, 342, 347, 351, 352, 354, 363, 365
immunity, 416
power, 79, 85, 96, 99, 111, 327
power density, 317
reduction, 344, 345
reduction systems, 341, 351, 353
reduction techniques, 347
spectral density, 112, 116
spectrum, 347

Noise (*cont.*)
threshold, 348
Noiseless channel, 58
Nominal peak-to-peak amplitude, 247
Nonadaptive, 363
Nonadaptive prediction, 361
Non-disc formats, 434
Noninterlaced system, 134
Noninterlacing, 135
Nonlinear distortion, 78
Nonlinear system, 69
Nonlinearities, 39, 234
Nonreturn-to-zero (NRZ), 230, 371, 376, 422
Nonreturn to zero-level (NRZ-L), 232
Nonreturn to zero-mark (NRZ-M), 232
Nonreturn to zero-space (NRZ-S), 232
Nonuniform quantization, 360, 365
Normalizing light, 155
Notch filter, 162, 179, 197, 199
NTSC
coder, 183
lines, 183, 188, 189
signal, 320
standard, 162, 225
system, 161, 218, 268, 329, 410, 411
TV standard, 136
NTSC-like, 37
NTSC-line demodulator, 195
Numerical aperture (NA), 18, 239, 240, 253, 290, 296
Nyquist bandwidth, 415
Nyquist criterion, 73, 82
Nyquist frequency, 357, 363
Nyquist rate, 360
Nyquist sampling, 363

Objective lens, 18, 23, 24, 53
Objective, 26, 27, 260
Objective fidelity criteria, 301
Object reflection, 153
Octal coded binary (OCB), 428
Offset, 224, 346
Offset QASK (OQASK), 77
On-off recording, 300
One-shot, 121
One-shot circuit, 239
One-tone modulation, 109
Operating controls, 278
Optic nerve, 302
Optical
card, 434
data storage, 2
depth, 296
diffraction, 54
digital disc, 373
digital recording, 370
disc, 17, 46, 50, 52, 213, 434
disc drive, 434
disc media, 49
disc pickup, 18
disc recording, 49
discs, 11, 39, 370
disc system, 51
edge detection, 20
edge detector, 21
illusion, 313
image, 127
interference, 43
lens system, 129
mass storage, 432
noise, 290
path length, 19
pickup, 16, 21, 23, 26, 27, 239, 240
readout method, 289, 431
recorders, 40, 53
recording, 376, 384, 413, 431
servo, 22
storage, 5
stylus, 239, 296
system, 18, 22, 23, 129
tape, 434
transfer function, 21
videodisc master, 39
videodisc systems, 29, 37
Optics, 51, 248
Optimum group codes, 423
Optimum receiver, 85
Optomechanical system, 40, 41
Optomodulator, 51
OQPSK, 77
Orthonormal functions, 325
Overcoating, 46
Overload noise power, 364
Overmodulation, 170
Oversampling index, 365
Overwrite, 58

Packaging, 39
Packing density, 39
PAL blanking, 186, 187
PAL Coding Process, 183
PAL decoding process, 186
PAL four-field sequence, 182
PAL lines, 176, 183
PAL-D receiver, 189
PAL-D system, 193
PAL-S, 191
PAL-S receiver, 191
PAL-system, 9, 82
Parallax stereograms, 402
Parasite AM, 256
Parseval's theorem, 108
Passband filter, 247
Patterning, 314
Pattern sensitivity, 416
Peak frequency deviation, 329
Peak power, 99
Perceptual distortion, 317, 367
Perceptual domain, 302
Perceptual phenomena, 280
Periodic signals, 67
Personal computer, 5
Phase
alternating line (PAL), 180
cancellation, 18
characteristics, 48, 74
comparator, 114
demodulator (PD), 68
detector, 26
deviation, 123
difference, 180
distortion, 144, 147, 178, 193
encoding, 230
error, 180, 251, 268
grating, 18
modulated, 125
modulation (PM), 17, 65, 76, 123, 268
offset, 87
quadrature, 164
response, 245
shift codes, 230
Phase-change optical media, 28
Phase-error condition, 196
Phase-locked-loop (PLL) demodulator, 114
Phase-lock-loop (PLL), 61, 405, 412
Phase-shift keying (PSK), 66
Phasor diagram, 174
Philips code, 226, 233
Phonographic recording, 4
Photodetector, 18, 23, 53
Photodiode, 20, 26, 218, 271
Photographic films, 49, 129
Photopolymerization, 45
Photoresist exposure, 44
Photoresist film, 42
Phototelegraphy, 140
Pickup devices, 129
Pickup stylus, 22, 299
Pickup transfer functions, 289
Pictorial clarity, 128
Picture, 128, 134, 269, 279, 324, 342, 390
carrier, 151
codes, 54

compression, 322, 324
content, 138, 180
detail, 148
element (pixel), 128, 133, 147, 161, 313, 323
frame, 274, 313
information, 315
interlace, 310
luminance signal, 247
number code, 217
numbers, 226, 229
quality, 3
repetition rate, 134
signal, 133, 138
stop codes, 54, 229
Picture-element, 154
Piezoelectric, 23
Pilot, 225, 269
carrier DSB, 89
signal, 225, 226, 236, 262, 268, 269, 415
tone, 225
PIN photodiode, 23, 218
Pipelining, 381
Pit
density, 40
depth, 18, 20, 28
length, 12, 17
pattern, 217, 413
width, 12, 17, 18
Pitch, 292, 343, 344
Pivoting mirror, 26, 27, 261
Pixels, 2, 310, 313
Plastic replica discs, 43
Plastics (PVC), 20, 51
Playback
beams, 51
decoder, 353
electronics, 412
process, 213
system, 33, 42, 49, 50, 309
video, 261, 262
Playback-only videodiscs, 39, 48
Player optics, 23
Playing modes, 11, 274, 275
Playing time, 4, 29, 240, 296, 297, 298, 307, 317
Polar signaling, 413
Polarization, 28
Polarizing filters, 402
Polarizing prism, 23
Polymethyl methacrylate (PMMA), 33, 40
Polyvinyl acetate (PVA), 48
Polyvinyl chloride (PVC), 33, 48
Positive-frequency energy spectrum, 137
Positive modulation, 221
Positive resist, 42
Positive spectrum, 140
Post-detection filter, 96, 98
Power spectral density (PSD), 96, 428
Power spectral density, 31, 111, 117
Preamplifier, 218
Prediction error, 323
Predictive compression, 322, 323
Predictive quantizing, 360, 361, 362
Predictive system, 323
Predictor network, 362
Preemphasis circuits, 217, 322
Pregrooved capacitive disc, 48
Pregrooved layer, 51
Pregrooved media, 56
Pregrooved optical discs, 3
Pregrooved systems, 3, 47
Premastering, 34, 39, 53, 272, 432
Preprocessing, 341, 342, 351
Prerecorded optical discs, 11
Primary color, 176
Primary sources, 153
Probability density function (PDF), 31, 358
Probability of error, 73, 85
Process control, 40
Product demodulation, 68
Product detectors, 122, 180, 410
Product FM demodulator, 236
Product modulators, 408
Program dump, 34, 392
Program time code, 226
Programmable discs, 8
Programmable optical disc systems, 49
Programmable ROM (PROM), 397
Programming mode, 392, 393
Projection objective, 294
Proportional asymmetry, 253
Psychophysics, 287
Pulse
AM, 77
amplitude modulation (PAM), 65, 69, 76
code modulation (PCM), 11, 65, 76, 214
count discriminator, 238
FM, 118, 125, 213, 220, 238
FM signal, 214, 224
frequency modulation (PFM), 11, 117, 213
instantaneous frequency, 234
PM, 125
phase modulation (PPM), 65, 76
pulse position modulation (PPM), 65, 69, 213
rate, 343
repetition frequency, 118, 119, 213, 221, 236, 239, 260
spectrum, 74, 119
train, 65, 118
transitions, 58
width modulation, 76
Pulse-code modulation (PCM), 357
Pulse-counting techniques, 321
Pulse-modulation, 117
Pulse-width (duty-cycle), 35
PWM/FM, 65

QAM chrominance signal, 184, 188
QAM, 165, 317, 410
QASK signals, 410
Q component, 166
Q signal, 411
Quadrature
amplitude modulation (QAM), 408
amplitude shift keying (QASK), 408
components, 196
crosstalk, 180
modulation, 164
PSK (APSK), 66, 76
Quadrature-modulated format, 37
Quality control, 53
Quality inspection, 39
Quantization, 67, 371
Quantization errors, 268, 358
Quantization noise, 361, 363
Quantizer, 357, 360, 365
Quantizing error, 357, 365
Quantizing noise, 367, 376, 383
Quarter-line offset, 182
Quarter-wavelength plate, 23, 51, 53
Quartz crystal, 24
Quartz delay lines, 264
Quartz oscillator, 264

Radial servo system, 23, 276
Radial tracking, 23
Radio communication, 127
Radio Electronic and Television Manufacturers' Association (RETMA), 161
RAM (random access memory), 268, 367, 392
Random access memory (RAM), 29, 391
Random data, 416
Random fields, 325
Random noise, 246, 247
Random-access memory, 381
Randomized NRZ, 422
Raster line, 197

Raster scanning lines, 143
Ratio code, 230, 232
Ratio detector, 236
Ratio of noise compression, 327
Read after write, 34
Read beam, 51
Read during writing, 8
Reading
  beam, 26, 256, 257, 261
  lens, 239, 246
  light, 290
  objective, 290
  process, 252
  spot, 18, 20
Read objective, 24
Read-only discs (RODs), 8, 12
Read-only memory (ROM), 397
Read-only videodisc system, 33, 49
Readout, 13, 16, 22, 28
  beam, 18
  mechanism, 20
  principles, 18
Read spot, 33
Read track, 294
Read while write, 44
Receiver power, 99
Receivers, 24, 147, 255, 322, 412, 414, 415
Receiving filters, 79
Recordable discs, 27, 28, 53, 54
Recordable laser videodisc (RLV), 51
Recordable optical discs, 11
Recordable optical videodiscs, 56
Recorded picture, 275
Recorded signal, 22
Recorder, 3, 40, 52, 64, 320
Recording
  area, 1
  beam, 18, 40
  channel, 415
  codes, 230
  film, 28
  layer, 51
  lens, 40
  material, 49
  mechanism, 28, 49
  media, 12, 39, 47, 56
  objective, 253
  process, 253, 260, 379
  processing, 39
  wavelength, 239
Redundancy, 230, 306, 362
Redundancy reduction, 311
Reed-Solomon (RS) codes, 380–383
Reference
  color equalization, 155
  frequency, 329, 330
  light, 155
  oscillator, 24
  source, 155
  subcarrier, 179, 189, 193
  white, 155, 156, 163
Reflective discs, 26, 46
Reflective index, 20, 42
Reflective optical disc, 17, 18
Reflective videodiscs, 18
Reflective videodisc systems, 373
Reflectivity, 54
Refraction index, 18
Remise en forme, 199
Remote control unit (RCU), 6, 391, 392
Removable optical storage media, 22
Replica disc, 376
Replicas, 1, 42, 43, 45
Replicated discs, 16, 54
Replicated videodiscs, 432
Replicated videodisc system, 379
Replication process, 34
Residual carrier, 239
Resist film, 42
Resistive matrix, 164
Resolution, 133, 134, 230, 249, 289, 312, 317, 376
Resolution cell, 2
Resolving elements, 145
Resolving power, 129
Resonances, 48, 348
Resonant frequency, 16
Retrace modes, 130, 131
Retracing time, 139
Return-to-bias (RB), 230, 232
Return-to-zero (RZ), 230, 232, 422
RF
  channels, 310
  preemphasis, 198
  signals, 220, 310
Rice code, 429
Ring modulator, 68
Rise time, 351
Rms amplitude, 247
Rms error, 301, 302
Rms signal-to-noise ratio, 301
Rms value, 358
ROM, 371, 377
Root-mean-square, 344
Root-mean-square (rms) error, 301
Rotational frequency, 24
Rotational speed, 27
Run-length-coded NRZ, 426
Run-length limited, 422
Run-length-limited codes, 230, 429
Sample-and-hold circuits, 371
Sampling
  frequency, 374
  period, 364
  pulse spectrum, 359
  rates, 357, 359, 367, 373, 402, 404
  theorem, 65, 67, 69, 357, 358, 360
Scan lines, 161
Scanning
  beam, 129, 133, 138, 139
  lines, 134, 138, 401
  mechanism, 129
  method, 131
  period, 139
  process, 129, 139, 303
  rate, 160
  velocity, 138
Scattering effects, 17
SECAM
  code, 198
  decoder, 199
  delay line, 199
  system, 198, 411
Second harmonic distortion, 215
Second harmonics, 40, 220, 245, 254, 384, 416
Second Nyquist criterion, 74
Second subcarrier harmonic, 188
Second-order sidebands, 234, 242, 384
Self-clocking, 376
Self-clocking code, 230
Semiconductor laser, 432
Sequential color a memoire (SECAM), 197
Sequential systems, 154
Serrated field pulses, 135
Servo controls, 31
Servo control system, 26
Servo systems, 22, 23, 49, 51, 268, 370, 371
Servomechanisms, 22
Setup level, 128, 134
Shift registers, 268, 380
Sideband components, 107
Sideband lines, 142
Sidebands, 105, 125, 221, 316
Signal
  elements, 11, 12, 13, 413
  encoding, 34
  energies, 408
  interference, 416
  parameters, 242
  playback, 18
  processing, 287, 299, 300, 301, 373
  recording, 341

spectral density, 116
spectrum, 415
tracks, 13
Signal coding, 365
Signal-quantizing noise ratio, 80
Signal-to-noise
degradation, 58
improvement ratio, 124
ratio (SNR), 12, 46, 245–247, 268–269, 347–348, 353, 358, 368, 370
Signal-to-quantization error rate, 365
Signal-to-quantizing noise ratio, 362
Signaling interval, 73, 74, 85
Signaling pulse waveform, 81
Silver halide materials, 39
Simultaneous systems, 154
Sinc pulses, 296
Sine and cosine transforms, 325
Single-sideband modulation (SSB), 67, 91
Single-tone modulation, 109
Skip-frame extended play, 309
Slide projector, 278
Slope overload, 363
Slot-loading player, 48
SNR improvement, 327, 332
Soft limiter, 253
Sound
carrier, 151, 161
over still-frame, 342
pressure levels (SPLs), 348
quality, 347
signals, 210
track, 273
Sound-over-still frame, 356
Spatial
cutoff frequency, 240, 253
frequencies, 40
frequency, 13, 307, 317, 325
frequency spectrum, 288
phase variations, 18
recording frequency, 322
resolution, 306, 322
resolution acuity, 308
resolution reduction, 306
Special effects, 26, 170, 274, 296
Special features, 288
Spectra, 35, 54, 74, 102, 213, 417, 419
Spectral
amplitude density, 137, 205
analysis, 26, 344
components, 69, 106, 107, 138, 142, 213, 224, 246, 303
densities, 67, 79, 112, 246, 416
lines, 105, 139, 207, 209
phase density, 137
power density, 231
zeros, 415
Speech information, 343
Speech perception, 342
Speed servo system, 47
Spherical aberration, 18
Spindle, 16, 27, 53
control, 22
motor, 24
Spingle servo, 24
Spiral track, 18, 42
Split detector, 296
Split photodetector, 20
Stamper replication, 39, 46
Stampers, 34, 45, 271
Standard
color television systems, 155
luminance curve, 163
signal, 210
TV, 36
TV receiver, 218
TV signal, 37
Start code, 212
Stereo sound, 2, 35
Stereo system, 371
Stereophonic sound, 342
Stereoscopic television, 401
Still frame, 11, 26, 367
Still-frame audio, 54
Still pictures, 127, 278
Stochastic signals, 66
Stop audio mode, 394, 398
Stop bit, 229
Stop code, 218
Stop-motion, 309
Storage density, 2
Stylus electrode, 15
Stylus tip, 13
Subcarrier, 160, 245, 255, 268, 278, 328, 407, 411
angular frequency, 183
burst, 186
frequency, 317, 406
reference signal, 178
spots, 162
Subjective fidelity criterion, 302
Subjective signal-to-noise ratio, 157
Substrates, 26, 288, 289
Switching element, 68
Symbol interleaving, 381
Synchronous demodulators, 188
Sync intervals, 210
Sync pattern, 422
Sync pulse level, 234
Sync pulses, 149
Sync signals, 34, 179, 199
Sync tip, 221, 242
Synchronization, 35, 74, 371
function, 371
pulses, 230
signals, 275
Synchronizing
interval, 74
level, 134
pulses, 128
signals, 220
Synchronous demodulation, 87, 95, 99, 179, 196
Synchronous detectors, 188
Synchronous systems, 72
Syndrome register, 380
System bandwidth, 260
System gamma, 157
System parameters, 200

Tachometer, 24
Tangential density, 11
Tangential servo system, 23
Tape mastering, 342
Tape noise, 351, 352
Teletex, 404
Television
cameras, 281
pictures, 128, 287
set, 310, 321, 390, 393
signal, 220, 303, 304, 307, 315, 316, 321
standards, 296, 308, 315, 406
systems, 297, 308, 401, 402, 407
Tellurium (Te), 40, 42
Temporal frequency, 240, 241, 307
Test colors, 279
Test signal, 34, 77
Thermal energy, 288
Thick videodiscs, 45
Thin metal films, 39, 42, 43
Thin videodiscs, 45
Third Nyquist criterion, 74
Third-order chroma sideband, 234
Three-color process, 152
Three-dimensional TV, 402
Three-phase modulation (3 PM) code, 423
Threshold effect, 99
Time
base, 22
constants, 222
errors, 74, 260, 261
multiplexer, 371
sharing, 37
window, 58
Time-base correction, 27, 264, 268

Time-base error (TBE), 27
Time-base error correction, 225, 260, 261, 321
Time-base jitter, 266
Time-lag errors, 46
Time-sharing, 37, 154
Timed frequency modulation (TFM), 77
Timing errors, 260, 261
Tone modulation, 86, 88, 91, 103, 123
Tone narrow-band FM, 283
Tone-modulated AM signals, 123
Tracking control, 376
Tracking error signal, 16
Tracking servo system, 16
Tracking signal elements, 13
Track kissing, 54
Track pitch, 12, 16, 26, 240, 289, 290, 373
Track surface, 16
Track-to-track pitch, 296
Track width, 29
Transducer, 261
Transfer characteristics, 243
Transfer function, 13, 15, 21, 68, 82, 115, 116, 222, 257, 350, 363
Transfer rate, 432
Transform compression, 322, 324
Transition density, 422
Transition distribution histogram, 61
Transition interval, 416
Transmission
  bandwidth, 94, 108, 113
  channel, 129
  line, 15
  medium, 343
Transmissive discs, 17, 26
Transmitter, 150, 412
Transmitting filter, 73
Transversal filter, 82
Transversal predictor, 367
Trapezoidal-cross-section, 47
Tree codes, 379
Triangular-cross section, 47
Triangular noise, 215, 223, 326, 347
Triangular noise spectrum, 246
Triangular random noise, 247
Tristimulus values, 279
Two-headed player, 4
Two-level AMI class II coding, 230
Two-sided power spectral density, 327
Type noise, 356

U and V signals, 183
UHF, 151
  band, 152
  modulator, 321
  signal, 16
Uncorrelated data, 325
Uniform quantization, 80, 377
Unipolar, 72
Unmodulated carrier, 236
Unweighted video noise, 247
User program, 391, 392

V modulator, 182
Varactor diodes, 262, 263
Variable-length coding, 428
Variance, 361, 362, 365
Varicap diode, 68
Velocity errors, 267
Vertical
  blanking interval, 34, 35, 201
  blanking period, 136
  blanking pulses, 142
  interval control, 226
  interval reference (VIR), 226
  interval test signals (VITS), 205
  interval time code (VITC), 35
  picture resolution, 318
  pulses, 135
  resolution, 134, 189, 210, 307, 321, 401
  scanning frequency, 134
  sync pulse, 135
Vestigial amplitude modulation, 168
Vestigial sideband characteristic, 95
Vestigial single-sideband modulation (VSB), 67, 94
Video
  bandwidth, 137, 226
  baseband, 226, 236
  breakup, 35
  camera, 5
  carrier, 160, 161, 224, 320, 346
  channel, 211, 368, 405, 407
  coding, 365
  deemphasis, 200
  domain, 178, 220
  FM signal, 224, 320
  frames, 367
  frequency modulation, 220
  group delay, 245
  high-density, 11
  information, 1, 220, 236
  lines, 226
  modulating signal, 234
  noise, 247
  picture, 384
  preemphasis, 37
  program, 394
  signal, 8, 9, 34, 52, 134, 248, 268, 269, 270, 272, 309, 315, 320, 367, 390
  signal recording, 341
  SNR, 268
Videocassette recorder, 5
Videodisc channel, 53
Videodiscs, 269, 273, 399, 412
  channel, 35, 37, 300, 303, 306, 346, 381, 407, 408, 411
  frequency spectrum, 220
  mastering, 39
  modulation system, 255
  noise, 356
  player, 5, 10, 22, 27, 278, 287, 390, 391, 399
  programming, 390
  recording, 225, 233, 239, 299
  stampers, 44
  system, 2, 4, 9, 16, 28, 64, 73, 211, 214, 249, 250, 253, 255, 256, 270, 274, 278, 279, 281, 287, 288, 299, 300, 309, 312, 322, 346, 356, 367, 379, 390, 393, 400, 404, 405
  technology, 1, 22, 58, 341, 370, 390, 434
  timing instability, 268
Video-signal domain, 249
Videotape, 34, 53, 402
Videotext, 5
Viewer, 128
Vinyl disc, 42
VIR (vertical interval reference), 212
Visible spectra, 156
Visible spectrum, 188
Vision carrier, 168, 170
Vision model, 300, 302, 303
Visual acuity, 401
Visual messages, 342
Visual phenomena, 311
Visual response, 280
Visual system, 287, 301, 302
Voice messages, 342
Voice recognition, 6
Voltage-controlled oscillator (VCO), 114, 264
Volumetric density, 434
Vowel pitch, 343
VSB filters, 95

Waveforms, 230, 260
Wavelength of light, 18

Weighted noise, 247
Weighting network, 247
White color, 155
White flag, 273, 274
White Gaussian noise, 110
White luminance levels, 234
White noise, 327, 345–348, 360
Wideband FM (WBFM), 109
Wideband modulation, 108
Wideband AM, 123
Window sliding, 60
Wobbling frequency, 26
Word processing, 400
Writable discs, 4, 8, 33, 52
Writable disc systems, 48
Write head, 53
Write-once discs, 1, 3, 12
Writing spot, 33

X-ray, 2

y factor, 158
Yields, 40

Zero-crossing detection, 76, 239
Zero crossings, 21, 60, 74, 101, 120, 224, 233, 283
Zero forcing equalizer, 85